Formal Description Techniques and Protocol Specification, Testing and Verification

Visit the IT & Applied Computing resource centre
www.IT-CH.com

IFIP – The International Federation for Information Processing

IFIP was founded in 1960 under the auspices of UNESCO, following the First World Computer Congress held in Paris the previous year. An umbrella organization for societies working in information processing, IFIP's aim is two-fold: to support information processing within its member countries and to encourage technology transfer to developing nations. As its mission statement clearly states,

> IFIP's mission is to be the leading, truly international, apolitical organization which encourages and assists in the development, exploitation and application of information technology for the benefit of all people.

IFIP is a non-profitmaking organization, run almost solely by 2500 volunteers. It operates through a number of technical committees, which organize events and publications. IFIP's events range from an international congress to local seminars, but the most important are:

- the IFIP World Computer Congress, held every second year;
- open conferences;
- working conferences.

The flagship event is the IFIP World Computer Congress, at which both invited and contributed papers are presented. Contributed papers are rigorously refereed and the rejection rate is high.

As with the Congress, participation in the open conferences is open to all and papers may be invited or submitted. Again, submitted papers are stringently refereed.

The working conferences are structured differently. They are usually run by a working group and attendance is small and by invitation only. Their purpose is to create an atmosphere conducive to innovation and development. Refereeing is less rigorous and papers are subjected to extensive group discussion.

Publications arising from IFIP events vary. The papers presented at the IFIP World Computer Congress and at open conferences are published as conference proceedings, while the results of the working conferences are often published as collections of selected and edited papers.

Any national society whose primary activity is in information may apply to become a full member of IFIP, although full membership is restricted to one society per country. Full members are entitled to vote at the annual General Assembly, National societies preferring a less committed involvement may apply for associate or corresponding membership. Associate members enjoy the same benefits as full members, but without voting rights. Corresponding members are not represented in IFIP bodies. Affiliated membership is open to non-national societies, and individual and honorary membership schemes are also offered.

Formal Description Techniques and Protocol Specification, Testing and Verification

FORTE X / PSTV XVII '97

IFIP TC6 WG6.1 Joint International Conference on
Formal Description Techniques for Distributed Systems
and Communication Protocols (FORTE X) and Protocol
Specification, Testing and Verification (PSTV XVII).
18–21 November 1997, Osaka, Japan

Edited by

Tadanori Mizuno
Shizuoka University
Hamamatsu
Japan

Norio Shiratori
Tohoku University
Miyagi
Japan

Teruo Higashino
Osaka University
Osaka
Japan

and

Atsushi Togashi
Shizuoka University
Hamamatsu
Japan

Published by Chapman & Hall on behalf of the
International Federation for Information Processing (IFIP)

CHAPMAN & HALL
London · Weinheim · New York · Tokyo · Melbourne · Madras

Published by Chapman & Hall, 2–6 Boundary Row, London SE1 8HN, UK

Chapman & Hall, 2–6 Boundary Row, London SE1 8HN, UK

Chapman & Hall GmbH, Pappelallee 3, 69469 Weinheim, Germany

Chapman & Hall USA, 115 Fifth Avenue, New York, NY 10003, USA

Chapman & Hall Japan, ITP-Japan, Kyowa Building, 3F, 2-2-1 Hirakawacho, Chiyoda-ku, Tokyo 102, Japan

Chapman & Hall Australia, 102 Dodds Street, South Melbourne, Victoria 3205, Australia

Chapman & Hall India, R. Seshadri, 32 Second Main Road, CIT East, Madras 600 035, India

First edition 1997

© 1997 IFIP

Printed in Great Britain by Athenæum Press Ltd, Gateshead, Tyne & Wear

ISBN 0 412 82060 9

A catalogue record for this book is available from the British Library

∞ Printed on permanent acid-free text paper, manufactured in accordance with ANSI/NISO Z39.48-1992 and ANSI/NISO Z39.48-1984 (Permanence of Paper).

CONTENTS

23 Modelling digital logic in SDL
   G. Csopaki and K.J. Turner   367

24 A methodology for the description of system requirements and the
   derivation of formal specifications
   A. Togashi, F. Kanezashi and X. Lu   383

25 On the influence of semantic constraints on the code generation
   from Estelle specifications
   R. Henke, A. Mitschele-Thiel and H. König   399

PART EIGHT   Industrial Usage Reports   415

26 Using a formal description technique to model aspects of a global air
   traffic telecommunications network
   J.H. Andrews, N.A. Day and J.J. Joyce   417

27 An experiment in using RT-LOTOS for the formal specification and
   verification of a distributed scheduling algorithm in a nuclear power
   plant monitoring system
   L. Andriantsiferana, J.-P. Courtiat, R.C. De Oliveira and L. Picci   433

28 Intelligent protocol analyzer with TCP behavior emulation for
   interoperability testing of TCP/IP protocols
   T. Kato, T. Ogishi, A. Idoue and K. Suzuki   449

29 Eight years of experience in test generation from FDTs using TVEDA
   R. Groz and N. Risser   465

Invited Talk III   481

   Distributed object consistency in mobile environments
   H. Chang   483

PART NINE   Concurrent Systems   485

30 Self-independent petri nets for distributed systems
   Y. Sun, S. Liu and M. Ohba   487

31 Combining CSP and object-Z: finite or infinite trace semantics?
   C. Fischer and G. Smith   503

32 Selective mu-calculus: new modal operators for proving properties on
   reduced transition systems
   R. Barbuti, N. De Francesco, A. Santone and G. Vaglini   519

33 On a concurrency calculus for design of mobile telecommunication
   systems
   T. Ando, K. Takahashi and Y. Kato   535

Index of contributors   547

Keyword index   549

PREFACE

This is the Proceedings of the 1997 IFIP TC6/WG6.1 Joint International Conference on FORMAL DESCRIPTION TECHNIQUES for Distributed Systems and Communication Protocols, and PROTOCOL SPECIFICATION, TESTING, AND VERIFICATION (FORTE/PSTV'97) which was held at NEC C&C Plaza Hall in Osaka Business Park (OBP) near Osaka Castle in the center of Osaka City, Japan on November 18-21, 1997.

After having been organized in Kaiserslautern (Germany, 1996), the two separate conferences FORTE and PSTV have been combined into a joint edition (FORTE/PSTV).

The aim of the conference is to be a meeting point between research and industry and between theory and practice of Formal Description Techniques (FDTs) applicable to Distributed Systems and Communication Protocols (such as Estelle, LOTOS, SDL, ASN.1, TTCN, Z, Automata, Process Algebra, Logic). The conference consists of the presentations of reviewed and invited papers, tutorials, panel session and tool demonstrations.

Totally 115 papers have been submitted, and the Program Committee has selected excellent papers among them. Many people have contributed to the success of the conference. We would like to express our thanks to all the PC members, reviewers and people participating in the local organization. We would like to express our appreciation to the invited/keynote speakers for accepting our invitation. Also our thanks go to Organization Committee Co-Chairs K. Suzuki (KDD, Japan) and K. Taniguchi (Osaka Univ.), and Tutorial Co-Chairs F. Sato (Shizuoka Univ.) and M. T. Liu (Ohio State Univ., USA).

Last, but not least, we gratefully acknowledge the support from Information Processing Society of Japan (IPSJ), The Telecommunication Advancement Foundation (TAF) and NEC Co. Ltd.

Tadanori Mizuno
Norio Shiratori
Teruo Higashino
Atsushi Togashi
Osaka, Japan, November 1997

Committee Members and Invited Speakers

General Co-Chairs
Tadanori Mizuno (Shizuoka Univ., Japan), Norio Shiratori (Tohoku Univ., Japan)

Organization Committee Co-Chairs
Kenji Suzuki (KDD, Japan), Kenichi Taniguchi (Osaka Univ., Japan)

Program Co-Chairs
Teruo Higashino (Osaka Univ., Japan), Atsushi Togashi (Shizuoka Univ., Japan)

Tutorial Co-Chairs
Fumiaki Sato (Shizuoka Univ., Japan), Ming T. Liu (Ohio State Univ., USA)

Program Committee
M. Aoyama (Nigata Inst. of Tech., Japan), G. v. Bochmann (Univ. of Montreal, Canada), T. Bolognesi (CNUCE, Italy), E. Brinksma (Univ. of Twente, Netherlands), H. Bowman (Univ. of Kent at Canterbury, UK), S. Budkowski (INT, France), A. Cavalli (INT, France), S. T. Chanson (Univ. of Sci. and Tech., Hong Kong), Y.Choi (Seoul National Univ., Korea), J.-P. Courtiat (LAAS-CNRS, France), R. Dssouli (Univ. of Montreal, Canada), S. Fischer (Univ. of Mannheim, Germany), R. Gotzhein (Univ. of Kaiserslautern, Germany), M. Gouda (Univ. of Texas at Austin, USA), R. Groz (CNET, France), D. Hogrefe (Univ. of Luebeck, Germany), E. Horita (NTT, Japan), Y. Kakuda (Osaka Univ., Japan), M. C. Kim (Korea Telecom, Korea), K. Futatsugi (JAIST, Japan), G. Leduc (Univ. of Liege, Belgium), D. Lee (Bell Lab., USA), S. Leue (Univ. of Waterloo, Canada), L. Logrippo (Univ. of Ottawa, Canada), N. Lynch (MIT, USA), J. de Meer (GMD FOKUS, Germany), E. Najm (ENST, France), N. Nakano (Mitsubishi, Japan), S. Obana (KDD, Japan), K. Ohmaki(ETL, Japan), J. Quemada (ETSI Telecom., Spain), H. Rudin (IBM, Switzerland), B. Sarikaya (Aizu Univ., Japan), R. Tenney (Univ. of Massachusetts, Boston, USA), K. Turner (Univ. of Stirling, UK), S. T. Vuong (Univ. of British Columbia, Canada), J. Wu (Tsinghua Univ., China)

Invited Speakers
Henry Chang (IBM T. J. Watson Center, USA), Shoichi Noguchi (Aizu Univ., Japan), Ichiroh Sakakibara (NTT, Japan)

Corporate Supporters
Information Processing Society of Japan (IPSJ), The Telecommunication Advancement Foundation (TAF)

List of External Reviewers
(PC members are not included in this list)

E. Abdeslam
F. Accordino
T. Ando
L. Andriantsiferana
M. Barbeau
H. Ben-Abdallah
A. Benzina
B. B. Bista
E. Boiten
R. Boubour
C. Bourhfir
J. Bredereke
P.-C. Brigitte
M. Caneve
R. Castanet
O. Catrina
G. Chakraborty
O. Charles
X. J. Chen
Z. Cheng
B.-M. Chin
P. R. D'Argenio
P. Dembinski
J. Derrick
R. Djian
J. Drissi
M. Faci
M. J. Fernández
V. François
D. de Frutos Escrig
H. Fukuoka
H. Fouchal
B. Ghribi
S. Gnesi
K. Go
J. Grabowski
T. Hasegawa
L. Heerink
O. Henniger
S. Heymer
M. Higuchi
G. Holzmann
H. Horiuchi

A. Idoue
S. Iida
Y. Isobe
C. Jard
T. Jéron
B. Joachim
I. Kang
D. Kang
S. Kang
T. Kato
Y. Kato
Z. Khasidashvili
F. Khendek
A. Khoumsi
S.-U. Kim
B. Koch
F. Kristoffersen
R. Langerak
D. Larrabeiti
D. Latella
M. Llamas
L. F. Llana Díaz
H. Lounis
T. Macavei
S. Maharaj
P. Maigron
K. Mano
G. Mansfield
C. Marylene
M. Massink
M. Matsumoto
F. Michel
N. Miyake
J.-F. Monin
E. Montes de Oca
S. Nagano
S. Naito
S. Nakajima
M. Nakamura
A. Nimour
F. Nitta
T. Ohta
K. Okada

L. Paula Lima Jr.
W. Penczek
A. Petrenko
S. Pickin
J.-L. Raffy
A. Rennoch
A. Rouger
T. C. Ruys
P. de Saqui-Sannes
F. Sato
M. Savi
M. Schmitt
H. van der Schoot
C. Shankland
J. Sincennes
R. Sinnott
M. Steen
B. Stepien
J.-F. Suel
K. Sugiyama
M. Tabourier
K. Takahashi
A. Takura
M. Tomono
M. Törö
T. Tonouchi
J. Tretmans
A. Umemura
D. Vincent
Y. Wakahara
D. Witaszek
I. Yahmadi
H. Yamaguchi
A. Yamanaka
K. Yamanaka
K. Yasumoto
N. Yevtushenko
S. Yoshida
S. Yuen
J. Zhu

New Generation Networks
and Applications

New Generation Networks and Applications

Ichiroh Sakakibara
Vice President, Network Access Service, Network Service Department
NIPPON TELEGRAPH AND TELEPHONE CORPORATION
TOKYO OPERA CITY TOWER, 3-20-2, Nishi-Shinjuku, Shinjuku-
ku, Tokyo 163-14 Japan
phone: +81-3-5353-9800   fax: +81-3-5353-5744
`bara@nws.hqs.ntt.co.jp`

Abstract

Recent technological progress in communications and information processing has been remarkable. As digital technology and optical fiber technology in telecommunications networks have become increasingly common and economical. We find personal computers on almost every desk nowadays. Although the hardware has become much more sophisticated, multimedia applications had not yet even started to mature a few years ago. Against this background, NTT proposed its "Joint Utilization Tests of Multimedia Communications" in order to stimulate new multimedia applications. These tests were carried out from September 1994 to March 1997, and involved more than one hundred participants in creating and verifying new applications. As the same time, progress in network technology has brought about a variety of network structures. Among them, CATV networks, mobile communications and satellite systems have become the new form of communications networks, as optical fiber has extended broadcasting capabilities to conventional communications networks. These networks are now the infrastructure on which many new and highly competitive network services such as Internet, ISDN and ATM are provided. This session presents an outline of the remarkable progress in networks and applications in recent years.

Formal Description Techniques and Protocol Specification, Testing and Verification
T. Mizuno, N. Shiratori, T. Higashino & A. Togashi (Eds.)   © 1997 IFIP. Published by Chapman & Hall

## PART ONE

Testing Theory for Concurrent Systems

1

Specification-based testing of concurrent systems

Andreas Ulrich[a], Hartmut König[b]
[a] Dept. of Comp. Science, Univ. of Magdeburg, PF 4120, 39016
Magdeburg, Germany; e-mail: ulrich@cs.uni-magdeburg.de
[b] Dept. of Computer Science, BTU Cottbus, PF 101344, 03013
Cottbus, Germany; e-mail: koenig@informatik.tu-cottbus.de

Abstract

The paper addresses the problem of test suite derivation from a formal specification of a distributed concurrent software system given as a collection of labeled transition systems. It presents a new concurrency model, called *behavior machine*, and its construction algorithm. Further, the paper outlines how test derivation can be based on the new concurrency model in order to derive test suites that still exhibit true concurrency between test events. A toolset is presented to support the generation of concurrent test suites from specifications given in the formal description technique LOTOS. Finally, some comments on requirements for the design of a distributed test architecture are given.

Keywords

Distributed concurrent software systems; conformance testing; test derivation; labeled transition systems; LOTOS; Petri nets.

1 INTRODUCTION

Testing is an important means in the development cycle of software. A challenging problem is the derivation of test suites that are able to detect faulty implementations

Formal Description Techniques and Protocol Specification, Testing and Verification
T. Mizuno, N. Shiratori, T. Higashino & A. Togashi (Eds.) © 1997 IFIP. Published by Chapman & Hall

of a system firmly. Driven by requirements in testing telecommunication systems, approaches were developed to assist the automatic derivation of test suites [ADL+91] [Fer96]. These approaches are usually based on a finite description of the behavior of the system, mostly the model of a finite state machine, that is also exploited in the verification phase of the system. However, current test derivation approaches only support the derivation of test suites for sequential systems. One reason is that they are faced with computational problems due to state explosion if they resolve specified concurrent behavior in an interleaving sequence of actions of a derived test suite. Furthermore, its execution in a standard black-box test architecture might be not sufficient to assess conformance of truly concurrent systems since the message exchange between components of the concurrent system must be observed and controlled by a tester in order to avoid nondeterministic test runs.

This paper continues work on the use of partial orders for test suite derivation of concurrent systems. It improves the previous work done in [UlCh95] and also other known work on this subject, e.g. [LSK+93], [KCK+96], by providing a sound concurrency model and an algorithm to construct it automatically from a collection of communicating labeled transition systems (LTSs). The new concurrency model, called *behavior machine* (BM), is an interleaved-free and finite description of concurrent and recursive behavior. The construction algorithm works as follows. First, the LTSs are mapped into a single Petri net representing the whole system. This Petri net is further used to construct its *unfolding*, another Petri net with a simpler structure, using an algorithm from [ERV96]. The behavior machine is then constructed from the finite prefix of a Petri net unfolding.

After the behavior machine is introduced as description model of concurrent behavior, the test derivation approach is extended to support the new model. It is shown how an extension of the transition tour can be derived from a behavior machine using algorithms already known from sequential systems. First results of the new testing approach that were obtained with a prototype implementation supporting specifications in LOTOS are discussed.

The paper is organized as follows. Section 2 introduces the model assumptions on a concurrent system. Section 3 sets up the new concurrency model. Some Petri net notions are explained in Section 4 that are needed for an easy understanding of the construction algorithm. Section 5 presents the algorithm to compute a behavior machine. In Section 6, the behavior machine serves as model for test derivation. An algorithm to derive a test suite from a behavior machine is presented. Section 7 discloses a first prototype implementation that supports LOTOS specifications, and finally, Section 8 explains concepts of the design of a distributed test architecture.

2 A MODEL FOR DISTRIBUTED CONCURRENT SYSTEMS

We consider distributed concurrent software systems consisting of a collection of software modules running on different host machines that are connected through a computer network. A module is implemented as a sequential unit realizing a certain function of the system. Modules communicate synchronously via interaction points. The synchronous communication pattern fits the properties of programming languages for concurrent systems, e.g. Ada, and function calls in high-level network programming, like remote procedure calls, which are exploited, for instance, in the middleware platform CORBA.

Starting point of our investigations is a formal specification that defines the desired behavior of the concurrent system. Sequential behavior of a module in a concurrent system is modeled as a *labeled transition system* (LTS). The model of an LTS is an abstraction that focuses on interactions of a module with other modules in the system and/or with its environment.

Definition 1 A *labeled transition system* (LTS) is defined by the quadruple $(S, A, \rightarrow, s_0)$, where S is a finite set of states; A is a finite set of actions (the alphabet); $\rightarrow \subseteq S \times A \times S$ is a transition relation; and $s_0 \in S$ is the initial state.

A concurrent system $\mathfrak{S} = M_1 \parallel M_2 \parallel \dots \parallel M_n$ is composed from a fixed number of communicating LTSs M_i. A composite machine $C_{\mathfrak{S}}$ of $\mathfrak{S}$ (also an LTS) is expressed by means of a composition operator $\parallel$ similar to that used in CSP. $P \parallel Q$ is the parallel composition of modules P and Q with synchronization of the actions common to both of their alphabets and interleaving of the others. The parallel composition $P \parallel Q$ of two LTSs $P = (S_1, A_1, \rightarrow_1, s_1)$ and $Q = (S_2, A_2, \rightarrow_2, s_2)$ is defined as a composite LTS $(S, A, \rightarrow, s)$, where $S \subseteq S_1 \times S_2$, $A \subseteq A_1 \cup A_2$, $s = (s_1, s_2)$, and the transition relation $\rightarrow$ is given as follows. If $P -a\rightarrow_1 P'$ then $(P \parallel Q) -a\rightarrow (P' \parallel Q)$ if $a \notin A_2$. If $Q -a\rightarrow_2 Q'$ then $(P \parallel Q) -a\rightarrow (P \parallel Q')$ if $a \notin A_1$. If $P -a\rightarrow_1 P'$ and $Q -a\rightarrow_2 Q'$ then $(P \parallel Q) -a\rightarrow (P' \parallel Q')$ if $a \in A_1 \cap A_2$.

3 A CONCURRENCY MODEL

The representation of concurrent behavior in a composite machine is accomplished by a tedious repetition of concurrent actions in order to construct all possible total orders. However, concurrent actions are independent to a certain extent from their occurrence in a total order. Instead of interpreting causality information in an interleaved-based model, we apply the notion of a *labeled partially ordered set* and its extension to a *partially ordered multiset*, which are interleaved-free representations of concurrent behavior [Pra86].

Definition 2 An *lposet* (*labeled partially ordered set*) is defined by the quadruple $(E, A, \leq, l)$, where E is a set of event names; A is a set of action names; $\leq$ is a partial order expressing the causality information between events, i.e. $e \leq f$ if event e precedes event f in time; $l: E \rightarrow A$ is a labeling function assigning action names to events. Each labeled event represents an occurrence of the action labelling it, with the same action possibly having multiple occurrences.

A *pomset* (*partially ordered multiset*) is an isomorphism class over event renaming of an *lposet*, denoted $[E, A, \leq, l]$. A *process* describing the behavior of concurrent system $\Im$ is a set of pomsets where each pomset describes a possible execution sequence of concurrent actions. Since the behavior of a system is frequently infinite due to recursive parts in the system description, the pomsets of a process are infinite, too. If branching occurs in a process, the set of pomsets forms an infinite *pomtree* [PLL+91], where an arc in the pomtree is an lposet or a concatenation of lposets, and a vertex is a branching point of the process (see Figure 3).

Since the construction of the composite machine from a set of communicating LTSs is not feasible in many cases due to state explosion, it follows that we need a new model that combines the advantages of both concepts: true concurrency between actions as preserved in an lposet and finiteness of the description as preserved in an LTS. This model is a *behavior machine* (BM), a similar model to the one introduced in [PLL+91].

Definition 3 The *behavior machine* of concurrent system $\Im$ is a quadruple $BM_\Im = (G, LPO, T, g_0)$ consisting of a finite set of global states G, where each element of G is an n-tuple of local states of all LTSs of $\Im$, i.e. $G \subseteq S_1 \times \ldots \times S_n$; a set of finite lposets LPO representing concurrent behavior in $\Im$; a concurrent transition relation $T \subseteq G \times LPO \times G$ that maps a start state to an end state when the actions of the corresponding lposet are executed; and an initial global state $g_0 = (s_1, \ldots, s_n) \in G$.

A global state of behavior machine $BM_\Im$, excluding its initial state, expresses always a branching point or a recurrence point within concurrent system $\Im$. A branching point is a global state where further behavior of the system branches off. A recurrence point is a global state where the behavior of the system repeatedly continues. An lposet in $BM_\Im$ is constructed in such a way that it connects always two global states of $BM_\Im$ by a concurrent transition $t \in T$. A pomtree can be obtained from $BM_\Im$ if its concurrent transitions are unrolled. In this case, branching points in the behavior machine correspond to branching points in the pomtree, whereas recurrence points diminish. Thus, unrolling a behavior machine is similar to the construction of a spanning tree from a directed graph.

Let t_1 an t_2 be two concurrent transitions of $BM_\Im$ with $t_1 = (g_1, lpo_1, g_2)$ and $t_2 = (g_3, lpo_2, g_4)$. The operation $t_1 \oplus t_2$ expresses concatenation of the two concurrent transitions. Concatenation is allowed if $g_2 = g_3$. It is carried out in such a way that each local state of the end state g_2 in t_1 is connected with the same local state of the start state g_3 in t_2. That means, the lposets lpo_1 and lpo_2 are merged according to the causal dependencies between their events.

Consider the system $\Im = A \parallel B$ whose LTSs are given in Figure 1. Under the assumption that actions a and c in each LTS synchronize, removal of the parallel operator by applying interleaved-based semantics rules yields the composite machine $C_\Im$. Figure 2 shows the behavior machine $BM_\Im$ of system $\Im$. It contains three global states $\{S_0, S_1, S_2\}$ and four concurrent transitions $\{t_1 - t_4\}$ and describes the same behavior of system $\Im$ as given in Figure 1. Each concurrent transition is described by an lposet that exhibits concurrency among actions (see transition t_4). If the behavior machine is unrolled, the pomtree in Figure 3 is obtained. The process of unrolling shows the full degree of concurrency between events. For instance, if transitions t_2 and t_4 are concatenated, we realize that event b is concurrent to e.

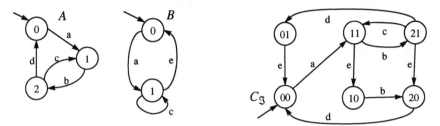

Figure 1 LTSs A and B, and the composite machine C_3 of system $\mathfrak{I} = A \parallel B$.

where S0 = (A0, B0), S1 = (A1, B1), S2 = (A2, B1) are global states, and the concurrent transitions are

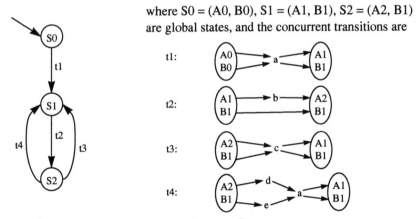

Figure 2 The behavior machine $BM_{\mathfrak{I}}$ of system $\mathfrak{I}$.

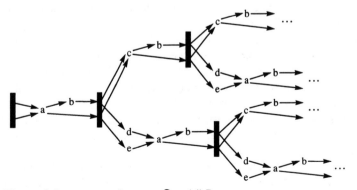

Figure 3 A pomtree of system $\mathfrak{I} = A \parallel B$.

Although the behavior machine in Figure 2 is not the smallest representation of con-current behavior due to its construction algorithm discussed below, it is still a very compact representation of concurrent behavior. In this specific example, action a is redundantly represented within concurrent transitions t_1 and t_4 that can be avoided in a minimal description. Still, a behavior machine is able to distinguish concur-rency from branching. This knowledge is lost in the composite machine C_3.

4   PETRI NET CONCEPTS

The construction algorithm of a behavior machine is based on a Petri net description of the concurrent system. In [McM95] and [ERV96], a verification approach was described that is based on the technique of net unfolding, a partial order semantics of Petri nets. The unfolding of a Petri net is another (usually infinite) net with a sim-pler structure. The proposed algorithms in both papers aim at constructing the initial part of the net unfolding that contains all reachable states of the original net, called the *finite complete prefix*.

A net is a triple (S, T, F), where S is the set of places, T is the set of transitions, $S \cap T = \varnothing$, $F \subseteq (S \times T) \cup (T \times S)$ is the flow relation, If $M: S \rightarrow \mathcal{N}$ is a marking of a net ($\mathcal{N}$ denotes the set of non negative integers), the 4-tuple $N = (S, T, F, M)$ is called a Petri net [Rei91]. The unfolding of a Petri net is a (unmarked) condition-event net (B, E, F), where B is the set of conditions, E is the set of events, with the properties: (1) it is an acyclic graph, (2) if two events (transitions) e_1, $e_2 \in E$ of the unfolding are in conflict, meaning that they are enabled from the same condition (place), then there exist two paths leading to e_1 and e_2 that start at the same condition and imme-diately branch off from another, (3) the nodes in the unfolding have a finite number of predecessors, and (4) no event is in self-conflict.

Figure 4 depicts the Petri net description of system $\mathfrak{I}$ and the initial part of its unfolding. Note that the unfolding is not finite. For instance, if event **E4** is per-formed, the unfolding continues by substituting place **B10** with **B3** and **B11** with **B4**, respectively. This applies similarly to event **E6**. Events **E4** and **E6** are also in conflict. The branching point where the paths leading to the events branch off from another is the marking {A1, B1}, represented by the set of conditions {**B5, B4**}.

A *local configuration* $[e]$ of event e in the unfolding describes a possible par-tially ordered run of the system which executes event e as its last event. It is a set of events satisfying the following two conditions: (1) if any event is in the local config-uration, then so are all of its predecessors, and (2) a local configuration is conflict-free. The local configuration captures the precedence relation between events. Any total order on these events that is consistent with the partial order is an allowed totally ordered run of the system. Throughout the paper, we use the notions *local configuration* and *configuration* interchangeable.

To compute the finite complete prefix of a Petri net, it is necessary to define a break-off condition to stop the construction of the unfolding. This is done by intro-ducing *cut-off events*. An event e is a cut-off event if the local configuration $[e]$

belonging to event e reaches a marking $Mark([e])$ in the unfolding that was reached before by a smaller local configuration of a different event.

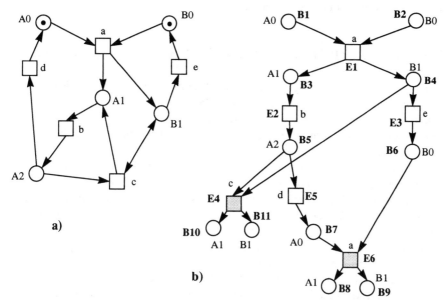

Figure 4 The marked Petri net of system $\mathfrak{S} = A \parallel B$ (a) and its unfolding (b).

Consider the unfolding in Figure 4b. The configuration of event **E6** is the set of events $[\textbf{E6}] = \{\textbf{E1, E2, E3, E5, E6}\}$. The reachable marking of this configuration is $Mark([\textbf{E6}]) = \{\text{A1, B1}\}$. This marking was reached before, however, by configuration $[\textbf{E1}] = \{\textbf{E1}\}$. We say, event **E6** corresponds to event **E1**. Since the configuration of **E1** has fewer elements than the configuration of **E6**, it follows that **E6** is a cut-off event. The second cut-off event is **E4**.

Proposition 1 Given the unfolding of a 1-save Petri net[*]. Any deadlock-free system is completely represented by the finite set of tuples of cut-off and corresponding events $\{(e_{\text{cutoff}_1}, e_{\text{corresp}_1}), (e_{\text{cutoff}_2}, e_{\text{corresp}_2}), ..., (e_{\text{cutoff}_n}, e_{\text{corresp}_n})\}$, where the two configurations of events in a tuple reach the same marking, $Mark([e_{\text{cutoff}_i}]) = Mark([e_{\text{corresp}_i}])$.

The proposition requires cut-off events for each execution branch in the net unfolding. Their existence was proven in [ERV96]. A deadlocking system, however, reaches a final marking that does not relate to a cut-off event. The construction algorithm of a behavior machine is currently restricted to concurrent systems without

---

[*]   A 1-save Petri net is a net whose places contain at most one token at a certain time. This type of net is obtained, for instance, if the net is constructed from a set of communicating LTSs.

deadlocks. This is, however, not a restriction for the purpose of test derivation since we usually require that the specification of a system has been verified to be dead-lock-free before it can be implemented.

5 CONSTRUCTION ALGORITHM OF A BEHAVIOR MACHINE

The first step in constructing a behavior machine from a set of communicating LTS is a transformation of the LTSs into a global Petri net. After the transformation, the unfolding algorithm is applied to unfold the Petri net and to construct the set of pairs of cut-off and corresponding events. Finally, the behavior machine is constructed from the unfolding.

Constructing a Petri net from a set of communicating LTSs is simple. The following algorithm is applied: first, each single LTS is transformed into a Petri net; then, all Petri nets are merged according to the synchronization constraints in order to obtain a global Petri net. If the same action name labels several transitions in several LTSs, then a net transition for each allowed way of synchronization has to be constructed in the global Petri net. The transformation was already presented in [GaSi90] and is used in the *Cæsar/Aldebaran* toolset that supports verification of specifications given in the formal description language LOTOS.

Figure 4a shows the Petri net constructed from the two LTSs in Figure 1. The next step is the construction of the finite complete prefix. This is done by applying the algorithm presented in [ERV96]. Figure 4b depicts the prefix of the example system. The last step, the construction of the behavior machine, is described below.

We assume that the finite complete prefix of a Petri net unfolding, including the set of cut-off and corresponding events, is given. The local configuration of an event describes an execution path through the behavior machine from the initial state to this particular event. The marking reached by the configuration of an event defines a global state in the behavior of a concurrent system. The reachable marking of a configuration can be identified with places in the unfolding that are reached if all events in the configuration are executed.

When the behavior machine is constructed, it is not necessary to compute all reachable markings in the unfolding. Instead, only those reachable markings have to be known that are recurrence or branching points. Cut-off events and events corresponding to them define recurrence points of the behavior machine. Yet, branching points have to be computed.

To identify the branching points, we do the following considerations. Given the finite complete prefix of an unfolding, each local configuration of a cut-off event or a corresponding event starts in the initial state of the system, i.e. the initial marking, and ends in a marking reached by the configurations of those events. Since the finite complete prefix covers all reachable states of the system, branching points exist only somewhere inside the configurations of cut-off and corresponding events. If we analyze any two configurations $[e_1]$ and $[e_2]$ with $e_1 \neq e_2$ from the same unfolding, we realize that the configurations start with a same subset of events and branch off from another after a certain event e_{branch} occurred in both configurations. Now, a branching point can be defined exactly by the reachable marking of the configuration formed by this event e_{branch} assuming that $e_{branch} \in [e_i]$, and $[e_{branch}]$ is the maximal configuration that holds the condition $[e_{branch}] \subseteq [e_i]$, with $i = \{1, 2\}$.

This observation leads to the construction algorithm. It takes the set of cut-off events and corresponding events that are contained in the finite complete prefix of an unfolding as input. The idea behind the algorithm is to construct the configurations of the given cut-off and corresponding events first. Then, the events in the configurations are analyzed in order to identify the branching points.

```
1    • let E be the set of cut-off and corresponding events in a finite prefix;
2    • let 𝓔 initially be the set of configurations from all events in E, i.e.
         𝓔 = {[e₁], [e₂], ...};
3    forall configurations [e] ∈ 𝓔 do
4        forall events d ∈ [e] with d ≠ e do
5            if (([d] ∉ 𝓔) AND (successors(d) are branching places)) then
6                • mark d as branching event;
7                𝓔 = 𝓔 ∪ {[d]};
8            end
9        end
10   end
11   forall configurations [e] ∈ 𝓔 do
12       forall configurations [d] ∈ 𝓔 do
13           if ((|[e]| < |[d]|) AND ([e] ⊆ [d])) then
14               • mark [e] if it is a maximal configuration contained in [d];
15           end
16   end
17   forall configurations [e] ∈ 𝓔 do
18       if ((e is a branching event) AND
             ([e] is marked as a maximal configuration less than twice)) then
19           𝓔 = 𝓔 \ {[e]};
20   end
21   return 𝓔;
```

Figure 5 Generation of configurations represented in the behavior machine.

As discussed above, a branching point is defined by the reachable marking of a maximum configuration contained within two or more other configurations. To identify these points, we analyze the successor places of an event e. If at least one of the successor places has more than one successor event, the reachable marking of the configuration $[e]$ might be a branching point in the behavior machine (static conflict). Since this result is obtained from a local analysis of a single event rather than from an analysis of the global system, not all events found in this way refer really to a branching point. The algorithm in Figure 5 takes into account this aspect and returns only those events and their configurations that will be finally considered in the construction of a behavior machine.

The initial set of configurations $\mathcal{E}$ is obtained from the configurations of cut-off and corresponding events contained in the prefix of the unfolding (line 2 in Figure 5). In the next step, further configurations of events are added to $\mathcal{E}$ if these events possess successor places that cause local branching (lines 3–10). The third step (lines 11–16) determines whether a configuration is contained in another one and marks the maximal configuration that fulfills this property. The final step (lines 17–20) deletes configurations of events added to $\mathcal{E}$ before if they are not marked as maximal configurations or if they are marked only once in another configuration. That means, configurations that do not determine a branch in the behavior are omitted in the construction of the behavior machine.

```
1     • let E be the set of local configurations;
2     global_states = ∅;
3     forall configurations [d] ∈ E do
4          • compute the reachable marking reachable_marks of [d];
5          global_states = global_states ∪ {reachable_marks};
6     end
7     conc_trans = ∅;
8     forall configurations [d] ∈ E do
9          • let [e] be the maximal configuration of [d];
10         conc_trans = conc_trans ∪ {[d] \ [e]};
11    end
12    return global_states, conc_trans;
```

Figure 6 Construction of global states and concurrent transitions in a BM.

The algorithm in Figure 5 returns the set of configurations $\mathcal{E}$ relevant in the construction of the behavior machine, i.e., the configurations have the property that they reach the marking of a recurrence point or a branching point. In the next algorithm (Figure 6), this knowledge is used to construct the global states and the concurrent transitions of the behavior machine of a concurrent system.

In line 4 of Figure 6, the reachable marking is computed. Note that the reachable markings are the same for the configuration of a cut-off event and the configuration of its corresponding event, thus the second computation is redundant. A concurrent transition in a behavior machine is computed in line 10. It is simply the set difference of configuration [d] and the maximal configuration [e] contained in [d]. This computation is correct since the configuration [e] is a subset of [d], and all events in [e] occur in the behavior machine in one or more other concurrent transitions. The behavior machine is now nearly complete. The missing initial global state of the behavior machine is computed from the reachable marking of the empty configuration, i.e., it is the initial marking in the Petri net.

The construction of a behavior machine is based on cut-off events and corresponding events in the finite complete prefix and the configurations belonging to them. If we assume that the events and conditions of the prefix are stored in doubly linked lists, local configurations and reachable markings can be computed in linear time. Thus, the highest computational complexity of the construction algorithm is contained in the identification of branching points (Figure 5, lines 11–16). The complexity of this algorithm is bound on $O(n^2 \cdot (\log_k n)^2)$, where n is the number of places in the prefix of the unfolding, and k is the largest number of successor places of any transition. All other parts of the construction algorithm are less complex.

Table 1 Results of the dining philosophers example, computed on a Sparc Station 5.

# philo.	# reachable states	# global states	# concurrent transitions	memory (kByte)	computation time (sec)
5	392	16	35	36	0.11
7	4,247	36	77	61	0.73
9	46,763	64	135	98	2.96
11	510,116	100	209	151	9.12
13	5,564,522	144	299	222	22.76
15	—	196	405	314	48.83

To demonstrate the feasibility of the construction algorithm, we compute the behavior machines for a variable number of processes of the *dining philosophers* example. The results are given in Table 1 for a varying number of philosophers. Note that this example contains a deadlock state whose configuration is not represented in the behavior machine due to Proposition (1). However, all other behavior parts are truly represented. The second column shows the number of reachable states computed by a traditional reachability analysis. The number of states grows clearly exponentially with the number of philosophers. The following two columns reveal the numbers of global states and concurrent transitions of the constructed behavior machines. We realize that the number of global states increases slightly worse than quadratic. Even though the computation time increases fast and seems to be a function of $n^{5.5}$, where n is the number of philosophers, the time is still reasonable small. This is also particularly true for the memory space used. The computation time of the finite complete prefix that is used as input for our construction algorithm was always less or around few seconds.

6 TEST GENERATION BASED ON BEHAVIOR MACHINES

The behavior machine is an appropriate model for test suite derivation. A test suite has to fulfill certain properties to be useful in software testing. Especially, it must

distinguish faulty implementations from correct ones according to a chosen conformance relation. A conformance relation commonly used in testing is the *trace*
equivalence between LTSs modeling the specification and the implementation. We
extend trace equivalence over LTSs to an equivalence over behavior machines. First,
equivalence of lposets is defined as follows borrowing ideas from [Bri88].

Definition 4 Lposet $lpo_1 = (E_1, A_1, \leq_1, l_1)$ *reduces* Lposet $lpo_2 = (E_2, A_2, \leq_2, l_2)$,
denoted $lpo_1 \sim lpo_2$, iff $A_1 \subseteq A_2$, and for all $e, f \in E_1$, if $e \leq_1 f$ then there exist $r, s \in$
E_2 such that $r \leq_2 s$, and $l_2(r) = l_1(e)$, $l_2(s) = l_1(f)$. Two lposets are *equivalent*, $lpo_1 \approx$
lpo_2, iff $lpo_1 \sim lpo_2$ and $lpo_2 \sim lpo_1$.

We define further a sequence *seq* of lposets as a concatenation of a matching
sequence of lposets according to the $\oplus$ operator (see Section 3): $seq = lpo_1 \oplus lpo_2$
$\oplus\ lpo_3 \oplus ....$ Thus, a sequence of lposets describes a pomset, i.e. an execution
branch of the concurrent system as depicted in its pomtree. Two sequences seq_1 and
seq_2 equal, $seq_1 = seq_2$, if their lposets are equivalent.

Definition 5 Given two concurrent systems $\mathfrak{S}$ and $\mathfrak{R}$ and their behavior machines
$BM_\mathfrak{S}$ and $BM_\mathfrak{R}$, respectively. Let *Seq(BM)* refer to the set of sequences of lposets of
which behavior machine *BM* is able to perform. $\mathfrak{S}$ and $\mathfrak{R}$ are *equivalent*, $\mathfrak{S} \approx_c \mathfrak{R}$, iff
$Seq(BM_\mathfrak{S}) = Seq(BM_\mathfrak{R})$.

A test suite consisting of a finite set of finite test cases is *sound* w.r.t. a fault model if
any conforming implementation passes the test suite. A test suite is *complete* w.r.t. a
fault model if any non-conforming implementation from the implementation
domain fails the test suite [PBY96]. An implementation domain that is often considered in conformance testing is the set of implementations with *acceptance faults*,
i.e., these implementations may not accept all actions by corresponding transitions
as required in the specification, thus they reduce the specification [Lan90].
      Acceptance faults can be detected by a *transition cover* of the concurrent sys-
tem. A transition cover is usually defined over an LTS as a set of traces covering all
transitions in the LTS. We extend now the notion of a *transition tour* [ADL+91]
over a behavior machine as the least sequence of lposets covering all lposets. Such a
sequence of lposets can be seen as a "transition cover" of a behavior machine. The
extended notion is called *concurrent transition tour* (CTT) [UlCh95].

Definition 6 (CTT) A *concurrent transition tour* through a behavior machine $BM_\mathfrak{S}$
of a concurrent system $\mathfrak{S}$ is the least pomset $CTT = [E_{CTT}, A_{CTT}, \leq, l]$ such that all
actions of $\mathfrak{S}$ are covered in the pomset, i.e. if $a \in A_1 \cup A_2 \cup ... \cup A_n$, then $a \in$
A_{CTT} and E_{CTT} is minimal.

Derivation of a CTT from a behavior machine is straightforward. Since the description of concurrent behavior is reduced to a finite directed graph, simple graph algorithms can be applied. To construct a CTT, an algorithm that solves the *Chinese*

postman problem is appropriate [ADL+91]. First, all strongly-connected components of maximum size contained in a behavior machine are computed. After that, a CTT is derived for each strongly-connected component. This approach assures full coverage of all transitions in the behavior machine. The complete algorithm is given in Figure 7. Note however that the size of CTTs computed in this algorithm might not be minimal if we assume, for example, a behavior machine consisting of two components where the second component is reachable through the first one. Further optimization strategies might become applicable in this case.

1	Find all strongly-connected subgraphs of maximum size bm^s_1, ..., bm^s_n in behavior machine $BM_\Im$.
2	For each bm^s_i find the shortest path p_i from the initial state of bm to bm^s_i.
3	For each bm^s_i find the Chinese postman tour pt_i.
4	A CTT of a subgraph of bm is found by concatenation of p_i and pt_i: $CTT_i = p_i \oplus pt_i$.
5	The test suite is the set of all CTTs found: $TS = \{CTT_1, ..., CTT_n\}$.

Figure 7 Test suite derivation.

Consider the behavior machine $BM_\Im$ of the system $\Im = A \parallel B$ in Figure 2. It contains one strongly-connected component consisting of the states S_1 and S_2. The initial path to reach this component is given by concurrent transition t_1. The Chinese postman tour through the component is the sequence of concurrent transitions $t_2 \oplus t_3 \oplus t_2 \oplus t_4$. The final test suite of system $\Im$ contains only a single CTT and is given in Figure 8 as a time-event sequence diagram where the gray-shaped arrows denote synchronization constraints between the modules A and B. This test suite describes the shortest path through the concurrent system fulfilling the requirements of a CTT.

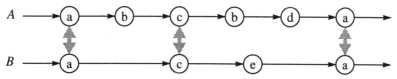

Figure 8 A concurrent transition tour for concurrent system $\Im = A \parallel B$.

7 A PROTOTYPE IMPLEMENTATION

The algorithms presented in Section 5 and Section 6 have been implemented as a first prototype tool to support concurrent test suite derivation from LOTOS specifications. In order to operate with Full LOTOS, we use the *Cæsar/Aldebaran* verification toolset [GaSi90]. *Cæsar* produces the Petri net of a LOTOS specification as it is the required input for the construction of a behavior machine. The produced Petri net has the advantage that values of variables are still represented symbolically. Thus, the Petri net description of the concurrent system remains small in size.

The Petri net of the specification serves then as input to the construction of the complete finite prefix of the net using the *PEP* tool [GrBe96]. Currently however, we do not support symbolic evaluation of data terms in LOTOS predicates or guard expressions. The behavior is considered to be executable under all cases instead. Consequently, the constructed prefix of the Petri net unfolding contains dead execution branches that cannot be executed by a correct implementation. At the moment, this code has to be removed manually in order to generate only executable test suites. Eventually, test suites according to the acceptance fault model are generated from the behavior machine.

A further drawback of the current prototype is the use of the verification tool *PEP*. Since *PEP* stops to unfold an execution branch of the Petri net if it has covered all reachable states (a sufficient condition in verification) instead of continuing to unfold until a cut-off event is reached, the resulting behavior machine contains uncompleted cycles. A new implementation of the unfolding algorithm is therefore necessary. If this work will be carried out, the construction of a behavior machine should then be integrated into a single tool.

8   DISTRIBUTED TESTER DESIGN

Although the derivation of test suites from concurrent systems appears to be manageable now by the approach presented before, there is still the problem to apply a concurrent test suite in a real test environment. Here we are faced with some unpleasant properties of concurrent systems, namely with the unpredictable progress in the concurrent modules when executing a test run and with hidden communication between modules that remain unobserved by the tester. Both issues result in nondeterministic behavior of the implementation under test (IUT). Whereas the second issue can be solved by applying a gray-box testing approach, the first one requires special means to ensure a deterministic test run, e.g. the instant replay technique presented in [TCO91].

In the simplest case of a test architecture, the tester is represented as a single module that implements a CTT of a given test suite. Since a tester module can exhibit only sequential behavior, the concurrent behavior of a CTT must be linearized in such a way that the causal relationships between events remain unchanged. For example, the CTT in Figure 8 can be implemented as the sequence of events "*a.b.c.b.d.e.a*" that fulfills the requirements. If the IUT is correctly implemented according to the chosen fault model, then it must be able to accept any sequence of events represented in the CTT and the selection of the actual sequence used in testing is arbitrary.

In case that the tester consists itself of distributed concurrent tester parts, further efforts are required to obtain reliable test results after a test run. Here, each tester part observes a sub-set of events of the IUT. The partial behavior containing those events visible to a tester part can be obtained from a projection from the complete CTT to a partial CTT containing only the visible events. Each tester part will then execute the behavior of the projected CTT. Yet, means are required to obtain a globally ordered test run of the IUT from the observed partial behavior, e.g. synchronization messages between tester parts.

9 CONCLUSIONS

The model of a behavior machine is used to support automatic test derivation for concurrent systems. The model has its merits as a finite description of concurrent behavior that still exhibits true concurrency among actions. Furthermore, a behavior machine distinguishes concurrency from execution branching. Vertices in the behavior machine refer usually to a small subset of the set of reachable global states in the system. The main contributions of this paper are the presentation of an algorithm that constructs a behavior machine from a Petri net unfolding as well as the application of the behavior machine to the realm of test derivation. For the purpose of test derivation, the notion of trace equivalence was extended to cope with concurrent behavior.

The presented approach to construct a behavior machine can be improved. Especially, the Petri net unfolding algorithm and the construction algorithm of the behavior machine should be combined into a single tool. Other extensions of a behavior machine may include the support of inputs and outputs and the supply of operations over behavior machines, e.g. the composition of two behavior machines, projections of behavior machines to submachines and other operations. Last but not least, more expanded fault models should be investigated and optimized test derivation algorithms should be elaborated for them.

Currently, other approaches are under investigation to construct a concurrency model that avoid the use of Petri nets [Hen97], though the applicability of the presented algorithms, i.e. their computational complexity, has still to be determined.

An extended version of the paper and other related work can be found on the web page *http://irb.cs.uni-magdeburg.de/~ulrich/*.

Acknowledgment The authors wish to thank Alex Petrenko for a very fruitful discussion and for his hints that helped to improve the quality of the paper.

REFERENCES

[ADL+91] A. V. Aho, A. T. Dahbura, D. Lee, M. Ü. Uyar: *An optimization technique for protocol conformance test generation based on UIO sequences and rural Chinese postman tours*; IEEE Transactions on Communications, vol. 39, no. 11 (Nov. 1991); pp. 1604–1615.

[Bri88] Ed Brinksma: *A theory for the derivation of tests*; 8th Int'l Symposium on Protocol Specification, Testing and Verification (PSTV'88), Atlantic City, USA; 1988.

[ERV96]   J. Esparza, S. Römer, W. Vogler: *An improvement of McMillan's unfolding algorithm*; 2nd Int'l Workshop on Tools and Algorithms for the Construction and Analysis of Systems; Passau, Germany; 1996.

[Fer96]   J.-C. Fernandez, C. Jard, Th. Jéron, César Viho: *Using on-the-fly verification techniques for the generation of test suites*; 8th Int'l Conference on Computer Aided Verification (CAV'96); New Brunswick, New Jersy, USA; 1996.

[GaSi90]   H. Garavel, J. Sifakis: *Compilation and verification of Lotos specifications*; 10th Int'l Symposium on Protocol Specification, Testing and Verification (PSTV'90); Ottawa, Canada; 1990; pp. 379–394.

[GrBe96]   B. Grahlmann, E. Best: *PEP – More than a Petri net tool*; 2nd Int'l Workshop on Tools and Algorithms for the Construction and Analysis of Systems (TACAS'96); Passau, Germany; 1996.

[Hen97]   O. Henniger: *On test case generation from asynchronously communicating state machines*; 10th Int'l Workshop on Testing of Communicating Systems (IWTCS'97); Cheju Island, Korea; Sep. 1997.

[KCK+96]   M. C. Kim, S. T. Chanson, S. W. Kang, J. W. Shin: *An approach for testing asynchronous communicating systems*; 9th Int'l Workshop on Testing of Communicating Systems; Darmstadt, Germany; Sep. 1996.

[Lan90]   R. Langerak: *A testing theory for LOTOS using deadlock detection*; 9th Int'l Symposium on Protocol Specification, Testing and Verification (PSTV'90); Enschede, The Netherlands; 1990.

[LSK+93]   D. Lee, K. K. Sabnani, D. M. Kristol, S. Paul: *Conformance testing of protocols specified as communicating FSMs*; IEEE INFOCOM'93; San Fransisco, CA, USA; 1993.

[McM95]   K. L. McMillan: *A technique of state space search based on unfolding*; Formal Methods in System Design, vol. 6, no. 1 (Jan. '95); pp. 45–65.

[Pra86]   V. Pratt: *Modelling Concurrency with partial orders*; International Journal of Parallel Programming, vol. 15, no. 1 (Feb. 1986); pp. 33–71.

[PLL+91]   D. K. Probst, H. F. Li, K. G. Larsen, A. Skou: *Partial-order model checking: a guide for the perplexed*; 3nd Int'l Conference on Computer-aided Verification (CAV'91); Aalborg, Denmark; 1991.

[PBY96]   A. Petrenko, G. v. Bochmann, M. Yao: *On fault coverage of tests for finite state specifications*; Special Issue on Protocol Testing, Computer Networks and ISDN Systems, vol. 29, 1996; pp. 81–106.

[Rei91]   W. Reisig: *Petri nets*; Springer Verlag, 1991.

[TCO91]   K. C. Tai, R. H. Carver, E. E. Obaid: *Debugging concurrent Ada programs by deterministic execution*; IEEE Transactions on Software Engineering, vol. 17, no. 1 (Jan. 1991); pp. 45–63.

[UlCh95]   A. Ulrich, S. T. Chanson: *An approach to testing distributed software systems;* 15th Int'l Symposium on Protocol Specification, Testing and Verification (PSTV'95); Warsaw, Poland; pp. 107–122; 1995.

2

Refusal testing for classes of transition systems with inputs and outputs

Lex Heerink and Jan Tretmans
Tele-Informatics and Open Systems group, Dept. of Computer Science
University of Twente, 7500 AE Enschede, The Netherlands
{heerink,tretmans}@cs.utwente.nl

Abstract

This paper presents a testing theory that is parameterised with assumptions about the way implementations communicate with their environment. In this way some existing testing theories, such as refusal testing for labelled transition systems and (repetitive) quiescence testing for I/O automata, can be unified in a single framework. Starting point is the theory of refusal testing. We apply this theory to classes of implementations which communicate with their environment via clearly distinguishable input and output actions. These classes are induced by making assumptions about the geographical distribution of the points of control and observation (PCO's) and about the way input actions of implementations are enabled. For specific instances of these classes our theory collapses with some well-known ones. For all these classes a single test generation algorithm is presented that is able to derive sound and complete test suites from a specification.

1 INTRODUCTION

An important aspect in the design and construction of systems is to validate whether an implementation operates as it has been specified. This can be done using conformance testing: experiments are conducted on an implementation under test (IUT) and from the observation of responses of the IUT it is concluded whether it behaves correctly. A formalisation of the conformance testing process hence requires formal models for the specification, for the implementation under test, and for experiments and observations, and the

Formal Description Techniques and Protocol Specification, Testing and Verification
T. Mizuno, N. Shiratori, T. Higashino & A. Togashi (Eds.)   © 1997 IFIP. Published by Chapman & Hall

formal definition of a correctness criterion, which is done by means of an implementation relation between models of implementations and specifications.

In formal conformance testing it is assumed that the formal specification is apriori known, and that the behaviour of an implementation can be formally modelled, but its model is not apriori known. The latter is called a test assumption. A well-known test assumption is that implementations can be modelled as labelled transition systems [2, 3, 14] that communicate in a symmetric and synchronous way with their environment; no notion of initiative of actions is present. However, it has been recognised that such symmetric communication is not very realistic. Most implementations communicate in practice with their environment via actions that are clearly initiated by one partner, and accepted by the other [9, 11, 13, 14]. This has triggered research in models that make an explicit distinction between actions that are controlled by the environment (input actions of the implementation) and actions that are controlled by the IUT (output actions of the implementation), e.g., input/output automata (IOA) [9, 13], input/output state machines [11], and input/output transition systems (IOTS) [15]. Many of these models additionally require that input actions are continuously enabled.

As indicated in [4, 15] many implementation relations for labelled transition systems can be defined extensionally in terms of a set of experimenters $\mathcal{U}$, a set of observations $obs(u, i)$ that experimenter $u \in \mathcal{U}$ causes when system i is tested, and a relation $\otimes$ between $obs(u, i)$ and $obs(u, s)$. Formally, this amounts to

$$i \text{ conforms-to } s \ =_{def} \ \forall u \in \mathcal{U} : obs(u, i) \otimes obs(u, s) \qquad (1)$$

One such testing relation is refusal testing [12], where experimenters are not only able to detect whether actions can occur, but also able to detect whether actions can fail. Another example is the extensional characterisation of quiescent trace preorder for input/output automata in [13].

This paper continues the track of extensionally defined implementation relations and testing for models that distinguish between input and output actions. What we add in this paper is the distinction between the different *locations* where these actions may occur on an interface, that is, we explicitly take the distributed nature of interfaces into account. Moreover, we weaken the requirement imposed on IOA and IOTS that input actions for implementations must always be enabled, thereby conciliating the criticism in [13] that this requirement is too restrictive. We obtain classes of models of implementations, one for each possible distribution of the interface. For these classes we apply refusal testing [12] where observers are able to observe the success and failure of actions conducted at each different location separately. We will show that refusal testing for transition systems without inputs and outputs [12] and refusal testing for IOTS [15] are just instances of our parameterised model (for the finest and the coarsest distribution of locations, respectively). In that way the worlds of testing transition systems with, and without, inputs and outputs are unified in a single testing framework. Furthermore, we define an intuitive correctness criterion in the same way as [15] which is manageable for test generation in realistic situations. A single test generation algorithm is given which can cope with each of these classes, and which derives tests that can distinguish between correct and incorrect implementations.

This paper is organised as follows. Section 2 fixes notation and recapitulates refusal testing for labelled transition systems [12]. In section 3 classes of implementation models, parameterised with assumptions about the distribution of the interfaces, are defined and co-related. Refusal testing theory for these classes is discussed in section 4 in a single framework. Section 5 presents a test generation algorithm which is proved to be sound and complete. Section 6 illustrates the operation of the algorithm, and section 7 wraps up with conclusions and further work.

2   REFUSAL TESTING FOR TRANSITION SYSTEMS

We use labelled transition systems to specify and model the behaviour of systems. This section recalls the basics of transition systems and refusal testing without distinguishing between inputs and outputs.

DEFINITION 1   *A (labelled) transition system over L is a quadruple $\langle S, L, \rightarrow, s_0 \rangle$ where S is a (countable) set of states, L is a (countable) set of observable actions, $\rightarrow \subseteq S \times L \times S$ is a set of transitions, and $s_0 \in S$ is the initial state.*

We denote the class of all transition systems over L by $\mathcal{LTS}(L)$. For the notation of transition systems, we use some standard process-algebraic operators, which are defined in the usual way (cf. LOTOS [5]). For this paper it suffices to use action-prefix $\alpha; B$, which can perform action α and then behave as B, and unguarded choice $\sum B$ which can behave as any of its members $B \in \mathcal{B}$. We abbreviate $\sum \{B_1, B_2\}$ by $B_1 + B_2$ and $\sum \emptyset$ by **stop**.

The behaviour of a labelled transition system, starting from a particular state (usually s_0), is expressed using sequences consisting of actions and sets of refused actions, i.e., sequences in $(\mathcal{P}(L) \cup L)^*$ ($\mathcal{P}(L)$ denotes the set of all subsets of L). Such sequences are called *failure traces* [1].

NOTATION 1   Let $p = \langle S, L, \rightarrow, s_0 \rangle$ be a transition system such that $s, s' \in S$, $\alpha, \alpha_i \in \mathcal{P}(L) \cup L$ and $\sigma \in (\mathcal{P}(L) \cup L)^*$. Then

$$s \xrightarrow{\alpha} s' \qquad =_{def} \begin{cases} (s, \alpha, s') \in \rightarrow & \text{if } \alpha \in L \\ s = s' \text{ and } \forall \mu \in \alpha : s \xrightarrow{\mu} \!\!\!/ & \text{if } \alpha \in \mathcal{P}(L) \end{cases}$$

$$s \xrightarrow{\alpha_1 \cdot \alpha_2 \cdots \alpha_n} s' =_{def} \exists s_0, s_1, \ldots, s_n : s = s_0 \xrightarrow{\alpha_1} s_1 \xrightarrow{\alpha_2} \ldots \xrightarrow{\alpha_n} s_n = s'$$

$$s \xrightarrow{\sigma} \qquad =_{def} \exists s' : s \xrightarrow{\sigma} s'$$

The self-loop transitions of the form $s \xrightarrow{A} s$ where $A \subseteq L$ are called *refusal transitions*; A is called a *refusal* of s. Such a refusal transition explicitly encodes the inability to perform any action in A from state s. Refusal transitions can be serialised: $s \xrightarrow{A_1 \cup A_2} s$ iff $s \xrightarrow{A_1} s \xrightarrow{A_2} s$.

DEFINITION 2 *Let $p \in \mathcal{LTS}(L)$, then*

1. $init(p) =_{def} \{\alpha \in L \mid \exists p' : p \xrightarrow{\alpha} p'\}$
2. $f\text{-}traces(p) =_{def} \{\sigma \in (\mathcal{P}(L) \cup L)^* \mid p \xrightarrow{\sigma} \}$
3. $traces(p) =_{def} f\text{-}traces(p) \cap L^*$
4. $p \textbf{ after } \sigma \textbf{ deadlocks } =_{def} \exists p' : p \xrightarrow{\sigma} p'$ and $init(p') \cap L = \emptyset$
5. $der(p) =_{def} \{p' \mid \exists \sigma \in L^* : p \xrightarrow{\sigma} p'\}$
6. $P \textbf{ after } \sigma =_{def} \{p' \mid \exists p \in P : p \xrightarrow{\sigma} p'\}$ where P is a set of states
7. *p is deterministic iff* $\forall \sigma \in L^* : |\ \{p\} \textbf{ after } \sigma\ | \leq 1$
8. *p has finite behaviour iff* $\exists N \in \mathbb{N} : \forall \sigma \in traces(p) : |\sigma| \leq N$

Observers are, just like specifications and implementations, modelled as transition systems. In order for an observer to observe the refusals of a system p we equip an observer u with a special deadlock detection label θ ($\theta \notin L$) [8]. The occurrence of θ indicates that the synchronised behaviour of u and p is not able to continue with any other action than θ, i.e., p refuses all other actions offered by u.

DEFINITION 3 *Let $p \in \mathcal{LTS}(L)$ and $u \in \mathcal{LTS}(L \cup \{\theta\})$, then $\rceil\!|: \mathcal{LTS}(L \cup \{\theta\}) \times \mathcal{LTS}(L) \to \mathcal{LTS}(L \cup \{\theta\})$ is defined by the following inference rules.*

$$\frac{u \xrightarrow{a} u', p \xrightarrow{a} p'}{u \rceil\!| p \xrightarrow{a} u' \rceil\!| p'} \ (a \in L) \qquad \frac{u \xrightarrow{\theta} u', init(u) \cap init(p) = \emptyset}{u \rceil\!| p \xrightarrow{\theta} u' \rceil\!| p}$$

Observations that can be made using an observer u interacting with p by means of $\rceil\!|$ now may include the action θ. This makes it is possible to detect when p was unable to perform any other action offered by u.

DEFINITION 4 *Let $p \in \mathcal{LTS}(L)$ and $u \in \mathcal{LTS}(L \cup \{\theta\})$.*

1. *The set of completed trace observations obs_c^{θ} is*

$$obs_c^{\theta}(u, p) =_{def} \{\sigma \in (L \cup \{\theta\})^* \mid (u \rceil\!| p) \textbf{ after } \sigma \textbf{ deadlocks } \}$$

2. *The set of trace observations obs_t^{θ} is*

$$obs_t^{\theta}(u, p) =_{def} \{\sigma \in (L \cup \{\theta\})^* \mid (u \rceil\!| p) \xrightarrow{\sigma} \}$$

Based on the ability to distinguish processes by means of observations a preorder on processes can be defined extensionally (cf. equation (1)). This preorder, called *refusal preorder*, is known to correspond to inclusion of failure traces [12].

DEFINITION 5 *Refusal preorder $\leq_{rf} \subseteq \mathcal{LTS}(L) \times \mathcal{LTS}(L)$ is defined by*

$$i \leq_{rf} s =_{def} \forall u \in \mathcal{LTS}(L \cup \{\theta\}) : \ \begin{aligned} obs_c^{\theta}(u, i) &\subseteq obs_c^{\theta}(u, s) \ \text{ and} \\ obs_t^{\theta}(u, i) &\subseteq obs_t^{\theta}(u, s) \end{aligned}$$

PROPOSITION 1 $i \leq_{rf} s$ *iff* $f\text{-}traces(i) \subseteq f\text{-}traces(s)$

3 CLASSES OF TRANSITION SYSTEMS

In [9] it is argued that the symmetric synchronisation mechanism between a system and its environment used in e.g., [5] exhibits the counter intuitive property that the system is able to block actions that are supposed to be controlled by its environment, and vice versa. Therefore, models have been developed that distinguish between the initiative (or direction) of actions. In these models either an action is initiated by the environment and accepted by the system (i.e., it is an *input action*), or an action is initiated by the system and accepted by its environment (i.e., it is an *output action*). Output actions for the environment are input actions for the system, and vice versa. By requiring that implementations must always be prepared to accept input actions and the environment must always be prepared to accept output actions, counter-intuitive blocking can no longer occur. Transition systems in which the labelset L is partitioned in a set of inputs L_I and a set of outputs L_U (i.e., $L = L_I \cup L_U$ and $L_I \cap L_U = \emptyset$), and which are always prepared to accept any input are called I/O automata (input/output automata) [9]. Figure 1 depicts the synchronisation between a coffee machine and its environment, both modelled as I/O automata ($L_I = \{coin\}$ and $L_U = \{coffee, tea\}$).

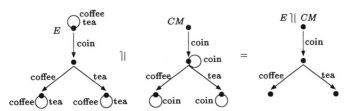

Figure 1 Coffee machine CM and environment E as I/O automata.

In [13, 15] testing theories based on the I/O automaton model (in terms of IOA and IOTS) are presented. These theories suffer from some deficiencies. First, the requirement imposed on IOA and IOTS to always accept input actions is quite strong. Secondly, [13, 15] implicitly assume that the environment is always capable of observing *every* output produced by the system, even if these outputs occur at geographically dispersed places (which is frequently the case if the system under test is distributed), and thereby ignores a possible distribution of the environment itself. In order to overcome these deficiencies we refine the I/O automaton model by taking the distribution of the interaction points (PCO's, points of control and observation [6]) on the interface of a system with its environment into account, and we weaken the requirement that inputs must be always enabled. This is accomplished by partitioning the inputs L_I in pairwise disjunct sets $L_I^1, \ldots, L_I^n$, and, similarly, L_U in pairwise disjunct sets $L_U^1, \ldots, L_I^m$. We shall refer to such sets as *channels*. The idea behind this partitioning is that each set L_I^j (or L_U^k) reflects the location on an

interface where these actions may occur. Furthermore, we weaken the requirement on input enabling to 'if some action in channel L_I^j can be performed,
then all actions in channel L_I^j can be performed".

DEFINITION 6 *A multi input-output transition system p over partitioning $\mathcal{L}_I$
of L_I and partitioning $\mathcal{L}_U$ of L_U is a transition system with inputs and outputs, $p \in \mathcal{LTS}(L_I \cup L_U)$, such that for all $L_I^j \in \mathcal{L}_I$*

$$\forall p' \in der(p), \; if \; \exists a \in L_I^j : p' \xrightarrow{a} \;\;\; then \;\; \forall b \in L_I^j : p' \xrightarrow{b}$$

*The universe of multi input-output transition systems over $\mathcal{L}_I$ and $\mathcal{L}_U$ is de-
noted by $\mathcal{MIOTS}(\mathcal{L}_I, \mathcal{L}_U)$.*

Each particular partitioning $\mathcal{L}_I$ and $\mathcal{L}_U$ induces a class of transition systems
$\mathcal{MIOTS}(\mathcal{L}_I, \mathcal{L}_U) \subseteq \mathcal{LTS}(L_I \cup L_U)$. In order to compare these we define an
ordering $\unlhd$ on partitionings of a set S, where we restrict to finite partitionings.

DEFINITION 7 *Let $Parts(S)$ be the set of all partitionings of S and let $\mathcal{X}, \mathcal{Y} \in
Parts(S)$, then $\mathcal{X} \unlhd \mathcal{Y} =_{def} \forall X \in \mathcal{X}, \exists Y \in \mathcal{Y} : X \subseteq Y$*

The relation $\unlhd$ reflects the ordering on the granularity of the partitioning
involved, and defines a lattice on partitionings. The minimal element is the
partitioning that contains only singleton sets consisting of elements of S:
$\min_{\unlhd}(Parts(S)) = \{\{s\} \mid s \in S\}$. The maximal element is the partitioning that
consists of a single set containing all elements of S: $\max_{\unlhd}(Parts(S)) = \{\{S\}\}$.

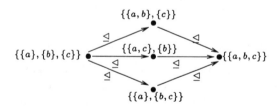

Figure 2 The partial order $\unlhd$ applied to partitionings of $\{a, b, c\}$.

The granularity of the partitioning $\mathcal{L}_I$ uniquely defines the class of potential
system models of implementations: the finer the partitioning of $\mathcal{L}_I$, the larger
the class $\mathcal{MIOTS}(\mathcal{L}_I, \mathcal{L}_U)$. The granularity of the partitioning of the set of
output actions L_U does not influence the class of potential system models of
implementations (cf. definition 6).

PROPOSITION 2 *Let $\mathcal{X}_1, \mathcal{X}_2$ be partitionings of L_I, and let $\mathcal{Y}$ be a partitioning
of L_U, then $\mathcal{X}_1 \unlhd \mathcal{X}_2$ implies $\mathcal{MIOTS}(\mathcal{X}_1, \mathcal{Y}) \supseteq \mathcal{MIOTS}(\mathcal{X}_2, \mathcal{Y})$*

Specific instances of these partitionings yield some well-known classes. In particular, for the finest partitioning on L_I and L_U the requirement imposed on multi input-output transition systems trivially becomes true, and the set $\mathcal{MIOTS}(\min_{\lhd}(Parts(L_I)), \min_{\lhd}(Parts(L_U)))$ equals the set $\mathcal{LTS}(L_I \cup L_U)$. The set $\mathcal{MIOTS}(\max_{\lhd}(Parts(L_I)), \max_{\lhd}(Parts(L_U)))$ collapses with IOTS [15] in case input actions are always enabled.

4   TESTING MULTI INPUT-OUTPUT TRANSITION SYSTEMS

We give an extensional comparison criterion, cf. equation (1), that decides which implementations, modelled as MIOTS, can be distinguished by external observers, and which cannot. The set of external observers $\mathcal{U}$ are assumed to be modelled as MIOTS, too: an observer is able to accept all outputs at a specific location that are produced by the implementation as long as the observer is able to accept only one of them. Furthermore, in order to observe the inability to accept an input action, or the inability to produce an output action, observers are (analogous to definition 3) equipped with deadlock detection labels. This time we use different deadlock detection labels for each channel: $\theta_i^j \notin L_I \cup L_U$ to observe the inability of the implementation to accept an input action in channel L_I^j, and $\theta_u^k \notin L_I \cup L_U$ to observe the inability to produce outputs in channel L_U^k.

Now, implementations that are modelled as members of $\mathcal{MIOTS}(\mathcal{L}_I, L_U)$ are observed by observers modelled in $\mathcal{MIOTS}(\mathcal{L}_U^\theta, \mathcal{L}_I^\theta)$, where $\mathcal{L}_U^\theta =_{def} \{L_U^1 \cup \{\theta_u^1\}, \ldots, L_U^m \cup \{\theta_u^m\}\}$ and $\mathcal{L}_I^\theta =_{def} \{L_I^1 \cup \{\theta_i^1\}, \ldots, L_I^n \cup \{\theta_i^n\}\}$. Communication between observer and system is modelled by operator $][$. The set Θ denotes the set of all deadlock detection labels: $\Theta =_{def} \{\theta_i^1, \ldots, \theta_i^n, \theta_u^1, \ldots, \theta_u^m\}$.

DEFINITION 8 *Let* $\mathcal{L}_I = \{L_I^1, \ldots, L_I^n\}$ *be a finite partitioning of* L_I *and let* $\mathcal{L}_U = \{L_U^1, \ldots, L_U^m\}$ *be a finite partitioning of* L_U, *then operator* $][$: $\mathcal{MIOTS}(\mathcal{L}_U^\theta, \mathcal{L}_I^\theta) \times \mathcal{LTS}(L_I \cup L_U) \to \mathcal{LTS}(L_I \cup L_U \cup \Theta)$ *is defined by the following inference rules.*

$$\frac{u \xrightarrow{a} u', p \xrightarrow{a} p'}{u\,][\,p \xrightarrow{a} u'\,][\,p'} \ (a \in L_I \cup L_U) \qquad \frac{u \xrightarrow{\theta_i^j} u', init(p) \cap L_I^j = \emptyset}{u\,][\,p \xrightarrow{\theta_i^j} u'\,][\,p} \ (j \in \{1, \ldots, n\})$$

$$\frac{u \xrightarrow{\theta_u^k} u', init(p) \cap L_U^k = \emptyset}{u\,][\,p \xrightarrow{\theta_u^k} u'\,][\,p} \ (k \in \{1, \ldots, m\})$$

An observer $u \in \mathcal{MIOTS}(\mathcal{L}_U^\theta, \mathcal{L}_I^\theta)$ that communicates with a system $p \in \mathcal{LTS}(L_I \cup L_U)$ may perform sequences of actions in $L_I \cup L_U$, possibly interleaved with deadlock detection labels θ_i^j and θ_u^k. Similar to definition 4, we define such sequences as the observations that can be made of such a system, thereby overloading the notations obs_c^θ and obs_t^θ.

DEFINITION 9 *Let* $p \in \mathcal{LTS}(L_I \cup L_U)$ *and* $u \in \mathcal{MIOTS}(\mathcal{L}_U^{\theta}, \mathcal{L}_I^{\theta})$.

1. The set of completed trace observations obs_c^{θ} *is*

$$obs_c^{\theta}(u, p) =_{def} \{\sigma \in (L_I \cup L_U \cup \Theta)^* \mid (u \rrbracket p) \textbf{ after } \sigma \textbf{ deadlocks }\}$$

2. The set of trace observations obs_t^{θ} *is*

$$obs_t^{\theta}(u, p) =_{def} \{\sigma \in (L_I \cup L_U \cup \Theta)^* \mid (u \rrbracket p) \xrightarrow{\sigma}\}$$

Now, following equation (1), and in the same line as definition 5, refusal preorder is defined under the assumption that implementations are modelled as members in $\mathcal{MIOTS}(\mathcal{L}_I, \mathcal{L}_U)$. However, we do not require this for specifications; specifications are just labelled transition systems over $L_I \cup L_U$.

DEFINITION 10 *The relation* $\leq_{mior} \subseteq \mathcal{MIOTS}(\mathcal{L}_I, \mathcal{L}_U) \times \mathcal{LTS}(L_I \cup L_U)$, *called multi input-output refusal preorder, is defined by*

$$i \leq_{mior} s =_{def} \forall u \in \mathcal{MIOTS}(\mathcal{L}_U^{\theta}, \mathcal{L}_I^{\theta}) : \; obs_c^{\theta}(u, i) \subseteq obs_c^{\theta}(u, s) \; and$$
$$obs_t^{\theta}(u, i) \subseteq obs_t^{\theta}(u, s)$$

Conceptually, when an observer experiments on an implementation that is modelled as MIOTS it can either provide inputs at an input channel (e.g., press a button), or observe outputs from an output channel (e.g., view a display). For each input channel the observer is equipped with a "finger" to perform a button-push experiment, and for each output channel the observer is equipped with an "eye" that notices the output actions occurring on the display (figure 3). By assumption, output actions at a specific location cannot be selectively perceived by observers: if one output can be observed, then all output actions at the same location can potentially be observed. Furthermore, it is assumed that unsuccessful input experiments and output experiments are noticed by the observer. Figure 3 depicts an interface for MIOTS.

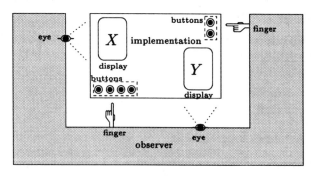

Figure 3 Observers of multi input-output transition systems.

A special class of observers are the *singular observers*. They consist of finite, serial compositions of providing a single input action at some channel L_I^j and detection of its acceptance or rejection, and observing some channel L_U^k and detection of the occurrence, or absence, of outputs produced at this channel. It turns out that it suffices to restrict to singular observers in order to establish whether implementations are $\leq_{mior}$-correct or not.

DEFINITION 11 *A singular observer u over $\mathcal{L}_U$ and $\mathcal{L}_I$ is a finite, deterministic multi input-output transition system $u \in \mathcal{MIOTS}(\mathcal{L}_U^\theta, \mathcal{L}_I^\theta)$ such that*

$$\forall u' \in der(u): \; init(u') = \emptyset \;\; or \;\; init(u') = L_U^k \cup \{\theta_u^k\} \;\; or \;\; init(u') = \{a, \theta_i^j\}$$

for some $j \in \{1, \dots, n\}, k \in \{1, \dots, m\}$ and $a \in L_I^j \in \mathcal{L}_I$. The set of all singular observers over $\mathcal{L}_U$ and $\mathcal{L}_I$ is denoted by $\mathcal{SOBS}(\mathcal{L}_U^\theta, \mathcal{L}_I^\theta)$.

PROPOSITION 3 *Let $i \in \mathcal{MIOTS}(\mathcal{L}_I, \mathcal{L}_U)$ and $s \in \mathcal{LTS}(L_I \cup L_U)$, then*

$$i \leq_{mior} s$$

iff $\forall u \in \mathcal{SOBS}(\mathcal{L}_U^\theta, \mathcal{L}_I^\theta) : obs_c^\theta(u,i) \subseteq obs_c^\theta(u,s)$ *and* $obs_t^\theta(u,i) \subseteq obs_t^\theta(u,s)$

iff $\forall u \in \mathcal{SOBS}(\mathcal{L}_U^\theta, \mathcal{L}_I^\theta) : obs_c^\theta(u,i) \subseteq obs_c^\theta(u,s)$

iff $f\text{-}traces(i) \cap (L_I \cup L_U \cup \mathcal{L}_I \cup \mathcal{L}_U)^* \subseteq f\text{-}traces(s)$

Since each singular observer is composed of actions that are able to detect whether an input at channel L_I^j is accepted or not, and observations that are able to detect whether outputs are produced at some channel L_U^k or not, it follows that execution of singular observers only ends in case no more actions can be conducted; the only way for a test execution $u]|i$ to deadlock is by deadlock of u.

PROPOSITION 4 *Let $u \in \mathcal{SOBS}(\mathcal{L}_U^\theta, \mathcal{L}_I^\theta)$ and $p \in \mathcal{MIOTS}(\mathcal{L}_I, \mathcal{L}_U)$, then*

$(u]|p)$ **after** σ **deadlocks** *implies* u **after** σ **deadlocks**

The observation that can be made from observer u communicating with system p uniquely determines the failure trace that was performed by p. This is possible because every observation of θ_i^j and θ_u^k in $u]|p$ corresponds to refusal of L_I^j and L_U^k, respectively. We denote with $\bar{\sigma}$ the trace σ where each occurrence of a refusals L_I^j or L_U^k is replaced by its detection label θ_i^j or θ_u^k, and vice versa.

PROPOSITION 5 *Let $p \in \mathcal{LTS}(L_I \cup L_U)$ and $u \in \mathcal{SOBS}(\mathcal{L}_U^\theta, \mathcal{L}_I^\theta)$, then for any $\sigma \in (L_I \cup L_U \cup \Theta)^*$*

$u]|p \xrightarrow{\sigma} u']|p' \quad iff \quad u \xrightarrow{\sigma} u' \; and \; p \xrightarrow{\bar{\sigma}} p'$

Yet another characterisation of the relation $\leq_{mior}$ exists that is based on the responses that the implementation can produce after having performed a specific trace. These responses consist of the output suspension labels (δ^k) indicating that the implementation is in a state that cannot produce an output at channel L_U^k, the input suspension labels (ξ^j) indicating that the implementation is in a state that cannot accept any input from channel L_I^j, and the outputs in L_U that the implementation can produce in the current state. All these responses are collected in the set *out*.

DEFINITION 12 *Let* $p \in \mathcal{LTS}(L_I \cup L_U)$ *and* $\sigma \in (L_I \cup L_U \cup \mathcal{L}_I \cup \mathcal{L}_U)^*$, *then the set* out(p after σ) *is defined by*

$$out(\,p\,\text{after}\,\sigma\,) =_{def} \quad \{x \in L_U \mid \exists p' : p \xrightarrow{\sigma} p' \xrightarrow{x} \}$$

$$\cup \{\xi^j \mid 1 \leq j \leq n, \exists p' : p \xrightarrow{\sigma} p' \text{ and } init(p') \cap L_I^j = \emptyset\}$$

$$\cup \{\delta^k \mid 1 \leq k \leq m, \exists p' : p \xrightarrow{\sigma} p' \text{ and } init(p') \cap L_U^k = \emptyset\}$$

The inability to accept input at channel L_I^j (i.e., input suspension) and the inability to produce output at channel L_U^k (i.e., output suspension) is now explicitly visible in terms of the input suspension labels ξ^j and the output suspension labels δ^k, respectively. It turns out that an implementations is $\leq_{mior}$-related to a specification in case all responses that the implementation can perform after a trace in $(L_I \cup L_U \cup \mathcal{L}_I \cup \mathcal{L}_U)^*$ are specified, i.e., an implementation is not allowed to suspend at some channel in case this is not specified, and the implementation is not allowed to produce unspecified outputs.

PROPOSITION 6 *Let* $i \in \mathcal{MIOTS}(\mathcal{L}_I, \mathcal{L}_U)$ *and* $s \in \mathcal{LTS}(L_I \cup L_U)$, *then*

$$i \leq_{mior} s \quad iff \quad \forall \sigma \in (L_I \cup L_U \cup \mathcal{L}_I \cup \mathcal{L}_U)^* : out(\,i\,\text{after}\,\sigma\,) \subseteq out(\,s\,\text{after}\,\sigma\,)$$

Checking the condition in proposition 6 for all traces in $(L_I \cup L_U \cup \mathcal{L}_I \cup \mathcal{L}_U)^*$ is too time consuming in practice. Therefore, we generalise this condition to an arbitrary (and possible finite) set $\mathcal{F} \subseteq (L_I \cup L_U \cup \mathcal{L}_I \cup \mathcal{L}_U)^*$, and define a corresponding implementation relation **mioco**$_\mathcal{F}$ in the same way as **ioco**$_\mathcal{F}$ in [15]. We will use this relation in the next section as the basis for deriving tests.

DEFINITION 13 *The implementation relation* **mioco**$_\mathcal{F} \subseteq \mathcal{MIOTS}(\mathcal{L}_I, \mathcal{L}_U) \times \mathcal{LTS}(L_I \cup L_U)$, *where* $\mathcal{F} \subseteq (L_I \cup L_U \cup \mathcal{L}_I \cup \mathcal{L}_U)^*$, *is defined by*

$$i \text{ mioco}_\mathcal{F} s =_{def} \forall \sigma \in \mathcal{F} : out(\,i\,\text{after}\,\sigma\,) \subseteq out(\,s\,\text{after}\,\sigma\,)$$

Furthermore, we define **mioco** $=_{def}$ **mioco**$_{f\text{-}traces(s) \cap (L_I \cup L_U \cup \mathcal{L}_I \cup \mathcal{L}_U)^*}$.

We remark here that, in general, observers in $\mathcal{SOBS}(\mathcal{L}_U^\theta, \mathcal{L}_I^\theta)$ are more

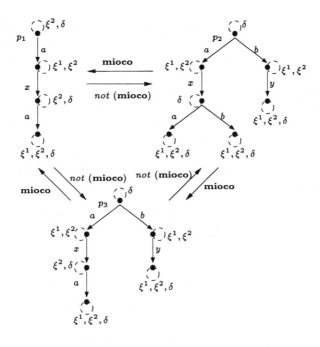

Figure 4 The relation **mioco** with $\mathcal{L}_I = \{\{a\}, \{b\}\}$ and $\mathcal{L}_U = \{\{x, y\}\}$.

powerful than observers for input/output automata or IOTS (cf. [13, 15]) due to their ability to observe output suspension at different output channels: singular observers can distinguish between systems that are unable to produce output actions at one channel, while at another channel the system is able to produce output actions. In terms of the relation $\trianglelefteq$ on partitions, this means that the finer the outputs are partitioned (i.e., the more output channels are present), the more selectively observers are able to observe. In particular, for the finest partitioning of inputs and the finest partitioning of outputs our relation $\leq_{mior}$ collapses with $\leq_{rf}$, while we claim that for the coarsest partitioning of the inputs and the outputs the relation $\leq_{mior}$ collapses with **ioco** [15] in case it is assumed that for implementations inputs are always enabled.

PROPOSITION 7 *Let* $i \in \mathcal{MIOTS}(\min_{\trianglelefteq}(Parts(L_I)), \min_{\trianglelefteq}(Parts(L_U)))$ *and* $s \in \mathcal{LTS}(L_I \cup L_U)$ *such that* $\min_{\trianglelefteq}(Parts(L_I))$ *and* $\min_{\trianglelefteq}(Parts(L_U)))$ *are finite, then* $i \leq_{rf} s \quad iff \quad i \leq_{mior} s$

5 TEST GENERATION FOR MIOTS

In this section we develop an algorithm to derive tests systematically from a specification such that these tests are able to reject implementations that are

mioco$_\mathcal{F}$-incorrect, and accept implementations that are **mioco$_\mathcal{F}$**-correct. The algorithm depends on the specification (modelled as a member of $\mathcal{LTS}(L_I \cup L_U)$), the correctness criterion (**mioco$_\mathcal{F}$** for some $\mathcal{F}$), and the test assumption (implementations are modelled as members of $\mathcal{MIOTS}(\mathcal{L}_I, \mathcal{L}_U)$).

Test cases need to incorporate some kind of verdict that can be used to give such an indication about the (in)correctness of implementations when running these test cases against implementations. We distinguish between two kinds of verdicts: **pass** to indicate that the implementation behaved as expected, and **fail** to indicate that the implementation behaved erroneously (cf. [6, 7]). We define a test as a member of $\mathcal{SOBS}(\mathcal{L}_U^\theta, \mathcal{L}_I^\theta)$ where the final states are identified with the verdicts **pass** or **fail**.

DEFINITION 14 *A test t over $\mathcal{L}_I^\theta$ and $\mathcal{L}_U^\theta$ is a singular observer $t \in \mathcal{SOBS}(\mathcal{L}_U^\theta, \mathcal{L}_I^\theta)$ such that for each $t' \in der(t)$*

$$init(t') = \emptyset \quad iff \quad t' = \pmb{pass} \ \ or \ \ t' = \pmb{fail}$$

The universe of tests over $\mathcal{L}_I^\theta$ and $\mathcal{L}_U^\theta$ is denoted by $TESTS(\mathcal{L}_U^\theta, \mathcal{L}_I^\theta)$.

Since tests will always end in a final state of the test (proposition 4) every test run is assigned a verdict, viz., the verdict of the final state of the test. Trace σ is a test run of $t][i$ iff $\sigma \in obs_c^\theta(t, i)$. An implementation i fails test t if there exists a test run of $t][i$ leading to a **fail** state, (i.e., i *fails* t $=_{def}$ $\exists \sigma \in obs_c^\theta(t, i), \exists i' : t][i \xrightarrow{\sigma} \pmb{fail}][i')$, and implementation i passes test t if it does not fail t. Implementation i fails a set of tests T if i fails a test $t \in T$, otherwise it passes T.

Soundness, exhaustiveness and completeness [7] are properties of test suites (i.e., sets of tests) that link the passing or failing of test suites to the correctness of the implementations. A test suite is called sound if this test suite will never reject **mioco$_\mathcal{F}$**-correct implementations, and a test suite is called exhaustive if each incorrect implementation always fails this test suite. In practice test suites are required to be sound, but not necessarily exhaustive; any error that is detected by a test suite indeed proves that the implementation under test was incorrect, but not finding an error does not mean that the implementation is error free! A test suite is called complete if it is both sound and exhaustive.

Figure 5 presents a test generation algorithm Π that produces tests that are able to distinguish between **mioco$_\mathcal{F}^\theta$**-correct and **mioco$_\mathcal{F}^\theta$**-incorrect implementations. The rationale behind the test algorithm is that it construct tests that check the condition

$$out(\, i \ \pmb{after} \ \sigma \,) \subseteq out(\, s \ \pmb{after} \ \sigma \,)$$

for $\sigma \in \mathcal{F}$ (cf. definition 13). The test generation algorithm takes a specification $s \in \mathcal{LTS}(L_I \cup L_U)$ and a set of failure traces $\mathcal{F} \subseteq (L_I \cup L_U \cup \mathcal{L}_I \cup \mathcal{L}_U)^*$, and produces tests in $TESTS(\mathcal{L}_U^\theta, \mathcal{L}_I^\theta)$. The variable S keeps track of the current states in the specification, which initially equals $\{s_0\} \ \pmb{after} \ \epsilon$, and the variable $\mathcal{F}$ keeps track of the failure traces that need to be investigated in order to establish correctness. Each time an action is performed the sets S and $\mathcal{F}$ are updated accordingly.

Input: set of states S
Input: set of failure traces $\mathcal{F} \subseteq (L_I \cup L_U \cup \mathcal{L}_I \cup \mathcal{L}_U)^*$
Output: test case $\Pi_{\mathcal{F},S} \in TESTS(\mathcal{L}_U^\theta, \mathcal{L}_I^\theta)$.

Initial value: $S = \{s_0\}$ **after** ϵ, where s_0 is the initial state of s.

Apply one of the following non-deterministic choices recursively.

1. (* terminate the test case if there are no more specified traces in $\mathcal{F}$ *)
   if $\mathcal{F} = \emptyset$ then
   $$\Pi_{\mathcal{F},S} := \mathbf{pass}$$

2. (* terminate the test case when a trace $\sigma \in \mathcal{F}$ has been performed *)
   if $\epsilon \in \mathcal{F}$ then take some $L_I^j \in \mathcal{L}_I$, and for some $a \in L_I^j$ (* supply input a *)
   $$\Pi_{\mathcal{F},S} := \begin{cases} a;\mathbf{pass} + \theta_i^j;\mathbf{fail} & \text{if } S \text{ after } L_I^j = \emptyset \\ a;\mathbf{pass} + \theta_i^j;\mathbf{pass} & \text{if } S \text{ after } L_I^j \neq \emptyset \end{cases}$$

3. (* terminate the test case when a trace $\sigma \in \mathcal{F}$ has been performed *)
   if $\epsilon \in \mathcal{F}$ then take some $L_U^k \in \mathcal{L}_U$, then (* observe channel L_U^k *)
   $$\Pi_{\mathcal{F},S} := \quad \sum\{x;\mathbf{pass} \mid x \in L_U^k \cup \{\theta_u^k\} \text{ and } S \text{ after } \overline{x} \neq \emptyset\}$$
   $$+ \sum\{x;\mathbf{fail} \mid x \in L_U^k \cup \{\theta_u^k\} \text{ and } S \text{ after } \overline{x} = \emptyset\}$$

4. (* supply an input for which you want to test deeper *)
   Take some $L_I^j \in \mathcal{L}_I$ and $a \in L_I^j$ such that $\{\sigma \mid a \cdot \sigma \in \mathcal{F}\} \neq \emptyset$, then
   $$\Pi_{\mathcal{F},S} := a; \Pi_{\mathcal{F}',S'} + \theta_i^j;\mathbf{pass}$$
   where $S' = S$ **after** a, $\mathcal{F}' = \{\sigma \mid a \cdot \sigma \in \mathcal{F}\}$

5. (* supply some input and continue if it is refused *)
   Take some $L_I^j \in \mathcal{L}_I$ such that $\{\sigma \mid L_I^j \cdot \sigma \in \mathcal{F}\} \neq \emptyset$, then
   $$\Pi_{\mathcal{F},S} := a;\mathbf{pass} + \theta_i^j; \Pi_{\mathcal{F}'',S''}$$
   where $a \in L_I^j$, $S'' = S$ **after** L_I^j, $\mathcal{F}'' = \{\sigma \mid L_I^j \cdot \sigma \in \mathcal{F}\}$

6. (* Find a channel L_U^k that produces an output for which to test deeper *)
   Take some $L_U^k \in \mathcal{L}_U$ such that $\{\sigma \mid \exists x \in L_U^k \cup \{L_U^k\} : x \cdot \sigma \in \mathcal{F}\} \neq \emptyset$, then
   $$\Pi_{\mathcal{F},S} := \sum\{x; \Pi_{\mathcal{F}',S'} \mid x \in L_U^k \cup \{\theta_u^k\} \quad \text{and } \mathcal{F}' = \{\sigma \mid \overline{x} \cdot \sigma \in \mathcal{F}\}$$
   $$\text{and } S' = S \text{ after } \overline{x}\}$$

Figure 5 Test generation algorithm.

Step 1 of the algorithm assigns **pass** in case no failure trace in $\mathcal{F}$ was performed (e.g., because the implementation responds with an output action that is not checked for in $\mathcal{F}$). Step 2 of the algorithm checks for all input channels whether the implementation is allowed to suspend input. Note that S **after** $L_I^j = \emptyset$ means that there is no state in S that can perform refusal transition L_I^j. Step 3 checks for all output channels whether all outputs that the

implementation can produce are indeed specified. Step 4 supplies an input to
the implementation at some channel L_I^j and continues if the implementation
is able to accept this input. Step 5 also supplies an input to the implementation at some channel L_I^j but now the algorithm recursively proceeds if the
input is refused. Finally, step 6 awaits an output action or observes an output suspension at output channel L_U^k after which the algorithm recursively
proceeds.

Note that the algorithm is guaranteed to finish in case the set $\mathcal{F}$ contains
a finite number of failure traces; in every step the length of the failure traces
in $\mathcal{F}$ are reduced, and since all failure traces in $\mathcal{F}$ are (by definition) finite
eventually step 2 or step 3 will always be applied.

PROPOSITION 8 *Let $\mathcal{F} \subseteq (L_I \cup L_U \cup \mathcal{L}_I \cup \mathcal{L}_U)^*$ and $s \in \mathcal{LTS}(L_I \cup L_U)$*
1. *Any test case obtained from algorithm Π for s and $\mathcal{F}$ is sound for s with
   respect to* **mioco$_\mathcal{F}$**.
2. *The set of all test cases that can be obtained from algorithm Π for s and $\mathcal{F}$
   is complete for s with respect to* **mioco$_\mathcal{F}$**.

REMARK 1 The algorithm presented in figure 5 can be seen as an extension
of the one presented in [15] in two ways. First of all, [15] considers implementations that are modelled as IOTS, so refusal of input is not considered.
Secondly, the algorithm in [15] is not able to deal with the different input
channels and different output channels on interfaces of implementations.

Although our algorithm is applicable to different classes of implementations,
the algorithm in [15] is (probably) more efficient in deriving tests for IOTS
than ours; it is likely that they need less tests to obtain a complete test suite
for these kind of systems than we do.

6 ILLUSTRATION OF THE ALGORITHM

Consider the coffee machine *CM* depicted in figure 6. After insertion of a
coin (*coin*) a user may press either the coffee button (*cb*) or the tea button (*tb*), which results in the production of coffee (*cof*) or tea (*tea*), respectively. There are two distinct input channels (a channel to insert coins and
a channel to push buttons) and a single output channel for providing coffee
or tea: $CM \in \mathcal{MIOTS}(\{\{coin\}, \{cb, tb\}\}, \{\{cof, tea\}\})$. The dashed arrows labelled ξ^1, ξ^2 and δ^1 denote refusal transitions for the sets $\{coin\}, \{cb, tb\}$ and
$\{cof, tea\}$, respectively.

Figure 6 also depicts some tests that are derived from *CM* for $\mathcal{F} = \{\epsilon, coin\cdot
cb\}$ using algorithm Π (see figure 5). For readability the steps of the algo-
rithm that were applied are indicated in the nodes of the tests. Tests (a) is
an immediate consequence of step (2) of the algorithm, and test (b) an immediate consequence of step (3). Test (a) checks that implementations initially
must accept a coin (refusal of a coin gives a **fail** verdict), and test (b) checks

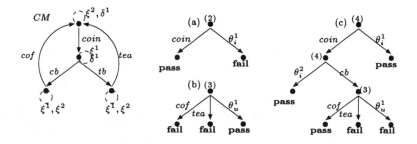

Figure 6 Some tests generated by Π from *CM*.

that implementations are initially not allowed to provide free drinks. Test (c) follows from the successive application of step (4), again step (4), and step (3). It checks that after *coin·cb* the production of tea, or the suspension of providing a drink, is considered incorrect.

Note that algorithm Π may produce tests that always return the verdict **pass** (e.g., *cb*·**pass** + θ_i^2·**pass**). Execution of such tests is not very sensible. The derivation of such meaningless tests indicates that the algorithm is not optimal and that there is room for improvement.

7 CONCLUSIONS AND FURTHER WORK

Conclusions   In this paper the theory of refusal testing [12] has been applied to several classes of transition systems that distinguish between the initiative of actions: either input or output. Each class is induced by the distribution of the locations through which these systems communicate with their environment. In this way a refusal testing theory is obtained that is parameterised by the distribution of the interface of implementations. Specific choices for the interfaces yield the seminal refusal testing theory of [12], and the (repetitive) quiescent trace testing theory for I/O automata [13] and for input/output transition systems [14, 15]. For the large variety of classes of transition systems that can be obtained, a correctness criterion **mioco**$_\mathcal{F}$ (definition 13) is defined that is explicitly parameterised by a set of failure traces $\mathcal{F}$. For all these classes of systems and the corresponding correctness criteria a single test generation algorithm (figure 5) is defined that is able to produce a sound and complete test suite from a specification. This algorithm is an extension of the one in [15]: that one is applicable to a smaller class of systems and is not parameterised over the distribution of the interface of implementations.

Further work   The test generation algorithm Π can produce a large, and possibly infinite, number of tests. Since it is not feasible to execute all of them, techniques have to be developed to measure the relevance of tests (coverage), to select the most relevant tests from a larger set of tests (test selection), or to avoid the generation of irrelevant tests. Furthermore, the test generation algorithm needs to handle data in a symbolic way in order to avoid explosion of the state space, and keep test generation manageable. Since the correctness criterion **mioco**$_\mathcal{F}$ is based on traces, i.e., linear sequences only, an explosion due to the branching structure of specifications (e.g., as in [2]) is avoided. Also, mechanisms to observe the suspension of input or output have to be developed, e.g., making use of timers: if no action occurs before the time-out it is assumed that no action can occur anymore. This requires techniques to carefully choose the timer values such that no incorrect suspension can be observed. Furthermore, the relation between MIOTS and input-complete Finite State Machines needs to be investigated (see, e.g., [10]).

8   REFERENCES

[1] J.C.M. Baeten and W.P. Weijland. *Process Algebra*. Number 18 in Cambridge Tracts in Theoretical Computer Science. Cambridge University Press, 1990.

[2] E. Brinksma. A theory for the derivation of tests. In S. Aggarwal and K. Sabnani, eds., *PSTV VIII*, p. 63–74. North-Holland, 1988.

[3] R. De Nicola. Extensional equivalences for transition systems. *Acta Informatica*, 24:211–237, 1987.

[4] R. De Nicola and M.C.B. Hennessy. Testing equivalences for processes. *Theoretical Computer Science*, 34:83–133, 1984.

[5] ISO. *LOTOS - A Formal Description Technique Based on the Temporal Ordering of Observational Behaviour*. IS-8807. ISO, Geneve, 1989.

[6] ISO. *Conformance Testing Methodology and Framework*. IS-9646. ISO, Geneve, 1991. Also: CCITT X.290–X.294.

[7] ISO/IEC JTC1/SC21 WG7, ITU-T SG 10/Q.8. *Proposed ITU-T Z.500 and Committee Draft on "Formal Methods in Conformance Testing"*. CD 13245-1. ISO – ITU-T, Geneve, 1996.

[8] R. Langerak. A testing theory for LOTOS using deadlock detection. In E. Brinksma et. al., eds., *PSTV IX*, p. 87–98. North-Holland, 1990.

[9] N.A. Lynch and M.R. Tuttle. An introduction to input/output automata. *CWI Quarterly*, 2(3):219–246, 1989.

[10] A. Petrenko, G. v. Bochmann, and R. Dssouli. Conformance relations and test derivation. In O. Rafiq, ed., *IWPTS VI*, p. 157–176. North-Holland, 1993.

[11] M. Phalippou. *Relations d'Implantation et Hypothèses de Test sur des Automates à Entrées et Sorties*. PhD thesis, L'Université de Bordeaux I, 1994.

[12] I. Phillips. Refusal testing. *Theoretical Computer Science*, 50(2):241–284, 1987.

[13] R. Segala. Quiescence, fairness, testing, and the notion of implementation. In E. Best, ed., *CONCUR'93*, p. 324–338. LNCS 715, Springer-Verlag, 1993.

[14] J. Tretmans. Conformance testing with labelled transition systems: Implementation relations and test generation. *Computer Networks and ISDN Systems*, 29:49–79, 1996.

[15] J. Tretmans. Test generation with inputs, outputs, and repetitive quiescence. *Software - Concepts and Tools*, 17:103–120, 1996.

3

A Framework for Distributed Object-Oriented Testing

*Alan C.Y. Wong, Samuel T. Chanson, S.C. Cheung
and Holger Fuchs**

*Department of Computer Science, Hong Kong University of Science
and Technology Clear Water Bay, Hong Kong*

Abstract

Distributed programming and object-oriented programming are two popular
programming paradigms. The former is driven by advances in networking
technology whereas the latter provides vigorous software principles needed in
developing complex software systems. While more and more distributed object-
oriented software has appeared, not much work exists on the testing of these
systems in an integrated manner. Instead, the distributed and object features have
been tested separately. In this paper, we propose an integrated framework known
as DOOT for incremental testing of distributed object-oriented software systems.
It combines various testing techniques to provide comprehensive test coverage at
four levels - class testing, intra-cluster testing, inter-cluster testing and system
testing. Each level uses a specific fault model, test strategy and test case
generation that build on the previous test level to reduce the overall test effort.
They are designed to handle the distinct requirements of the two paradigms at
each level. Moreover, a reduction algorithm for testing the inherited class is also
included in the framework. The approach is illustrated using a real life example
of a conferencing system.

1. INTRODUCTION

The popularity of the Internet coupled with advances in local area network and
high speed network technologies have introduced many new distributed
applications. Object-oriented techniques are often used to cope with the
complexity of developing these software systems which have come to be known as
distributed object-oriented software systems (DOOSS). Like other software these
systems must be thoroughly tested before use. Software testing deals with
checking the correctness of the implementation against its formal specification
[10, 20]. Much work has been done on verifying [13] or testing [6, 7, 9, 15, 16,
19, 23, 24] object-oriented software. However, the validation [8] or testing [11,

---

* Holger Fuchs is currently with the Brandenburg University of Technology at Cottbus, Germany.

Formal Description Techniques and Protocol Specification, Testing and Verification
T. Mizuno, N. Shiratori, T. Higashino & A. Togashi (Eds.) © 1997 IFIP. Published by Chapman & Hall

12, 25, 26, 27] of distributed systems have received less attention. Moreover, few existing work has specifically addressed the issue of combining the two types of testing methodologies in testing DOOSS.

This paper reports a framework for distributed object-oriented testing known as DOOT. The objective is to integrate the testing techniques for distributed systems and object-oriented software into an integrated framework while minimizing the testing effort. DOOT unifies the results at different test levels and reduces the global test space to a manageable size. The test levels are class testing, intra-cluster testing, inter-cluster testing and system testing. Each level has its own fault model, test strategy and test case generation scheme that addresses the specific requirements of the distributed and object paradigms at their level of abstraction. Each level utilises the test results from the previous levels so that the overall test effort is minimized.

The rest of this paper is organised as follows. Section 2 describes the general structure of DOOSS and their advantages as well as the problems in testing them. The DOOT framework is outlined in Section 3. Sections 4, 5, 6 and 7 present the details of each level of testing, namely class testing, intra-cluster testing, inter-cluster testing and system testing respectively. Section 8 concludes the paper, and the application of our approach is illustrated using a conferencing system in the Appendix.

2.   DISTRIBUTED OBJECT-ORIENTED SOFTWARE SYSTEMS

Due to the many useful software construction features such as encapsulation, abstraction, inheritance, genericity and dynamic binding, the object model is widely adopted in both academic research and in the industry in recent years. These features also help to promote software reusability, maintainability, reliability and performance [5]. Distributed systems have also become increasingly common as more and more organisations use networks to share resources, enhance communication and increase performance. Examples of these systems range from the Internet, to workstations in a local area network within a building, to processors within a single multiprocessor [14]. They are characterised by the presence of independent activities, loosely coupled parallelism, heterogeneous software and hardware.

Increasingly, the object model and distributed technologies are being amalgamated [1, 4]. The advantage is obvious: the complexity and dependencies of the entities can make use of the object model in a distributed system to break down the intensive design process into efficient constructs. Moreover, the techniques contributed by CORBA, OLE and WWW have greatly promoted the approach of distributed object technology [22].

However, DOOSS presents a major challenge for testing and maintenance. Many problems known to object-oriented systems are compounded in the distributed system environment, especially those related to concurrency. Some solutions have been proposed in each area individually. In general, these approaches adopt traditional testing and analytic techniques that work well in either sequential programs [10, 23] or communication protocols [8, 12]. The major problem of state space explosion in DOOSS remains unsolved. Moreover, the solutions for object-oriented testing are not handled satisfactorily. These include the efficient use of inheritance to reduce test effort for derived classes.

Our approach combines existing analysis and test techniques with new solutions specifically oriented towards a new set of conditions and requirements in DOOSS. The selected test cases result in an acceptable level of confidence on the correctness of our case study implementation (see Appendix).

3.   THE FRAMEWORK OVERVIEW

The DOOT framework distinctly separates the testing of object-oriented and distributed properties. To handle the various types of interactions found in distributed object-oriented software, DOOT is driven by four test levels, each associated with a different fault model and test strategy. Table 1 relates the four test levels to those commonly adopted in traditional testing approaches.

Traditional System	Distributed Object-Oriented Software System		
approaches	object-oriented properties		distributed properties
unit testing	*class testing*	*method testing* *object testing*	---
Integration testing	*intra-cluster testing*		*inter-cluster testing*
System testing	*System testing*		

Table 1. Classification of test Levels

Class testing comprises two procedures: method testing and object testing. The goal of class testing is to validate the class definition of an object. Method testing is essentially unit testing on each method in the class. During the test, a value table (VT) and a control flow diagram (CFG) are derived for each method. While method testing focuses on the details of each method, object testing examines the interactions among the member functions (methods) of a class. Integration testing is realised in two levels of testing: intra- and inter-cluster testing. A cluster is formed by a collection of objects executed within a single control thread. Intra-cluster testing constructs from the CFGs an execution tree (ET). While intra-cluster testing aims at checking the interaction among objects within a cluster, the inter-cluster testing focuses on the interactions between clusters. The last level of DOOT follows the traditional approach of system testing. Algorithm 1 presents an overview of the entire framework. The input and output elements will be described in subsequent sections.

```
1    input:   CLIB - class library for each cluster
2             CPC, FSM - pseudo-code and finite state machine for each class
3    output:  VT, CFG - value table and control flow graph for each method
4             LIG, ET - list of interaction graphs and execution tree per cluster
5             TT - Transaction tree of the system
6    -------------------------------------------------------------------------
7    LIG = empty;                        // initialise LIG
8    for every cluster do {              // for each single thread
9       for every concrete class in CLIB do {   // no testing for abstract classes
10         for every new method do {    // no redundant methodTesting
11            apply methodTesting(CPC);  // also produce VT and CFG (algorithm 2)
12            derive IG from CFG;        // form IG for each method (algorithm 3)
13            add IG to LIG;             // to be used in intraClusterTesting
14         }
15         apply objectTesting(FSM,CFG); // testing causal order of methods
16      }
17      apply intraClusterTesting(LIG);  // also generate ET (algorithm 4)
18   }
19   combine ET to TT;                   // produce the composite TT (algorithm 5)
20   apply interClusterTesting(TT);      // testing the cluster communications
21   apply systemTesting;                // random walk and/or requirement testing
```

Algorithm 1. Distributed object-oriented testing

DOOT supports an integrated and incremental approach for software testing. It is an integrated approach because each level utilises the test results from the previous levels. For example, the CFGs derived from the pseudo-code are reused by object testing (lines 11 and 16 in Algorithm 1). Hence the overall test effort is reduced. DOOT is also an incremental framework since the set of elements tested in one level is considered as a basic unit of interaction in the next level. For example, object testing examines method interactions whereas intra-cluster testing uses objects as test units. This is an important concept of DOOT. First, the complexity of state space is reduced by this folding technique. Furthermore, given the well-tested paths in the lower level, the next level needs only transverse one of these paths for a comprehensive test coverage. Finally, the same or similar test modelling techniques can be employed at different levels of abstraction in the framework to improve understandability and reduce the complexity of the approach. Each level of testing will be explained in more detail in the following sections.

4.   CLASS TESTING

In DOOT, a class is assumed not to contain internal concurrency or non-determinism[1]. Thus no special effort to deal with concurrency or non-determinism is made in class level testing which consists of two procedures: method testing and object testing. Method testing is a structural testing of individual class methods whose behaviours are specified in pseudo-code. While method testing focuses on the internal details of each method, object testing

---

[1] Current components within a class should be discouraged and replaced by multiple interacting objects.

validates the interaction among these methods against the allowable object behaviour. Like other object-oriented testing techniques [7, 23], the allowable object behaviour is assumed to be given in the form of finite state machine (FSM). Class testing is only the first step of our testing framework. The testing of class relationships is done in other steps in the framework. Class testing can also be further optimised to take advantage of the inheritance hierarchy of an object-oriented system which has been discuss in by others [16, 17, 24, 25]. Class testing aims at identifying the following types of faults:

- Data-anomalies such as misspelling, name confusion, deletion of statements and missing initialisation. These faults can be detected by static data flow analysis, compilation [2, 28].
- State errors associated with object behaviour. These include the selection of a wrong transition, invalid execution of a transition and missing transitions.
- Errors that are associated with object instantiation, such as memory allocation, invalid object name and missing initialisation of attributes.

4.1 Method Testing

Method testing comprises of analysing and collecting test information. The analytical phase scans the pseudo-code of each method, and generates a predicate-used [10] variable table VT with its values and its corresponding CFG. This information is collected for controlling method execution. Method testing is summarised in the following algorithm:

```
1    input:   CPC - pseudo-code for each class
2    output:  VT, CFG - value table and control flow graph for each method
3    ----------------------------------------------------------------
4    for each new method in CPC do {
5        create a VT with all attributes and method variables;
6        fill in the VT with the values from the transition predicates;
7        create a CFG from the VT and label the predicates;
8        generate test cases based on the values in the VT;
9        execute every path of the CFG at least once;
10   }
```

Algorithm 2. Method testing

Let us consider a simple class *Account* (bank account) that consists of three attributes (*accountNo, customerName* and *balance*) and several methods (four of which are relevant to this discussion: *open, deposit, withdraw* and *close*). Figure 1 illustrates the mechanism of method testing using a segment of the *Account* class specification. Variables other than the three class attributes are formal parameters of the methods. As indicated in Algorithm 2, a VT and a CFG are generated for each method. The node column of each VT denotes the statement number and the variable column lists all variables accessed by that method. There are three types of values in VT. A *don't care* means the initial value of that variable is not important, e.g., *accountNo* in node 3 of the *open* method. A *random* entry means a random value is provided for the variable in testing that method, e.g., a random integer for *accNo* in the same node. Lastly, each predicate generates a set of specific values. For example, two values of *acctNo* are provided in the *deposit* method: one is equal to *acccountNo* and the other does not. They are also labelled (*a* and *b* in this case) for identification. In this testing, a method is considered to be an imperative program and its interactions with other methods are treated as interaction to black boxes, e.g., *error(accNo)* in node 5 of the *deposit* method. Each VT is then transformed to a CFG where the predicates are shown as parameters associated with the transitions (arrows). The 'n_0' and 'n_e' symbols in the CFG denote the start and the end nodes respectively, whereas the labels reference the conditions given in the VT for the associated transitions to occur.

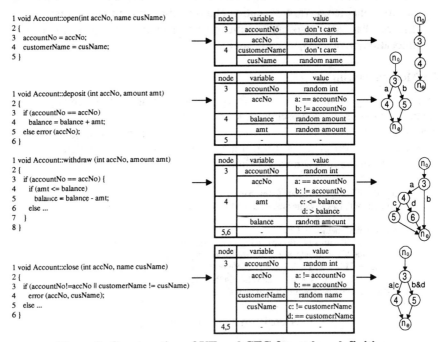

Figure 1. Construction of VT and CFG from class definition

The test coverage of method testing is governed by the CFGs of a class. Each predicate-use combination generates a separate test case. For instance, the three test cases of the *withdraw* method are given by the parameter set {*ac, ad, b*}. This test coverage guarantees that every specification statement of a method (or every control flow in a CFG) is executed at least once. The *close* method demonstrates the combined parameters, where the 'I' and '&' stand for the 'or' and the 'and' boolean operator respectively. This method also generates three test cases which are defined by the set {*a, c, b&d*}.

4.2 Object Testing

The second part of class testing is object testing. The objective is to identify faults in the implementation that violate the predefined permissible object behaviour. The permissible object behaviour is specified in an FSM. Each path in the FSM describes a legitimate method execution sequence. A shortest test sequence is generated using transition tours (such as the Chinese postman algorithm) that covers all transitions in the automaton. Figure 2 illustrates object testing by showing an FSM of the *Account* class. The correct causal order of method invocations is examined. For example, the *open* method must be invoked before the first *deposit* method; and a *withdrawAll* method must be followed by a *close* method.

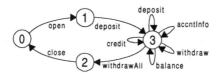

Figure 2. Finite state machine for the *Account* class

After a test sequence is generated, the values of test data are set to ensure the execution path is followed. This is achieved by using a simple static data flow analysis [18, 28]. For each invoked method in the sequence, the predicate-use values of each path in the CFG are examined. The largest matching parameter set is chosen for generating the test cases. For instance, in order to execute the sequence [*open deposit withdraw*] of the *Account* class in Figure 1, the three methods must use the same value of *accNo*. Therefore, while a random value is allowed in the *open* method, the parameter *a* of the *deposit* method must match that of the *withdraw* method. On the other hand, if no matching parameter is needed, the longest sequence is used and the unreachable transitions are marked for use in the generation of the next test sequence. This procedure is applied until all transitions in the FSM are covered. Give the test data to enforce the execution of the following sequence [*open deposit accntinfo deposit withdraw credit balance withdrawAll close*] in the Account class example.

5.   INTRA-CLUSTER TESTING

A cluster is a collection of objects to be executed in a single thread. Intra-cluster testing focuses on finding faults related to object interaction within a cluster. The difficulties of this testing lie in the complexity and dynamic structure of a cluster [16, 19]. As such, traditional techniques like data flow and control flow testing are not adequate [23]. However, by combining several techniques, it is possible to reduce the complex structure of object interactions into something more manageable. DOOT tackles this by localising all dependencies of interactions in the cluster. Intra-cluster testing consists of two steps: static and dynamic analyses. In static analysis, an interaction graph (IG) is constructed from each CFG derived in the method testing phase. The interaction graph of a method *m* shows all the methods that can be invoked within *m* and the conditions for their invocations. The IGs in the same cluster are then linked up to form an execution tree (ET). In dynamic analysis, test cases are generated by traversing the ET.

Static analysis searches for faults that are related to unreachable codes and attribute anomaly. The unreachable codes may reveal errors due to the dynamic binding feature of object-oriented systems. Data anomaly may occur within a single method or in attributes that are used by more than one method. For example, errors may exist when an attribute is used before being defined or is redefined before use. On the other hand, dynamic analysis concentrates on wrong method invocation and missing instantiation of objects. The former may be caused by confusion on method names, and the latter may occur if the type of the invoked method is incorrect or the object is not yet instantiated when called.

5.1 Static Analysis

From the predicate-used variables in the VT, the information on all method invocation is collected and specified in the CFG. Additional control flows (V-headed arrows) are inserted to indicate method dependence. The simplified model is a tree of depth one and is known as an IG. The root node of an IG contains the method name whereas the leaf nodes contain the names of invoked methods. An edge from the root node to a leaf node represents a method invocation. These leaf nodes are depicted from left to right in the order of sequential execution. Each edge may contain a guarded condition that has to be satisfied to invoke the corresponding method. The condition is obtained by a trace of parameters on the path defined in the CFG. Algorithm 3 shows the generation of a list of IGs within a cluster.

Figure 3 illustrates the transformation of three IGs (for methods A, B and C) from the models in class testing. The VTs only show the variables (x, y, z, v) that are used in the predicates. The labels of predicate (a, b, c, d) specify the conditions for which the corresponding transitions in the CFG will take place. An extra column is added to show possible method invocation. For instance, $m2(b)$ in node 4 of method B indicates that $m2$ is invoked if condition b holds. A grey node in the CFG depicts the point of method invocation. DOOT classifies the methods within a cluster into four mutually disjoint categories:

main method: every cluster has only one method of this type

entry method: an interaction point where calls between clusters are made, i.e., can be called from other clusters and may call other methods in the local cluster (e.g. method C)

agent method: a non-entry method that may invoke other methods (e.g. method B)

service method: a non-entry method that does not invoke any methods (e.g. method A)

```
1    input:    CLIB - class library of the cluster
2              CPC, CFG - pseudo-code for each class and control flow graph per method
3    output:  LIG - list of interaction graphs
4    --------------------------------------------------------------------------------
5    for each class in CLIB do                          // each class in the cluster
6       for each method in CPC do {                     // each method in a class
7          create new root node to LIG;                 // create a new IG
8          for each path in CFG do                      // obtained from method testing
9             while (not end of path) do {
10               follow the path;                       // ignore non-interaction statements
11               collect parameter on predicate-use;    // form trace of parameters
12               if (invocation of other method) {      // find new method invocation
13                  create new node with the method;    // create leaf node
14                  record parameter on edge;           // create condition on edge
15                  if (!first node at this path)
16                     connect to last node found;      // create links between leafs
17               }
18            }
19    }
```

Algorithm 3. Interaction graph generation

An entry method is depicted as a black node in IG. There are two more points to note in Figure 3. First, an interaction without parameter is an *automatic invocation* (e.g. node 3 in method *B*). Second, there may be causal relationships among the variables used in the predicates (e.g. in method *B*, *x* depends on *y* and *y* depends on *z*). These method types can be easily determined by scanning the pseudo-code.

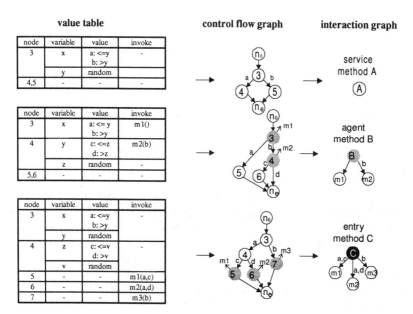

Figure 3. Transformation of models from class testing into interaction graph

After the IGs of each method are constructed, an execution tree (ET) is formed by linking and merging the IGs in the cluster. An ET is a finite tree whose root node corresponds to an initial method and the leaf nodes are the terminal services in the cluster. The conditions on the edges from each IG are also preserved in the ET. The original control flow model of the cluster can be very complex, but this reduced model only captures the interaction information for intra-cluster testing. The construction of the execution tree is shown in Algorithm 4.

```
 1   input:   CLIB, LIG - class library of the cluster and a list of interaction graphs
 2   output:  ET - execution tree for the cluster
 3   -------------------------------------------------------------------------
 4   LLIG = empty;                           // set LLIG for construction phase
 5   for every concrete class in CLIB do     // no abstract classes
 6      for each method in LIG do {          // get method of the class
 7         if the method not in LLIG          // new method found
 8            create new node and add to LLIG; // create node with method, connect to list
 9         if method type is not service      // main or agent or entry method
10            for each method invocation do { // found new link
11               create edge to invoked method; // connect invoked methods
12               copy parameters to the edge;   // store conditions for the call
13            }
14      }
15   for every method in LLIG do             // expansion phase
16      for each edge to other method do     // deal with a link each time
17         if invoked method != the method {  // not recursion in ET
18            add node, edge and parameters;  // clone the branch
19            mark the invoked method is called; // record the subtree is tested
20            handle the node at the same way; // depth first expansion
21         }
22   for each marked method in LLIG do       // delection phase
23      delete the subtree start from the method; // redundant subtree
24   ET = LLIG;                              // final execution tree
```

Algorithm 4. Constructing an execution tree from interaction graphs

The example used class testing is expanded to demonstrate the construction of ET in static analyse and test case generation in dynamic analysis. The automatic teller machine (ATM) network model[2] from Rumbaugh's book [21] is used. Figure 4 shows the components in a banking system. The four types of cluster are shown by the rectangular boxes. Each cluster may have multiple instances with the exception of *Consortium*[3]. A solid arrow depicts a communication link between two clusters whereas a V-headed arrow denotes method invocation among the objects in the cluster. Also, each method is labelled for identification. In order to simplify the discussion, only the *BankComputer* cluster contains multiple objects in this example.

---

[2] The system supports a computerised banking network including both human cashiers and ATMs to be shared by a consortium of banks. Each bank provides its own computer to maintain its own accounts and process transactions against them.

[3] ATMs communicate with the Consortium which clears transactions with the appropriate banks. The bankInfo method serves to select the right instance of BankComputer to interact.

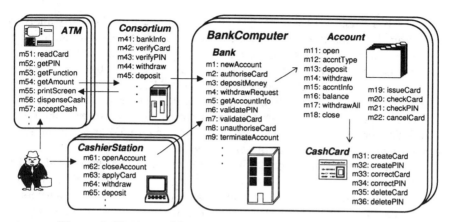

Figure 4. Clusters, objects and methods in a banking system

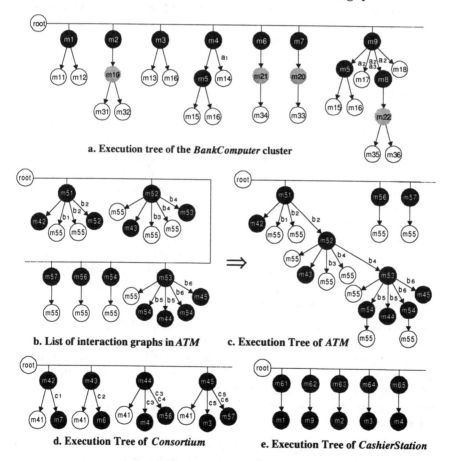

a. Execution tree of the *BankComputer* cluster

b. List of interaction graphs in *ATM*

c. Execution Tree of *ATM*

d. Execution Tree of *Consortium*

e. Execution Tree of *CashierStation*

Figure 5. Execution tree of the banking system example

A list of IGs is derived from the VTs and CFGs of each method. By applying Algorithm 4, an execution tree of each cluster is constructed as shown in Figure 5. All entry methods of a cluster are connected to the root node. They serve as an interface of that cluster. For example, all methods of the *bank* object (m1 to m9 in Figure 5.a) are considered as entry methods in the BankComputer cluster. An agent method or an entry method is refined until all leaf nodes are replaced by service methods or entry methods from other cluster. For instance, in the ET of *CashierStation* (Figure 5.e), the *applyCard* method (m63) triggers an entry method *authoriseCard* (m2) in the *BankComputer* cluster. On the other hand, this method in the *Bank* object (Figure 5.a) invokes an agent method (m19) *issueCard* in the *Account* object, which instantiates a new *CashCard* by invoking the *createCard* and *createPIN* (m31, m32) service methods. The parameters in each edge specify the conditions of method invocation. This is illustrated in Figures 5.b and 5.c which show the generation of the *ATM* execution tree from the interaction graphs. In static analysis the execution tree is examined for erroneous structures. A path is unreachable if an agent method or a service method connects directly to the root node. Since only the main method and entry methods can be initiated by the root node, a fault is detected.

5.2 Dynamic Analysis

Test cases in intra-cluster testing are presented by sequences of method invocations. Since we have assumed no concurrent execution within a cluster, every method execution sequence has pre-determined start and end points. Starting from a root node of an ET, a backtracking algorithm derives the path to every leaf node. Each path specifies a sequence of method invocations from top to bottom and from left to right of a subtree. The conditions in each edge are noted and they must be satisfied by the previous invoked methods. Static data flow analysis is employed to derive the values of test data required for the sequence. These values define a particular path of execution in a method such that the desired conditions are fulfilled. In intra-cluster testing, an invocation of an entry method in another cluster is considered as an interaction to a black-box (denoted as 'X' in the test cases shown below). The faults caused by this type of interactions are handled by inter-cluster testing.

Every execution sequence generates a test case. The conditions associated to each invocation is placed in the brackets next to the method (similar to that of VT shown in Figure 3). By eliminating the sequences that are completely contained in some other test sequences, the number of test cases is significantly reduced. The complete set of test cases of the *ATM* cluster (shown in Figure 6a) can be reduced to that shown in Figure 6b. In this example, the reduction ratio is approximately 5 to 2.

The faults such as type confusion of objects, name confusion of methods, missed instantiations may result in wrong or shorten execution paths, or even stopped execution. These incorrect interactions can be identified by examining the ET. The static analysis detects every unreachable code of the cluster whereas the dynamic analysis guarantees that every interaction in a cluster is executed at least once.

a. complete set of test cases
1: m51 X
2: m51 X m55(b1)
3: m51 X m55(b2)
4: m51 X m55(b2) m52(b2)
5: m51 X m55(b2) m52(b2) m55
6: m51 X m55(b2) m52(b2) m55 X
7: m51 X m55(b2) m52(b2) m55 X m55(b3)
8: m51 X m55(b2) m52(b2) m55 X m55(b4)
9: m51 X m55(b2) m52(b2) m55 X m55(b4) m53(b4) m55
10: m51 X m55(b2) m52(b2) m55 X m55(b4) m53(b4) m55 m54(b5) m55
11: m51 X m55(b2) m52(b2) m55 X m55(b4) m53(b4) m55 m54(b5) m55 X(b5)
12: m51 X m55(b2) m52(b2) m55 X m55(b4) m53(b4) m55 m54(b6) m55
13: m51 X m55(b2) m52(b2) m55 X m55(b4) m53(b4) m55 m54(b6) m55 X(b5)
14: m56 m55
15: m57 m55

b. reduced set of test cases
2: m51 X m55(b1)
7: m51 X m55(b2) m52(b2) m55 X m55(b3)
11: m51 X m55(b2) m52(b2) m55 X m55(b4) m53(b4) m55 m54(b5) m55 X(b5)
13: m51 X m55(b2) m52(b2) m55 X m55(b4) m53(b4) m55 m54(b6) m55 X(b6)
14: m56 m55
15: m57 m55

Figure 6. Intra-cluster test cases for the *ATM* cluster

6. INTER-CLUSTER TESTING

DOOT verifies the functionality and reachability of a cluster by exhaustive class testing and intra-cluster testing. After the first two levels of testing, the correctness and consistency of individual cluster is assumed. Therefore inter-cluster testing only focuses on faults that are related to the interactive calls between entry methods among clusters. The success of this test ensures the reachability of all possible communication links within a distributed system. This is achieved by constructing a transaction tree of the entire system. Since there is no concurrency within a cluster, each execution path in a cluster can be considered as a transaction. The condition for the successful execution of a transaction is the aggregate of all the conditions denoted in the edges of the path.

Algorithm 5 describes the construction of a transaction tree from the ETs of each cluster. The algorithm comprises of three phases. The first phase removes all non-entry methods but keeps the necessary conditions. The second phase expands each leaf entry method with its subtree structure. The last phase deletes all the redundant subtrees.

```
1    input:   ET - execution tree per cluster
2    output:  TT - transaction tree of the whole system
3    ---------------------------------------------------------------------
4    LRET = empty;                                    // form an empty list of reduced ET
5    for each ET in the system do                     // reduction phase
6      for each subtree do {                          // handle a subtree at a time
7        create new node in LRET;                     // this can be entry method or main
8        for each path do                             // search all paths
9          while (not end of path) do {               // ignore non-entry methods
10           follow path and collect conditions on edge;  // form trace of parameters
11           if (invoked an entry method)             // find new entry method invocation
12             create new node and record conditions; // produce interaction point with edge
13         }
14      }
15   for each subtree in LRET do                      // combination phase
16     while (a leaf is defined as subtree) do        // the transaction can be extended?
17       substitute the leaf with the subtree and mark it;  // copy related transaction and record tested
18   for all marked subtree do                        // deletion phase
19     delete from LRET;                              // remove redundant subtree
20   TT = LRET;                                       // final transaction tree
```

Algorithm 5. Formation of transaction tree

Figure 7.a shows the transaction tree that starts from the entry method *m51* of the *ATM* cluster (refer to the ETs in Figure 5). Figure 7.b shows some transaction sequences in the system[4]. Since the reachability of all possible paths within each cluster has been checked in the previous test levels, these paths are not the focus in this test. For instance, the sequence $\alpha_1, \alpha_2, \alpha_3$ in the *ATM* cluster[5] is represented by the transaction flow β. It is only necessary for each transaction to connect two entry methods. For example, the transaction flow β connects the entry methods *m51* and *m52* in the *ATM* cluster, and the transaction flow γ links the entry method *m52* to the entry method *m43* in the *Consortium* cluster. Moreover, a transaction sequence terminates if it invokes an entry method that does not have any outgoing transaction flow. For example, the transaction flow δ terminates the sequence on invoking the entry method *m7*.

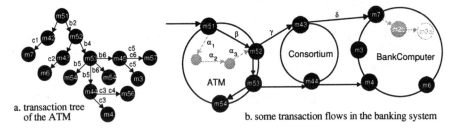

a. transaction tree of the ATM   b. some transaction flows in the banking system

Figure 7. Inter-cluster testing

---

[4] All transaction flows have been recorded in the transaction tree. This incomplete diagram is shown only for description purpose.

[5] The shaded parts are not supposed to be in the diagram, they are shown for clarity reason.

In the transaction tree, every path from the root node to a leaf node forms a possible test case. The execution sequences between two entry methods are taken from the test cases generated by the intra-cluster testing. Similar to other test levels, all test cases that are completely embedded in other test cases are removed. The transaction tree in Figure 7.a generates two test sequences given below. In this case, they are the expanded version of test cases 11 and 13 from Figure 6.

new 11: m51 m42 m41 m7(c1) m20 m33 m55(b2) m52(b2) m55 m43 m41 m6(c2) m21 m34
m55(b4) m53(b4) m55 m54(b5) m55 m44(b5) m41 m4(c3) m5 m15 m16 m14(a)
m56(c3,c4) m55
new 13: m51 m42 m41 m7(c1) m20 m33 m55(b2) m52(b2) m55 m43 m41 m6(c2) m21 m34
m55(b4) m53(b4) m55 m54(b5) m55 m45(b5) m41 m3(c5) m3 m13 m16 m57(c5,c6) m55

Again, all the conditions in the transaction flow must be satisfied for the execution. The results of data flow analysis are reused for deriving test data. This test level guarantees the coverage of all coupled entry methods between clusters. In other words, every communication link of the entire system is executed at least once.

7. SYSTEM TESTING

Due to the interleaving and non-deterministic behaviour of DOOSS, traversing all possible paths is not practical (and often impossible). Various techniques have been proposed to leave out a particular interleaving in the state space search or to reduce the global states by partial ordering reduction [8, 25]. DOOT employs two common approaches for system testing. The first one checks user specified requirements and the second one employs the random walk technique.

The former detects context constraints that are specific to the application domain. This enhances the exhaustive testing in the previous levels with compositional reachability analysis [3]. The important system invariants are declared as test requirements in the domain context. Test cases are specifically generated with the proper parameter values to enforce the execution of pre-determined paths. With this test in DOOT, we can guarantee the essential constraints of the system are satisfied. The latter performs a random simulation that walks through various classes and clusters of the system. Due to the scale of DOOSS, the global states can neither be represented nor found exhaustively. In this controlled search based on random selections of successor states, no effort is made to predict where errors are likely to occur in the state space. Some researchers argue that the approach is not only the simplest technique to implement, but is also likely to produce the highest quality search in testing [17].

8. CONCLUSION

We have proposed a framework for testing distributed object-oriented software systems known as DOOT. DOOT employs an integrated incremental testing approach at four levels - class level, intra-cluster level, inter-cluster level, and system level. An exhaustive reachability test is adopted in the fine grain class level testing so that the test effort is significantly reduced in the coarse grain cluster and system level testing. Each level addresses the requirements at its level and builds on the results of testing at the previous levels. We also provide guidelines for system testing to cover the user-specified requirements and a final random-walk error detection. The approach was tested on a conferencing system given in the Appendix. More substantial work on inter-cluster testing is under development.

9. REFERENCES

[1] Andleigh P.K. and Gretzinger M.R., *Distributed object oriented data-systems design*, Prentice Hall, 1992.

[2] Chanson S.T. and Zhu. J., Automatic protocol suite derivation, *Proceedings of INFOCOM '94 Conference on Computer Communications*, Vol. 2, pp. 792-799, 1994.

[3] Cheung S.C. and Kramer J., Contextual local analysis in the design of distributed systems, *International Journal of Automated Software Engineering*, Vol. 2, No. 1, pp. 5-32, 1995.

[4] Chin R.S. and Chanson S.T., Distributed object-based programming systems, *ACM Computing Surveys*, Vol. 23, No. 1, pp. 91-124, 1991.

[5] Graham I., *Object-Oriented Methods*, Addison-Wesley, 1994.

[6] Hayes J.H., Testing of object-oriented programming (OOPS): A fault-based approach, *Proceedings of 14th ICSE*, IEEE Press, pp. 205-220, 1992.

[7] Hoffman D., A case study in class testing, *PROC/CASON'93*, Vol.1, pp.472-82, 1993.

[8] Holzmann G.J., *Design and validation of computer protocols*, Prentice-Hall, 1991.

[9] Jorgensen P.C. and Erickson C., Object-oriented integration testing, *Communications of the ACM*, Vol. 37, No. 9, pp. 30-33, 1994.

[10] Jorgensen P.C., *Software testing - a craftsman's approach*, CRC Press, 1995.

[11] Kim M., Chanson S.T. and Kang S., An approach or testing asynchronous communicating systems, *Proceedings of IWTCS'96*, pp. 141-155, 1996.

[12] Kim M.C., Chanson S.T. and Kim G.H., Concurrency model and its application to formal specifications of asynchronous protocols, *Proceedings of IEEE GLOBECOM*, Vol. 3, pp. 1580-4, 1995.

[13] Kirani S. and Tsai W.T., *Specification and verification of object-oriented programs*, Technical report, University of Minnesota, 1994.

[14] Lamport L. and Lynch N., Distributed computing: models and methods, *Handbook of theoretical computer science.* pp. 1157-1199, Elseiver Science, 1990.

[15] Marick B., *The craft of software testing - subsystem testing including object-based and object-oriented testing*, Prentice Hall, 1995.

[16] McGregor J.D. and Korson T.D., Integrating object-oriented testing and development processes, *Communications of the ACM*, Vol. 37, No. 9, pp. 59-77, 1994.

[17] Mihail M., Papadimitriou C.H. and Dill D.L., On the random walk method for protocol testing, *Proceedings of Computer Aided Verification*, pp. 132-41, 1994.

[18] Mueller F., Whalley D.B. and B. Le Charlier, Efficient on-the-fly analysis of program behavior and static cache simulation, *Proceedings of First International Static Analysis Symposium*, SAS '94, Springer-Verlag, pp. 101-15, 1994.

[19] Murphy G.C., Townsend P. and Pok S.W., Experiences with cluster and class testing, *Communications of the ACM*, Vol. 37, No. 9, pp. 48-58, 1994.

[20] Poston R.M., *Automating specification-based software testing*, IEEE Press, 1996.

[21] Rumbaugh J., Blaha M., Premerlani W., Eddy F. and Lorensen W.: *Object-Oriented Modelling and Design*, Prentice Hall, 1991.

[22] Ryan T.W., *Distributed object technology: concepts & applications*, Prentice Hall, 1997.

[23] Shel S., *Object oriented software testing*, John Wiley & Sons, 1996.

[24] Smith M.D. and Robson D.J., A framework for testing object-oriented programs, *Journal of object-oriented programming*, Vol. 5, No. 3, pp. 45-53, 1992.

[25] Tai K.C. and Carver R.H., Testing of distributed programs, *Handbook of parallel and distributed computing*, pp 955-978, McGraw Hill, 1995.

[26] Ulrich A. and Chanson S.T., An approach to testing distributed software systems, In *Proceedings of PSTV'95*, Chapman & Hall, pp. 121-136, 1995.

[27] Ulrich A., A Description model to support test suite derivation for concurrent systems, *Kommunikation in verteilten systemen (KiVS'97)*, Springer Verlag, pp. 151-166, 1997.

[28] Ural H., Testing sequence selection based on static data flow analysis, *Computer communication*, 10(5), 1987.

[29] Wong C.Y., Chanson S.T., Cheung S.C. and Fuchs H., *Testing distributed object-oriented system*, Technical report, Hong Kong University of Science and Technology, April 1997.

Appendix - Case Study

A simple conferencing system was implemented and tested using the framework. The system was built using the configuration language DARWIN and the REGIS-runtime environment and follows the client-server architecture. A conference client is required at each terminal while a conference server runs as a perpetual process. The conference server can handle more than one conference simultaneously, and a user can join more than one conference by starting different instances of the client program. In this exercise, class testing and intra-cluster testing concentrate on individual clusters which are the client and server programs. Inter-cluster testing deals with the concurrency within the conference client and the interaction between the the client and server programs. Different test levels are illustrated as given in the framework. To save space, the following description only focuses on the testing of the conference client (please refer to our technical report [29] for full details).

Conference Client
The conference client consists of a number of interacting objects organised in different class hierarchies. Some of these are static objects that exist throughout the lifetime of the client program. Others are created and destroyed dynamically. The two threads are described by the *console* and *c_client* objects. *Console* is a predefined REGIS keyboard class which interacts with the user. During user input, this object is blocked instead of the whole process of the conference client. In this way, the *c_client* object can also wait for incoming messages from the conference server. *C_client* contains the following basic objects:

server: is responsible for all interactions to and from the server through ports and entries;
interface: is responsible for all interactions of the user with the client-program;
user: is responsible for getting and maintaining data from user for identification;
screen: is responsible for all formatted outputs of the conference client;
overview: is an abstract class that presents an overview for selecting functions, records,
  users, conferences etc. It manages the display by using the *maskhandler* or *pullup* object.
conference: manages an active conference by controlled actions, such as input messages
  from message window, whiteboard or chairperson menu;
confpart: is responsible for the display and control of the conference participants;
pullup, maskhandler: are responsible for some special functionalities in the client program.

Figure A depicts a simplified architecture of the conference-client. A small circle denotes
a port or an entry point whereas an arrow represents a 'used-by' relationship between two
objects.

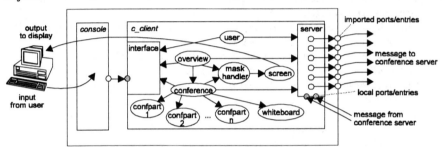

Figure A. The conference client cluster with use-relationships

Class Testing - The following table shows the number of test cases generated in method
testing and object testing in each object class. It also includes the size of each class in
terms of number of lines of code, attributes, methods and number of nodes in the CFG and
IG. (# stands for 'number of' whereas % stands for 'number of test cases')

class	c_client	interface	server	user	screen	overview	conference	confpart	maskhandler	total
# lines in header	118	44	89	47	42	70	44	35	36	525
# lines in body	73	276	327	141	191	92	244	189	90	1623
# attributes	-	4	8	9	6	13	11	9	7	67
# methods	1	13	25	8	18	7	16	5	9	102
# nodes in CFG	14	187	188	104	96	68	139	108	53	957
# nodes in IG	12	40	97	67	40	46	91	55	39	487
% method testing	1	77	42	25	27	27	44	46	12	301
% object testing	0	8	10	6	10	16	22	6	5	83

Intra-Cluster Testing - By applying the intra-cluster testing, the global state transitions
are transformed to interactions in the execution tree. The test case generation employs a
reduction algorithm to eliminate repeated execution sequences. In the conference client
example, the overall number of test cases generated for this test level is found to be 359.

Inter-Cluster Testing - This exercise only considers the interactions between the *console*
and *c_client* objects. Since the *c_client* object is tested exhaustively in the previous levels,
the internal execution sequences are not of interest. Therefore the overall number of test
cases generated for inter-cluster testing is around 100 compared to the traditional approach
(no abstraction in test levels) where the total number of test cases is well over 100,000.

4

Interoperability Test Suite Derivation for Symmetric Communication Protocols

Sungwon Kang and Myungchul Kim
Korea Telecom Research & Development Group
Sochogu Umyundong 17, Seoul 137-792, Korea
e-mail: kangsw@sava.kotel.co.kr, mckim@sava.kotel.co.kr

Abstract

Communication protocols are commonly designed in such a way that implementations of the same protocol can be used as peers for communication. Such a protocol is said to be symmetric. When two or more entities are employed to perform a certain task as in the case of communication protocols, the capability to do so is called interoperability and considered as the essential aspect of correctness of communicating systems. This paper deals with the problem of deriving interoperability test suite for control part of symmetric protocols. A new approach to efficient interoperability testing is described with justifications and the method of interoperability test suite derivation is shown with the example of the ATM Signaling protocol.

Keywords

Interoperability testing, test suite generation, symmetric protocol

1. INTRODUCTION

When more than one objects are employed to perform a certain function, there arises the problem of whether they together behave correctly. This is the problem of

Formal Description Techniques and Protocol Specification, Testing and Verification
T. Mizuno, N. Shiratori, T. Higashino & A. Togashi (Eds.) © 1997 IFIP. Published by Chapman & Hall

interoperation in a general sense. Here (1) involvement of more than one objects and (2) the objects together behaving as expected are the two key characteristics of (correct) interoperation. Thus it can be said that two or more objects *interoperate* if they together behave as expected. For interoperability of communication protocols, such expectations are documented in specifications. Usually the expectations about interoperation are not described in a single dedicated document. Rather they should be inferred or derived from the relevant specifications.

Within the context of communication networks, an object mentioned above can be a network node, a layer of a node or a component of a layer or a plane or even a network, i.e. anything we decide to view as a whole. These notions of interoperation and object allow us to view various kinds of interacting behavior within a network and between networks as special kinds of interoperation. Thus *internetworking* can be seen as interoperation of objects which are networks. *Interworking* of two nodes or two networks utilizing an interworking function unit becomes interoperation of the three objects, i.e. two nodes or networks together with the interworking function unit in between.

An abstract view of communication can be conceived when protocols of network nodes are layered and underlying layers are regarded as the service provider for the layer above. Then by similarly abstracting from the underlying layers, we can focus on the behavior of a certain layer and think about interoperation of the objects which realize the particular layer under consideration. In this way, the above definition of interoperation remains valid even when we take an abstract view of network nodes and networks.

Once the notion of interoperation is clearly understood, there arises the problem of verifying interoperation for target implementations which interact with each other. This is the task of interoperability testing. By virtue of the two characteristics of interoperation noted at the beginning of this section, interoperability testing differs from conformance testing. Because of the first characteristic, more varied test architectures are possible than with conformance testing. Because of the second characteristic, test suite derivation become a more challenging task for which the expected behavior need be inferred first.

1.1 Related Work

In the past, research on protocol testing mainly concentrated on conformance test sequence generation (cf. [Chow 78] [Sidh 90] and the bibliographies therein). Accordingly, it didn't take long before the international standard for the methodology and framework for conformance testing has come into existence [ISO/IEC 9646]. Although conformance testing is regarded as a necessary step on the way to achieving interoperation, it is agreed that it is insufficient to ensure interoperation of communication network entities. Some sort of direct testing of interoperation is considered indispensable. Work on interoperability testing can be classified into two categories depending on whether it is more geared to practical things such as clarification and implications of interoperability testing or to systematic generation of interoperability test suite. As work along the former line are [Bonn 90] [GRSS 90] [Cast 91] [VerB 94]. [Bonn 90] [VerB 94] present interoperability testing experiences. [GRSS 90] gives a comprehensive discussion on various aspects related to interoperability testing.

For the latter line of work, there are [RafC 90] [AraS 92] [APRS 93] [CasK 94] [LuBP 94] [KanK 95]. All these base interoperability test suite derivation on some sort of reachability analysis. For interoperability test architecture, [RafC 90] uses upper testers as well as lower testers. [AraS 92] [LuBP 94] introduces notions of stable state to reduce the size of relevant state space. [CasK 94] develops interoperability test suite method for synchronous models. [KanK 95] shows how to derive test suites for dynamic testing of interoperability.

The previous work, however, did not provide a coherent framework for interoperability testing in that the notions of interoperability, interoperability testing, interoperability test case and interoperability test architecture were not presented in an integrated manner nor were interrelated for the purpose of interoperability test suite development. In particular, there is no work specially treating symmetric protocols. So the consequences and possible optimizations for interoperability test derivation that may arise from a communication protocol being symmetric remained unexplored.

This paper, which belong to the second line of work on interoperability testing, addresses these issues. Starting from the general definition of interoperability which we already gave in the previous section, we carefully select an interoperability test architecture. The chosen architecture, combined with natural assumptions and inherent limitations for testing, is shown to induce a notion of interoperability test case. And this notion of interoperability test case allows us to focus on genuine interoperability aspect and at the same time to derive interoperability test suite in a cost-effective manner. It is shown that the test suite derivation and the actual testing itself can be made very efficient in particular when the protocol under consideration is symmetric. The remainder of this paper describes in detail this approach to efficient interoperability testing and the method of interoperability test suite derivation with the example of the ATM Signaling protocol.

1.2 Our Approach

We consider interoperability testing of two interacting implementations as the most basic type of interoperability testing. When more than two objects are involved, interoperability testing requires a more complicated test architecture. Also the test derivation, the test execution and the test result analysis become more complicated.

In this paper, we restrict our attention to control part of symmetric communication protocols. A communication protocol is said to be *symmetric* if it is designed in such a way that the implementations of the same protocol can be used as communication peers. In order to be symmetric, a protocol should be such that its peers have exactly the same functional features. Peers for a symmetric protocol need not be placed adjacent. Commonly protocols are symmetric or near symmetric (with asymmetry resulting only from differences in some features). Examples of asymmetric protocol are master-slave protocol and client-server protocol.

In our approach to interoperability test suite derivation, it is assumed that (1) the communication protocol is structured as a finite state machine (FSM) and that (2) a complete set of conformance abstract test cases has been already developed (elsewhere) based on the FSM structure. Because of this second assumption, our approach will yield an interoperability test suite which has no overlapping with conformance test suites as we will see later. Furthermore, (3) we adopt the test architecture in Figure 1. We call an individual object involved in interoperation an Implementation Under Test (IUT). In spite of the term IUT, it is important to note that the target of testing here is not the individual objects (which are the targets of conformance testing) but the system as a whole which consists of those objects. Still, IUT is a convenient term and will be used throughout the paper.

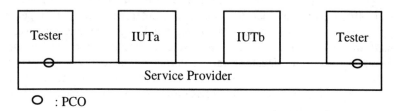

Figure 1. Test architecture for interoperability testing.

Note that there are only two Points of Control and Observation (PCO's) in Figure 1. Often in practice a monitoring point or Point of Observation (PO) is set between IUT's. We do not set such PO. For with such PO, (1) it would be more costly to generate test suite, (2) it would be more costly to perform testing and (3) interoperability testing would have much overlapping with conformance testing.

As with conformance testing, interoperability testing is restricted by observability and controllability that is allowed in a chosen test architecture. In addition, it is usually assumed that the slowness of the environment [LuBP 94] places practical limit on observability and controllability. That is, transient states cannot be observed or controlled in a predictable way and hence are considered useless for the purpose of testing. This justifies basing any notion of test case on stable states as we do in this paper.

The interoperability test approach of this paper is to *test interoperability of implementations against interoperability of specifications*. In this approach, interoperability testing is a (yet another) conformance testing, i.e. conformance testing of the system which is composed of the actual subsystem implementations. Any notion of interoperation, if any, should be expressed in principle in a given specification or should be agreed upon in advance by specification writers and implementers. Ascertaining correctness of specifications in this respect is the task of verification and validation, which goes beyond the proper scope of testing. For the target of testing is the relation between specification and implementation rather than between specifications. However, if specific interoperability requirements are subject to validation and verification at the specifications level, then the same validation and verification method can be applied to a system of implementations for testing.

As the result of interoperability testing of implementations, problems residing in specifications may be found and reported. Then the additional ingredient interoperability testing provides in addition to verification of interoperability of specifications is to fill the gap left by usual conformance testing. The real work of interoperability assurance should come at the stage of specification development. But still the fine points that may have been missed at the time of specification writing (including validation and prototyping) can be augmented through testing.

2. COMMUNICATING SYSTEMS OF IOSM'S

In this paper, we take the view that specifications and implementations of communicating entities can be modeled as some sorts of FSM's. This makes possible rigorous discussion on conformance and interoperability issues which would provide the basis for automating test suite derivation process. In this section, we make precise the FSM model to be called IOSM and its communication behavior.

Definition 1 An IOSM is a 5-tuple $<St, s_0, L_{in}, L_{out}, Tr>$ where:
(1) $St = \{s_0, ...,s_{n-1}\}$ is a set of states,
(2) $L_{in} = \{v_1, ...,v_m\}$ is a set of input symbols,
(3) $L_{out} = \{u_1, ...,u_p\}$ is a set of output symbols,
(4) $Tr \subseteq \{s-v/U \to s' \mid s,s' \in St \wedge v \in L_{in} \wedge U \in P(L_{out})\}$ and
(5) $s_0 \in St$ is the initial state.

In the definition, $P(X)$ denotes the power set of the set X. L_{in} and L_{out} are respectively called *input alphabet* and *output alphabet*. L_{in} and L_{out} can be further subdivided into two classes: one prefixed with '*i_*' and the other without the prefix. '*i_*' is used to indicate internal messages as opposed to external messages. So for example when two IOSM's M_1 and M_2 are communicating, the messages between M_1 and M_2 are internal messages and are prefixed with '*i_*' but messages between M_1 (or M_2) and their environment are external messages and are not prefixed. Tr is a set of transitions. '*v/U*' is called a *label*. The message before '*/*' is the received message and the set of messages after '*/*' are messages sent. The set notation is used because upon receiving a message IOSM can send zero or more messages. In the set Tr of transitions of a *deterministic* IOSM, for any state there is only one transition with the same input symbol. Note that Definition 1 does not restrict IOSM to be deterministic.

Definition 2 Let M be an IOSM. Let $v,v_i \in L_{in}$, $u,u_i \in L_{out}$, $v \in L_{in}^*$ and $s, s' \in St$. Then:
(1) $\sigma(s,v) = \{(s',v/u) \mid (s-v/U \to s') \in Tr \wedge u \in U \}$
(2) $\sigma(s,\mathbf{v}) = \{(s',v_1/u_1...v_k/u_k) \mid \mathbf{v} = v_1...v_k \wedge$
   $(s-v_1/U_1 \to s_1) \in Tr \wedge ... \wedge (s_{k-1}-v_k/U_k \to s') \in Tr \wedge u_1 \in U_1 \wedge ... \wedge u_k \in U_k\}$
(3) $M(\mathbf{v}) = \sigma(s_0,\mathbf{v})$

A member of $M(\mathbf{v})$ is a sequence of labels and is called a *run* (or *trace*) of M for the input sequence $\mathbf{v}$. Next we define communicating system of IOSM's.

Definition 3 A (communicating) system Σ of n IOSM's is $<\{(M_i,Q_i) \mid 1 \leq i \leq n\}, L_{\Sigma,in}, L_{\Sigma,out}, s_{\Sigma,0}>$ where:
(1) $M_i, 1 \leq i \leq n$, is a IOSM $<St_i, s_{i,0}, L_{i,in}, L_{i,out}, Tr_i>$ as in Definition 1.

(2) Q_i, $1 \leq i \leq n$, is an input queue for M_i.

(3) $L_{\Sigma,in}$ is a set of external input symbols.

(4) $L_{\Sigma,out}$ is a set of external output symbols.

(5) The initial state of the system is $s_{\Sigma,0} = <(s_{1,0},Q_0), ...,(s_{n,0},Q_n)>$ where Q_i is empty and $s_{i,0}$ is the initial state of M_i, $1 \leq i \leq n$.

In this model, there is one explicitly defined input queue for each IOSM and implicit bi-directional communication channels between each pair of IOSM's. In the next definition, we give a precise description of the global behavior of a communicating system of IOSM's. It is assumed that each message contains enough information to identify the receiving IOSM.

Definition 4 (Communicating system of IOSM's) Let Σ be as in Definition 3. Let $v \in L_{\Sigma,in}$, $U \in P(L_{\Sigma,out})$, $v \in L_{\Sigma,in}^*$, $\mathbf{u},\mathbf{u_1},\mathbf{u_2} \in L_{\Sigma,out}^*$, s, s' be states of Σ. Then

(1) $\sigma(s,v) = \{(s',v/\mathbf{u_1}.\mathbf{u_2}) \mid s=<(s_{1,j1},\epsilon),...,(s_{i,ji},\epsilon),...,(s_{n,jn},\epsilon)> \wedge$

$\qquad \exists 1 \leq i \leq k : (s_{i,j}-v/U \rightarrow s_{i,j'}) \in \mathrm{Tr}_i \wedge \mathbf{u_1} \in \varphi(U) \wedge$

$\qquad (s',\mathbf{u_2}) \in \sigma'(\mu(s,\{(s_{i,j'},\epsilon)/(s_{i,j},\epsilon)\} \cup \{(s_{i,j},Q_i)/(s_{i,j},i\_w.Q_i) \mid i\_w \in U\})) \}$

where $\sigma'(s) = \{(s',\mathbf{u_1}.\mathbf{u_2}) \mid \mathbf{u_1} \in \varphi(U) \wedge (s',\mathbf{u_2}) \in \sigma'(\mu(s,\{(s_{k,j},Q_i)/(s_{k,j},i\_w.Q_k) \mid i\_w \in U\}))\}$

$\qquad$ if $\exists 1 \leq i \leq n : s = <...,(s_{i,j},Q_i.i\_w),...> \wedge (s_{i,j} - i\_w/U \rightarrow s_{i,j'}) \in \mathrm{Tr}_i$

$\quad \sigma'(s) = \{(s,\epsilon)\}$ otherwise

$\quad$ where $\mu(s,\varnothing) = s$

$\qquad \mu(<...,(s,Q),...>, P \cup \{(s',Q')/(s,Q)\}) = \mu(<...,(s',Q'),...>,P)$

$\qquad \varphi(\varnothing) = \{\epsilon\}$

$\qquad \varphi(\{i\_w\} \cup U) = \varphi(U)$

$\qquad \varphi(\{u\} \cup U) = \{u.u \mid u \in \varphi(U)\}$

(2) $\sigma(s,vv) = \{(s'',v_1/u_1...v_k/u_kv/u) \mid$

$\qquad v = v_1...v_k \wedge (s',v_1/u_1..v_k/u_k) \in \sigma(s,v) \wedge (s'',v/u) \in \sigma(s',v) \}$

$\quad \sigma(s,\epsilon) = \{(s,\epsilon)\}$

(3) $\Sigma(v) = \sigma(s_{\Sigma,0},v)$

A system state in which input queues are all empty is called a *stable state* [LuBP 94]. The communicating system as defined above takes a sequence of inputs from the environment one by one. The next input from the environment is processed only when the system is in a stable state. This definition of stable state makes it unnecessary in later sections to describe input queues to show (stable) states of a communicating system. In (1), σ with a single external input defines a set of new states reached by processing of a single input from the environment followed by sequences of messages (including sequences of internal state changes) sent by the receiving IOSM. The reception of a message and sending messages as its response constitutes an atomic action. Thus inputs from the environment or IOSM's is instantly placed into the input queues of the receiving IOSM's. μ is a function that updates states. φ takes a set of messages and returns the set of all possible orderings of external messages excluding internal messages. (2) extends σ to a sequence of external inputs. (1)-(3) for communicating systems are similar to (1)-(3) for IOSM's of Definition 2. In a communicating system of IOSM's, *deadlock* is said to occur when any combination of (external and internal) inputs cannot change the system state. *Livelock* is said to occur when system state changes forever without reaching a stable state. Note that Definition 4 can also be regarded as defining a composition of n IOSM's, denoted $M_1 \times M_2 \times ... \times M_n$, for which

(1) the states are stable states of Σ

(2) between its states s and s' there is a transition with the label 'v/**u**' if and only if $(s', \mathbf{u}) \in \sigma(s, v)$

(3) its initial state is the same as the initial state of Σ.

Therefore we will use Σ and $M_1 \times M_2 \times ... \times M_n$ in an interchangeable way.

In conformance testing, conformance test suite is developed in such a way that for the given specification with FSM structure, absence of operation errors (or correctness of input/output behavior) and absence of transfer errors (or correctness of the state reached after the transition) are examined with respect to each transition of the FSM [Chow 78]. For this purpose, a conformance test case consists of three parts: preamble, test body and postamble. *Test body* is the part of a conformance test case that checks operation errors and transfer errors. Test body ends by reaching a stable state. New test cases are needed to examine the behavior of the machine M at the stable state reached after the transition. *Preamble* is the part that takes M to the stable state at which the test body can be started. *Postamble*, which is often empty, is the part that takes M to a stable state. Methods to generate such a conformance test suite are well-known [Chow 78][Sidh 90]. Let *TS*(M) be the set of all such conformance test cases, *conformance test suite* in short, for M. We say that a conformance test suite is *complete* if each and every transition in the specification modeled as IOSM has a corresponding conformance test case as defined above.

3. INTEROPERABILITY TEST SUITE DERIVATION

To illustrate our method we use as an example the ATM Signaling Protocol for the ATM switch. The behavior of the protocol at the user-network interface is specified in [ITU-T Q.2931] and also in ATM Forum UNI specification [AF UNI]. The two specifications have much in common. The behavior of the protocol at the network-network interface differs from that of user-network side. Among ATM Forum specifications, PNNI specification [AF PNNI] is the one that gives the most complete description of protocol behavior at the network-network interface. We use ATM Forum UNI 3.1 and PNNI specifications as the definitive specifications of the interface behavior of the ATM switch which together constitute the ATM signaling protocol for the ATM switch (to be subsequently abbreviated as Signaling Protocol).

3.1 Pruning Transitions

The first step is to derive an IOSM, say *S*, from the relevant informal specifications. Assuming that we already derived *S*, let *pr*(*S*) be the IOSM which is the same as *S* except that *pr*(*S*) does not contain the transitions in *S* which (1) do not change state and (2) do not contain '*i_*' prefixed messages. For example, upon receiving

STATUS_ENQUIRY message at any state, the signaling protocol entity should send STATUS message reporting its current state [AF UNI][AF PNNI]. Such transitions would not appear in $pr(S)$. In this way, we come up with an IOSM depicted in Figure 2 from Signaling Protocol.

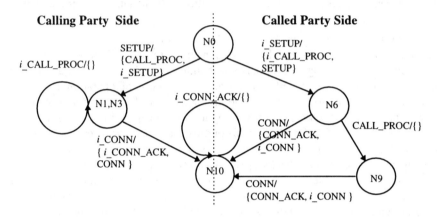

Figure 2. Call establishment part of the pruned IOSM for Signaling Protocol.

Conceptually an IOSM is one connected graph. With this example, the picture in one graph would be severely cluttered due to its complexity. Therefore we present the graph partitioned with respect to functionalities. Figure 2 describes state transitions for call establishment. At any moment, signaling protocol entity functions either as a calling party or as a called party (but not both). In the figure, it is made clear with the dotted line in the middle. There are two kinds of messages (or PDU's): one appearing at the UNI interface and the other appearing at the NNI interface. In Figure 2, NNI messages are prefixed with '$i\_$'. The reason for the prefix is the interactions occurring at NNI are internal events between two ATM switches. As explained with Figure 1 in Section 1.2, we take the view that, by not monitoring NNI, interactions at NNI do not directly affect test results.

When combined with the conformance test suites for S_a and S_b, the interoperability test suite for $pr(S_a) \times pr(S_b)$ covers all the test cases for $S_a \times S_b$. The following theorem states this completeness formally.

Theorem 1 Let S_a, S_b be two IOSM's. Then
$$TS(S_a \times S_b) \subseteq \{TS(pr(S_a) \times pr(S_b)) \cup TS(S_a) \cup TS(S_b)\}$$

Note that this step of optimization is applicable regardless of whether the protocol is symmetric or not. For symmetric protocols, the behavior of the global system can be described as the composition of $pr(S_a)$ with itself, i.e. $pr(S_a) \times pr(S_a)$.

3.2 Interoperability Test Case

Figure 3 depicts a typical message interaction sequence for successful call establishment. The terminal equipment TEa initiates a call to TEb by sending

SETUP through the switch IUTa. TEb accepts the call by sending CONN through the switch IUTb. The shaded area represents a black-box whose internals we decide not to observe.

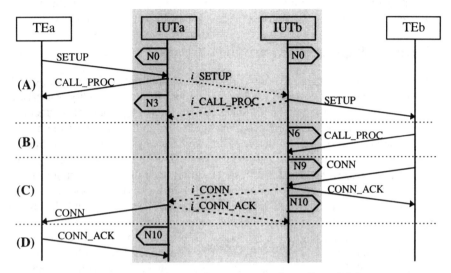

Figure 3. Message interaction sequence for call establishment.

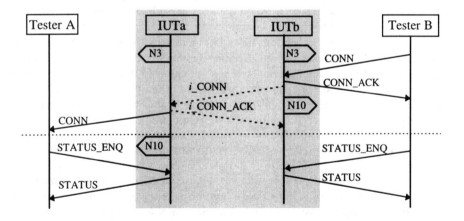

Figure 4. An interoperability test case for subsequence (C).

As with the conventional conformance testing, unstable states are not available for control purposes and can be utilized to reduce the size of state space of protocol entities logically built from specifications and at the same time the number of test cases. There are 4 *stable states*, (N0,N0), (N3,N6), (N3,N9) and (N10,N10) in the interaction sequence of Figure 3. Since we assume that test bodies are from a stable state to a stable state, there are 4 subsequences which are candidates for testing as marked with (A), (B), (C) and (D). However, (B) and (D) are actually conformance test cases for IUTb and IUTa, respectively, since only one IUT is involved. (A) and (C) are genuine interoperability test cases. Figure 4 is one of the interoperability test cases derivable from the interaction sequence of Figure 3.

3.3 Optimizing Test Suite Derivation Process

In the previous sections, in the course of optimizing interoperability testing, we (1) pruned transitions of IOSM and (2) gave definition of interoperability test case in such a way that it is based on stable states and abstracts from internal interactions. In this section, we consider further optimizations. To do that, we need the notions of initiator and responder. An active tester which always initiates a test body is called *initiator*. A tester which may be active or passive but never initiates a test body is called *responder*. An active responder may send messages during test case execution. For example, it may initiate preamble part by sending a message.

In the test architecture of Figure 1, suppose that IUTa and IUTb are implementations of the same protocol specification. One of Tester A or Tester B must send a message to initiate a test body. If Tester B were to send a message to initiate a test body, then by making Tester A initiate the test body with the positions of Tester A and Tester B interchanged the mirror image of the same sequence of interactions would occur. This observation can be stated as follows:

> **Theorem 2**   Given the test architecture of Figure 1, for interoperability testing of symmetric protocols, it is sufficient to have one initiator and one responder.

For example, the test execution shown in Figure 4 can be achieved by switching the positions of IUTa and IUTb and keeping Tester A as the initiator and Tester B as the responder. An implication of the observation is that the same test suite should be run twice, the second run after switching the positions of Initiator and Responder. However, the necessary test suite size is now reduced approximately to half. In general Theorem 2 is not true of asymmetric protocols.

3.4 An Efficient Algorithm to Generate Stable States

Not only can we reduce the test suite size approximately to half as shown in the previous section, but also it is possible to optimize the test generation process itself. Relying on Theorem 2, we show below an algorithm to generate all stable states which form the starting and the ending states of interoperability test cases. Actual test cases generation is realized by decorating the algorithm with the step to store the applicable sequence of events between stable states. In our approach to interoperability testing, the number of stable states is small enough that it can be used as a suitable basis for manual calculation as well as for automatic generation of interoperability test suites.

input: IOSM's S_1 and S_2
output: Stable[j] contains all the stable states of $S_1 \times S_2$
j :=0;
Stable[j]   := {$(s_{1,0}, s_{2,0})$};   /* All stable states generated up to Stage j */
New   := {$(s_{1,0}, s_{2,0})$};   /* Stable states to be expanded at Stage j+1 */
while New $\neq \varnothing$ **do begin**
    <New, (s_1, s_2)> := *delete-one-element*(New);
    OldFrontier := $\varnothing$;   /* Already expanded nodes at Stage j */
    NewFrontier := {(s_1, s_2)};   /* Nodes to be expanded at Stage j+1 */
    while (NewFrontier $\neq \varnothing$) **do begin**
        <NewFrontier, (s_1, s_2)> := *delete-one-element*(NewFrontier);
        OldFrontier := OldFrontier $\cup$ {(s_1, s_2)};
        Event_Seq_Set := *interaction-sequences*((s_1, s_2));
        while Event_Seq_Set $\neq \varnothing$ **do begin**
            <Event_Seq_Set, Event_Seq> := *delete-one-element*(Event_Seq_Set);
            if Event_Seq *ends with a stable state* (s_1, s_2)
            then begin
                        if $s' \notin$ (Stable[j] $\cup$ New $\cup$ OldFrontier)
                        then NewFrontier := NewFrontier $\cup$ {(s_1, s_2)};
                end
            else begin
                        There is an error in the specifications.
                        Stop and fix the error.
                end
        end while;
    end while;
    j := j+1;
    Stable[j] := Stable[j−1] $\cup$ OldFrontier;
    New := {(s_2, s_1) | $(s_1, s_2) \in$ Stable[j] $\wedge$ $(s_2, s_1) \notin$ Stable[j] };
end while;

Figure 5. An efficient algorithm to generate stable states.

In the algorithm, (s_1, s_2) denotes the global system state where s_1 and s_2 are stable states of S_1 and S_2, respectively. *delete-one-element*(*Set*) is a function that choose an arbitrary element from the set *Set* and removes it. The output of this function is a pair of which the first argument is the chosen element and the second argument is the resulting *Set* without the chosen element. *interaction-sequences*((s_1, s_2)) generates all possible sequences of interactions which are initiated by the IOSM which is in state s_1. Let $TS'(S_1, S_2)$ denote the test suite generated by the algorithm.

An application of the algorithm is shown in Figure 6. The global states in boxes are stable states and those not in boxes are transient states. Underline for instance in $\underline{s}$ indicates that the tester for the IUT with the underlined state is the initiator. After the first stage is over, the only node of which the full expansion has been postponed is (**N3,N6**) marked in bold face. Therefore the 2nd stage would begin with the expansion of (**N6**,N3). In general, there can be more than one such nodes. Similarly, in the 2nd stage the only node which would be not fully expanded is (**N9**,N3). Hence in the 3rd stage (**N3**,N9) should be expanded.

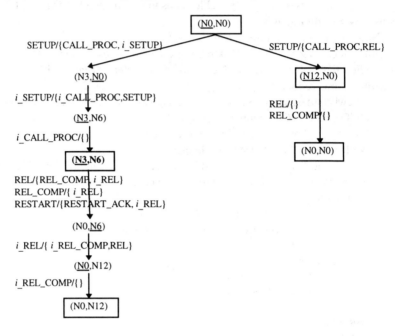

Figure 6. The 1st stage of test suite generation algorithm.

Let $tc_{S1, S2}$ be one of the test cases generated from the algorithm. Let $tc_{S2, S1}$ denote the test case with exactly the same structure as $tc_{S1, S2}$ except that in the former the roles of S_1 and S_2 are exchanged. For example, if $tc_{Sa, Sb}$ to be obtained from the 2nd stage is stated as:
    The initial state is (N6,N3).
    After IUTa receives CALL_PROC from LTa, IUTa does not respond.
    The final state is (N9,N3).
then $tc_{Sb, Sa}$ is as follows:
    The initial state is (N3,N6).
    After IUTb receives CALL_PROC from LTb, IUTb does not respond.
    The final state is (N3,N9).
It is the characteristic of the algorithm that it only generates one of these two test cases. But neither of the test cases can be dispensed with. The following states an obvious fact about the algorithm.

Lemma 3 Let S_1, S_2 be IOSM's and $tc_{S1, S2}$ and $tc_{S2, S1}$ be as defined above. Then
$$tc_{S1, S2} \in TS'(S_1, S_2) \text{ if and only if } tc_{S2, S1} \in TS'(S_2, S_1)$$

Thus the algorithm can be said to be complete in the sense that if it is applied twice, the second time with the positions of S_1 and S_2 interchanged, all the test cases generated by the conventional method would be covered. The next theorem states this completeness.

Theorem 4 Let S_1, S_2 be IOSM's. Then
$$TS(S_1 \times S_2) \subseteq TS'(S_1, S_2) \cup TS'(S_2, S_1)$$

Symmetry of $TS'(S_1, S_2)$ and $TS'(S_2, S_1)$ implies that for actual application the effect of applying the two test suites would be achieved simply applying one of the suites twice, the second time with the positions of two IUT's interchanged.

When it is known that the two IUT's are identical, possible nonconforming behavior which would be detected with the positions of IUT's interchanged would be detected without interchanging their positions. Let $exec(tc, I_a \times I_b)$ denote the result of applying the test case tc to the system of IUT's $I_a \times I_b$. For test suite TS, define
$$exec(TS, I_a \times I_b) = \{exec(tc, I_a \times I_b) \mid tc \in TS\}.$$

Corollary 5 Let S_1, S_2 be IOSM's and I_a, I_b be IUT's . Then
$$exec(TS(S_1 \times S_2), I_a \times I_b) \subseteq exec(TS'(S_1, S_2), I_a \times I_b)$$

This means that if $I_a = I_b$ it is sufficient to apply $TS'(S_1, S_2)$ (or $TS'(S_2, S_1)$) once for interoperability testing. Note that Lemma 3, Theorem 4 and Corollary 5 are stated generally and they remain valid in the special case when the relevant IOSM's are pruned machines.

4. FURTHER DEVELOPMENT

When the approach shown in this paper is fully applied to Signaling Protocol, we obtain 19 test case skeletons (3 for call establishment, 11 for call clearing, 5 for restarting). In [KanC 96], they were called *test case skeletons* because actual abstract test cases can be derived from them by embellishing PDU's. In general, more than one abstract test cases may be constructed depending on the chosen test approach. With the test architecture used in this paper (Figure 1), interoperability test cases can be adequately described in TTCN.

Signaling Protocol is structured in such a way that depending on whether each IUT supports CALL_PROC, there are four different possible sets of test cases as follows:

		IUTa	IUTb
Upon receiving SETUP	(1)	Yes	Yes
Does IUT send CALL_PROC	(2)	No	Yes
to the user?	(3)	Yes	No
	(4)	No	No

We used the case (1) for the example in this paper. The case (4) can be handled in the exactly the same manner shown in this paper.

5. CONCLUSION

In this paper, first we clarified the notions of interoperation and interoperability testing and showed a specific interoperability test architecture and testing approach in accordance with those notions. We applied the approach to symmetric protocols and showed how various optimizations can be achieved. The details of optimizations were exhibited with the example of the ATM Signaling Protocol. In order to do that, firstly, we defined as the conceptual basis interoperability test case such that it utilizes stable states, abstracts from interactions internal to the two communicating protocol entities and is disjoint from conformance test. Secondly, we showed how to prune IOSM's so that, provided that the relevant conformance test suites already exist, the interoperability test suite derived from the pruned IOSM has no overlapping with them and covers all the proper interoperability test cases. Thirdly, we developed an efficient algorithm to generate all stable states to optimize test suite generation process itself. As the result, the test suite size can be reduced approximately to half in the case of symmetric protocols. Moreover, it was shown that when it comes to interoperability testing of two identical IUT's the number of test cases to be actually executed can be reduced to half.

There are some preliminary steps necessary before the interoperability testing can be applied. It is usually taken for granted that interoperability testing should be done *after* IUT's pass conformance testing [ISO 9646-1]. The first reason for this is that if IUT's are non-conforming most likely they would not interoperate. Moreover, when a system of implementations fails, we would not know which one is responsible for non-interoperation or whether there is an interoperation problem residing in specifications themselves. In addition, before the interoperability testing can be applied, implementations of the underlying service need to be thoroughly tested and should be known to be conforming. These assumptions are in line with the established conformance test methodology [ETSI 96].

Since the purpose of the B-ISDN signaling protocol is to provide ATM layer connection, one may think that one should check that traffic does indeed go through ATM layer connection after a successful call setup. This involves not only the behavior of the Signaling layer but also the behavior of the ATM layer. Thus testing of such interoperation behavior requires a more complicated test architecture with additional PCO's for the User Plane above the ATM layer. Being based on the test architecture shown in Figure 1, our interoperability test cases do not cover such cases. A similar view has been taken in ATM Forum signaling user-side ATS [AF Sig U-ATS] and network-side ATS [AF Sig N-ATS] and also CCITT ISDN Layer 3 D-channel testing [ITU-T Q.931bis], where data transport through B-channel and ATM connection, respectively, are not examined. Verdicts for nondeterministic test cases should be given by adopting *all-weather conditions assumption* as in other work [LuBP 94][Brin 88], i.e. run them sufficiently many times to expose any nonconformance. In the example in Section 3, such test cases are those beginning at (N0,N0) with SETUP and at (N6,N3) with CALL_PROC.

The basic idea of interoperability testing framework and approach developed in this paper was partially proposed and was used for interoperability test suite development inside the ATM Forum [KKCS 96][KanC 96]. In Section 4, we mentioned the cases where the protocols are asymmetric. Our approach would not directly work for those cases. However, if protocols are *almost* symmetric then it is very likely that modification of our approach would work for them. We also plan to extend our application target to include point-to-multipoint signaling feature. Furthermore, we plan to extend our method for testing protocol data incorporating data flow analysis.

Acknowledgment We thank Dr. David Su, David Cypher and Leslie Collica at NIST, USA, for having numerous discussions with us on the subject of interoperability testing. Special thanks are due to David Cypher who clarified many subtle issues related to the ATM signaling protocol.

6. REFERENCES

[APRS 93] Arakawa, N., Phalippou, M., Risser, N. and Soneoka, T., "Combination of conformance and interoperability testing", *Formal Description Techniques V (C-10)*, M. Diaz and R. Groz (Eds.) Elsevier Science Publishers B. V. (North-Holland), 1993.
[AraS 92] Arakawa, N. and Soneoka, T., "A Test Case Generation Method for Concurrent Programs", *Protocol Test Systems, IV*, J. Kroon, R. J. Heijink and E. Brinksma (Eds.), Elsevier Science Publishers B. V. (North-Holland), 1992.
[AF UNI] ATM Forum, *ATM User-Network Interface Specification, Version 3.1*, 1994.
[AF PNNI] ATM Forum, *ATM Forum PNNI Draft Specification*, 1996.
[AF Sig U-ATS] ATM Forum, "ATM Forum 96-0979: ATS for the UNI3.1 Signaling - User Side", 1996.
[AF Sig N-ATS] ATM Forum, "ATM Forum 95-1145R1: ATS for the UNI3.1 Signaling - Network Side (Part I)", 1996.
[Bonn 90] Bonnes, G., "IBM OSI Interoperability Verification Services", IFIP TC6/WG6.1 The 3rd International Workshop on Protocol Test Systems, 1990.
[Brin 88] Brinksma, E., "A Theory for the Derivation of Tests", *Protocol Specification, Testing and Verification, VIII*, S. Aggarwal, K. Sabnani (eds.), North-Holland, Amsterdam, pp. 63-74. 1988.
[CasK 94] Castanet, R. and Kone, O., "Deriving Coordinated Testers for Interoperability", *Protocol Test Systems, VI(C-19)*, O. Rafiq(Ed.), Elsevier Science B. V. (North-Holland), IFIP, 1994.
[Chow 78] Chow, T. S., "Testing Software Design Modeled by Finite-State Machines", *IEEE Trans. on SE, Vol. SE-4, No. 3*, May 1978.
[ETSI 96] ETSI, *Making Better Standards: Practical Ways to Greater Efficiency and Success*, ETSI MTS, 1996.
[GRSS 90] Gadre, J., Rohre, C., Summers, C., and Symington, S., "A COS Study of OSI Interoperability", *Computer Standards and Interfaces, Vol. 9, No. 3*, pp 217-237, 1990.
[ISO 95] ISO/IEC 9646-3, "Information Technology - OSI Conformance testing methodology and framework - Part 3: The Tree and Tabular Combined Notation (TTCN)", 1995

[ITU-T Q.2931] ITU-T, "ITU-T draft Recommendation Q.2931: B-ISDN User-Network Interface Layer 3 Specification for Basic Call/Bearer Control", 1994.

[ITU-T Q.931bis] ITU-T, "ITU-T Rec. Q.931bis: Abstract Test Suite for Basic Call Control Conformance Testing", 1994.

[KKCS 96] Kang, S., Kim, M., Cypher, D., and Su, D., "A Proposal for ATM Signaling Protocols Interoperability Test Suite Development", ATM Forum/96-0167, February 1996.

[KanC 96] Kang, S., and Cypher, D., "Test Case Skeletons for Interoperability Testing of ATM Signaling Network-Side Protocols", ATM Forum/96-0336r2, Baltimore, August 1996.

[KanK 95] Kang, S., and Kim, M., "Test Sequence Generation for Adaptive Interoperability Testing", IFIP TC6/WG6.1 The 8th International Workshop on Protocol Test Systems, Evry, France, September 1995.

[LuBP 94] Luo, G., Bochmann G. and Petrenko, A., "Test Selection Based on communicating Nondeterministic Finite-State Machines Using a Generalized Wp-Method", *IEEE Transactions on S.E., Vol 20, No. 2*, February 1994.

[RafC 90] Rafiq, O. and Castanet, R., "From Conformance Testing to Interoperability Testing", The 3rd International Workshop on Protocol Test Systems, 1990.

[Sidh 90] Sidhu, D. P., "Protocol Testing: The First Ten Years, The Next Ten Years", *Protocol Specification, Testing and Verification, X*, L. Logrippo, R. R. Probert & H. Ural (Eds.), Elsevier Science Publishers B. V. (North-Holland), 1990.

[VerB 94] Vermeer, G.S., and Blik, H., "Interoperability Testing: Basis for the Acceptance of Communication Systems", *Protocol Test Systems, VI (C-19)*, Elsevier Science Publishers B. V. (North-Holland), 1994.

Sungwon Kang received B.A. degree from Seoul National University in Korea in 1982 and M.S. and Ph.D. in computer science from the University of Iowa in U.S.A in 1989 and 1992. Since 1993, he has been a senior researcher at Korea Telecom. From 1995 to 1996, he was a guest researcher at National Institute of Standards and Technology of U.S.A. In 1997, he was the co-chair of the 10th International Workshop on Testing of Communicating Systems and currently is the head of Protocol Testing Technology Team at Korea Telecom R&D Group. His research interests include communication protocol testing, program optimization and programming languages.

Myungchul Kim received B.A. in electronics engineering from Ajou Univ. in 1982, M.S. in computer science from the Korea Advanced Institute of Science and Technology in 1984, and Ph.D. in computer science from the Univ. of British Columbia in 1992. Since 1984, he has been working for Korea Telecom. In 1997, he was co-chair of the 10th International Workshop on Testing of Communicating Systems. Currently he is managing director of Testing Technology Research Section at Korea Telecom R&D Group and chairman of Profile Test Specifications - Special Interest Group of Asia-Oceania Workshop. His interests include protocol engineering on multimedia and telecommunications.

## PART TWO

MSC and ODP

5
A Hierarchy of Communication Models for Message Sequence Charts

A. Engels, S. Mauw, M.A. Reniers
Department of Mathematics and Computing Science,
Eindhoven University of Technology
P.O. Box 513, 5600 MB Eindhoven, The Netherlands
{engels|sjouke|michelr}@win.tue.nl

Abstract

In a Message Sequence Chart (MSC) the dynamical behaviour of a number of co-operating processes is depicted. An MSC defines a partial order on the communication events between these processes. This order determines the physical architecture needed for implementing the specified behaviour, such as a FIFO buffer between each of the processes. In a systematic way, we define 50 communication models for MSC and we define what it means for an MSC to be implementable by such a model. Some of these models turn out to be equivalent, in the sense that they implement the same class of MSCs. After analysing the notion of implementability, only ten models remain, for which we develop a hierarchy.

Keywords

Message Sequence Charts, semantics, implementation, validation, buffering, communication models, hierarchy

1 INTRODUCTION

In recent years much attention has been paid to graphical languages for the visualisation of communication traces in distributed systems. One of the most popular classes of formalisms for this purpose is the class of sequence charts. Of those, Message Sequence Chart (MSC) has been standardised by the International Telecommunication Union (ITU) as Recommendation Z.120 (ITU-TS 1996). Two important reasons for

Formal Description Techniques and Protocol Specification, Testing and Verification
T. Mizuno, N. Shiratori, T. Higashino & A. Togashi (Eds.) © 1997 IFIP. Published by Chapman & Hall

the popularity of MSCs are that they provide a clear intuition to both engineers and designers and at the same time posses a well-defined semantics.

Although MSC is primarily concerned with presenting the asynchronous communication between processes in a distributed system, no information is given as to the way in which these communications are supposed to be realized in an implementation. The only assumption about the implementation of communication is that an output precedes its corresponding input.

This impossibility to specify the communication model becomes a problem when a specific communication model is presupposed, for example due to hardware requirements. Whenever MSC is used to specify the communication behaviour, the question arises whether the behaviour defined by an MSC is feasible with respect to the desired communication model. It may be the case that all traces defined by the MSC are feasible, that at least one trace is, or that none of the traces is feasible. For example, an MSC with two inherently crossing messages cannot be implemented with an architecture containing one single global FIFO buffer for message exchange.

There are two approaches to deal with this under-specification in MSC. The first is to select a single preferred model and revise the semantics of MSC accordingly. Keeping in mind the broad context in which MSC is used in practice, this option is not realistic. The only acceptable choice would be the most general random-access buffer model that has been chosen in the current standardised semantics of MSC.

The alternative would be to allow the user of MSC to indicate the desired communication model explicitly. This can be done by extending the syntax of MSC with a means to specify the intended model and by developing dedicated tools for the analysis of MSC with respect to certain implementation models. We propose to study this second alternative and it is our aim in this paper to provide a solid and formal basis for defining the relation between a communication model and an MSC.

For a given MSC we define the notions of strong and weak implementability. Strong implementability of an MSC in a given communication model means that all traces of the MSC can be realized with the given communication model and weak implementability means that there is a trace that can be realized.

In this way, we attach to each implementation model the class of MSCs that are strongly or weakly implementable with respect to that model. A natural question to ask is whether there are communication models that define the same class of MSCs. This means that for a given MSC one has a choice of communication model for implementation. It turns out that the initial number of fifty MSC classes can be reduced to a hierarchy of ten different models.

Acknowledgements We would like to thank Thijs Cobben, Loe Feijs, Herman Geuvers and Bart Knaack for their valuable input.

2   MESSAGE SEQUENCE CHARTS

In this section we explain the semantical foundations of Message Sequence Chart (MSC). We use a partial order on the events of an MSC to express the semantics. In literature several ways to define the semantics of MSC are proposed (Mauw and

Reniers 1994, Ladkin and Leue 1995, Grabowski, Graubmann and Rudolph 1993). The process algebra approach (Mauw and Reniers 1994) has been standardised as Annex B to ITU recommendation Z.120 (ITU-TS 1995). The partial order representation (Alur, Holzmann and Peled 1996) used in this paper coincides with most of these proposals for the class of Basic Message Sequence Charts. We also define the traces expressed by an MSC.

The MSCs studied here consist of a collection of instances (or processes) with a number of messages attached to them. These are known as Basic Message Sequence Charts, but in this paper we use the term MSCs to denote them.

Some examples of MSCs can be seen in Figure 3. MSCs consist of vertical lines, denoting the various communicating processes, which we call 'instances' and arrows between these instances, denoting exchanged messages.

We allow messages from an instance to itself, but we only consider closed systems, that is, we do not consider messages to the environment. Neither do we consider any other specific features such as local actions and recursion. We assume that the names of the instances and messages are unique. Therefore, the instances to which a message is attached are determined uniquely by the message name.

The easiest way to express the semantics of such a simple MSC is by using a partial order on the events that are comprised in an MSC. Depending on the particular dialect of the MSC language, one can assign different classes of events to an MSC. For example, in Interworkings (Mauw, Wijk and Winter 1993) every message is considered to be a single event. There is no buffering, and thus communication is synchronous.

In MSC (ITU-TS 1996), messages are divided into two events, the output and the input of the message. The output of message m is denoted by $!m$ and the input by $?m$. The only assumption about the implementation of communication is that an output precedes its corresponding input. This corresponds to the most general implementation model in which processes communicate via unbounded random-access buffers.

In this paper we go one step further, and add a third event, denoted by $!!m$, that we call *transmit m*. The basic idea is that a message passes two buffers before arriving at its destination. The intuition here is that $!m$ denotes the putting of a message into an output buffer, $!!m$ is the transmission of the message from the output buffer to the appropriate input buffer, and $?m$ is the removal of the message from the input buffer. We assume these events to be instantaneous. Furthermore, we concentrate on FIFO-buffers only.

Although the intermediate transmit events $!!m$ play a crucial role in our description of the communication models, we do not encounter them in the definition of an MSC, nor in the partial order describing the formal semantics of an MSC. An MSC still describes a partial order on output and input events only.

Definition 1 (MSC) An MSC is a quintuple $\langle I, M, from, to, \{<_i\}_{i \in I} \rangle$, where I is a finite set of instances, M is a finite set of messages, *from* and *to* are functions from M to I, and $\{<_i\}_{i \in I}$ is a family of orders. For each $i \in I$ it is required that $<_i$ is a total order on $\{!m \mid from(m) = i\} \cup \{?m \mid to(m) = i\}$.

In the above definition, *from(m)* denotes the instance which sends message m. Likewise, $to(m)$ denotes the instance which receives message m. Given an instance i, the order $<_i$ denotes in which order the events attached to instance i occur. The order $<_i$ is lifted in the trivial way to the set $\{!m, ?m, !!m \mid m \in M\}$.

The partial order denoting the semantics of an MSC is derived from two requirements. First, the order of the events per instance is respected, and second, a message can only be received after it has been sent. The first requirement is formalised by defining the partial order $<^{inst} := \bigcup_{i \in I} <_i$, and the second requirement is formalised by the *output-before-input* order $<^{oi} := \{(!m, ?m) \mid m \in M\}$.

Now, we define the partial order induced by the MSC as the transitive closure (denoted by $^+$) of the instancewise order and the output-before-input order. For an MSC k, we denote this order by $<^{msc}_k$ or by $<^{msc}$ if k is known from the context.

Definition 2 For a given MSC $k = \langle I, M, from, to, \{<_i\}_{i \in I}\rangle$, the relation $<^{msc}_k$ is defined by $<^{msc}_k := (<^{inst} \cup <^{oi})^+$.

We define similar notions for 3-traces. We define the *output-before-transmit-before-input* order by $<^{oti} := \{(!m, !!m), (!!m, ?m) \mid m \in M\}$, and the relation $<^{m3}$ by adding the instancewise ordering on the MSC.

Definition 3 For a given MSC $k = \langle I, M, from, to, \{<_i\}_{i \in I}\rangle$, the ordering $<^{m3}$ is defined by $<^{m3} := (<^{inst} \cup <^{oti})^+$.

It is easy to see that $<^{msc}$ is the restriction of $<^{m3}$ to output and input events. It may be the case that $<^{msc}$ does not define a partial order, due to cyclic dependencies of the events. Such an MSC is said to contain a *deadlock*, or is called *inconsistent*. In Z.120 (ITU-TS 1996) inconsistent MSCs are considered illegal, and in (Ben-Abdallah and Leue 1997) an algorithm is described for determining whether a given MSC is consistent. In the remainder of this paper we consider consistent MSCs only, which implies that both $<^{msc}$ and $<^{m3}$ are partial orders.

From an operational point of view, one can say that an MSC describes a set of traces. We distinguish 2-traces and 3-traces. A 2-trace denotes the ordering of output and input events ($!m$ and $?m$), a 3-trace those of transmit events ($!!m$) as well.

Definition 4 (2-traces, 3-traces) A *2-trace* t over a set of messages M is a total ordering $(e_1, e_2, \ldots, e_n)$ of the set $\{!m, ?m \mid m \in M\}$. This ordering is denoted by $<^{trace}_t$. A trace $(e_1, e_2, \ldots, e_n)$ is denoted $e_1 e_2 \ldots e_n$. A *3-trace* is equal to a 2-trace, except for the fact that it contains transmit events as well.

Definition 5 (MSC-trace) A 2-trace t is said to be a trace of the MSC k iff it is defined over the messages M of k, and $<^{msc}_k \subseteq <^{trace}_t$. A 3-trace t is said to be a trace of the MSC k iff it is defined over the messages M of k, and $<^{m3}_k \subseteq <^{trace}_t$.

A 3-trace can be turned into a 2-trace by removing all transmit events ($!!m$). If, for a 3-trace t this results in a 2-trace t', then t is said to be an extension of t'. It is not hard to see that a 3-trace t is a trace of an MSC k iff the 2-trace of which it is an extension is a trace of the MSC and $<^{oti}_k \subseteq <^{trace}_t$.

For MSC 2a in Figure 3 the following orderings hold: $!a <^{msc} ?a$, $!b <^{msc} ?b$, and $?a <^{msc} ?b$. The first two are implied by the $<^{oi}$-order, the third by the $<^{inst}$-order. The MSC has exactly three 2-traces: $!a\,?a\,!b\,?b$, $!a\,!b\,?a\,?b$, and $!b\,!a\,?a\,?b$. These 2-traces can be extended to ten 3-traces, such as $!a\,!!a\,?a\,!b\,!!b\,?b$ and $!a\,!b\,!!b\,!!a\,?a\,?b$.

3   IMPLEMENTATION MODELS

We discuss possible architectures for realizing an MSC. We consider only implementation models consisting of FIFO buffers for the output and input of messages. For MSC traces, we define what it means to be implementable on some architecture.

3.1   Locality of buffers

The particular implementation models which we are interested in are constructed of processes that communicate with each other via FIFO buffers. We assume that the buffers have an unbounded capacity. We discern two uses of buffers, namely for the output and for the input of messages.

A second distinction can be made based on the locality of the buffer. From most global to most local we distinguish the following types:

- *global*: A global FIFO buffer: All messages from all instances pass this buffer.
- *inst*: A FIFO buffer, local to an instance: All messages sent (or received) by one single instance go through the same buffer.
- *pair*: A FIFO buffer, local to two instances: All messages that are sent from one specific instance to another specific instance go through this buffer.
- *msg*: A FIFO buffer, local to a message: There is one buffer for every message.

This last model, a buffer per message, is a specific architecture to catch up the cases in which the buffers do not behave like FIFO queues. Taking into account the assumption that messages are unique, it can easily be seen that it is equivalent to a global random-access buffer. A communication model with only a random-access buffer represents the model of the MSC standard: the only assumption made about the implementation of communication is that output precedes input.

Finally, we consider the following possibility of using no buffers at all, denoted by *nobuf*. In this case communication is synchronous.

We assume that the transmission from an instance to its output buffer, from one buffer to another buffer, or from an input buffer to the instance it belongs to, is synchronous. We also assume that all output buffers are of the same type, and similarly that all input buffers are of the same type. This results in four possibilities for the output as well as for the input. Adding the possibility of using no buffer at all, we have a total of 25 possible architectures. To denote the different architectures, we use the notation (X,Y), where X denotes the type of output buffer, and Y the type of input buffer.

3.2   Examples of communication models

In Figure 1 we give examples of a physical architecture of three communication models. A circle denotes an instance, a rectangle denotes a buffer, and an arrow denotes a communication channel. Each example contains three instances. The first example illustrates the (*nobuf,global*) model. There is no output buffer, and one universal input buffer. As there is no output buffer, the messages go straight into the input buffer. This single buffer could be regarded as an output buffer as well, so this example is an illustration of (*global,nobuf*) too. The second example shows the (*global,inst*) model. There is one general output buffer and every instance has a local input buffer. The third architecture is an example of the (*pair,pair*) model.

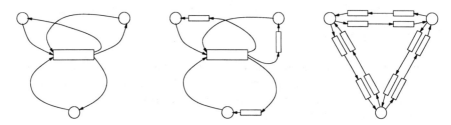

Figure 1 Some models: (*nobuf,global*), (*global,inst*) and (*pair,pair*).

Many of these architectures occur in practice as either the underlying communication architecture of a programming language or as a physical architecture. We give some examples of languages. The model (*nobuf,nobuf*) is typical for process algebraic formalisms based on synchronous communication, such as LOTOS and ACP. The specification language SDL, which is closely related to MSC, has as a general communication model (*pair,msg*), but if we leave out the *save* construct we obtain (*pair,inst*) and if we also do not consider the possibility of delayed channels, we have (*nobuf,inst*). Some examples of physical architectures are: an asynchronous complete mesh has a (*nobuf,pair*) architecture, and an Ethernet connection with locally buffered input and output behaves like (*inst,inst*).

3.3   Implementability

The main question of this paper is, whether a given MSC can be the behaviour of a given implementation model. To answer this question, we first give a formal definition of what it means for a trace to have a certain implementability property. The definitions below can be seen as a formalisation of the notions introduced in Section 3.1.

Definition 6 (Output-implementability)

- *nobuf-output*: Every output event is directly followed by the corresponding transmit event. Thus, output and transmit event may be combined into one new event.

A 3-trace t is *nobuf*-output implementable iff $\forall_{m \in M} \neg \exists_{e \in \{!m, !!m, ?m | m \in M\}} !m <_t^{trace}$
$e <_t^{trace} !!m$.

- *global-output*: The order of two output events is respected by the corresponding transmit events. A 3-trace t is *global*-output implementable iff $\forall_{m,m' \in M} !m <_t^{trace}$
$!m' \Leftrightarrow !!m <_t^{trace} !!m'$.

- *inst-output*: The order of any two output events from the same instance is respected by the corresponding transmit events. A 3-trace t is *inst*-output implementable iff $\forall_{m,m' \in M} from(m) = from(m') \Rightarrow (!m <_t^{trace} !m' \Leftrightarrow !!m <_t^{trace} !!m')$.

- *pair-output*: The order of two outputs with the same source and the same destination, is respected by the corresponding transmit events. A 3-trace t is *pair*-output implementable iff $\forall_{m,m' \in M} from(m) = from(m') \wedge to(m) = to(m') \Rightarrow (!m <_t^{trace} !m' \Leftrightarrow !!m <_t^{trace} !!m')$.

- *msg-output*: A 3-trace t is always *msg*-output implementable.

The input implementabilities are defined analogously.

Definition 7 (Input-implementability) A 3-trace t is

- *nobuf*-input implementable iff $\forall_{m \in M} \neg \exists_{e \in \{!m, !!m, ?m | m \in M\}} !!m <_t^{trace} e <_t^{trace} ?m$;
- *global*-input implementable iff $\forall_{m,m' \in M} !!m <_t^{trace} !!m' \Leftrightarrow ?m <_t^{trace} ?m'$;
- *inst*-input implementable iff $\forall_{m,m' \in M} to(m) = to(m') \Rightarrow (!!m <_t^{trace} !!m' \Leftrightarrow ?m <_t^{trace} ?m')$;
- *pair*-input implementable iff $\forall_{m,m' \in M} from(m) = from(m') \wedge to(m) = to(m') \Rightarrow (!!m <_t^{trace} !!m' \Leftrightarrow ?m <_t^{trace} ?m')$;
- always *msg*-input implementable.

Having defined formally the notions of output- and input-implementability, we now combine them and obtain our notion of communication model.

Definition 8 A 3-trace is (X,Y)-*implementable* (for $X, Y \in \{nobuf, global, inst, pair, msg\}$) iff it is X-output implementable and Y-input implementable. A 2-trace is (X,Y)-*implementable* iff it can be extended to a 3-trace that is (X,Y)-implementable.

4 CLASSIFICATION OF IMPLEMENTABILITY OF TRACES

To each of the implementation models defined in the previous section we can associate the set of all traces that are implementable in the model. Based on the subset relation on these sets of traces, we can order implementation models. We consider two models equivalent if they have the same set of implementable traces.

In Lemma 9 we give a classification of the notions of output-implementability. It states that a trace that is implementable on a certain architecture is also implementable on an architecture where these buffers are partitioned into buffers with a more restricted locality. For example, if a trace can be implemented on an architecture with one output buffer per instance, it can also be implemented on an architecture with an output buffer per pair of instances (provided the input buffers remain the same).

Lemma 9 (Classification of output-implementability) Every *nobuf*-output implementable trace is *global*-output implementable. Every *global*-output implementable
trace is *inst*-output implementable. Every *inst*-output implementable trace is *pair*-
output implementable. Every *pair*-output implementable trace is *msg*-output implementable.

For the proof of this lemma, and of the other lemmas and theorems for which no
proof is given in this paper, we refer to (Engels, Mauw and Reniers 1997).

The following lemmas give the orderings between the implementation models.

Lemma 10 Every (*global, global*)-implementable 2-trace is (*global, nobuf*)-imple-
mentable. Every (*inst,global*)-implementable 2-trace is (*inst,nobuf*)-implementable.
Every (*pair,pair*)-implementable 2-trace is (*pair,nobuf*)-implementable. Every (*msg,
msg*)-implementable 2-trace is (*msg,nobuf*)-implementable.

For the previous lemmas the analogue obtained by switching output buffers and
input buffers is equally true. Next, we describe how the above lemmas are useful in
ordering the models. Lemma 9 provides us with a partial ordering on the various implementations: Any (*X,Y*)-implementable trace is implementable by all implementation models located to the right of or below (*X,Y*) in Figure 2. 10, together with the
order provided by Lemma 9, gives us the equivalences as expressed in Figure 2 by
means of the clustering of implementation models.

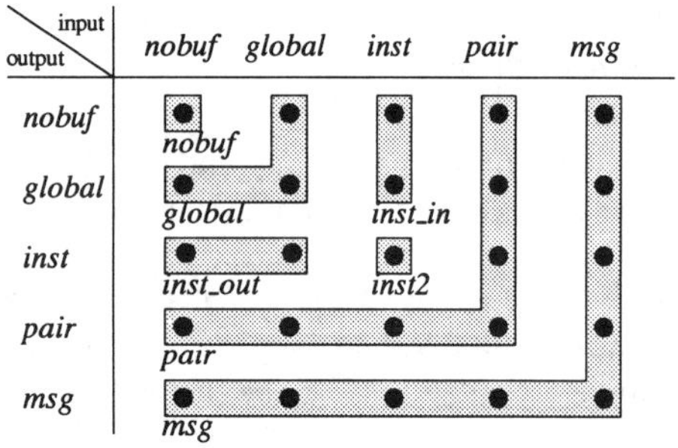

Figure 2 Equivalence of implementation models for traces.

For example, the models from the last column are equivalent. This can be seen as
follows. Because of the analogue of Lemma 10, any (*msg,msg*)- implementable 2-
trace is (*nobuf,msg*)-implementable, while Lemma 9 gives that any (*nobuf,msg*)-im-
plementable 2-trace is (*X,msg*)-implementable, and every (*X,msg*)-implementable 2-
trace is (*msg,msg*)-implementable.

Now we have brought down the number of implementation models to only seven
different classes. Of course some of these could still be equivalent for other reasons.
That this is not the case, will be seen in Theorem 12 below. We name the equivalence
classes as follows: *nobuf, global, inst_out, inst_in, inst2, pair, msg* (see Figure 2).

Note that of these seven cases only *inst2* is not of the form (*X, nobuf*) or (*nobuf, X*).

As these forms imply that there is respectively no input buffer or no output buffer, of these seven cases only the case *inst2* needs two buffers, all other cases can be modelled such that each message goes through at most one buffer.

5 CLASSIFICATION OF MSCS

There are two principal ways to lift the definition of implementability from the level of traces to the level of MSCs. The first is to define that an MSC can be implemented in a certain communication model iff every 2-trace of the MSC can. The second is to define that an MSC can be implemented in a certain implementation model iff some 2-trace can. We call these notions strong and weak implementability. We first focus on the strong implementability, then on weak implementability. After this we consider the relation between classes from the strong and the weak spectrum.

5.1 Strong implementability

Definition 11 An MSC k is said to be *strongly X-implementable*, notation X_s-implementable, iff all 2-traces t of k are X-implementable.

From this definition it follows immediately that the ordering of the implementation models for traces also holds for MSCs as far as strong implementability is concerned (see the left part of Figure 5). Next, we demonstrate that the implementation models, obtained by lifting them from the trace level to MSCs in the strong way, are indeed different. This is achieved by finding examples of MSCs that are in one class but not in another.

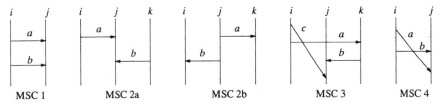

Figure 3 MSCs to distinguish the implementation models: strong case.

MSC 1 in Figure 3 shows an example that is *global_s*-implementable, but not *nobuf_s*-implementable. It is not *nobuf_s*-implementable, because the trace $!a\,!b\,?a\,?b$ is not. The inputs necessarily have to be ordered in the same way as the outputs, so it is *global_s*-implementable.

MSC 2a is *inst_out_s*-implementable, but not *global_s*-implementable due to the trace $!b\,!a\,?a\,?b$. That MSC 2a is *inst_out_s*-implementable can be seen as follows: All messages go through a different output buffer, so there is no problem with the output buffers at all. Similarly, MSC 2b is *inst_in_s*-implementable, but not *global_s*-implementable due to the trace $!a\,!b\,?b\,?a$.

MSCs 2a and 2b show the difference between *inst_out$_s$* and *inst_in$_s$*. MSC 2a is *inst_out$_s$*-implementable, as mentioned before, but not *inst_in$_s$*-implementable. The trace !b !a ?a ?b is not *inst_in*-implementable, because the inputs of instance j do not reach the input buffer in the order in which they are to be manipulated. For MSC 2b the reverse is the case: It is *inst_in$_s$*-implementable, but not *inst_out$_s$*-implementable. MSC 2a is *inst_out$_s$*-implementable and therefore also *inst2$_s$*-implementable. We have already established that it is not *inst_in$_s$*-implementable. Similarly, MSC 2b is *inst_in$_s$* and *inst2$_s$*-implementable, but not *inst_out$_s$*-implementable. Together, these show that *inst_out$_s$*, *inst_in$_s$* and *inst2$_s$* are all different.

MSC 3 is an example of an MSC that is *pair$_s$*-implementable, but not *inst2$_s$*-implementable. It is easy to see that it is *pair$_s$*-implementable, because each message goes through a different buffer. Its only 2-trace is !c !a ?a !b ?b ?c. If we try to extend this to an *inst2*-implementable 3-trace t', we need to have !!c $<_t^{trace}$!!a $<_t^{trace}$!!b $<_t^{trace}$!!c, which is impossible (the first $<_t^{trace}$ is because of the *inst*-output implementability and !c $<_t^{trace}$!a, the second is clearly true for every 3-trace of the MSC, and the third is because of the *inst*-input implementability together with ?b $<_t^{trace}$?c).

Finally, MSC 4 shows the difference between *pair$_s$*- and *msg$_s$*-implementability. All other implementation models are also pairwise different. This result is obtained due to the transitive closure of the ordering as presented in Figure 5.

Together the examples show that if we look at strong implementability, the seven remaining implementation models are indeed different for MSCs, and thus that they are also different for 2-traces.

Theorem 12 The classes *nobuf*, *global*, *inst_out*, *inst_in*, *inst2*, *pair*, and *msg* are different for implementability of traces and for strong implementability of MSCs, and for strong implementability they are ordered as shown in the left part of Figure 5.

5.2   Weak implementability

Definition 13 An MSC k is said to be *weakly X-implementable*, notation X_w-implementable, iff there is an X-implementable 2-trace t of k.

As was the case for strong implementability, for weak implementability we also have the ordering for traces as a starting point. However, using weak implementability, we do not have anymore that all implementation models differ. To see this, we first give an alternative way to characterise some of the implementations and prove that these are equivalent to the original definition.

Definition 14 Let k be an MSC over the set of messages M. Then we define the relations $<_k^{io}$ and $<_k^{ii}$ on $\{!m, ?m \mid m \in M\}$ and the relation $<_k^{i2}$ on $\{!m, !!m, ?m \mid m \in M\}$ as follows:

$$<_k^{io} := (<_k^{msc} \cup \{(?m, ?m') \mid m, m' \in M \wedge from(m) = from(m') \wedge !m <_k^{msc} !m'\})^+,$$

$$<_k^{ii} := (<_k^{msc} \cup \{(!m, !m') \mid m, m' \in M \wedge to(m) = to(m') \wedge ?m <_k^{msc} ?m'\})^+,$$

$$<_k^{i2} := (<_k^{m3} \cup \{(!!m, !!m') \mid m, m' \in M \land from(m) = from(m') \land !m <_k^{m3} !m'\}$$
$$\cup \{(!!m, !!m') \mid m, m' \in M \land to(m) = to(m') \land ?m <_k^{m3} ?m'\})^+.$$

We explain the definition of the ordering $<_k^{io}$ which is defined in order to check the *inst_out*-property. The ordering is obtained from $<_k^{msc}$ by adding pairs of input events to it. More specifically, if two outputs are defined on the same instance of the MSC, and thus are ordered in some way, then we add their corresponding input events in the same order. This is motivated as follows. For a trace to be *inst_out*-implementable it is required that the input events are ordered in this way anyway. Thus by adding this pair explicitly we construct an ordering representing the MSC given that it has to be implemented on an architecture with one output buffer per instance.

The *inst_out*-implementable traces of the MSC are also traces of the ordering $<_k^{oi}$ as they respect the requirements for *inst_out*-implementability by definition, and vice versa. Basically this is what is expressed in Theorem 15.

Theorem 15 Let t be a 2-trace of an MSC k. Then, t is *inst_out*-implementable iff $<_k^{io} \subseteq <_t^{trace}$, t is *inst_in*-implementable iff $<_k^{ii} \subseteq <_t^{trace}$, and t is *inst2*-implementable iff there exists a 3-trace t' which is an extension of t such that $<_k^{i2} \subseteq <_{t'}^{trace}$.

Proof. We only give the proof for the last proposition. The proofs for the first two propositions follow the same line. Suppose that t is *inst2*-implementable. Then we must prove that $<_k^{i2} \subseteq <_{t'}^{trace}$ for some 3-trace t' which is an extension of t. For t' we choose any *inst2*-implementable 3-trace of t. It suffices to prove that $e <_{t'}^{trace} e'$ for an arbitrary pair of events $e, e' \in \{!m, !!m, ?m \mid m \in M\}$ with $e <_k^{i2} e'$. Since $e <_k^{i2} e'$ we have the existence of $e_1, \dots, e_n$ such that $e \equiv e_1 < e_2 < \cdots < e_n \equiv e'$ where for all $1 \le i < n$ we have one of the following:

- $e_i <_k^{m3} e_{i+1}$;
- $e_i \equiv !!m, e_{i+1} \equiv !!m', from(m) = from(m')$ and $!m <_k^{m3} !m'$ for some $m, m' \in M$;
- $e_i \equiv !!m, e_{i+1} \equiv !!m', to(m) = to(m')$ and $?m <_k^{m3} ?m'$ for some $m, m' \in M$.

For all of these cases we can conclude that $e_i <_{t'}^{trace} e_{i+1}$, and hence, $e <_{t'}^{trace} e'$.

Next, suppose that $<_k^{i2} \subseteq <_{t'}^{trace}$ for some extension t' of t. We must prove that t is *(inst,inst)*-implementable. Thereto it suffices to show that t' is *(inst,inst)*-implementable, i.e., that t' is *inst*-output implementable and *inst*-input implementable. We prove that t' is *inst*-output implementable, the proof that t' is *inst*-input implementable is analogous. Let $m, m' \in M$ such that $from(m) = from(m')$. Then it suffices to show that $!m <_{t'}^{trace} !m' \Leftrightarrow !!m <_{t'}^{trace} !!m'$. Thereto, suppose that $!m <_{t'}^{trace} !m'$. Since $from(m) = from(m')$, we have $!m <_k^{msc} !m'$. So $!!m <_k^{i2} !!m'$. Because $<_k^{i2} \subseteq <_{t'}^{trace}$ we therefore have $!!m <_{t'}^{trace} !!m'$. Suppose that $!m \not<_{t'}^{trace} !m'$. Then $!m' <_{t'}^{trace} !m$. With similar reasoning as before we obtain $!!m' <_{t'}^{trace} !!m$. Therefore, $!!m \not<_{t'}^{trace} !!m'$. $\square$

Thus far, we have seen that the ordering $<_k^{io}$ contains all *inst_out*-implementable traces of MSC k. An MSC k is *inst_out_w*-implementable iff it has a trace t that is *inst_out*-implementable. Clearly, such a trace exists iff there is a trace for the ordering $<_k^{io}$, in other words, iff $<_k^{io}$ is cycle-free. Therefore, we have the following characterisation theorem, which follows directly from Theorem 16.

Theorem 16 An MSC k is *inst_out$_w$*-implementable iff $<_k^{io}$ is cycle-free. An MSC k is *inst_in$_w$*-implementable iff $<_k^{ii}$ is cycle-free. An MSC k is *inst2$_w$*-implementable iff $<_k^{i2}$ is cycle-free.

We use the alternative characterisations provided by the previous theorem in the proof of the equivalence of the classes *inst_out$_w$*, *inst_in$_w$*, and *inst2$_w$*.

Theorem 17 The implementation models *inst_out$_w$*, *inst_in$_w$*, and *inst2$_w$* are equal.

Proof. We show that each *inst2$_w$*-implementable MSC is *inst_out$_w$*-implementable. The reverse implication is trivial, and the proofs with *inst_in$_w$* are analogous. From Theorem 16 we see that it suffices to prove that $<^{io}$ is cycle-free if $<^{i2}$ is cycle-free. We prove this using contraposition, so we assume that $<^{io}$ has a cycle. Let $e_1 <^{io} e_2 <^{io} \cdots <^{io} e_n <^{io} e_1$ be an arbitrary largest cycle. For every ordering in the cycle, say $e_i <^{io} e_{i+1}$ either $e_i <^{msc} e_{i+1}$, and hence $e_i <^{i2} e_{i+1}$, or $e_i \equiv ?m$, $e_{i+1} \equiv ?m'$ for some $m, m' \in M$ such that $!m <^{msc} !m'$. If the first is always the case, then we have a cycle in $<^{msc}$, so certainly in $<^{i2}$. Now assume we have the second at least once in the cycle. In that case we have at least two inputs in the cycle, say $?m$ and $?m'$. Then $?m <^{io} ?m'$ and $?m' <^{io} ?m$. As is shown in (Engels et al. 1997), this implies that $!!m <^{i2} !!m'$ and $!!m' <^{i2} !!m$. Thus clearly $<^{i2}$ has a cycle. $\square$

Theorem 17 establishes that the classes *inst_out$_w$*, *inst_in$_w$*, and *inst2$_w$* are equivalent. In the remainder we denote this class by *inst$_w$*. The remaining models are all different. The MSCs 3 and 4 in Figure 3 show the difference between *inst$_w$* and *pair$_w$*, and *pair$_w$* and *msg$_w$* respectively in the weak case too (these MSCs have only one 2-trace, so their weak implementability equals their strong implementability). MSC 5 in Figure 4 is *global$_w$*-implementable, but not *nobuf$_w$*-implementable. The trace $!a\ !b\ ?a\ ?b$ is *global*-implementable, but because both outputs must have been executed before any input can be processed, there is no *nobuf*-implementable trace.

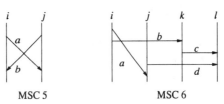

MSC 5   MSC 6

Figure 4 MSCs to distinguish the implementation models: weak case.

MSC 6 is *inst$_w$*-implementable, but not *global$_w$*-implementable. It is not *global$_w$*-implementable, as can be seen thus: $!a <^{msc} !b$, so for every *global*-implementable

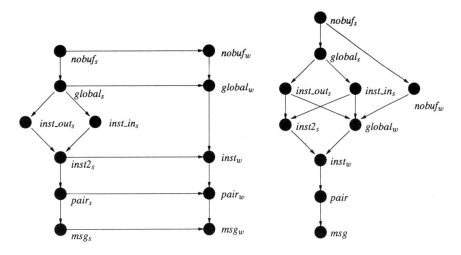

Figure 5 Incomplete hierarchy.   **Figure 6** Final hierarchy.

trace t we must have $?a <_t^{trace} ?b$. Because $!d <^{msc} ?a$ and $?b <^{msc} !c$, we get $!d <_t^{trace}$ $!c$. But we also have $?c <^{msc} ?d$, and thus $?c <_t^{trace} ?d$, from which it follows that t cannot be *global*-implementable. On the other hand, the trace $!a\ !b\ !d\ ?a\ ?b\ !c\ ?c\ ?d$ is *inst_out*-implementable, so the MSC is $inst_w$-implementable.

Theorem 18 *The implementation models for weak implementability of the right part of Figure 5 are all different.*

5.3 Combining the strong and weak hierarchy

The relations between classes in the strong implementability hierarchy and the relations between classes in the weak hierarchy have been studied extensively in the previous sections. In this section we focus on the relations between implementation models from the different hierarchies. From the definitions of strong and weak implementability it is clear that any X_s-implementable MSC is also X_w-implementable. These orderings are also depicted in Figure 5.

Theorem 19 establishes that the classes $pair_s$ and $pair_w$, and msg_s and msg_w are equivalent. In the remainder we denote these by *pair* and *msg* respectively.

Theorem 19 *An MSC is $pair_s$-implementable iff it is $pair_w$-implementable. An MSC is msg_s-implementable iff it is msg_w-implementable. Every $inst_out_s$-implementable or $inst_in_s$-implementable MSC is $global_w$-implementable.*

In Figure 6 we give all communication models that remain after the identifications obtained until now. The arrows between these models follow also from the previous theorems and lemmas. Finally, we have to prove that the arrows between models from the strong and weak hierarchy are strict and that there are no additional arrows necessary. It suffices to show that the following arrows do not exist: $global_s$ to $nobuf_w$, *nobuf$_w$* to *inst2$_s$*, and *inst2$_s$* to *global$_w$*. The rest then follows because of transitivity. For example, the nonexistence of an arrow from *global$_s$* to *nobuf$_w$* implies the nonexistence of an arrow from *inst_out$_s$* to *nobuf$_w$*, because if the second arrow exists then, by transitivity, also the first must exist. Similarly we obtain the nonexistence of arrows from *inst_in$_s$* and *inst2$_s$* to *nobuf$_w$*. We use the MSCs in Figure 7 to indicate that the first two arrows do not exist. MSC 7 is *global$_s$*-implementable, but not *nobuf$_w$*-implementable. On the other hand MSC 8 is *nobuf$_w$*-implementable, but not *inst2$_s$*-implementable. The trace !*a* ?*a* !*b* ?*b* !*c* ?*c* is *nobuf*-implementable, while the trace !*b* !*c* ?*c* !*a* ?*a* ?*b* is not *inst2*-implementable.

Figure 7 Distinguishing MSCs: comparing strong and weak.

The non-existence of an arrow from *inst2$_s$* to *global$_w$* is taken care of by MSC 6 in Figure 3. It has already been shown not to be *global$_w$*-implementable. It is *inst2$_s$*-implementable because every 2-trace of this MSC can be extended to an *inst2*-implementable 3-trace by adding !!*a* and !!*b* immediately after !*a* and !*b*, and !!*c* and !!*d* immediately before ?*c* and ?*d*.

Theorem 20 The implementation models from Figure 6 are all different, and they are ordered as expressed in Figure 6.

6   CONCLUDING REMARKS AND FUTURE RESEARCH

We have considered implementation models for asynchronous communication in Message Sequence Chart. These models contain of FIFO buffers for the sending and reception of messages. By varying the locality of the buffers we have arrived, in a systematic way, at 25 models for communication. With respect to traces, consisting of putting a message into a buffer and removing a message from a buffer, there are seven different models.

By lifting this implementability notion from traces to Message Sequence Charts in two ways, strong and weak, we obtain fourteen models. After identification, ten essentially different models on the level of Message Sequence Charts remain.

For defining the models we have used the notion of 3-traces; these are a natural extension of normal MSC-traces if a message can pass two buffers on its way from source to destination.

In this paper, we have only considered Basic Message Sequence Charts. An interesting question is how to transfer the notions and properties defined for this simple language to the complete language MSC. As many of our theorems rely on the fact that the events on an instance are totally ordered, an extension to MSC with more sophisticated ordering mechanisms (e.g., coregion and causal ordering) will imply a revision of the hierarchy. Another interesting question is whether the implementation properties are preserved under composition by means of the operators of MSC.

Furthermore, we have restricted ourselves to the treatment of architectures in which each message has exactly one possible communication path and where each such path contains at most two buffers. The extension to more flexible architectures is non-trivial and is expected to lead to an extension of the hierarchy.

Finally, our assumption of infinite FIFO buffers may be relaxed, allowing other types of buffers and buffers with finite capacity.

The results obtained in this paper form a solid base for several applications. First, they allow us to discuss the relation between different variants of MSC, such as Inter-workings (Mauw et al. 1993). Interworkings presuppose a synchronous communication mechanism. An Interworking can be considered as the restriction of the semantics of an MSC to only the *nobuf*-implementable traces. Thus, an MSC can be interpreted as an Interworking if and only if there is at least one such trace, i.e., the MSC is *nobuf$_w$*-implementable. We also envisage more practical applications. Consider a tool in which a user can select a communication model, draw an MSC and invoke an algorithm to check if the MSC is implementable with respect to the selected model. Alternatively, the user can provide an MSC and use a tool to determine the minimal architecture, according to our hierarchy, which is needed for implementation.

Often a user is interested in the question whether all traces of his MSC are im-plementable with respect to a certain architecture. We can also envisage two possi-ble uses relying on the implementability of a single trace. First, MSCs are often used to display one single trace, for example if it is the result of a simulation run. In this case, the question is not whether the MSC is strongly or weakly implementable, but whether the implied trace is implementable (as defined in Section 4). Second, given an MSC, a user may want to know if at least one trace is implementable and if so, which trace that is. He is interested in a *witness*. Both applications can easily be derived from the results on weak implementability. The algorithms (see below) can easily be mod-ified to check implementability of a given trace and to produce a witness.

A more involved application would be to use a selected communication model to reduce the set of traces defined by a given MSC to only those traces that are imple-mentable on the given model. In this way, the semantics of an MSC would be relative to some selected model.

For most of these applications computer support would be useful. Based upon the definitions presented in this paper, it is feasible to derive efficient algorithms. All models in the weak-spectrum can be characterised in terms of the cycle-freeness of an extended ordering relation, as is shown in (Engels et al. 1997). An example of such a characterisation is given in Theorem 16. There it is stated that an MSC k is *inst_out$_w$*-implementable iff the ordering $<_k^{io}$ (which is an extension of $<_k^{msc}$) is cycle-free. Thus checking if an MSC is *inst_out$_w$*-implementable boils down to checking cycle-freeness of this relation. This immediately gives a wide range of efficient im-plementations for checking class-membership as many algorithms are known in literature for determining whether a given ordering is cycle-free. For the strong spectrum characterisations are given in (Engels et al. 1997) as well.

There are two papers in which a similar subject is discussed. In (Charron-Bost, Mattern and Tel 1996) four different implementations for MSC-like diagrams are discussed: *RSC* (Realizable with Synchronous Communication), *CO* (Causally Ordered), *FIFO* and *A* (Asynchronous). They find that there is a strict ordering $RSC \subset CO \subset FIFO \subset A$. As shown in (Engels et al. 1997), the implementations *RSC*, *FIFO* and *A* correspond to our implementations $nobuf_w$, *pair* and *msg*, while *CO* is positioned strictly between the implementations $inst_w$ and *pair*.

Another paper in which different communication models for MSC have been studied, is (Alur et al. 1996). The models from our hierarchy are incomparable with their models, because the ordering of certain combinations of events on an instance is subject to a chosen communication model, thereby relaxing our fundamental total ordering of events on an instance.

REFERENCES

Alur, R., Holzmann, G. J. and Peled, D.: 1996, An analyzer for Message Sequence Charts, *Software - Concepts and Tools* **17**(2), 70–77.

Ben-Abdallah, H. and Leue, S.: 1997, Syntactic detection of process divergence and non-local choice in Message Sequence Charts, *in* E. Brinksma (ed.), *Tools and Algorithms for the Construction and Analysis of Systems*, number 1217 in *Lecture Notes on Computer Science*, Springer Verlag, pp. 259–274.

Charron-Bost, B., Mattern, F. and Tel, G.: 1996, Synchronous, asynchronous and causally ordered communication, *Distributed Computing* **9**(4), 173–191.

Engels, A., Mauw, S. and Reniers, M. A.: 1997, A hierarchy of communication models for Message Sequence Charts, *Technical Report CSR 97-11*, Eindhoven University of Technology, Department of Computing Science.

Grabowski, J., Graubmann, P. and Rudolph, E.: 1993, Towards a Petri net based semantics definition for Message Sequence Charts, *in* O. Færgemand and A. Sarma (eds), *SDL'93 - Using Objects*, Proceedings of the Sixth SDL Forum, North-Holland, pp. 179–190.

ITU-TS: 1995, *ITU-TS Recommendation Z.120 Annex B: Algebraic semantics of Message Sequence Charts*, ITU-TS, Geneva.

ITU-TS: 1996, *ITU-TS Draft Recommendation Z.120: Message Sequence Chart 1996 (MSC96)*, ITU-TS, Geneva.

Ladkin, P. and Leue, S.: 1995, Interpreting message flow graphs, *Formal Aspects of Computing* **7**(5), 473–509.

Mauw, S. and Reniers, M. A.: 1994, An algebraic semantics of Basic Message Sequence Charts, *The Computer Journal* **37**(4), 269–277.

Mauw, S., Wijk, M. v. and Winter, T.: 1993, A formal semantics of synchronous Interworkings, *in* O. Færgemand and A. Sarma (eds), *SDL'93 - Using Objects*, Proceedings of the Sixth SDL Forum, North-Holland, pp. 167–178.

6
Timing Constraints in Message Sequence Chart Specifications

Hanêne Ben-Abdallah and Stefan Leue
University of Waterloo
Waterloo, ON N2L 3G1, Canada, telephone +1 (519) 888 4567,
email [hanene|sleue]@swen.uwaterloo.ca

Abstract
When dealing with timing constraints, the Z.120 standard of Message Sequence Charts (MSCs) is still evolving along with several proposals. This paper first reviews proposed extensions of MSCs to describe timing constraints. Secondly, the paper describes an analysis technique for timing consistency in iterating and branching MSC specifications. The analysis extends efficient current techniques for timing analysis of MSCs with no loops nor branchings. Finally, the paper extends our syntactic analysis of process divergence to MSCs with timing constraints.

Keywords
Message Sequence Charts, timing constraints, timing consistency analysis

1 INTRODUCTION

Various flavours of Message Sequence Charts (MSCs) have been used in software engineering of telecommunications systems as well as object-oriented analysis and design notations, e.g. (Selic, Gullekson & Ward 1994, Algayres, Lejeune, Hugonment & Hantz 1993, Jacobson & et al. 1992, Ichikawa, Itoh, Kato, Takura & Shibasaki 1991). The increasing popularity of MSCs recently motivated a standardization effort that produced the ITU-T Recommendation Z.120 (ITU-T 1996). The Z.120 standard defines two main concepts: *basic MSCs* (bMSCs) and *High-Level MSCs* (hMSCs). A bMSC consists of a set of processes that run in parallel and exchange messages in a one-to-one,

Formal Description Techniques and Protocol Specification, Testing and Verification
T. Mizuno, N. Shiratori, T. Higashino & A. Togashi (Eds.) © 1997 IFIP. Published by Chapman & Hall

asynchronous fashion. An hMSC combines references to basic MSCs to describe parallel, sequential, iterating, and non-deterministic execution of basic MSCs. In addition, an hMSC can describe a system in a hierarchical fashion by combining hMSCs within an hMSC.

To facilitate the specification of real-time systems, a few extensions to MSCs have been proposed to express timing constraints: timers (ITU-T 1996), interval delays (Alur, Holzmann & Peled 1996, Meng-Siew 1993) and timing markers (Booch, Jacobson & Rumbaugh 1996). The proposed extensions evolved independently and differ in terms of their expressiveness and support for formal analysis. Further, all proposed analysis of MSCs with timing constraints have been so far limited to basic MSCs (Alur et al. 1996, Meng-Siew 1993).

In an effort to help consolidate the proposed timing extensions possibly within the standard, in this paper we first review the various proposed syntactic annotations of basic MSCs with timing constraints. For each of the proposed timing extensions, we highlight the syntactic features, expressiveness and limitations, and we discuss ambiguities that must be addressed when bMSCs are composed within an hMSC.

Another motivation for this paper is to extend the timing consistency analysis for bMSCs to deal with iterating and branching MSC specifications. The analysis technique we present has been implemented within our prototype toolset for requirements engineering based on MSCs (Ben-Abdallah & Leue 1996). In addition, in this paper we extend our syntactic analysis of the process divergence problem in MSC specifications (Ben-Abdallah & Leue 1997*b*) in the presence of timing constraints. We use the example of an automatic teller machine system to illustrate our selected timing extension and the presented timing analysis.

2   TIMING CONSTRAINTS IN BASIC MSCS

There are essentially four classes of syntactic constructs to express timing constraints in MSCs and MSC reminiscent notations: timers (ITU-T 1996, Alur et al. 1996), delay intervals (Alur et al. 1996, Meng-Siew 1993), drawing rules (Booch et al. 1996), and other timing markers (Booch et al. 1996).

Timers.   Recommendation Z.120 provides timers to express timing constraints in a basic MSC. Within a single process, a timer can be *set* to a value, *reset* to zero, and observed for *timeout*. A timer cannot be shared among concurrent processes in a basic MSC. Figure 1 (a) shows an example of a basic MSC with timing constraints expressed through two timers. In this example, for instance if the timer T3.1 is set to 5, then P3 must exchange its messages within at most five time units relative to the timer setting event. Process P1, on the other hand, first sets the timer T1.1 say to three time units, receives message a, then resets its timer. Since process P1 does not explicitly use a timeout event for the timer T1.1, the *implicit* assumption here is that the timer T1.1 does not expire before it is reset. As this example illustrates, a timer can be used to express a *maximal* delay between two or more consecutive events in one process. In addition, a timer can be also used to express a *minimal* delay between two consecutive events in one process.

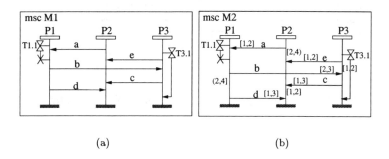

(a)   (b)

Figure 1 Timing constraints expressed through: (a) Z.120 timers; (b) Z.120 timers and delay intervals

Delay Intervals. Depending on how delay intervals annotate a bMSC, they express three types of timing constraints: 1) *event-associated* timing constraints (Meng-Siew 1993) which are denoted as an interval that is associated with an event in the basic MSC; 2) *message delivery* delays (Alur et al. 1996, Meng-Siew 1993) which are expressed as a time interval over a message arrow; and 3) *Processor's speed* constraints (Alur et al. 1996, Meng-Siew 1993) which are expressed as time intervals between two consecutive events in a process.

An event-associated timing constraint is a *global* constraint on the timed occurrence of an event: the event must occur within the specified minimal and maximal time delays with respect to *any* previous event, whenever it occurs in an execution trace of the bMSC. Figure 2 (a) shows sample event-associated timing constraints.

In the message delivery and processor's speed constraints, a delay interval is delimited with respect to the occurrence of the two consecutively, visually ordered events it constrains. Figure 1 augments the timing constraints in Figure 1 with message delays (i.e., intervals on message arrows) and processor's speed constraints (i.e., intervals on vertical lines). In this version, message **b** takes between two and three time units from the time it is sent by process **P1** to the time it is received by process **P3**. In addition, process **P3** requires that message **b** be received between one and two time units from the time it sends message **e**.

In (Meng-Siew 1993), the author generalizes the message delivery and processor's delay intervals (called *trace-associated* timing constraints) by using

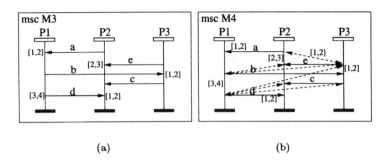

(a)   (b)

Figure 2 (a) Event-associated and (b) trace-associated timing constraints

a semantic notion of *consecutive* events: two events are consecutive if they can be executed one after the other. In addition, this work extends the use of trace-associated timing constraints to express timing constraints between events that are not related. For this, the syntax of bMSCs is extended with *precedence* edges that connect unrelated events. The user can then annotate the extended bMSC with timing constraints to impose on unrelated events. Figure 2 (b) shows sample trace-associated timing constraints where the precedence edges are drawn with dashed-line, bidirected edges. As this example illustrates, while precedence edges allow the expression of more timing constraints, they may result in a cumbersome graph.

Drawing rules and timing markers.   Sequence diagrams within the Unified Modeling Language (Booch et al. 1996) extend the Z.120 MSCs with additional information, e.g., focus of control to show the time when a process has a thread of control. Timing constraints are represented in a sequence diagram in two ways: the drawing rules of message arrows and *timing markers*. A horizontal message arrow indicates the *simultaneous* occurrence of the send and receive events of the message. A downward slanted message arrow, on the other hand, indicates a required delay between the send and receive events of the message. In addition, within each object outgoing message arrows can be drawn at a single point to indicate the simultaneous sending of a message. (Incoming message arrows are not allowed to meet at the same point within an object.)

To describe more quantitative timing constraints, timing markers are attached to a sequence diagram. Timing markers are boolean expressions placed in braces and attached to the diagram (Booch et al. 1996). The boolean expressions can constrain particular events or the whole diagram. However, since neither the precise syntax of timing markers nor their formal semantics is defined, we cannot completely assess their expressiveness. In addition, no formal analysis of timing constraints has been proposed within the Unified Modeling Language.

Timing Analysis Based on Timers and Delay Intervals. Timing analysis consists of validating a timing assignment and verifying timing consistency. A timing assignment is essentially a time-stamp function that associates with the MSC events occurrence times with respect to a global clock. A timing assignment is valid if it respects the timing constraints in the MSC. A bMSC is timing consistent if there is at least one valid timing assignment that allows the MSC to have a behavior where the events occur according to the specified timing constraints.

For bMSCs extended with timers and timing delays, one can use the temporal constraint network techniques in (Dechter, Meiri & Pearl 1991) to reduce timing analysis to computing all-pairs-shortest-paths in a labeled directed graph. In the worst case, this can be computed in $O(n^3)$ time where n is the number of events in the bMSC. We will discuss timing consistency analysis with this technique in detail in Section 4.

The MSC analyzer tool (Alur et al. 1996) offers in addition to the above timing analysis for bMSCs, timed analysis based on a semantics that accounts for the queuing strategies in a bMSC. To analyze MSC specifications within this tool, the user would have to select the various bMSCs that compose one sequential path in the hMSC and analyze each path separately. However, in the presence of loops in the hMSC, this tool offers no hints on how many times the user is supposed to unfold a loop to conclude timing consistency of the loop. Further, analysis based on path selection should resolve certain issues about the usage of timer events and the interpretation of the timing constraints in the MSC specification as we discuss in the next section.

3 INTERPRETING TIMING CONSTRAINTS IN MSC SPECIFICATIONS

The hMSC in an MSC specification connects basic MSCs to describe sequential, possibly iterating and non-deterministic behavior. The presence of timing constraints as described in the previous section in MSC specifications requires attention for one essential reason: timing constraints can be spread across sequentially connected basic MSCs. We next illustrate how the Z.120 standard syntax (ITU-T 1995) is ill-defined when timers are used in hMSCs, and outline possible choices of interpreting timing consistency in MSC specifications with branchings.

Interpreting Iterations
Current analyses of iterations in an hMSC rely on unfolding loops a finite number of times and analyzing resulting basic MSCs (Alur et al. 1996). In the case of timed behavior, this technique raises several questions about: 1) interpreting multiple occurrences of the set event of the same timer, 2) resolving the correspondence between several timers' set and timeout events, and 3) the syntactic well-formedness of MSC specifications with timers.

Consider the MSC specification of Figure 3 (a) where the timer T1.1 is

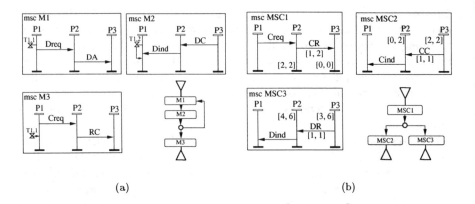

Figure 3 (a) Timers in a loop; (b) Timing constraints and branching

set in the basic MSC M1 and its timeout is detected in the basic MSC M3. As control iterates through the basic MSCs M1 and M2, it is unclear whether the system generates a new instance of the timer, or uses the same timer during all iterations. Both interpretations can be justified by the common interpretation of a loop in an hMSC through a finite unfolding to a basic MSC. More specifically, Consider the following execution scenario: M1, M2, M1, M2, and then M3. Using loop unfolding, this sequence of basic MSCs represents a syntactically legal basic MSC. It then seems that a *new* timer is generated each time M1 is executed. Therefore, we need to resolve the association of the one timeout event in M3 with the two timer setting events. The choice obviously affects the timing consistency analysis.

Let us try to derive the semantics of repeatedly invoked timers from that of repeatedly sent messages. Ladkin and Leue (Ladkin & Leue 1995) use a finite state argument to interpret multiple sends of a message as deactivated by one receive of the message. (At the time of writing Z.120 does not deal with the semantics of iterating system behavior.) In the case of timers, this interpretation translates into associating the timeout event with *any* of the two timer setting events. A more reasonable choice is to associate it with the first timer since it would time out first. For T1.1 set to 5 and T1.2 set to 3, this interpretation makes the above execution scenario timing inconsistent. Another alternative is to associate the timeout event with the last timer setting event. This coincides with the interpretation where the same timer is reset then reused during all iterations. In this case, the same execution scenario for the example of Figure 3 (a) is timing consistent in the sense that the last timer set does not expire before sending the event Creq; however, such an analysis is misleading when the overall behavior is considered: the first time the message Dreq was sent and sending the event Crq are separated by at least 6 time units and thus one timeout event, i.e., deadline was obviously missed.

The above ambiguity in interpreting timers within loops results from the ill-defined syntax of hMSCs when timers are involved. The Z.120 syntax of an hMSC assumes the well-formedness of the bMSCs used in the hMSC. The Z.120 syntax of bMSCs only restricts the usage of timers such that a reset or timeout event may occur only after a timer is set (ITU-T 1995); that is, this syntax neither forces the reset or timeout event to occur after a timer set event, nor does it restrict multiple occurrences of a timer set event prior to its reset or timeout event. As the above example illustrates, this relaxed syntax of bMSCs can lead to ambiguities when a loop in an hMSC contains bMSCs where one timer is set but neither its timeout nor reset event occurs in the loop.

In a broader context, the above example raises a fundmental question about what an MSC specification means: does it describe *all* behaviors of a system, or does it describe a set of *sample* behaviors of a system? In the first case, the standard syntax must be further restricted to disallow the above example. In the second case, the above example should be allowed and interpreted according to the second alternative; that is, timers may expire without explicitly being modeled in the MSC specification. However, this interpretation may create practical difficulties since timers' expirations are usually implemented as interrupts and thus can not be ignored in some occasions and handled at other times.

Interpreting Branchings

When an MSC specification contains branchings, we can determine whether its timing constraints are satisfiable in two ways:

•*Local semantics:* select one path at a time and analyze its timing constraints, independently of other paths that may branch out of the selected path (Alur et al. 1996). This interpretation of timing constraints allows the derivation of several timing assignments, one for each path in the hMSC. In other words, any particular basic MSC that is shared by different paths may have different timed behavior depending on both the *past* and *future* behavior of the system.

•*Global semantics:* all paths must be analyzed simultaneously. This analysis technique assumes that any timing assignment for the hMSC must be valid along all shared portions of all paths in the hMSC. In this approach, each basic MSC will have the same timed behavior independently of the execution path on which it resides, hence independently of the future behavior of the system. This approach produces tighter timing constraints than the local semantics.

To illustrate the differences between the two approaches, consider the timed MSC specification show in Figure 3 (b) and which describes a simple connection establishment. The example is locally timing consistent. However, it is not globally timing consistent since there is no timing assignment for the common prefix of its two paths and that allows both paths to be simultaneously timing consistent. Timing consistency of the path leading to MSC2 requires that receive CR occurs at most 1 time unit after it is sent, whereas timing consistency of the path leading to MSC 3 requires that receive CR not to occur before 2 time units after it is sent.

4   TIMING ANALYSIS OF MSC SPECIFICATIONS

In this section, we first define the syntax of timed MSC specification we will use. Second, we augment the timing analysis for bMSCs presented in (Meng-Siew 1993, Alur et al. 1996) to handle the possibility that a timer is set in a bMSC but no reset nor timeout event follows the timer setting in the bMSC. We then extend it to analyze MSC specifications with branchings and iterations. For the proofs of the lemmas and theorems in this and the following Sections see (Ben-Abdallah & Leue 1997*a*).

Timed MSC Specifications

In remainder of this paper, we assume that the untimed bMSCs contain only message exchanges drawn in accordance to Z.120. To express timing constraints we use the Z.120 timer events together with (non-standard) timing delay intervals. A timer can be set to a positive integer value, and rest to zero. In compliance with the Z.120 standard (ITU-T 1995), a reset and time-out event must be preceded by a timer setting event. In addition, a timer is private to a process and thus its events can be used by a single process only. Further, to distinguish between timers, we assume that each timer has a unique identifier associated with it.

A timing delay interval is a label over either a message arrow, or a control flow segment, i.e., a portion in a process's line that is delimited by two consecutive events. (We use the generic term *event* to denote one of the following types of events: the start of a process, the end of a process, sending a message, receiving a message, or a timer event.) A delay interval labeling a message arrow denotes the relative minimal and maximal delays between the events of sending the message and receiving it. A delay interval labeling a control flow segment denotes the relative minimal and maximal delays between the events delimiting the control flow segment. Delay intervals can be of the form $[a, b]$, $[a, b)$, or $(a, b]$ where $a \in \mathbf{N}$ and $b \in \mathbf{N} \cup \{\infty\}$.

An *MSC specification* is a structure $S = (B, V, suc, ref)$ where: B is a finite set of bMSCs; V is a finite set of nodes partitioned into the singleton-set of *start* node, the set of *intermediate* nodes, and the set of *end* nodes; *suc* is the relation which reflects the connectivity of the hMSC of S such that all nodes in V are reachable from the start node; and *ref* is a function that maps each intermediate node to a bMSC in B^*.

A *path* in an MSC specification $S = (B, V, suc, ref)$ is a sequence of intermediate nodes (i.e., bMSCs), $b_1, b_2, \cdots$, such that $(b_i, b_{i+1}) \in suc$ for $i \geq 1$. A path is *simple* if all its nodes are distinct. A *loop* in S is a path $b_1, b_2, \cdots, b_n$ with $(b_n, b_1) \in suc$, and a loop is called *simple* if all its nodes are distinct.

In compliance with the Z.120 standard we allow timer events to be split

---

*We assume that an MSC specification contains one level of nesting; however, the definitions and results presented in this paper can be easily extended to deal with MSC specifications where nodes refer to *other* hMSCs.

across bMSCs in an hMSC. The Z.120 standard restriction that each timeout and timer resetting event must be preceded by a timer setting event in bMSCs is extended to paths in the MSC specifications, i.e., in every path timeout and timer resetting events must be preceded by a timer setting. However, to avoid the ambiguities described earlier, we require that every simple loop in an MSC specification has *matched* timer events: 1) every timer setting event in the loop must be followed by either a timeout or reset event; and 2) every timeout event must be preceded by a timer setting event in the loop. The first restriction disallows the example in Figure 3 (a). Note that this restriction is for loops only; that is, it does not force the use of a timeout or reset event in non-looping paths. As we see in the next section, our timing analysis will ensure that the absence of these events does not mean the specification is incomplete. The second restriction disallows the possibility that time ellapes in the loop making the timeout event obsolete.

Timing Consistency of bMSCs

To determine the timing consistency of a basic MSC, we adopt an approach similar to the ones presented in (Meng-Siew 1993, Alur et al. 1996). First the bMSC is translated into a directed, labeled graph that we call *temporal constraint graph*. The vertices and edges in the graph reflect the control flow and message exchanges in the bMSC. The edge labels represent the timing constraints in the bMSC. Once a temporal constraint graph is constructed, to verify that the bMSC is timing consistent, we just check that the temporal constraint graph has no cycles with a negative cost (Dechter et al. 1991, Meng-Siew 1993, Alur et al. 1996).

Since it is unclear whether the informal translation presented in (Alur et al. 1996) handles the possibility that a timer is set but no reset nor timeout event is explicitly included in the bMSC, we next present the translation we assume in our analysis of timing consistency of MSC specifications.

From a bMSC to a temporal constraint graph. An edge in the temporal constraint graph is labeled with either the lower or upper bound of the delay interval imposed on the corresponding "edge" in the bMSC. To extract the bounds of a delay interval I with bounds $a, b \in \mathbf{N} \cup \{\infty\}$ ($a \leq b$), we use the functions $\mathcal{L}(I)$ and $\mathcal{U}(I)$ defined as follows: $\mathcal{L}([a,b]) = \mathcal{L}([a,b)) = a$, $\mathcal{L}((a,b]) = \mathcal{L}((a,b)) = a^-$;

It is straightforward to represent a bMSC as a directed, labeled graph. A node in the graph represents one of the following events: the start of a process, the end of a process, sending a message, receiving a message, or setting, resetting or timeout a timer. An edge in the graph can have one of three types that represent the dependencies between events: 1) a "signal" edge (x, y) represents sending a message from one process to another; 2) a "next event" edge (x, y) represents the control flow within a process where event x appears before event y on the vertical line of the process; and 3) a "temporal" edge (x, y) connects a timer setting event x with a timer reset or timeout event y. The label of an edge is the timing delay and for a signal edge the message type in addition. (For details, the reader is referred to (Ladkin & Leue 1995) where basic MSCs without timing constraints are represented as Message Flow Graphs. This translation is easily augmented with the temporal edges to represent timing constraints.) Given a bMSC M, its temporal graph $\mathcal{T}_g(M)$ is obtained as follows:

- Each node in M is represented by a node in $\mathcal{T}_g(M)$.
- Each next event edge, each signal edge and each temporal edge (e, e') with timing label I in M is represented by two labeled edges in $\mathcal{T}_g(M)$: edge (e, e') with label $-\mathcal{L}(I)$, and edge (e', e) with label $\mathcal{U}(I)$.
- For each process P_i in M, for each of its set timer event e_i with value t and no matching reset or timeout event, the following two edges are added in $\mathcal{T}_g(M)$: edge (e_i, e_i') with label $-t$, and edge (e_i', e) with label t, where the node e_i' is the node that corresponds to the end node of process P_i in M.

The above translation could differ from previous translations in the last step. As mentioned earlier, this additional step allows us to cover the lose Z.120 syntax (ITU-T 1995) which does not force a set timer to have a reset or timeout event in the same bMSC. The analysis of the temporal constraint graph as constructed above ensures that those timers that were not explicitly reset or timeout, in fact, did not expire. Hence, the analysis results of the extended temporal constraint graph are coherent with the implicit assumption that a missing timeout/reset event in a process is interpreted as the set timer not having expired prior to the process's end of execution.

A bMSC is timing consistent if and only if its temporal constraint graph has no cycles with a negative cost (Dechter et al. 1991, Meng-Siew 1993, Alur et al. 1996)*. Detecting cycles in the temporal constraint graph can be done through the Floyd-Warshall's algorithm (Papadimitriou & Steiglitz 1982) which computes all-pairs-shortest paths in the graph in a worst case time of $O(n^3)$ where n is the number of events in the graph.

Timing Consistency of hMSCs

Our notion of timing consistency of an MSC specification relies on a local interpretation of the timing constraints.

Definition 4.1 An MSC specification S is *timing consistent* if *every* path that starts from the start node in S is timing consistent. S is *partially* timing consistent if *some* of its paths are timing consistent. S is *timing inconsistent* if *none* of its paths is timing consistent.

The above definition of timing consistency is impractical since in the presence of iterations in the MSC specification, the number of paths is infinite. We

---

*Addition over the natural numbers $\boldsymbol{N}$ is extended over $\boldsymbol{N} \cup \boldsymbol{N}^- \cup \{\infty, -\infty\}$ in a straightforward way.

next present a syntactic approach to determine the consistency of an MSC specification based on a finite subset of its paths. In the sequel, we adopt the following notation to ease readability: given two bMSCS, M_1 and M_2, we denote the MSC specification that consists of the sequential composition of M_1 followed by M_2 as $M_1 \bullet M_2$.

Lemma 4.1 If the MSC specification $M_1 \bullet M_2 \bullet M_1$ is timing consistent with $M_1 \bullet M_2$ having matched timer events, then the MSC specification $M_1 \bullet M_2 \bullet M_1 \bullet M_2$ is also timing consistent.

The above Lemma allows us to deduce the timing consistency of a loop from the timing consistency of an augmented version of the simple path that the loop contains, without unfolding the loop. Thus, to decide the timing consistency of an MSC specification, we can focus on simple paths and augmented simple paths that represent loops in the specification. We next define these paths.

Definition 4.2 Let $S = (B, \top \cup I \cup \bot, suc, ref)$ be an MSC specification. A *sequential component* in S is a simple path $n_1, n_2, \cdots, n_k$ in S such that: $(s, n_1) \in suc$ for the start node $s \in \top$; and either $(n_k, e) \in suc$ for an end node $e \in \bot$, or there exists $n_j \in \{n_1, n_2, \cdots, n_k\}$ such that $(n_k, n_j) \in suc$.

Informally, a sequential component is either a simple path from the start node to an end node, or a path that starts with a simple path and ends with a simple loop. We call the first type *finite* sequential components, and the second *infinite* sequential components. In addition, we say that a path $n_1, n_2, \cdots, n_j, \cdots, n_k, n_j$ in an MSC specification S is a *closed* infinite sequential component if $n_1, n_2, \cdots, n_j, \cdots, n_k$ is an infinite sequential component in S.

Theorem 4.1 An MSC specification S is timing consistent if 1) each finite and each closed infinite sequential component in S is timing consistent; and 2) each infinite sequential component in S has matched timer events.

The condition on the infinite sequential components is stronger because the loop in the component allows time to progress an arbitrary amount, and thus possibly missing a timeout event. If a set timer is missing a reset or timeout event inside the loop, our analysis can not conclude from one iteration that the timer will never expire. In fact, if a timing consistent, infinite component has unmatched timer events, then the specification is partially timing consistent is the best we can predict.

Timing consistency algorithm. To examine the timing consistency of an MSC specification, we have implemented within our MSC tool (Ben-Abdallah & Leue 1996) the algorithm shown in Table 1. Step 1 is carried out through a depth-first-search algorithm of the the hMSC of S. To construct the temporal constraint graph of a sequence of bMSCs, step 3 extends the bMSC to temporal graph translation of Section 4 in a straightforward way based on the

Table 1 Timing consistency analysis algorithm

Input: an MSC specification S
Output: for each timing inconsistent sequential component in S, the
        events involved in a negative cost cycle
1.   Find the finite and closed infinite sequential components in S
2.   For each sequential component L:
3.   construct the temporal constraint graph $\mathcal{T}_g(L)$
4.   compute all-pairs-shortest paths in $\mathcal{T}_g(L)$
5.   report all events involved in a negative cost cycle
End

---

following fact: the behavior of $M_1 \bullet M_2$ is equivalent to the behavior of the
bMSCs obtained by gluing M_2 after M_1, with the timing delays at the end of a
process in M_1 added to the timing delays of the same process at the beginning
of M_2. Step 4 uses the Floyd-Warshall's algorithm on a matrix representation
of the temporal constraint graph. In step 5 an event is in a cycle with a negative cost if its corresponding diagonal element in the all-pairs-shortest-path
matrix is negative.

Process Divergence in Timed MSC Specifications
In the presence of loops, an MSC specification may suffer from process divergence: a system execution where a process sends messages an unbounded
number of times ahead of the messages being received (Ben-Abdallah &
Leue 1997*b*). An MSC specification with a process divergence can be either
unimplementable as it requires message queues with an infinite size, or it can
be implementable with discrepancies, e.g., unexpected deadlocks since message queues are finite and messages must be dropped or over-written.

In (Ben-Abdallah & Leue 1997*b*), we have syntactically characterized process divergence in untimed MSC specifications by examining the communication patterns of its processes. Informally, we proved that an (untimed) MSC
specification suffers from process divergence if and only if it has a loop where
at least one of its processes does not depend on, i.e., sends to but never receives
directly or indirectly from another concurrent process in the loop.

In the presence of timing constraints through timers and/or delay intervals, our syntactic characterization of process divergence can be extended as
follows.

Theorem 4.2   Given an MSC specification S that has untimed process
divergence through the processes $P_1, \cdots, P_n$ in a loop L which can jointly race
ahead of the remaining processes. S has timed process divergence iff either 1)
the sum of all minimal delays in the processes $P_1, \cdots, P_n$ within L is equal
to zero; or 2) the minimum of all maximal delays of the processes receiving
messages from $P_1, \cdots, P_n$ including delays on the received messages is equal
to ∞.

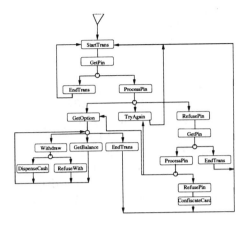

Figure 4 High-level MSC for the ATM example

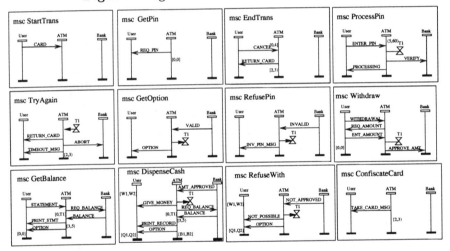

Figure 5 Basic MSCs in the ATM example

5 ATM EXAMPLE

To illustrate our timing analysis, consider the MSC specification of an automatic teller machine (ATM) system shown in Figures 4 and 5. The ATM system consists of three components: potential customers (process **User**), the ATM controller (process **ATM**), and a host computer in a bank (process **Bank**).

Initially, the ATM controller waits to receive the customer's bank card. Then, it either receives a request to cancel the transaction within $[0,4]$ seconds (bMSC **EndTrans**), or receives the customer's pin number within $[5,60]$ seconds (bMSC **ProcessPin**). If the ATM receives a request to cancel the transaction (bMSC **EndTrans**), it returns the customer's card and takes between $[2,3)$ seconds to return to its initial state. The ATM expects a reply from the bank within T_1 seconds. This timing constraint is expressed through the T1 timer as well as the $[0, T_1]$ delay interval. In (bMSC TryAgain), when T1 times out, the card is returned, an appropriate message is then displayed, and the ATM takes again between $[2, 3)$ seconds to return to its initial state. Our specification also describes the following constraints: a customer expects a withdraw request to be processed within $[W_1, W_2]$ seconds relative to the time of entering an amount; a customer takes $[Q_1, Q_2]$ seconds to decide whether to make another transaction while the ATM has the card; the ATM takes $[B_1, B_2]$ seconds for book-keeping after dispensing cash; the ATM takes $[3, 5]$ seconds to print a receipt after receiving the balance information from the bank. Each ATM-customer communication has a delay of $[0, 2)$ seconds and each vertical line without a time delay has a default delay of $[0, \infty)$, which we do not explicitly represent in the chart. Note that when timing constraints extend over more than just one bMSC it is necessary to use auxiliary $[0, 0]$ constraints.

Table 2 Sample results of timing consistency analysis

Case #	(1)	(2)	(3)	(4)	(5)	(6)
$[W_1, W_2]$	$[0, \infty)$	$[0, 3]$	$[0, 4]$	$[0, 4]$	$[0, 4]$	$[0, 4]$
$[Q_1, Q_2]$	$[0, \infty)$	$[0, \infty)$	$[0, \infty)$	$[0, 2]$	$[0, 2]$	$[0, 1]$
$[B_1, B_2]$	$[0, \infty)$	$[0, \infty)$	$[0, \infty)$	$[5, 6]$	$[4, 6]$	$[4, 6]$
Consistent?	yes	no	yes	no	yes	no

Timing Analysis.   It is easy to check that our MSC specification of the ATM satisfies the syntactic conditions of Theorem 4.1. We used our analysis tool (Ben-Abdallah & Leue 1996) to verify automatically the timing consistency of the specification for $T_1 = 10$ and various values of $W_1, W_2, Q_1, Q_2, B_1,$ and B_2. Table 2 shows sample results. The tool generated 43 infinite closed sequential components whose temporal graphs were then examined for cycles with a negative cost. For the case (1) in Table 2, the user does not impose any timing constraints on the system, which in fact makes any value acceptable for the remaining variables. In the case (2), the user expects the ATM to process their withdraw request within $[0, 3]$ seconds which leads to a timing inconsistency. The tool detects the event send ENT_AMOUNT as being in a cycle with a negative cost of -1. This gave us the hint to increase the upper-bound of the delay to 4 which gave us timing consistency (case (3)). In the cases (4) and (5), we examine the effects of the book-keeping time $[B_1, B_2]$. Due to the implicit $[0, \infty)$ bounds we only needed to vary the lower bound. Case (6) proves the dependency between the minimum book-keeping delay B_1 and the delays between consecutive customer requests while the ATM holds the customer's card, interval $[Q_1, Q_2]$. In the above cases, when a timing inconsistency is detected, the cost of the cycle and the involved events reported by the tool helped us to focus on which variables to adjust by which amount.

6 CONCLUSION

We have reviewed four proposed extensions of MSCs to express timing constraints and available analysis techniques for timing consistency of basic MSCs. The Z.120 standard timers together with delay intervals as suggested in (Meng-Siew 1993, Alur et al. 1996) can describe timing constraints for events within a process and events that are directly related, i.e., via the control edge or message arrow. To express more general timing constraints, e.g., to relate events within different basic MSCs or processes, these extensions must be further augmented with temporal predicates (Booch et al. 1996).

Following the analysis techniques of temporal constraint networks (Dechter et al. 1991), timing consistency of a basic MSC is reduced to checking cycles with negative cost in a directed graph (Meng-Siew 1993, Alur et al. 1996). To extend this analysis to MSC specifications with iterations and branchings, we highlighted syntactic issues that the Z.120 standard syntax must address. Based on specific syntactic recommendations, we have extended the analysis of timing consistency of bMSCs with timers and delays to the analysis of MSC specifications with branchings and iterations. To deal with branchings, we adopted a local interpretation of the timing constraints. To handle iterations, we showed that, under a reasonable assumption, a loop in the MSC specification can be analyzed by analyzing a simple extension of it, hence eliminating the need to unfold the loop to examine its timing consistency. Furthermore, we have extended in this paper our syntactic analysis of process divergence in iterating MSCs in the presence of timing constraints.

Acknowledgements.   This work was supported by ObjecTime Limited and the Information Technology Research Centre of the Province of Ontario.

REFERENCES

Algayres, B., Lejeune, Y., Hugonment, F. & Hantz, F. (1993), The AVALON project: a validation environment for SDL/MSC descriptions, *in* O. Faergemand & A. Sarma, eds, 'Proceedings of the 6th SDL Forum, SDL'93: Using Objects'.

Alur, R., Holzmann, G. J. & Peled, D. (1996), An analyzer for message sequence charts, *in* T. Margaria & B. Steffen, eds, 'Tools and Algorithms for the Construction and Analysis of Systems, Lecture Notes in Computer Science, Vol. 1055', Springer Verlag, pp. 35–48.

Ben-Abdallah, H. & Leue, S. (1996), Architecture of a requirements and design tool based on message sequence charts, Technical Report 96-13, Department of Electrical & Computer Engineering, University of Waterloo. 18 p.

Ben-Abdallah, H. & Leue, S. (1997*a*), Expressing and analyzing timing constraints in message sequence chart specifications, Technical Report 97-

04, Department of Electrical & Computer Engineering, University of Waterloo. 24 p.

Ben-Abdallah, H. & Leue, S. (1997*b*), Syntactic detection of process divergence and non-local choice in message sequence charts, *in* E. Brinksma, ed., 'Tools and Algorithms for the Construction and Analysis of Systems, Lecture Notes in Computer Science, Vol. 1217', Springer Verlag, pp. 259–274.

Booch, G., Jacobson, I. & Rumbaugh, J. (1996), *Unified Modeling Language for Object-Oriented Development (Version 0.91 Addendum)*, RATIONAL Software Corporation.

Dechter, R., Meiri, I. & Pearl, J. (1991), 'Temporal constraint networks', *Artificial Intelligence* **49**, 61–95.

Ichikawa, H., Itoh, M., Kato, J., Takura, A. & Shibasaki, M. (1991), 'SDE: Incremental specification and development of communications software', *IEEE Transactions on Computers* **40**(4), 553–561.

ITU-T (1995), 'Recommendation Z.120, Annex B: Algebraic Semantics of Message Sequence Charts', ITU - Telecommunication Standardization Sector, Geneva, Switzerland.

ITU-T (1996), 'Recommendation Z.120', ITU - Telecommunication Standardization Sector, Geneva, Switzerland. Review Draft Version.

Jacobson, I. & et al. (1992), *Object-Oriented Software Engineering - A Usecase Driven Approach*, Addison-Wesley.

Ladkin, P. B. & Leue, S. (1995), 'Interpreting Message Flow Graphs', *Formal Aspects of Computing* **7**(5), 473–509.

Meng-Siew, N. (1993), Reasoning with timing constraints in Message Sequence Charts, Master's thesis, University of Stirling, Scotland, U.K.

Papadimitriou, C. & Steiglitz, K. (1982), *Combinatorial Optimization: Algorithms and Complexity*, Prentice-Hall, Englewood Cliffs, NJ.

Selic, B., Gullekson, G. & Ward, P. (1994), *Real-Time Object-Oriented Modelling*, John Wiley & Sons, Inc.

Hanêne Ben-Abdallah   received the B.S. degree in computer science and mathematics from the University of Minnesota, Minneapolis, in 1989, the M.S.E. degree in 1991 and Ph.D. degree in 1996 in computer science from the University of Pennsylvania, Philadelphia. From 1996 to 1997 she was a postdoctoral research fellow in the Dept. of Elect. & Comp. Eng. at the University of Waterloo. Her research interests include real-time systems and formal methods in software engineering.

Stefan Leue   received his Diplom-Informatiker degree from the University of Hamburg (Germany) in 1990 and his Ph.D. from the University of Berne (Switzerland) in 1995. He is currently an Assistant Professor in the Department of Electrical & Computer Engineering of the University of Waterloo (Canada).

7
Consistent Semantics for ODP Information and Computational Models

Joubine Dustzadeh, Elie Najm
Ecole Nationale Supérieure des Télécommunications
46, rue Barrault, 75634 PARIS CEDEX 13, France
{joubine, najm}@email.enst.fr

Abstract
We tackle two important ODP viewpoints, namely the information and the computational. We first provide formal semantics for object diagrams of some popular application development methodologies (such as OMT and Fusion) and show how these notations support ODP information modeling. We also formalize an essential part of the ODP computational viewpoint including the concepts of distributed objects and concurrent serializable activities. We then complement the two semantics by providing rules for consistent mappings between the two models.

Keywords
ODP, viewpoint consistency, information modeling, OMT, actions, computational modeling, serializability

1 INTRODUCTION

Distributed systems are inherently complex and their description in one single large specification is not desirable in practice. To deal with this complexity, system specifications need to be decomposed through a process of separation of concerns, and based on different levels of abstraction. Specification decomposition is a recurrent concept that is found in many architectures for distributed systems. For instance, the reference model of Open Distributed Processing – ODP considers five viewpoints: *enterprise, information, compu-*

Formal Description Techniques and Protocol Specification, Testing and Verification
T. Mizuno, N. Shiratori, T. Higashino & A. Togashi (Eds.) © 1997 IFIP. Published by Chapman & Hall

tational, engineering and *technology*. Specifications of systems in ODP can be made from any of these viewpoints and different – formal or informal – specification languages may be used.

The ODP information viewpoint provides a highly abstract model of real world entities and their relationships along with the constraints that govern their behavior. ODP does not prescribe specific languages to be used for information modeling partly due to the high level of abstraction required to capture information structures and due to the possible reflective nature of information. Informal object models (such as OMT [2]) are largely used for information modeling. In addition to the problems related to use of informal specification languages, the correspondence between specifications from different viewpoints (and particularly between the information and computational viewpoints) poses a considerable challenge in ODP. To better address these issues, the semantics of information models merits particular attention in ODP. The semantics of information models in ODP includes: (i) the semantics of the information specification (when considered to be isolated from other viewpoints) and (ii) the examination of how the information specification relates to the computational structure of the system.

How this Paper is Organized

Section 2 discusses various notations used for information modeling in ODP. Section 3 introduces the object model of OMT and provides a formal semantics for it. Section 4 introduces and formalizes the concept of information actions. Section 5 introduces a model of the ODP computational viewpoint. Section 6 provides a framework for valid mappings between the information and the computational models. Finally, some concluding remarks are given in section 7.

2   NOTATIONS FOR INFORMATION MODELING IN ODP

There exist several notations and methods that are candidates for information modeling within ODP. Three important requirements should be considered carefully: (i) the ability to directly capture the information model, (ii) abstractness, and (iii) friendly graphical user interface. An ideal notation for the ODP information model should support the principles of object-orientation and should also offer an easy-to-use interface to the service specifier for the task of mapping real world entities into information objects. Yet, this notation should have a formally defined semantics for the sake of analyzability, tool support, and compatibility between the information specification and the specifications made in other viewpoints. Information structures and relationships can be naturally captured by a user friendly graphical notation.

Graphical interfaces are also well suited to support incremental specification. As an information specification progresses, one can modify and refine objects and relations in an iterative fashion.

Z is being considered as highly appropriate for information modeling. Although Z is not fully object oriented, it has those object features that match the needs of information modeling. Static, invariant and dynamic schemas can be directly mirrored in a Z specification. Moreover, Z provides a good basis for relating specifications in other viewpoints or languages [10]. There is a large community of Z users and researchers. The object model of some object-oriented analysis methods, like e.g. OMT or Fusion [3], are also well suited for ODP information modeling. Although they are not really formal, they have been used because of their appealing graphical syntax and because they are part of fully integrated development methodologies.

3 FORMAL OBJECT-ORIENTED INFORMATION MODELING

Various surveys indicate that the *object model* of OMT is the most common object model that is practiced for the purpose of information structuring in ODP. However, as mentioned earlier, the object model of OMT lacks a formal semantics. The example illustrated in figure 1 shows how an OMT information specification may lead to ambiguity and therefore to multiple interpretations. The dangers of such an ambiguous information specification include:

- several possible (and semantically different) implementations of the same specification.
- impossible realization (implementation) of the specification (figure 1).

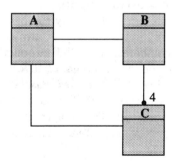

Figure 1 Example of an Incorrect Object Diagram in OMT

This OMT diagram uses some basic OMT concepts such as objects and binary associations (and their cardinality constraints). It defines three classes

A, B, C and three associations between these classes.

We have run a survey in the relevant mailing list* and news group* on the Internet and requested that OMT experts examine the semantical correctness of the diagram illustrated in figure 1. The post included the following questions:

```
- for each instance a1, is there exactly one instance b1?
  Let us represent it as following: <a1,b1>
- for b1, are there exactly four instances c1,...,c4?
  <b1,c1> <b1,c2> <b1,c3> <b1,c4>
- What about a1 and c-i? (since there is a one-to-one association
  between A and C)?
```

Surprisingly enough, this simple diagram resulted in several different interpretations and a long debate was initiated on the semantical correctness of OMT diagrams in general.

3.1   Formalization of OMT Object Diagrams

We choose the *object model* of OMT as a representative of all object models that are appropriate for information modeling in ODP. Despite this choice, the formal rules outlined in our approach remain applicable to other object models with minor modifications. The challenge of formalizing the OMT object model is twofold: definition of an abstract syntax of the language and definition of the semantical function for the elements of the language. Thus, our formalization method consists of the following steps:

Definition of a Formal Abstract Syntax : we first need to provide an abstract linear syntax that captures exactly the permissible OMT diagrams. This syntax should exist independently of any graphical OMT editor. It will be defined by a combination of BNF productions and a set of static semantics rules. We introduce the domain $\mathcal{U\_CRS}$ of *Unflattened Class Relation Systems* consisting of *unflattened classes** (translated from OMT classes) and *relations* (reflecting OMT basic associations and aggregation relations). Let $\mathcal{D}$ represent all OMT diagrams and $\mathcal{U\_CRS}$ represent the set of unflattened class relation systems. The linearization function $\mathcal{L}$ can be represented as following:

$$\mathcal{D} \xrightarrow{\mathcal{L}} \mathcal{U\_CRS}$$

---

* `otug@rational.com`
* `comp.object`
* *unflattened* means that classes may contain inheritance links to other classes

Flattening Process : in order to completely linearize the OMT diagrams translated into unflattened class relation systems, we use a flattening function $\mathcal{F}$ to remove all inheritance links from unflattened classes. This process leads to a new domain $\mathcal{CRS}$ whose elements are called *Class Relation Systems*. The flattening function can be represented as following:

$$\mathcal{U\_CRS} \xrightarrow{\mathcal{F}} \mathcal{CRS}$$

The flattening process also allows the definition of some static semantics rules for unflattened class relation systems (for instance, inheritance may not occur recursively).

Instance Systems : at this stage, we define the semantical domain $\mathcal{IS}$ for class relation systems. Elements of this domain are called *Instance Systems*. A generic element of $\mathcal{IS}$ is composed of instances of objects and instances of associations. Each instance system represents a possible state of the system that is specified by the OMT diagram. The semantics of an OMT diagram is defined in terms of the set of allowable configurations of instance systems.

3.2 Definition of a Formal Abstract Syntax

Our first concern here is to give an abstract syntax to the OMT notation. Specification of a syntax for a model implies the definition of an alphabet and the permissible combinations of its symbols. Unflattened class relation systems are defined by a BNF grammar rule as described below. U_CLASS is intended to be the linear representation of a class in an OMT object diagram, and RELATION a linear representation of both OMT associations and aggregations. U_CLASS and RELATION are defined below:

- U_CRS ::= U_CLASS | U_CRS U_CLASS | U_CRS RELATION
- U_CLASS ::= $\langle name : \aleph, parset = \pi, abstract : \mathbb{B}, attlist = \eta \rangle$
- RELATION ::= $\langle class_1 : \aleph, role_1 : \tilde{\aleph}, card_1 : \mathbb{F}\,N, name : \tilde{\aleph}, card_2 : \mathbb{F}\,N, role_2 : \tilde{\aleph}, class_2 : \aleph, attlist = \eta \rangle$

Notations and Conventions
- since we are not concerned here with syntactic constraints of different names and types, we introduce the given set $\aleph$ whose internal structure is transparent in our model. All the permissible names by OMT for classes, attributes, operations, operation arguments, associations and roles are drawn from this set. However, since in OMT role names and relation names may be omitted, we introduce the following notation to reflect absence of role names or association names.

$$\tilde{\aleph} ::= \aleph \mid \varepsilon \mid \kappa$$

where ε is a special name to represent absence of a name and κ indicates aggregation. Thus, for an aggregation relation, $r.role_1 = \kappa$ and $r.name = \kappa$.

- *attlist* $= \eta$ represents a list of attributes tagged by their name and their type. $\eta ::= \langle x_1 : t_1, \ldots, x_n : t_n \rangle$.
- *name* is the class name (resp. relation name) in CLASS (resp. in RELATION).
- *parset* is the set of names of the immediate parents (superclasses). If a class does not have any superclasses, *parset* will be an empty set. For simple inheritance, *parset* contains only one element (the name of the superclass) and for multiple inheritance it is composed of as many names as there are immediate parents. We will restrict our study to simple inheritance.
- *abstract* is a boolean. it holds the value true for an abstract class and false otherwise.
- *class*$_1$ and *class*$_2$ are the names of the classes that are related to each other via a relation.
- *role*$_1$ and *role*$_2$ are the role names in an association. For an aggregation, the corresponding role name (next to the aggregation diamond) is set to κ. If there is no role name specified in an association, *role*$_i$ will be ε.
- *card*$_1$ and *card*$_2$ are the multiplicity constraints defined for an association and they take their values in $\mathbb{F}\,\mathbb{N}^*$. For a one-to-one association, both *card*$_1$ and *card*$_2$ are the set $\{1\}$. In the case of a multiplicity *many* placed on the *class*$_i$ side of an association, if no set is numerically specified, *card*$_i$ is $\mathbb{N}$.

Well-formedness of Unflattened Class Relation Systems

Now that we have defined a formal syntax for unflattened class relation systems, we can provide formal rules to ensure their well-formedness. The BNF syntax allows construction of elements that may not correspond to any OMT diagrams. Rules of well-formedness should be defined to eliminate inconsistent unflattened class relation systems. These rules of well-formedness are also needed for a correct construction of OMT object diagrams. For example, two attributes within the same class may not have the same name, or two classes may not have the same name.

Definition 1 *An unflattened class relation system ucrs is well-formed iff it satisfies rules UCRS.WF defined below*[*].

- UCRS.WF.1 : a binary association can exist only between classes already defined.
- UCRS.WF.2 : the class name should be unique.
- UCRS.WF.3 : the attribute names should be unique within a class.
- UCRS.WF.4 : roles are mandatory for associations between two objects of the same class.

---

[*]$\mathbb{F}$ indicates a finite set. Thus, for instance, $\mathbb{F}\,S$ is the type whose elements are finite sets of elements of type S.

[*]A formal representation of these rules is given in the appendix (figure 5).

- UCRS.WF.5 : aggregation is anti-symmetric (it implies non-symmetry between the roles).
- UCRS.WF.6 : no role name should be the same as an attribute name of the source class.
- UCRS.WF.7 : all role names on the far end of associations attached to a class must be unique.
- UCRS.WF.8 : inheritance must not be cyclic.
- UCRS.WF.9 : only concrete classes may be a leaf class in an inheritance tree.

3.3 Flattening Process

Once unflattened class relation systems have been checked against the above well-formedness rules, a flattening process $\mathcal{F}$ may begin in order to remove inheritance links and abstract classes. In the appendix, we propose an algorithm describing $\mathcal{F}$. Note that the flattening process maybe achieved based on different parsing algorithms used for trees and forests in graph theory.

The flattening process leads to a new domain $\mathcal{CRS}$ whose elements are called *Class Relation Systems*. The BNF representation of $\mathcal{CRS}$ is the following:

- CRS ::= CLASS | CRS CLASS | CRS RELATION
- CLASS ::= $\langle name : \aleph, attlist = \eta \rangle$

3.4 Instance Systems

In our model, an instance system is viewed as a collection of allowable interrelated object instances. A valid instance system is one where all the constraints and rules expressed in the object model are reflected in the object instances and in the dependencies between object instances. Multiplicity constraints in associations between object instances are examples of such consistencies.

Syntax of Instance Systems

At a given point in time, an instance system can be considered as a collection of record structures composed of values (attributes) of given types. The effect of operations (defined in OMT classes) is reflected by an evolution of the state space of an instance system, that is, a transformation of values in the record structures and of the collection of records (e.g. create and delete records). Instance systems are defined by the following BNF grammar rule:

- InstanceSystem ::= information_element | information_element InstanceSystem
- information_element ::= object_instance | association_instance
- object_instance ::= $\langle id : \delta, name : \aleph, attset = \beta \rangle$
- association_instance ::= $\langle id : \delta, name : \aleph, id_1 : \delta, role_1 : \tilde{\aleph}, id_2 : \delta, role_2 : \tilde{\aleph}, attset = \beta \rangle$

Where $attset = \beta$ represents a list of attributes tagged by their name and their value. $\beta ::= \langle x_1 = v_1, \ldots, x_n = v_n \rangle$.

Well-formedness of Instance Systems

Similarly to a class relation system, an instance system must satisfy a number of well-formedness and static semantics rules. A non-exhaustive list of these rules are given below (a formal representation of these rules may be found in the appendix, figure 6):

- IS.WF.1 : instances must have a unique identity.
- IS.WF.1 : an association between two object instances requires that both object instances exist.
- IS.WF.3 : attribute names should be unique within an instance.
- IS.WF.4 : aggregation is anti-symmetric.

Semantics of Class Relation Systems

Now that we have given the well-formedness rules for instance systems, we can define the semantics of a class relation system in terms of the set of allowable configurations of instance systems.

Notations
Let crs be a well-formed class relation system, c a class in crs and r a relation in crs. Let is be a well-formed instance system, ω an object instance in is and ρ an association instance in is. For $i \in \{1, 2\}$, we introduce the following notations to facilitate the expression of the semantics rules for class relation systems.

Meaning
- $crs.r.i$ denotes the unique class referenced by $r.class_i$
- $is.\rho.i$ denotes the unique object referenced by $\rho.id_i$
- $\omega : c$ means that object instance ω is of class c
- $\rho : r$ means that association instance ρ is of class r
- $is.r.i.id$ denotes the set of object instances related to ω (with identifier id) via associations of type r

Notation		Formal Definition
$crs.r.i$	$=$	$\{c \in crs \mid c.name = r.class_i\}$
$is.\rho.i$	$=$	$\{\omega \in is \mid \omega.id = \rho.id_i\}$
$\omega : c$	$\Leftrightarrow$	$((\omega.name = c.name) \wedge \forall j :$ $(\omega.attset_j.name = c.attset_j.name) \wedge$ $(\omega.attset_j.type = c.attset_j.type))$
$\rho : r$	$\Leftrightarrow$	$((\rho.name = r.name) \wedge \forall j :$ $(\rho.attset_j.name = r.attset_j.name) \wedge$ $(\rho.attset_j.type = r.attset_j.type))$
$is.r.i.id$	$=$	$\{\omega \in is \mid \exists (\rho : r) \in is$ such that: $(\rho.id_i = id) \wedge (\rho.id_{\bar{i}} = \omega.id)\}$

- $\bar{i} = 2$ if $i = 1$ and $\bar{i} = 1$ if $i = 2$.

Definition 2 *We define valid correspondence between class relation systems and instance systems by relation $\mathcal{VC}$ such that $\mathcal{VC} \subset \mathcal{CRS} \times \mathcal{IS}$ and $(crs \ \mathcal{VC} \ is)$ iff the rules VC.1, VC.2, VC.3 hold:*

(VC.1): Any object instance in is is an instance of a class in crs.
(VC.2): Any association instance in is is an instance of a relation in crs.
(VC.3): The multiplicity constraints defined for any relation r in crs are respected in is.

These rules are formally expressed in figure 2.

(VC.1):	$\forall \omega \in is, \exists c \in crs \mid \omega : c$
(VC.2):	$\forall \rho \in is, \exists r \in crs \mid (is.\rho.i : crs.r.i))$
(VC.3):	$\forall r \in crs, \forall id, \ card\{is.r.i.id\} \in r.card_{\bar{i}}$

Figure 2 Semantics Rules for Instance Systems

Definition 3 *Semantics of class relation systems is given by the following rule:*

$$[\![crs]\!] \mathrel{\hat{=}} \{is \mid crs \; \mathcal{VC} \; is\}$$

where $\mathcal{VC}$ is a valid correspondence between a class relation system (translated from an OMT diagram) and an instance system.

Definition 4 *A sufficient condition* for the equivalence between two class relation systems crs_1 and crs_2 is:*

$$[\![crs_1]\!] = [\![crs_2]\!]$$

Now let us examine again the example of figure 1. In our semantics, the diagram of figure 1 denotes either an empty set of instances or a set of an infinity of instances of classes A, B and C. Thus, this diagram denotes only degenerated cases and should be discarded.

4   ACTIONS IN INFORMATION MODELING

In many realistic case studies, one may have to group a certain number of objects that participate simultaneously in the same action. While most object models of popular object-oriented methodologies such as OMT and Booch offer powerful constructs to staticly structure information entities, they fall short in expressing collective behaviors for groups of objects. As an example, the OMT object model which is one of the most commonly used specification languages for the information viewpoint does not support the concept of *multi-object actions*.

While one of the most commonly recognized advantages of the object approach is to improve the modularity and incrementability of a specification, and to allow encapsulation of data and individual behavior of objects, traditional object-oriented specification tools mostly focus on the specification of individual behavior of objects and usually neglect collective behaviors of objects. This approach could be accepted at a more program-oriented abstraction level, that is, at a stage of specification where computational objects and their interfaces have already been identified. In contrast, a multi-object action approach allows for a more expressive specification of collective behaviors in the sense that decisions on how the participating objects in an action make joint decisions on whether to commit in joint actions, elementary computations or exchange messages to coordinate them can be left to later stages of design [4]. For this reason, we choose a multi-object action approach, à la Fusion, for modeling dynamic schemas in the ODP information viewpoint.

---

*Class relation systems can be further equated whereby attributes and relations are interchangeable.

4.1 Semantics of Information Actions

Let us consider a class relation system *crs* (resulting from the flattening of some OMT diagram). We extend the information model of this *crs* with the following elements:

- *Inv* represents an invariant of the system. When an instance system *is* satisfies the invariants, we have: $Inv(is) = $ true.
- $[\![crs]\!]_{Inv}$ denotes a restriction to $[\![crs]\!]$ such that: $\forall\, is \in [\![crs]\!]_{Inv} \Rightarrow Inv(is) = $ true
- *A* denotes an information action that is defined on information objects. An action *A* embodies a pre-condition predicate A_{Pre}, a body part A_B and a post-condition predicate A_{Post}.

$$A = (A_{Pre}, A_B, A_{Post})$$

where: $A_B : \mathcal{IS} \longrightarrow \mathcal{IS}$, $A_{Pre} : \mathcal{IS} \longrightarrow \mathbb{B}$, $A_{Post} : \mathcal{IS} \longrightarrow \mathbb{B}$

Note that in this semantics, information actions are considered as atomic. In other words, the body part A_B of an action *A* operates on the whole instance system without being interrupted by or interleaved with other actions. Moreover, A_B may be considered as an imperative language whose syntax need not be explicitly given in our study.

Definition 5 (Valid Information Actions) *An information action A is valid iff:*

$$\forall\, is \in [\![crs]\!]_{Inv}, A_{Pre}(is) = true \;\Rightarrow$$
$$A_{Post}(A_B(is)) = true \;\wedge\; Inv(A_B(is)) = true$$

5   THE COMPUTATIONAL MODEL

As seen earlier, RM-ODP allows for a multi-viewpoint specification of distributed systems. In the current state of the RM-ODP, the least natural transition seems to lie in the passage from the information model to the computational model.

An action (considered at the information viewpoint level) corresponds to the execution of a computational activity which can involve several computational objects. Also, because of the multi-thread property of computational objects, several activities (relative to different actions) may run in the same computational object in parallel. The global state of the system changes due to the execution of computational activities. One can then observe, by taking *snapshots* at given moments in time, the changes of the global state of the system due to executions of computational activities. These snapshots – taken at *relevant* moments – will be recorded in *object logs* and will contain enough information to allow checking for the consistency of the global state of the system against the invariant schemas defined at the information viewpoint level.

Actions are specified in an atomic way in the information model. Designers need not worry about concurrency control at this stage. However, the corresponding computational activities could run in an interleaved fashion within computational objects. In other words, computational activities may have to access and share the system resources in a concurrent way. More specifically, variables and attributes of objects may be accessed for *read* or *write* operations concurrently. Also, during the execution of an action, new objects may be created. That leaves us with a concern for the consistency checking of the global state of the system at the end of execution of a given action. A consistent global state is one where the cardinality constraints, the invariants and the post-conditions remain satisfied.

We consider a computational model that is a subset of the computational model defined by RM-ODP. We consider only implicit bindings plus some new concepts such as object logs and activities. By introducing these new concepts, we hope to bring some clarity into the passage from the information model and to make a contribution in extending the RM-ODP computational model. A description of the computational terms that we consider in our model is given in the following:

Computational Elements : a computational element can be either an object or a message.
Objects : an object has multiple interfaces. Objects and interfaces within objects can be created and deleted dynamically.
Messages : messages can be interrogation requests, announcement requests and interrogation replies. A generic type of message is proposed to represent all kinds of messages. Each message is sent on the account of an activity and contains the identifier of this activity. Messages can be created and deleted dynamically.
Object Logs : in order to ensure that current executions of actions do not violate the consistency of the system, we build execution logs for activities. To each object, an execution log is associated and the log gets updated: (i) every time an attribute of the object is updated (write) and (ii) every time an attribute of the object is read. The snapshots recorded in the object logs must always satisfy the following rules:

- a computational activity may not run while the preconditions of the corresponding action are not satisfied.

- after the execution of a computational activity, the postconditions of the corresponding action as well as all invariants schemas defined at the information viewpoint must be verified.

5.1 Syntax of Computational Terms

A distributed computation Ω is a collection of *distributed elements* which may evolve in parallel. The parallel operator is denoted as $\|$. A distributed element can be either an object or a message μ. This abstraction does capture the understanding that a distributed program, from the point of view of the computational model, is merely a set of asynchronously interacting objects. Figure 3 defines a formal abstract syntax for computational terms.

$$\Omega \quad ::= \quad \theta \mid \mu \mid \Omega \parallel \Omega$$
$$\theta \quad ::= \quad \langle id : \delta, attset = \beta, class = c, actset = \tau, loglist = \lambda \rangle$$
$$\mu \quad ::= \quad \langle tgt = u, src = v, act = \alpha, cont = m \rangle$$

Figure 3 Syntax of Computational Terms

In figure 3, the following conventions apply:

- θ is a computational object.
- $\theta.id$ is the unique identifier of a computational object. δ is a generic type for all identifiers.
- $\theta.attset = \beta$ is the set of all attributes within an object. β takes the form of a list of attributes tagged by their name and their value:
$\beta ::= \langle x_1 = v_1, \ldots, x_n = v_n \rangle$.
- $\theta.class = c$ represents the corresponding class name in the information viewpoint. c takes its value in $\aleph$.
- $\theta.actset = \tau$ is the set of all activities that are present in an object. τ takes the form of a list of activity identifiers: $\tau ::= \langle \alpha_1 \rangle \cdots \langle \alpha_n \rangle$.
- $\theta.loglist = \lambda$ is the list of all log entries that have been recorded for an object. λ takes the form of a list $\lambda ::= \langle l_1 \diamond l_2 \diamond \cdots \diamond l_n \rangle$ of log entries l_k where :

$$\diamond ::= \begin{cases} \langle \rangle \diamond l = \langle l \rangle \\ \langle l_1 \diamond \cdots \diamond l_k \rangle \diamond l = \langle l_1 \diamond \cdots \diamond l_k \diamond l \rangle \end{cases}$$

Each entry l_k (interchangeably denoted $[\lambda]_k$) contains an activity identifier

α_j, the type of the operation performed by the activity (read or write), and an attribute x_i whose value v_i has been read or updated by α_j. In case of a read operation, we use the notation $l_k = \alpha_j^r(x_i = v_i)$ where v_i holds the value read by α_j, and in case of a write operation, we use the notation $l_k = \alpha_j^w(x_i = v_i)$ where v_i holds the new value written by α_j. When there is no confusion, v_i may be omitted.

- $\mu.tgt$ is the target interface of message μ. In case of an interrogation request (or an announcement) $\mu.tgt$ holds a value u which is the interface reference of the target interface.
- $\mu.src$ is the source interface of message μ. In case of an interrogation response, $\mu.src$ holds a value v which is the reference of an interface previously invoked.
- $\mu.cont = m$ denotes the remaining content of the messages. This part is not relevant in our study (see [5]).

5.2 Serializability of Computational Activities

Definition 6 (Serializability) *Let Ψ be a set of computational activities. An interleaved execution ξ of Ψ is said to be serializable if and only if ξ generates the same result as some serial (sequential) execution of Ψ.*

Definition 7 (Causal Precedence) *For a given object θ, we define a causal precedence between two activities α_i and α_j – denoted $\alpha_i \overset{\theta}{\rightsquigarrow} \alpha_j$ – if there exists an attribute x of θ such that :*

- *x is read by α_i before x is updated (written) by α_j, OR*
- *x is updated by α_i before x is read by α_j, OR*
- *x is updated by α_i before x is updated by α_j.*

This definition can be formalized as following:

$$(\alpha_i \overset{\theta}{\rightsquigarrow} \alpha_j) \iff \exists x \mid \alpha_i^r(x) \overset{\theta}{\rightsquigarrow} \alpha_j^w(x) \lor \alpha_i^w(x) \overset{\theta}{\rightsquigarrow} \alpha_j^r(x) \lor \alpha_i^w(x) \overset{\theta}{\rightsquigarrow} \alpha_j^w(x)$$

Object Logs can serve to track precedence relations between activities:

$$(\alpha_i \overset{\theta}{\rightsquigarrow} \alpha_j) \iff \exists l_n, l_m \in \theta.loglist \mid n < m, \exists x \in \theta.attset \mid$$
$$(l_n = \alpha_i^r(x) \land l_m = \alpha_j^w(x)) \lor$$
$$(l_n = \alpha_i^w(x) \land l_m = \alpha_j^r(x)) \lor$$
$$(l_n = \alpha_i^w(x) \land l_m = \alpha_j^w(x))$$

Let Ψ be a set of computational activities and let $\rightsquigarrow$ be the associated dependency relation. The serializability theorem can then be expressed as following:

Theorem 1 (Serializability) *An execution ξ of Ψ is a serializable execution iff the transitive closure of the dependency relation $\rightsquigarrow$ is a partial order.*

Thus, we can define computational consistency: a distributed computation Ω and a collection of activities ξ are computationally consistent iff ξ is serializable.

5.3 Computational Consistency

Let Ω be a distributed computation and ξ an execution of a set Ψ of computational activities. In order for Ω to be consistent (from a computational viewpoint) ξ must be serializable. The serializability of ξ implies that ξ generates the same result as some sequential execution of Ψ. Let us represent the serialized (sequence of) activities by Ψ_s:

$$\Psi_s = \{\alpha_1, \ldots, \alpha_i, \ldots, \alpha_n\}$$

For this serialized execution (of Ψ_s), we use $\longrightarrow_s$ to reflect the serialized nature of the system transitions. The evolution of the global state of the system can therefore be represented by:

$$\Omega_{init} \xrightarrow{\alpha_1}_s \Omega_1 \cdots \xrightarrow{\alpha_i}_s \Omega_i \xrightarrow{\alpha_{i+1}}_s \cdots \xrightarrow{\alpha_n}_s \Omega_n$$

Or, more simply: $(\Omega_{init}, \longrightarrow_s, \Psi_s)$.

At the end of each activity α_i, we have: $\Omega_i = \Omega_{init}.\alpha_1 \ldots \alpha_i$

Note that within this serialization, at the end of execution of an activity α_i, there are no messages left (since there are no present activities), and the system is composed of instances of computational objects only.

6 ARTICULATION BETWEEN THE TWO MODELS

In order to carry on in our study of articulation between the information and computational models, we need to make a choice for the mapping between information objects and computational objects. We choose to map each computational object to an information object. However, our results could be obtained also with more general forms of mappings. In our mapping, associations between information objects are represented by special attributes. In order to capture this mapping, we can extend our formal abstract syntax of computational objects (figure 3) and add a new label $\theta.relset$ as following:

$$\theta ::= \langle id : \delta, class = c, attset = \beta, relset = \varphi, actset = \tau, loglist = \lambda \rangle$$

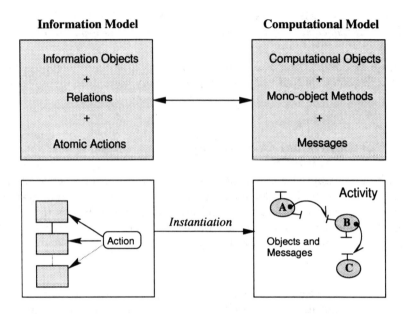

Figure 4 Relation between the Information and Computation Models

- $\theta.relset = \varphi$ represents a list of relation attributes tagged by a relation name r, a role name and a list of identifiers of computational objects to which θ is associated via relation r. Each relation attribute $[\varphi]_i$ may be represented by: $[\varphi]_i = \langle relname = r, role : \aleph, idlist : \mathbb{F}\,\delta \rangle$

We can now consider a function $\mathcal{H} : \Theta \longrightarrow \mathcal{IS}$ defined on instances of computational objects, that would yield elements of *instance systems*, i.e. instances of information objects and instances of information relations.

An Information/Computation Consistency Property

Besides being consistent (serializable) from a computational viewpoint, distributed computations must also satisfy the invariants defined in the information viewpoint. More specifically, (i) for each computational activity α_i, the pre-conditions of the corresponding information action A^{α_i} must be satisfied before the activity is able to be executed, and (ii) at the end of execution of each activity, the post-conditions of the corresponding information action as well as the global information invariants must be satisfied. These rules are formalized in the following:

Definition 8 *We define valid correspondence between an information representation* $(cs, Inv, \Gamma = \{A_1, \ldots, A_k\})$ *and a computational representation* $(\Omega_{init}, \longrightarrow_s, \Psi^* = \{\alpha_1, \ldots, \alpha_n\})$ *iff :*

$$\forall \alpha_i \in \Psi_s, \mathcal{H}(\Omega_{init}.\alpha_1 \ldots \alpha_i) \in [\![cs]\!]_{Inv} \quad \wedge$$
$$A^{\alpha_i}_{Pre}(\mathcal{H}(\Omega_{init}.\alpha_1 \ldots \alpha_{i-1})) = \text{true} \quad \wedge$$
$$A^{\alpha_i}_B(\mathcal{H}(\Omega_{init}.\alpha_1 \ldots \alpha_{i-1})) = (\mathcal{H}(\Omega_{init}.\alpha_1 \ldots \alpha_i)) \quad \wedge$$
$$A^{\alpha_i}_{Post}(\mathcal{H}(\Omega_{init}.\alpha_1 \ldots \alpha_i)) = \text{true}$$

7 CONCLUSION

In the study presented here, we aimed at providing a formal basis for the ODP information and computational models. In particular, we first provided a formal semantics for a hybrid OMT/Fusion object model. OMT is a widely used graphical notation for information modeling in ODP. We used OMT's class/relation paradigm as a notation to describe the structure of information of systems. Much research is now being carried out to endow these models with formal semantics, either directly (like our approach) or through a translation to Z [9] or through a translation to an Abstract Data Types language [8]. As for the dynamic aspects of the information model, we choose the Fusion model of actions, which we find more suited for the description of multiple object interactions.

We then considered the computational model for which we extended an existing formalization with the concepts of activities and object logs. Computational activities may run concurrently on collections of computational objects and thus, we provided consistency rules which define acceptable interleaving of activities. For that purpose, we choose the well known serializability property to characterize the valid interleavings.

The computational model, endowed with its intra-consistency rules, can then be checked against the information model specification. It is important to observe the state changes of a distributed system and to make sure that the system never reaches a state where the rules and invariant schemas defined in higher levels of abstraction could be violated. Thus, in our study, we observe a system at the computational level to ensure its consistency (i) from a computational viewpoint, and (ii) with respect to the rules and invariants imposed by its information specification.

A future direction to our study would be the provision of Information to Computational transformations that preserve the consistency of the source and target models. Such transformations would need to cover two main issues: (i) mapping the global state, assignment oriented, approach of programming into a distributed, message passing style; and (ii) introducing mechanisms, such as transactional processing monitors, in order to ensure the intra-computational consistency for multiple, concurrent, and interleaved activities.

REFERENCES

[1] UIT-T X.901 | ISO/IEC 10746-1 Part 1 Overview: "Reference Model of Open Distributed Processing" – July 1994.
[2] Object-Oriented Modeling Technique – J. Rumbaugh, M. Blaha, W. Premerlani, F. Eddy, W. Lorenson: Prentice Hall, Englewood Cliffs, N.J. 1991.
[3] Object Oriented Development: The Fusion Method – D. Coleman, P. Arnold, S. Bodoff, C. Dollin, H. Gilchrist, F. Hayes and P. Jeremaes: Prentice Hall, 1994.
[4] Fundamentals of Object-Oriented Specification and Modeling of Collective Behaviors – Reino Kurki-Suonio: (Eds. H. Kilov, W. Harvey), Kluwer Academic Publishers 1996.
[5] A Formal Semantics for the ODP Computational Model – E. Najm and J-B. Stefani: Computer Networks and ISDN Systems, Volume 27, 1995.
[6] Formal Support to ODP – Joubine Dustzadeh, Elie Najm: INDC'96, Sixth IFIP/ICCC Conference on Information Network and Data Communication, June 1996. Chapman & Hall.
[7] Semantics of Information Models for Open Object-Based Distributed Systems – Joubine Dustzadeh: Ph.D. thesis, Dec. 1996, ENST Paris.
[8] Robert H. Bourdeau, Betty H. C. Cheng: "A Formal Semantics for Object Model Diagrams" – IEEE Transcations on Software Engineering, October 1995.
[9] B. W. Bates, J-M. Bruel, R. B. France, M. M. Larrondo-Petrie: "Formalizing Fusion Object Oriented Analysis" – FMOODS'96, March 1996.
[10] H. Bowman, E.A. Boiten, J. Derrick and M.W.A. Steen: "Viewpoint Consistency in ODP, a General Interpretation" – FMOODS'96, March 1996.

APPENDIX

In figure 5, *ucrs* denotes a generic unflattened class relation system, uc, uc_1 and uc_2 denote classes and r, r_1 and r_2 denote relations.

In figure 6, *is* represents an instance system, ie, ie_1 and ie_2 represent information elements (instances of object or association) within an instance system, ω_1 and ω_2 represent object instances, and ρ is an association instance.

In figure 7, F represents a forest composed of inheritance trees. T denotes an inheritance tree. $\sigma(T)$ is a boolean predicate on T which returns true if T is a singleton tree and false otherwise. R_T is the root of tree T. N_j^T denotes the j-th (immediate) subclass (node) of the root R_T.

(UCRS.WF.1):	$\forall\, r \in ucrs,\ \exists\, uc_1, uc_2 \in ucrs$ such that : $(r.class_1 = uc_1.name) \wedge (r.class_2 = uc_2.name)$
(UCRS.WF.2):	$\forall\, uc_1, uc_2 \in ucrs, (uc_1 \neq uc_2) \Rightarrow (uc_1.name \neq uc_2.name)$
(UCRS.WF.3):	$\forall\, uc \in ucrs, \forall\, a_1, a_2 \in uc.attlist,$ $(a_1 \neq a_2) \Rightarrow (uc.a_1.name \neq uc.a_2.name)$
(UCRS.WF.4):	$\forall\, r \in ucrs, (r.class_1 = r.class_2) \Rightarrow$ $(\{r.role_1, r.role_2\} \neq \{\varepsilon, \varepsilon\})$
(UCRS.WF.5):	$\forall\, r \in ucrs, (r.role_i = \kappa) \Rightarrow (r.role_{\bar{\imath}} \neq \kappa)$
(UCRS.WF.6):	$\forall\, r \in cs, \exists\, c \in cs \wedge \exists\, i \in \{1,2\}$ such that: $((r.class_i = c.name) \wedge (\forall\, a_j \in c.attributes, r.role_i \neq c.a_j.name))$
(UCRS.WF.7):	$\forall\, r_1, r_2$ such that: $\{r_1.class_1, r_2.class_2\} = \{r_2.class_1, r_2.class_2\}$ (if $(r_1.class_1 = r_2.class_1) \Rightarrow$ $(r_1.role_1, r_1.name, r_1.role_2) \neq (r_2.role_1, r_2.name, r_2.role_2))$ (if $(r_1.class_1 = r_2.class_2) \Rightarrow$ $\{r_1.role_1, r_1.name, r_1.role_2\} \neq (r_2.role_{\bar{1}}, r_2.name, r_2.role_{\bar{2}}))$
(UCRS.WF.8):	There is no sequence $uc_1, \ldots, uc_n \in ucrs$ such that: $(\forall\, i < n, (uc_{i+1}.name \in uc_i.inherits)) \wedge$ $(uc_1.name \in uc_n.inherits)$
(UCRS.WF.9):	$\forall\, uc \in ucrs, (uc.abstract = \text{true}) \Rightarrow$ $(\exists\, uc' \in ucrs \mid uc.name \in uc'.inherits)$

Figure 5 Some Well-formedness Rules for Unflattened Class Relation Systems

(IS.WF.1): $\forall\, ie_1, ie_2 \in is, (ie_1 \neq ie_2) \Rightarrow (ie_1.id \neq ie_2.id)$

(IS.WF.2): $\forall\, \rho \in is,\ \exists\, \omega_1, \omega_2 \in is$ such that : $(\rho.id_1 = \omega_1.id) \wedge (\rho.id_2 = \omega_2.id)$

(IS.WF.3): $\forall\, ie \in is, \forall\, a_1, a_2 \in ie.attlist,$
$$(ie.a_1 \neq ie.a_2) \Rightarrow (ie.a_1.name \neq ie.a_2.name)$$

(IS.WF.4): $\forall\, \rho \in is, \{\rho.role_1, \rho.role_2\} \neq \{\kappa, \kappa\}$

Figure 6 Some Well-formedness and Static Semantics Rules for Instance Systems

let F be the forest of all inheritance trees;

begin
    while $\exists\ T_i \in F \mid \sigma(T_i) =$ false **do**
        $\forall j \in \{1, \ldots, k\}$ **do**
          $N_j^{T_i}.attlist := N_j^{T_i}.attlist \bigcup R_{T_i}.attlist;$
          create a relation to N_i^T for each relation of R_{T_i};
          remove parent link $N_j^{T_i}.parset$;
        od
        if $(R_{T_i}.abstract =$ false) **then**
          $F := (F - T_i) \bigcup R_{T_i} \bigcup \{N_1^{T_i}, \ldots, N_k^{T_i}\};$
        fi
        if $(R_{T_i}.abstract =$ true) **then**
          remove relations of R_{T_i};
          $F := (F - T_i) \bigcup \{N_1^{T_i}, \ldots, N_k^{T_i}\};$
        fi
    od
end

Figure 7 A Flattening Algorithm for Inheritance Trees

8
Specifying the ODP Trader: An Introduction to E-LOTOS

Giovanny F. Lucero
Dept. of Informatics, Federal University of Pernambuco
P.O. Box 7851, CEP 50732-970 Recife – Brazil
gflp@di.ufpe.br

Juan Quemada
Dept. of Telematics Engineering, Technical University of Madrid
ETSI Telecomunicación (B-202), E-28040 Madrid – Spain
quemada@dit.upm.es

Abstract
An E-LOTOS specification of the computational viewpoint of the ODP trader
is presented. E-LOTOS is a new FDT being standardized in the ISO ODP
framework. The trader specification is on one hand an assessment of the capabilities of E-LOTOS in specifying ODP elements, but is on the other hand an
introduction to the most relevant features of E-LOTOS. E-LOTOS maintains
the spirit and the most fundamental operators of LOTOS. It has includes a
new functional data typing language, as well as a rich set of new elements
such as time, exception handling, a generalized sequential operator, a more
general parallel composition operator, gate typing, partial synchronization,...

Keywords
LOTOS, E-LOTOS, ODP trading.

1 INTRODUCTION

LOTOS (ISO89 1989) is an ISO standard which has been successfully used
to produce formal description of a great variety of applications such as communication protocols, reactive systems, phone systems, ..., as well as to de-

Formal Description Techniques and Protocol Specification, Testing and Verification
T. Mizuno, N. Shiratori, T. Higashino & A. Togashi (Eds.)   © 1997 IFIP. Published by Chapman & Hall

sign real implementations (Turner 1993, Boumezbeur *et al.* 1993, Faci *et al.* 1993, Fernández *et al.* 1992). Its success is due to two main facts: 1) It is mathematically well-defined, but has nevertheless a very high expressive power; 2)It supports many parts of the software/system design cycle such as abstraction, a variety of design styles, a stepwise refinement approach, a mathematical framework for verification and testing, tools that support most of the parts of the life cycle, ...

A question raised in 1991 asking to enhance the existing LOTOS Standard led to the creation of a Work Item (WI) on "Enhancements to LOTOS" in 1993. The reasons for enhancement were the lack of some technical features, such as user friendly and executable data types, modules, time, gate typing, ... The WI has analyzed during its life time a large number of enhancement proposals. A decision was taken in May 1996. The resulting language has a reasonable increase in complexity (compared to LOTOS). It includes the following enhancements: executability of data expressions, user friendly data types with predefined types, modules, subtyping, partial functions, time, exceptions and exception handling, gate typing and partial synchronization, ... The rich variety of specification styles and the abstraction capabilities of LOTOS is even enlarged with the new enhancements. The WI has proposed the progression of the work to an independent standard which is usually referenced as E-LOTOS (ISO/IEC JTC1/SC21/WG7 1997). The definition has passed the CD Ballot with the support of 11 countries and will be submitted for FCD balloting by the end of the Summer 1997. The basic structure of E-LOTOS has been defined, but some adaptation may occur before E-LOTOS reaches its final status.

In this paper, we present an E-LOTOS specification of the ODP trader from the computational viewpoint. The ODP trader is a paradigm of a broker object which links clients and service providers in an open and dynamically changing distributed system. A trader models the procedures to be followed, on one hand by service providers to announce the services offered and on the other hand by users to find service offers. The trader specification is therefore a good example of how E-LOTOS can be applied to specify real ODP systems. The specification given follows the computational description of the trading function given in ISO/IEC DIS 13235 (1995) which is a non formal textual description. The specification describes the most relevant parts of the functionality given in the informal description. Most features not considered in the formal specification can be added directly without difficulty.

In the remainder of the paper we assume that the reader is familiar with LOTOS. We give explanations only of the new features which do not exist in LOTOS. The paper can be used therefore as an introduction for LOTOS users of the most relevant new features of E-LOTOS.

The paper is structured as follows: Section 2 summarizes the changes and enhancements introduced in E-LOTOS with respect to the original version of LOTOS. Section 3 provides an informal introduction of the trading function.

Section 4 includes the most relevant parts of the E-LOTOS specification of the ODP trader complemented with more generic descriptions of the most important parts of E-LOTOS. This explanations should provide newcomers a deeper insight into the capabilities of the new language. Section 5 describes the conclusions obtained by the authors about this specification experiment and about E-LOTOS as a specification language. Finally, section 6 acknowledges the rest of the regular members of the E-LOTOS WI which have participated in the design of E-LOTOS.

2 E-LOTOS

E-LOTOS (ISO/IEC JTC1/SC21/WG7 1997) is formed of a behavioral process algebra part, which generalizes the basic LOTOS algebra in several directions but which keeps most LOTOS operators as particular cases of the new ones, and an executable and user friendly functional data definition language.

A list is given now which tries to highlight from the user point of view the most notable new features from which an E-LOTOS user should benefit when using the language.

- A loop construct which avoids the need for implemented cyclic behaviours with recursive processes has been introduced. This is the only place where multiple assignment of a variable is allowed in each iteration of the loop.
- Output variables in processes and functions (Garavel *et al.* 1996). The use of output variables in processes and functions together with the new sequential composition operator provides a much more readable and concise way for value communication in sequential compositions. The following example illustrates the way assignment, functions and processes communicate through input and output values

```
?x:= 2 ;                      (* a write-once assignment *)
phase1 [...] (x,?result) ;    (* a process *)
compute(result,?res1,?res2) ; (* a function *)
lphase2 [..] (x,res1) ||| rphase2 [..] (x,res2)
                      (* parallel comp. of two processes *)
```

- A general parallel n-ary operator which supports synchronization of J processes among K processes $(J \leq K)$ (Garavel *et al.* 1996). Synchronization patterns of 2 processes among N are possible. The new parallel operator is clear and readable and has the LOTOS operators as particular cases.
- A Suspend/Resume operator which generalizes disabling by allowing a disabled behaviour to be resumed by a specific action of the disabling behaviour (Logrippo *et al.* 1994, Hernalsteen 1996).
- Exceptions and exception handling in the behavioural and data parts with a

uniform approach. Exception mechanisms permit new ways of structuring and are demanded by ODP (Quemada *et al.* 1992, Garavel *et al.* 1996, Jeffrey 1996).

- Explicit renaming operator for observable actions or exceptions (Jeffrey *et al.* 1996). The renaming operator allows not only to change the name of the events occurring but to add and remove fields from the structure of events, or to merge and to split gates.
- Gates must be explicitly typed with a given record structure. Partial synchronization can be achieved when the gate types satisfy a subtyping relation in addition to value matching on the common part. Partial synchronization allows very elegant constraint oriented specifications (Jeffrey 1996, Garavel 1994, Mañas 1994, Najm *et al.* 1995).

3   AN OVERVIEW OF THE ODP TRADER

In order to use services in a open distributed systems, users need to know which services are available and who are their providers. Since sites and applications are frequently changing in large distributed systems, it seems compulsory to have a mechanism which enables software components to find appropriate services providers. This mechanism, called Trading Function, is supplied in ODP (ISO/IEC DIS 10746 1995) by the *trader object*.

Following the philosophy of ODP, the trader is specified through several viewpoints. In the next sections, we gives an informal description of these viewpoints.

3.1   Enterprise viewpoint

From the enterprise viewpoint, a trader is an object that enables clients to find dynamically suitable servers in an ODP system. A trader can be viewed as an advertiser where objects can announce their capabilities and become aware of capabilities of other objects.

An announcement in a trader is called *service offer*. It describes the characteristics or properties satisfied by the service. In addition to service properties, a service offer also contains the interface where the service is available.

Advertising a service offer is called *export*. When a trader accepts an export request, it stores the exported service offer in a centralized or distributed database. This database is often termed *service offer space*.

On the other hand, an importer can require knowledge about adequate service providers. In this case, the trader accepts a request, called *import*, containing an expression of service requirements desired by the importer. The trader matches the importer's service request with its database of service offers and selects a list (possibly empty) of appropriated service offers which satisfy the requirements made by the importer.

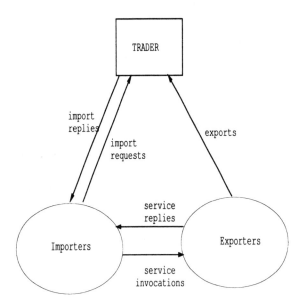

Figure 1 Interactions of a trader and its clients

The list of matched services offers is returned to the importer which may then interact directly with any service described in the list. Figure 1 summarizes the interactions of a trader and its clients.

Export and import activities are governed by a *trading policy*, which comprises trader policies, importer policies and exporter policies. Where an activity involves interactions between objects, the resulting policy will be a compromise between the wishes of the interacting objects. Therefore, a trader's behaviour is limited by the policies established for these activities. In other words, trader policies determine and guide a trader's behaviour. For example, a trader policy can restrict resources used by an individual import request.

Several autonomous traders can be "linked" in order to share their service offer spaces. Thus, a trader also can play the role of exporter or importer with respect to other trader(s). Such a group of autonomous traders is termed *interworking group*. A trader within an interworking group enlarges the service offer space for its users by including offers of other traders in the group. This enlargement of the service offer space is made indirectly when a trader propagates import requests to neighbor traders.

3.2   Information viewpoint

The trader information viewpoint defines the information elements and the relationships between them which are manipulated by the ODP trading function. In the standard ISO/IEC DIS 13235 (1995), this viewpoint is formally described using the formal specification notation Z. The specification includes basic concepts for information and static invariant and dynamic schemata for the ODP trading function. In this paper we do not consider this viewpoint.

3.3   Computational viewpoint

The trading function computational viewpoint describes an object template for a trader. This object has interfaces for service and management operations. Service operations are related to import and export activities whereas management operations are provided to add, delete or modify links to other traders.

In the standard ISO/IEC DIS 13235 (1995), this viewpoint description comprehends:

- signature templates for the service interface and management interface (defined in CORBA IDL),
- types used in the operations parameters (defined in CORBA IDL), and
- informal descriptions of the trader's behaviour.

The trader's behaviour is given by the behaviour of every service operation and management operation, plus a set of constraints on interleaving of actions performed by these operations. In the next section, we give in E-LOTOS a formal description of the trader's behaviour. This formal specification could be viewed as a complement of the informal one given in ISO/IEC DIS 13235 (1995).

4   E-LOTOS SPECIFICATION OF THE TRADER

This section outlines our E-LOTOS specification of the ODP trader's computational viewpoint. We present only the most relevant parts of the specification. In particular, we pay attention to the definition of the trader's behaviour. The reader interested in the complete specification is referred to the annex A of ISO/IEC JTC1/SC21/WG7 (1997).

4.1   The structure of a trader community

A trader is modeled by a process which can interact with its environment through two ports. On one port, the trader receives operation invocations from clients and management objects. On the other port, it returns the results to the respective invoker. Following the ODP terminology, we will call *termination* the action of returning a result.

In a trading community, every object (trader or client) has an interface identifier. An object in this community uses these identifiers in order to refer to the other objects. Thus, invocations and terminations are represented by events having respectively the following schemes

```
inv (!id !origId !...operation parameters...)
ter (!id !origId !...termination results...)
```

where **id** identifies the invoked trader and **origId** identifies the object which requests the operation. The trader receives invocations on the gate **inv** and sends the respective terminations on the gate **ter**.

A trading community can be described using the new E-LOTOS operator **par**. This operator allows to specify structurally and clearly architectures of systems which are difficult or, sometimes, impossible to express in LOTOS. Thus for example, the following E-LOTOS code shows a scenario where a trader communicates directly with an importer object and indirectly, through a binding object, with an importer-exporter object. The binding object redirects invocations and terminations between objects.

```
par inv#2, ter#2, bind#2
    [inv,ter] -> Trader[inv,ter](...)
 ||[inv,ter] -> importer[inv,ter]
 ||[bind] -> importerExporter[bind]
 ||[inv,ter,bind] -> bindingObject[inv,ter,bind]
endpar
```

The code above specifies that communication in the gates tagged with **#2** is one-to-one, i.e., processes synchronize by pairs on the gates listed in the square brackets before the respective "**->**". Notice that the style of specification used in this piece of E-LOTOS code is resource oriented. The **par** operator provides better flexibility and expressiveness than the three original parallel operators joined (| [...] |, | | and | | |). Specification styles like constraint oriented and resource oriented, which use parallel composition extensively, greatly benefit by this E-LOTOS extension.

4.2   Type declarations

In the specification, the greatest part of the type declarations comprises types for the operation parameters. The standard (ISO/IEC DIS 13235 1995) defines these types in CORBA IDL and their translation to E-LOTOS is very easy. Also, we translate exception declarations given in CORBA IDL to type declarations in the E-LOTOS notation. In order to make more concise the specification, we have removed the tailing string "Type" from all type names in ISO/IEC DIS 13235 (1995).

Ports are typed in E-LOTOS, therefore we need to specify types for data exchanged (offered) by E-LOTOS processes. In particular, we need to specify which are the types involved in trader interactions.

Data types in E-LOTOS are declared as in SML but using records rather than tuples in the constructor arguments. Thus, the signature for operation invocations is defined by the following type:

```
type InvocationSig is
   ExportOfferInv ( serviceDescription => Servicedescription,
                    servicePropValues => PropertyValueList,
                    offerPropValues => PropertyValueList,
                    serviceInterfaceId => InterfaceId)
 | ImportOfferInv
     ( serviceDescription => ServiceDescription,
       matchingCriteria => Rule,
       preferenceCriteria => Rule,
       orderingRequirementList => OrderRequirementList,
       servicePropertiesOfInterest => PropertiesOfInterest,
       offerPropertiesOfInterest => PropertiesOfInterest )
 | AddLinkInv ( newLinkName => name,
                linkPropValues => PropertyValueList,
                targetInterfaceId => InterfaceId )
 | ... (* other invocation signatures *)
endtype
```

This declaration introduces a union type and its constructors. Each branch separated by "|" defines one constructor and the type of its arguments. The type of the arguments is given by a record type using the "=>" notation to define named fields. The name of the field and its type are separated by the "=>" symbol.

An invocation signature states the input parameters of the respective operation. Thus, the signature for export operation invocations has a description of the service which will be exported, a list with the properties of the service, a list with the properties of the service offer, and a identifier of the interface where the service will be provided. A property of a service might be the cost of the service. A property of a service offer might be the date when the offer expires.

An import operation invocation contains a service description of the desired service, a *rule* to constrain the matching of suitable servers, a rule to indicate preferred services and some indications for the formatting of results. Rules are used to define politics of search in an import operation. A rule is expressed by a predicate over the properties of the desired service and offer service. For example, a rule expressing matching criteria might be "the cost of the service is less than \$10". Rules are modeled using recursive data types. By lack of space we do not include their definition here.

The operation **AddLink**, which allows to create links between traders, has a invocation signature with the name of the link, a list of properties of the link and a identifier of the target trader.

For each kind of invocation, a proper kind of termination is needed. We define termination signatures for trader operations by

```
type TerminationSig is
   ExportOfferTer (offerId => ServiceOfferId)
 | ImportOfferTer
      (detailsOfServiceOffers => ServiceOfferDetailList)
 | AddLinkTer (linkId => LinkId)
 | ... (* other termination signatures *)
endtype
```

A termination signature states the output parameters for a given operation. The export offer operation returns a new identifier for the recently exported service offer. The import operation returns a list detailing the properties of suitable (matched) services. An add link operation returns a new identifier for the link added by the performance of this operation.

The data exchanged for trading are invocations and terminations. The type of this data is defined below using record types. We define the type **Invocation (termination)** for typing gates that exchange invocations (terminations) of trader operations. The type **Invocation (Termination)** is a record with three fields: the invoked trader identifier, the originator client identifier and the input (output) parameters of the operation.

```
type Invocation is
  (interfaceId => InterfaceIdentifier,
   originatorId => InterfaceIdentifier,
   invocation => InvocationSig)
endtype

type Termination is
  (interfaceId => InterfaceIdentifier,
   originatorId => InterfaceIdentifier,
   termination => TerminationSig)
endtype
```

4.3   Function declarations

In this section, we only show some simple examples of function declarations. The examples illustrate the functional language used in E-LOTOS to write data expressions.

The following declaration defines a function which allows to see if a (property) name occurs in lists of name-value pairs.

```
function existProperty ( pName: PropertyName,
                         props: PropertyValueList ) : Boolean
   case props is
     Nil -> false
   | Cons((name => !pName, etc), any:PropertyValueList) -> true
   | Cons(any:PropertyValue, ?tl) -> existProperty(nm, tl)
   endcase
endfunc
```

Notice some details in the above the definition. The case constructor uses pattern matching. In patterns, binding occurrences of variables are decorated with "?" and expressions decorated with "!" denote values offered. This syntax is in accord with the syntax of events. The occurrence of "etc" is not meta-linguistic. The **"etc"** notation completes the record pattern and denotes the remaining components of the record.

The following specification takes advantage of the exception mechanism introduced by E-LOTOS in order to define some needed partial functions. For example, the following **propValueOf** function receives a property name and a list of property name–value pairs. When the name appears in the list, it returns the value associated to the name in the list. In the other case, the function raises an exception.

```
function propValueOf (pName: PropertyName,
                      props: PropertyValueList) : Value
                      raises [err: (TraderErrorExceptions)]
   case props is
     Nil -> raise err(UndefinedProperty(pName))
   | Cons ((!pName, ?v), any:PropertyValueList) -> v
   | Cons (any:PropertyValue, ?tl) -> propValueOf(pName, tl)[err]
   endcase
endfunc
```

4.4   Computational behaviour of the trader

Like LOTOS, E-LOTOS makes it possible to describe the behaviour of systems in a stepwise fashion, moving from one abstraction level to another. Each level can be specified in different specification styles. We follow this methodology in our specification. At the most abstract level, the trader object is a parallel composition of three resources (processes). The **ServiceInterface** and **ManagementInterface** processes provide functionality for service operations and management operations, respectively. The process **StateProc** represents the trader state.

```
process Trader [inv: Invocation, ter: Termination]
                ( interfaceId: InterfaceIdentifier,
                  properties : Properties,
                  offerSpace : ServiceOfferSpace,
                  linkSpace :  linkOfferSpace ) : exit(none) is
  hide sa:stateAccess in
      ( ServiceInterface [inv, ter, sa] (interfaceId)
      |||
        ManagementInterface [inv, ter, sa] (interfaceId)
      )
    |[sa]|
        StateProc [sa] ( properties, offerSpace, linkSpace )
  endhide
endproc (*Trader *)
```

Service operations are related to export and import activities and they only write in the trader service offer space. Management operations manipulate connections between interworking traders updating trader links. Therefore, service operations and management operations can be performed in parallel without destroying the consistency of the trader state.

The process **StateProc** encapsulates three elements which conform the trader state: a set of trader properties, the service offer space and the link space. Service and management operations use the trader state through the port **sa** which is hidden within the trader.

In order to illustrate how the operation's behaviour is defined we show the definition of the **ServiceInterface** process. In this process, all service operations are offered to clients in parallel. However, the availability of operations is constrained in such way that accessing operations (as import offer) and modifying operations (as export offer) can not be overlapped in time. Availability of operations is defined in a constraint oriented style by composing the processes representing operations with the **OrderingConstraints** process.

```
process ServiceInterface [inv: Invocation, ter: Termination,
                          sa: stateAccessCh ]
                          (interfaceId: InterfaceIdentifier)
                          : exit(none) is
    ( ImportOffer [inv, ter, sa] (interfaceId)
      |||
      ExportOffer [inv, ter, sa] (interfaceId)
      |||
      .... (* other service operations *)
    )
  |[inv, ter]|
    OrderingConstraints [inv, ter]
endproc (*ServiceInterface *)
```

In one more step of refinement we define below the **ExportOffer** and
ImportOffer processes. The **ExportOffer** process enables clients to perform
offer exports. For each invocation, the process interacts with the state to update the offer database adding the new offer and then, in the termination, it
replies with the fresh identifier assigned to the offer. Afterwards, when the
offer expires, the management object is notified about the expiration.

```
process ExportOffer [inv: Invocation, ter: Termination,
                     sa: stateAccess ]
                     (id: InterfaceId) : exit(none) is
    inv (!id, ?origId,
         ExportOfferInv(?sd, ?sProps, ?oProps, ?interfId)) ;
    sa NewServiceOfferId(?oId) ;
    sa ReadOffers(?offers) ;
    sa !WriteOffers(Cons((oId,sd,interfId,sProps,oProps),
                         offers)) ;
    ter (!id, !origId, !ExportOfferTer(oId)) ;
    trap
      exception
        err(UndefinedProperty("exipiration")) is exit
      endexn
    in
      TimeValue(?t) := propValueOf("expiration",oProps)[err] ;
      inv (!mangmtId, !id, !Expired(oId)) @ t1 [t1=t]
    endtrap
  |||
    exportOffer [inv,ter,sa] (id)
endproc
```

This process defines a behaviour which replicates infinitely the behaviour of one export. This infinite process represents the possibility of simultaneously accepting of several exports. However, the availability of exports is constrained by the `OrderingConstraints` in the service interface (see the process `ServiceInterface` above).

The behaviour representing one export sequentially perform the following actions: accepts an export offer invocation, gets a fresh offer identifier, writes the offer in the offer space, replies to the exporter and finally, if the offer has an expiration property, notifies the management object at time of expiration.

In this last definition above we use some new concepts introduced by E-LOTOS: assignment, exception trapping and explicit time restrictions. The left part of the assignment ":=" is a pattern which is matched with the evaluation of the right part.

The trap constructor allows to capture and to handle exceptions raised when expressions are evaluated. Thus for example, the evaluation of the function `propValueOf` will raise an exception when the expiration property is undefined. In this case, the exception is trapped and handled by an `exit`.

The `@ t1 [t1=t]` construction in the last `inv ...` action occurrence restrict the performance of the action at a given time. Thus, when the invocation is performed, `t1` is bound to the current time hence the guard `[t1=t]` constraint its realization at time of the offer expiration.

The definition of the process `ExportOffer` exemplifies the use of the real time capabilities of E-LOTOS which are strongly based on the proposal for real-time LOTOS described in Léonard *et al.* (1997). Here, we use time to define timeouts to signal offer expirations. Real time is important also in other applications. In ODP, for example, it can be used to describe features like QoS of binding objects.

Notice that the actual deletion of expired offers will depend on the management object policy therefore it is not specified here. For example, it can dictate that expired offers must be deleted immediately. Another policy could be to leave expired offers in the offer database until new exports demand the space occupied by them.

In addition, it is worth to note the readability of the definition. It is reached through the use of E-LOTOS proper constructs like assignment and traps. The reader who would attempt to write a similar specification in LOTOS would perceive a considerable loss of readability.

Another significant service operation is the import of an offer. The behaviour of the operation is given below by the `ImportOffer` process definition.

Informally, for each invocation, the `ImportOffer` process performs the following activities: receive the invocation, access the state to get the offer database, trader links and trader properties, match the offers against arguments and propagate the import invocation to interoperate with "neighbor" traders and, when some trader finds the right matches, they are submitted in the termination.

```
process ImportOffer [ inv: Invocation, ter: Termination,
                    sa: stateAccess ]
                (id: InterfaceId) : exit(none) is
    inv ( !id, ?origId,
         ImportOfferInv ( ?sd, ?matchingC, ?preferenceC,
                        ?ordering, ?spOfInt, ?opOfInt ) ) ;
    sa Read( ?traderProp, ?offers, ?links ) ;
    trap
      exception X ( offers: ServiceOfferDetailList ) is
         ter ( !id, !origId, ImportOfferTer(offers) )
      endexn
      exit is
         ter ( !id, !origId, ImportOfferTer(emptyOffers) )
      endexit
    in
      ... (* match local offers *)
      | | |
      ... (* interoperate with other traders *)
    endtrap
  | | |
  ImportOffer [ inv, ter, sa ] ( id )
endproc
```

The above definition is a good example to illustrate a non-orthodox use of the trap constructor. In general, the trap is used as a tool for exception handling. However, it also increase the expression power allowing to define new behaviour schemes. Here, for example, we use this constructor to express that, when several traders are interoperating in an import operation, the matched offers returned to the original importer are those of the "first" trader which was successful in the matching activity.

5 CONCLUSIONS AND FURTHER WORK

This paper provides a brief introduction to E-LOTOS and to its usage in specifying the ODP Trader. This is one of the first applications of this new FDT to a medium size example. E-LOTOS and the new enhancements have proved to be suitable for the description of the trader. The use of time, new parallel, exceptions and new typing facilities makes the specification shorter, more readable, more concise and allows a better structuring. The usage of the enhancements introduced has only started to be understood, but the results are very encouraging.

The definition of E-LOTOS is now in its final stages and examples as this one provide the basis for the last tuning. The standard is now in the last stages of the process where the last (small) changes can and probably will be introduced. Of course the core of E-LOTOS should be stable in the future. Changes in the syntax or in the static semantics may still occur. The main motivation for such changes will be, the alignment with ODP standards such as CORBA or IDL, the production of a uniform syntax which can be properly parsed, the removal of inconsistencies or errors in the present definition,...

6 ACKNOWLEDGEMENTS

The development of E-LOTOS has been a collective effort in which many contributors have participated. The work in ISO is performed within Work Items (WI) where national experts contribute and progress the work. One of the co-authors, Juan Quemada, has coordinated and chaired the "Enhancement to LOTOS" WI rapporteur and editor and he would like to acknowledge the work performed by many other national experts which have also contributed with his work, ideas and contributions to the definition of E-LOTOS. Among all the participants in the WI, the ones which have had a regular participation with contributions during some phase of the work are: Arnaud Février, Hubert Garavel, Alan Jeffrey, Guy Leduc, Luc Leonard, Luigi Logrippo, Jose Manas, Elie Najm, Mihaela Sighireanu and Jacques Sincennes.

This work has been jointly supported by Spanish National Research Program – the CICYT project DISC and CNPq/Brazil.

REFERENCES

ISO 8807 (1989) LOTOS—A formal description technique based on the temporal ordering of observational behaviour, International Standard.

ed. K.J. Turner (1993), Using Formal Description Techniques, an introdution to Estelle, LOTOS and SDL, Jonh Wiley & Sons.

Boumezbeur, R. and Logrippo, L. (1993) Specifying telephone systems in LOTOS, in *IEEE Communications Magazine*, August.

Faci, M. and Logrippo, L. (1993) Specifying hardware systems in LOTOS, in *Proc. of Computer Hardware Description Languages and their Applications XI (CHDL'93)*, North–Holland.

Fernández, A. and Miguel, C. and Vidaller, L. and Quemada, J. (1992) Development of Satellite Communication Networks based on LOTOS, in *Protocol Specification, Testing and Verification XII*, North Holland, Amsterdam.

ISO/IEC JTC1/SC21/WG7 1.21.20.2.3 (1997) Working Draft on Enhancements to LOTOS, ISO Working Group 7, Also available by ftp at ftp://ftp.dit.upm.es /pub/lotos/elotos/Working.Docs/cd.ps.gz.

ISO/IEC DIS 13235 (1995) ODP Trading Function.

Garavel, H. (1994) On the Introduction of Gate Typing in E-LOTOS, VERIMAG, France.

Garavel, H. and Sighireanu, M. (1996) A Wish List for the Behaviour Part of
    LOTOS, VERIMAG, France.
Logrippo, L. and Steppien, B. (1994) Specifying the Suspend/Resume Oper-
    ator, University of Ottawa.
Hernalsteen, C. (1996) Introduction of a Suspend/Resume Operator in E-
    LOTOS, ENST, France.
Quemada, J. and Azcorra, A. (1992) Structuring Protocols with Exception in
    a LOTOS Extension, in *PSTV XII*, Orlando.
Jeffrey, A. (1996) A Core Data and Behaviour Language for E-LOTOS, Uni-
    versity of Sussex.
Jeffrey, A. and Leduc, G. (1996) E-LOTOS Core Language, University of
    Sussex.
Mañas, J. (1994) Typed Gates, DIT-UPM, Spain.
Najm, E. and Février, A. (1995) Extending Gate Typing to Mobile LOTOS,
    ENST, France.
ISO/IEC 10746 (1995) Open Distributed Processing – Part 1-4.
Léonard, L. and Leduc, L. (1997) An introduction to ET-LOTOS for the
    description of time-sensitive systems, Computer Networks and ISDN
    Systems, **29 3**, 271–292.

7   BIOGRAPHY

Giovanny Lucero is currently a Guest Professor at Department of Computer
Science of the Federal University of Sergipe and a Ph. D. student at the
Federal University of Pernambuco (UFPE) at Brazil. He obtained a M.Sc.
in Computer Science at UFPE in 1993 and his main research interest is the
semantics of programming languages, and especially the overlap between functional programming and theories of objects and concurrency.

Juan Quemada is currently Professor at the Telematics Engineering Department of the Universidad Politecnica de Madrid, from where he achieved
his Telecommunication Engineering Degree and a Phd in Telecommunication
Engineering in 1981. He gives courses and conferences regularly and has au-
thored over a hundred technical publications, including contributions in over
30 co-authored books. He is a regular member of the program committee of scientific conferences such as FORTE, PSTV, FMOODS or ICODP. With about
twenty years of experience in the area of telecommunications and distributed
systems his present focus of activity is in the use of CSCW in new service
creation and in the use of formal methods in the design of communication
and distributed systems. As rapporteur and editor of the WI on "Enhancements to LOTOS" he has been coordinating the definition of the E-LOTOS
Standard in ISO/IEC.

## PART THREE

LOTOS and Extension

9

A Computer Aided Design of a Secure Registration Protocol

F. Germeau, G. Leduc
Research Unit in Networking, Université de Liège
{germeau,leduc}@montefiore.ulg.ac.be

Abstract

We use the formal language LOTOS to specify a registration protocol between a user and a Trusted Third Party, that requires mutual authentication. We explain how a model-based verification method can be used to verify its robustness to attacks by an intruder. This method is also used to find a simpler protocol that remains secure.

Keywords

Security, Authentication, Registration protocol, Guillou-Quisquater, Trusted third party, Formal verification, LOTOS specification

1 INTRODUCTION

With the development of the Internet and especially with the birth of electronic commerce, the security of communications between computers becomes a crucial point. All these new applications require reliable protocols able to perform secure transactions. The environment of these operations is very hostile because no transmission channel can be considered safe. Formal descriptions and verifications can be used to obtain the assurance that a protocol cannot be threatened by an intruder.

In this paper, we will show that it is possible to make a formal verification of a security protocol. We can certify that an intruder cannot break a protocol with different kinds of attacks. We will also show how the verification process is able to give useful information to correct the protocol if necessary. The verification technique we have developed is based on the LOTOS (Bolognesi *et al.* 1987)(ISO8807 1989) language and the CADP package (Fernandez *et al.* 1996) included in the Eucalyptus toolbox (Garavel 1996).

Formal Description Techniques and Protocol Specification, Testing and Verification
T. Mizuno, N. Shiratori, T. Higashino & A. Togashi (Eds.) © 1997 IFIP. Published by Chapman & Hall

We use a model-based approach that, until recently, was not felt adequate to tackle the verification of security protocols (Leduc *et al.* 1996)(Lowe 1996).

We will illustrate the method on a registration protocol. The Equicrypt protocol (Lacroix *et al.* 1996) is a conditional access protocol under design in the European ACTS OKAPI project (Guimaraes *et al.* 1996). It allows a user to subscribe to multimedia services such as video on demand. Equicrypt is an open protocol where the user must first register with a Trusted Third Party (TTP) using a challenge-response exchange. After a successful registration, this third party issues a public-key certificate which allows the user to subscribe to a service with different service providers. The subscription part has been studied in (Leduc *et al.* 1996) and some possible attacks have been reported. In this paper, we will focus on the design and verification of the registration protocol which must provide the authentication of the user by the TTP and authentication of the TTP by the user. The protocol is also used to transmit the user's public key to the TTP.

The paper is organized as follow. The section 2 describes the registration protocol that we want to verify and possibly correct. In section 3 we present the formal specification of the protocol written in LOTOS and the section 4 is dedicated to the properties we want to verify. The verification itself is explained in section 5 and concludes this paper.

2   THE REGISTRATION PROTOCOL

2.1   Notation

The protocol involves several cryptographic operations, for which we give an abstract view only. Each scheme uses peer encryption and decryption keys K_E and K_D and functions $E(\_,\_)$ and $D(\_,\_)$ such that $D(K_D, E(K_E, m)) = m$ for any message m. In public key cryptography, the encryption key is the public key and the decryption key is the private key for ciphering operations. For signature operations, the encryption key is the private key and the decryption key is the public key. We also use the more compact notation $\{m\}K_E$ to denote the message m encrypted with the key K_E. That is $\{m\}K_E = E(K_E, m)$.

K_A^P denotes the public key of the user A and K_A^S the private secret key of the user A. The expression $\{m\}K_E$ where K_E is a public key represents the message m encrypted with the key K_E. The same expression where K_E is a private key represents both the message m in clear and a hash of the message m encrypted with the key K_E.

We widely use the concept of nonce (i.e. a number used only once). A nonce is a random number that must be used during only one instance of the protocol. This prevents an intruder from replaying outdated messages and is an abstract model of the pair "time stamps, random number".

All the messages have the following structure:

$Number : Source \rightarrow Destination : Message\ Id\ < Message\ Fields >$

2.2   Principles

The following is a presentation of the Equicrypt system and its registration protocol. The aim of the Equicrypt system is to control the access to multimedia services proposed by service providers. To avoid requiring different access systems for every service provider, a unique decoder uses a public-key cryptography protocol to subscribe to and decode different services. An independent entity known as the Trusted Third Party (TTP) acts as a registering authority trusted by both users and providers. The registration protocol must achieve the mutual authentication of the user and the TTP. The TTP must be sure that the claimed identity of the user is the right one and the user must be sure that it registers with the right TTP. The TTP must also receive the good user's public-key to issue a public-key certificate similar to X.509 certificates (ITU-T X.509 1993). This kind of certificate is the user's public key signed with the TTP's private key.

The authentication of the user by the TTP uses the Guillou-Quisquater zero-knowledge identification scheme (Guillou *et al.* 1988). When the user buys his decoder, he receives secret personal credentials derived from its real-life identity. These credentials will help the user to prove who he is. The goal of the Guillou-Quisquater (GQ) algorithm is to convince the TTP that the user has valid credentials without revealing them. The authentication of the TTP by the user uses a challenge based on a nonce similar to the 3-way authentication protocol (Schneier 1996). When the user receives his credentials, he also receives the TTP's public-key that will allow him to perform the required checks on the messages and to authenticate the TTP.

The transmission of the user's public key is the third purpose of the registration protocol. The TTP must be sure that the received public key is really the user's one. He must make a link between the user's identity and his public key. An improved version of GQ algorithm proposed in Lacroix *et al.* (1996) can be used to check this.

2.3   The Guillou-Quisquater identification scheme

The cryptographic details of the GQ algorithm are beyond the scope of this paper but the principles will be exposed. Basically, the credentials the user receives are mathematically related to its identity. Let the user act as the prover P and the TTP act as the verifier V in the following protocol.

$1 : P \rightarrow V : Request \ < ID, K_P^P, T(K_P^P, r) >$
$2 : V \rightarrow P : Challenge \ < d >$
$3 : P \rightarrow V : Response \ < t(r, d, B) >$

The prover generates a random number r and computes a function T of this number and of his public key. He sends the verifier his identity ID, his public key K_P^P and the result of the function T. As a response, the verifier sends back another random number d. Then the prover computes a function t with the two random numbers r and d and his credentials B and sends it to the verifier. When he receives the response, the verifier can check that the credentials used to compute t correspond to the identity claimed in the first message, thanks to the existing mathematical relationship between ID and B. The user's credentials B must be kept secret so that the only one who could have computed a right function t is the real user. Thus the TTP has obviously received a fresh response from the right user and has authenticated him.

In message 1, the user's public key has also been scrambled (by the function T) with the random number r. When the verifier received message 3, he gains the mathematical ability to check that the public key received in message 1 is also the one used to compute T. Although the public key is transmitted in clear in message 1 and is thus known to an intruder, this intruder cannot forge a fake message 1 with another public key. This is because he does not know the random number r used again in message 3 and so he cannot generate a valid function $T(K_P^P, r)$.

2.4   Abstract model of the Guillou-Quisquater algorithm

In fact, the GQ algorithm can be seen as a general encryption/decryption scheme. This will be very useful for our formal description. We can consider the user's identity together with its public key as a public decryption key and the credentials as a corresponding secret encryption key. Then, the GQ algorithm looks like an authentication scheme based on a nonce and works as follows. The prover sends his decryption key and receives back the random number d from the verifier. The random number d acts as the nonce. Then he encrypts it with his encryption key. The verifier can check that the nonce he sent has a good signature.

This scheme resists to the "man-in-the-middle" attack because the decryption key is mathematically linked to the prover's identity: the identity itself being a part of the decryption key. When this authentication scheme is used with the classical public key cryptography, not the GQ algorithm, the verifier must receive the prover's public key in another way by a secure channel.

The real algorithm also involves the random number r. As said previously, its main purpose is to scramble the user's public key in the function T. If the intruder generates such a fake function in the first message, the credentials computation performed by the verifier when he receives the third message will fail. We will obtain the same result if the intruder changes the user's public key. This behaviour is exactly transposed in our model because both the user's identity and its public key are used to check the credentials. The second purpose of r is to prevent the TTP from guessing B. Our specification does not take these cryptographic attacks into account. Thus we do not need to consider the random number r and we can ignore it in our model. To avoid confusion, we use the special notation $F(B, d)$ to express the encryption of the nonce d with the credentials B. This will help the reader to keep the modelling in mind.

2.5   Protocol description

The complete registration protocol is as follows. The protocol comprises the authentification of the user by the TTP with the GQ algorithm. We have added the authentication of the TTP by the user with a challenge based on a nonce. Finally, we have added a fourth message to carry the registration result and we use the abstraction of the GQ identification scheme depicted above. This first version of the protocol has a flaw. We will see in section 5 that the formal verification has revealed it and has given information to correct the protocol and to produce new versions.

(a) Initial knowledge of the user

- An identity: $UserID$.
- A pair of public/private keys: K_U^P and K_U^S.
- Credentials: B.
- The public parameters of the GQ algorithm.
- The public key of the TTP: K_{TTP}^P.

(b) Initial knowledge of the TTP

- A pair of public/private keys: K_{TTP}^P and K_{TTP}^S.
- The public parameters of the GQ algorithm.

(c) Message exchanges

The user generates a random nonce n and sends the message 1.

$1: User \rightarrow TTP: Register\ Request\ < UserID, K_U^P, \{n\}K_{TTP}^P >$

When the TTP receives message 1, he decrypts the nonce n and signs it, generates a random number d and sends them to the user. The TTP can handle several registrations at a time. So he maintains an internal table with one entry for each user who has a registration in progress and he records the tuple $< UserID, K_U^P, n, d >$.

$2: TTP \rightarrow User: Register\ Challenge\ < d, \{n\}K_{TTP}^S >$

When the user receives message 2, he checks the signature. If the signature is correct, he performs the GQ calculation and sends the result to the TTP.

$3: User \rightarrow TTP: Register\ Response\ < F(B, d) >$

When the TTP receives message 3, he checks the GQ authentication using this message and the data found in his internal table. Then, he sends a response according to the result. The response is signed and includes both the user's identity and the nonce n. The user's entry in the internal table is deleted. If the response is positive, the TTP registers the tuple $< UserID, K_U^P >$.

$4^+: TTP \rightarrow User: Register\ Ack\ < \{Yes, UserID, n\}K_{TTP}^S >$
$4^-: TTP \rightarrow User: Register\ Ack\ < \{No, UserID, n\}K_{TTP}^S >$

Now that we have presented the registration protocol, we will continue with its specification, its verification.

3 FORMAL SPECIFICATION

The formal specification has been written in LOTOS which is a standardized description language suitable for the description of distributed systems.

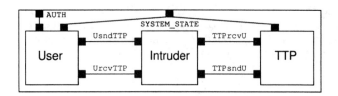

Figure 1 Structure of the LOTOS specification

3.1   Behaviour

The LOTOS specification models both the authentication system and the environment. The authentication system is composed of the user, the TTP and the intruder. Figure 1 shows the general structure of the processes and their interaction points.

The communication channel between the user and the TTP is replaced by the intruder. He intercepts all messages and transmits them or not, with or without modification. We give more details about the intruder in section 3.3. Gates UsndTTP and UrcvTTP are used by the user for its communication in both directions. The TTP uses the gates TTPsndU and TTPrcvU.

The environment is responsible for the management of specific events. Firstly, he plays the role of the real user who asks his decoder to register with an interaction at the gate AUTH. Secondly, he receives messages that give information about the internal state of the user and about the internal state of the TTP. These messages will help us to perform the formal verification. In this paper, we call them the special events. We have defined six of them received through the gate SYSTEM_STATE:

$1 : User \rightarrow Environment : USER\_START\_REG < UserID >$

This message notifies the environment that the user whose identity is $UserID$ has received the order to register. The user generates this message before sending a valid registration request to the TTP. In our specification, the user and the TTP always behave correctly.

$2 : TTP \rightarrow Environment : TTP\_START\_REG < UserID >$

With this message, the TTP informs the environment that he has received a valid registration request from the user who claims that his identity is $UserID$.

$3 : TTP \rightarrow Environment : TTP\_REG\_SUCCEEDED < UserID, K >$

When the TTP sends this message, this means that he has successfully registered the user $UserID$ with the public key K. This message occurs when the TTP owns a valid response to his GQ verification. He will then send a message 4^{+}.

$4 : TTP \rightarrow Environment : TTP\_REG\_FAILED < UserID, K >$

This message corresponds to the previous one but when the GQ verification has failed. The TTP will send a message 4^{-}.

$5 : User \rightarrow Environment : USER\_REG\_SUCCEEDED < UserID >$

The user informs the environment he has received a valid successful registration acknowledgement from the TTP.

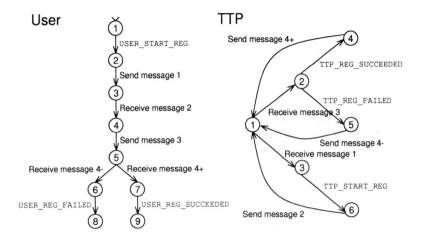

Figure 2 Behaviour of the user and the TTP

$6: User \rightarrow Environment : USER\_REG\_FAILED < UserID >$

The user informs the environment that he has received a valid refused registration acknowledgement from the TTP. That is, the user has received a message 4^- where the TTP's signature is valid but his response is negative.

Finally, the third task of the environment is to receive error messages. The user and the TTP perform several checks when they receive a message. If one of these checks fails, a message indicating the reason of the error is generated. It is very important to understand the difference between the two kinds of interruptions a registration can encounter. The registration can fail because the TTP has decided that the user does not own good credentials. That is what we will call a failure. The other cases are errors. An error occurs when the registration protocol stops due to a badly formed message: wrong signature, wrong nonce, ... We obviously focus on failures because we want to defeat the intruder when he generates good messages. An intruder can always create errors by sending garbage in the transmission channel.

Figure 2 sketches the main behaviours of the user and the TTP. Each transition is labelled with the transmission of a message, the reception of a message or the generation of a special event. Error cases and data manipulation are not shown for simplicity.

3.2 Data types

This specification has been written using data type language extensions, as offered by the APERO tools (Pecheur 1996) included in the Eucalyptus toolbox. The original text has to be processed by the APERO translator to get a valid LOTOS specification. This provides for a smaller and more readable specification.

The abstract data types are composed of:

- Base values: identifiers, keys, credentials described as explicit enumerations.
- Cryptographic functions: Encryption and decryption are modelled as abstract operations that are the reverse of each other. If a decryption is performed with a bad key, the result is not the encrypted message but a special junk value.

```
type EncryptedMessage is Message, PublicKey, PrivateKey
sorts EncryptedMessage
opns
 E (*! constructor *) : PublicKey, Message -> EncryptedMessage
 D : PrivateKey, EncryptedMessage -> Message
eqns
forall msg : Message,
       pubkey : PublicKey
       prvkey : PrivateKey
ofsort Message
  Match(pubkey,prvkey) => D(prvkey,E(pubkey,msg))=msg;
  not(Match(pubkey,prvkey)) => D(prvkey,E(pubkey,msg))=Message_Junk;
endtype
```

- Set of values: They are specially used to model the knowledge of the intruder. For example, to form a message, the intruder will pick a value in each of his sets non determinatically.
- Tables: Needed for storing information about registrations. The TTP can manage several registrations simultaneously so he must store the values received in the messages to make the authentication.

3.3   The Intruder

The intruder replaces the channel between the user and the provider. We want him to mimic any attack a real-world intruder can realize. Thus our intruder must be able to:

- Eavesdrop on and/or intercept any message exchanged among the entities.
- Decrypt parts of messages that are encrypted with his own public key and store them.
- Introduce fake messages in the system. A fake message is an old message replayed or a new one built up from components of old messages including components he was unable to decrypt.

The LOTOS process that models the intruder is always ready to interact at the four gates UsndTTP, TTPsndU, UrcvTTP and TTPrcvU. When the user, respectively the TTP, sends a message to the gate UsndTTP, respectively TTPsndU, the intruder catches the message and tries to decrypt its encrypted parts. Then he stores each part of the message in separate sets of values. These sets constitute the intruder's knowledge base that increases each time a message is received. When the user, respectively the TTP, expects a message on the gate UrcvTTP, respectively TTPrcvU, the intruder builds a new message with values stored in his sets. With this method, the intruder tries every message it can create.

The intruder is parameterized with some initial knowledge which gives him a certain amount of power. This power includes the capabilities to act as a user with the real TTP and to act as a TTP with the real user. Thus the intruder owns a valid identity, valid credentials and a valid pair of public/private keys. To give the intruder the capability of generating nonces, his initial knowledge also contains nonces that are distinct from those used by the entities. The system we modelled only includes one real user and one real TTP. With his knowledge, the intruder can be seen as a second user and a second TTP. So, our specification incorporates the case where a second valid user tries to cheat and the case where a second valid TTP tries to catch the registration.

The initial knowledge of the intruder is as follows:

- An identity: $IntruderID$.
- The identity of the user: $UserID$.
- A pair of public/private keys: K_I^P et K_I^S.
- Valid credentials: B_I.
- The public parameters of the GQ algorithm.
- The public key of the user K_U^P and the public key of the TTP K_{TTP}^P.
- Nonces.

We assume that our intruder cannot break the public key cryptosystem. That is, he cannot get a message in clear from an encrypted message and he cannot forge a signature without the private key. Note that LOTOS easily provides processes that transgress this rule. Care must be taken to avoid these kinds of unrealistic behaviours. A more detailed description of the intruder can be found in Germeau *et al.* (1997)

3.4   Labelled Transition System

To gain confidence into the specification, it has been simulated with the XSimulator tool from the Eucalyptus toolbox in step-by-step execution mode. This allows us to get a LOTOS specification which is likely to behave correctly without the intruder. Then we have used the CADP package to carry out the verification. The first step consists of using the Caesar tool to generate from the LOTOS specification a graph called Labelled Transition System (LTS). To be able to generate a finite-state LTS of reasonable size, some limitations were required. The exponential growth of states we meet forces us to limit the user to only one registration and the TTP to only two registrations. This has no effect on the generality of our result because the intruder is still able to perform a registration aside the user's one.

The size of the resulting graph greatly depends on the version of the protocol we study. The generated LTS of the protocol presented previously was composed of 487446 states and 2944856 transitions. But the corrected version that will be used in section 5.2 raises to 973684 states and 7578109 transitions. All the computations were performed on a Sun Ultra-2 workstation running Solaris 2.5.1 with 2 CPUs and 832 Mb of RAM. The CPU time required for the generation went up to six hours.

The second step in the process consists of using the Aldebaran tool to minimize the resulting graph. The minimization is always done modulo the strong bisimulation equivalence that preserves all the properties of the graph. This phase is generally carried out in less than fifteen minutes of CPU time. The reduction factor obtained is very important. The minimized LTS of the first protocol is made of 3968 states and 37161 transitions. This clearly shows that our biggest problem is the generation of the brute LTS with the Caesar tool.

As we will see in the next section, all the properties we want to verify are safety properties. Thus the minimization could have been improved modulo the safety equivalence which preserves all the properties expressible in Branching time Safety Logic (Bouajjani *et al.* 1991). This was not mandatory because the graphs were already small enough to make the verification.

4 SAFETY PROPERTIES TO BE VERIFIED

Our goal is to verify that the user always correctly authenticates the TTP, that the TTP always correctly authenticates the user and that the TTP receives the right user's public key. We are going to reach it with the combination of the following safety properties.

- P1: When the TTP successfully registers the user, the user must have started a registration with the TTP before.
- P2: When the TTP successfully registers the user, it must have started a registration with this user before.
- P3: When the TTP refuses to register the user, it must have started a registration with this user before. This refusal is what we called a failure.
- P4: The verdict given by the TTP (i.e. registered or failed) must always be correct and consistent with the acknowledgement received by the user. This property will be further explained below.
- P5: The TTP always registers the user with its real public key.

Each of these properties can be expressed with the special events managed by the environment. For instance, property P1 is translated to "All TTP_REG_SUCCEEDED with a particular user identifier must be preceded by a USER_START_REG with the same user identifier". This kind of condition can be easily written in the language of our verification tools as a reference graph composed of 3 states and 3 transitions.

If we consider the user whose identity is USERID_A and whose public key is USERPKEY_A, the graph is as follows:

```
des(0,3,3)
(0, "SYSTEM_STATE !USER_START_REG !USERID_A",1)
(1, "SYSTEM_STATE !TTP_REG_SUCCEEDED !USERID_A !USERPKEY_A", 2)
(2, "SYSTEM_STATE !TTP_REG_SUCCEEDED !USERID_A !USERPKEY_A", 2)
```

This is a small graph that requires a USER_START_REG event before any TTP_REG_SUCCEEDED event. Property P1 will be verified if the LTS of our system where events other than these two have been turned into internal events is related to this LTS by the safety preorder (Bouajjani *et al.* 1991). Informally,

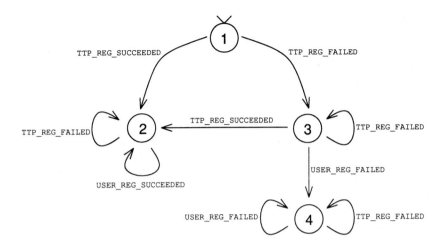

Figure 3 Labelled transition system modelling property P4

the LTS of a system is related to the LTS of a safety property by the safety preorder if and only if the behaviour of the system is allowed by the property. The comparison of two graphs modulo a particular relation is performed by the Aldebaran tool.

Property P4 can be best expressed by the graph shown on figure 3. It shows the temporal orderings that we authorize among the TTP_REG_SUCCEEDED, TTP_REG_FAILED, USER_REG_SUCCEEDED and USER_REG_FAILED events. In particular, a USER_REG_SUCCEEDED must always be preceded by one TTP_REG_SUCCEEDED because, when the user learns that he has successfully registered, the TTP must have successfully registered him. A USER_REG_FAILED must always be preceded by at least one TTP_REG_FAILED and no TTP_REG_SUCCEEDED because, when the user learns that his registration failed, the TTP must have refused to register him at least once and the TTP must not have registered that user successfully. A USER_REG_FAILED must never follow a TTP_REG_SUCCEEDED.

Properties P1 and P4 achieve the mutual authentication of the user and the TTP. The authentication of the user by the TTP is considered successful only if the TTP registers the user when the user wants to be registered. Thus we need to be sure that the user has started a registration with the TTP when the TTP registers the user. This is provided by property P1. We also need to be sure that the intruder is unable to perform a new registration of the user. Hence, property P4 allows only one successful registration. The authentication of the TTP by the user is considered successful if the user receives the right response from the TTP. This is guaranteed by property P4.

Properties P2 and P3 ensure that the TTP has really started a registration with the user when he gives a verdict. We need this check because the TTP can manage several registrations simultaneously. Finally, property P5 ensures that the user is always registered with its own public key (and not e.g. the intruder's one). To do so, the TTP_REG_SUCCEEDED event has two parameters: the user's identity and its public key. We must verify that these two fields always match for every TTP_REG_SUCCEEDED event in the LTS of our system.

5   VERIFICATION OF THE PROTOCOL

This section is the core of our study. We will show how the registration protocol can be certified using the Eucalyptus toolbox.

5.1   A flaw

When checking our properties, Aldebaran discovered that property P4 was not satisfied. We use the Exhibitor tool of the CADP package to produce a diagnostic sequence of 19 steps that exhibits one scenario that leads to the undesirable state. This sequence of transitions comprises an event USER_REG_FAILED before an event TTP_REG_SUCCEEDED. Thus the TTP successfully registers the user after the user has learned that his registration failed. This clearly does not fulfil property P4.

The diagnostic sequence is the following:

```
<initial state>
 1: "AUTH !USERID_A"
 2: "SYSTEM_STATE !USER_START_REG !USERID_A"
 3: "USNDTTP !USERID_A !USERPKEY_A !E (TTPPKEY, NONCE_A)"
 4: "TTPRCVU !USERID_A !USERPKEY_A !E (TTPPKEY, NONCE_A)"
 5: "SYSTEM_STATE !TTP_START_REG !USERID_A"
 6: "TTPSNDU !RANDOM1_TTP !S (TTPSKEY, NONCE_A)"
 7: "TTPRCVU !USERID_A !S (CERT_I, RANDOM1_TTP)"
 8: "SYSTEM_STATE !TTP_REG_FAILED !USERID_A !USERPKEY_A"
 9: "TTPSNDU !S (TTPSKEY, NO, NONCE_A, USERID_A)"
10: "TTPRCVU !USERID_A !USERPKEY_A !E (TTPPKEY, NONCE_A)"
11: "SYSTEM_STATE !TTP_START_REG !USERID_A"
12: "TTPSNDU !RANDOM2_TTP !S (TTPSKEY, NONCE_A)"
13: "URCVTTP !RANDOM2_TTP !S (TTPSKEY, NONCE_A)"
14: "USNDTTP !USERID_A !S (CERT_A, RANDOM2_TTP)"
15: "URCVTTP !S (TTPSKEY, NO, NONCE_A, USERID_A)"
16: "SYSTEM_STATE !USER_REG_FAILED !USERID_A"
17: "TTPRCVU !USERID_A !S (CERT_A, RANDOM2_TTP)"
18: "SYSTEM_STATE !TTP_REG_SUCCEEDED !USERID_A !USERPKEY_A"
<goal state>
19: "TTPSNDU !S (TTPSKEY, YES, NONCE_A, USERID_A)"
```

At line 1, the environment asks for a registration of user A. The user's decoder receives the order and begins the registration with a USER_START_REG event. It sends a register request message to the TTP at step 3 (see section 2.5).

$User \rightarrow Intruder : Register\ Request\ < A, K_A^P, \{N_A\}K_{TTP}^P >$

The intruder intercepts the message and replays it without alteration to the TTP at line 4.

$Intruder \rightarrow TTP : Register\ Request\ < A, K_A^P, \{N_A\}K_{TTP}^P >$

When the TTP receives this message, he starts the registration and sends back a message 2 with a random number R_1 at step 6.

$TTP \rightarrow Intruder : Register\ Challenge\ < R_1, \{N_A\}K_{TTP}^S >$

The intruder learns the random number required by the GQ verification when he receives this message. He immediately generates a fake response: that is line 7.

Intruder → *TTP* : *Register Response* $< F(B_I, R_1) >$

Obviously, the GQ verification fails because the intruder does not own the user's credentials. The TTP declares a failed authentication and sends a negative response.

TTP → *Intruder* : *Register Ack* $< \{No, A, N_A\}K_{TTP}^S >$

At this point, the TTP knows that he has refused the user A's registration but this user is still waiting for a response to his registration request. The intruder goes on with the attack by replaying the register request at line 10. The TTP starts a second registration of the user A and sends back a new challenge with a random number R_2 different from the previous one. The intruder still intercepts the message but this time he forwards it to the user (steps 12 and 13).

Intruder → *TTP* : *Register Request* $< A, K_A^P, \{N_A\}K_{TTP}^P >$
TTP → *Intruder* : *Register Challenge* $< R_2, \{N_A\}K_{TTP}^S >$
Intruder → *User* : *Register Challenge* $< R_2, \{N_A\}K_{TTP}^S >$

The user receives the so long awaited response and answers to it.

User → *Intruder* : *Register Response* $< F(B_A, R_2) >$

The intruder immediately replies by replaying the previous negative register acknowledgement message recorded at stage 9.

Intruder → *User* : *Register Ack* $< \{No, A, N_A\}K_{TTP}^S >$

This acknowledgement is considered valid by the user though it does not belong to the right registration. The user closes by declaring a failed registration with the event USER_REG_FAILED at step 16. Meanwhile, the intruder forwards the user's response to the TTP.

User → *Intruder* : *Register Response* $< F(B_A, R_2) >$

This response is valid, so the TTP successfully registers the user and sends a positive response.

TTP → *Intruder* : *Register Ack* $< \{Yes, A, N_A\}K_{TTP}^S >$

Both the user and the TTP have finished their exchange but they have not the same view of the registration.

For this attack to succeed, the intruder does not even need valid credentials. It only needs to create a fake response to the first registration to obtain a negative acknowledgement from the TTP. When he owns it, he replays the user's request and inserts the negative response in the exchange at the right place. Hopefully, this attack does not allow the intruder to authenticate himself as the user. So the TTP still authenticates correctly the user. But the authentication of the TTP by the user failed. The intruder can obtain a denial of service by performing this attack systematically.

The strength of our technique is that the analysis of the sequence immediately brings us the reason of the failure. The acknowledgement of the TTP is too general because it can be considered valid in two distinct registrations.

5.2   A corrected version

A way to prevent the attack is to add to the acknowledgement a unique identifier of the registration. The random number used in the GQ verification is the right candidate. This number is meant to be different at each registration. Its integration into the signature of the fourth message will allow the user to check its freshness. Here is the corrected version of our registration protocol:

$1 : User \rightarrow TTP : Register\ Request\ < UserID, K_U^P, \{n\}K_{TTP}^P >$
$2 : TTP \rightarrow User : Register\ Challenge\ < d, \{n\}K_{TTP}^S >$
$3 : User \rightarrow TTP : Register\ Response\ < F(B, d) >$
$4^+ : TTP \rightarrow User : Register\ Ack\ < \{Yes, UserID, n, d\}K_{TTP}^S >$
$4^- : TTP \rightarrow User : Register\ Ack\ < \{No, UserID, n, d\}K_{TTP}^S >$

Aldebaran states that all our properties are fulfilled with this version. Hence, the mutual authentication and the transmission of the public key succeed despite the attempts of the intruder. We conclude that this is a secure registration protocol provided that the cryptographic computations cannot be broken.

5.3   The simplest protocol

Section 5.2 demonstrates that the signature of the registration acknowledgement message is very important. It can certainly not be removed as it performs the authentication of the whole registration. We have found that the addition of the random number d in the signature of the fourth message makes the nonce n useless. It was used at first for the user to authenticate the TTP but the TTP's signature of the acknowledgement is sufficient to perform this authentication. The authentication of d with a signature in the registration challenge message is not anymore mandatory. These two simplifications lead to a very simple protocol with only one signature:

$1 : User \rightarrow TTP : Register\ Request\ < UserID, K_U^P >$
$2 : TTP \rightarrow User : Register\ Challenge\ < d >$
$3 : User \rightarrow TTP : Register\ Response\ < F(B, d) >$
$4^+ : TTP \rightarrow User : Register\ Ack\ < \{Yes, UserID, d\}K_{TTP}^S >$
$4^- : TTP \rightarrow User : Register\ Ack\ < \{No, UserID, d\}K_{TTP}^S >$

All the five properties are satisfied. This version is as robust as the previous one from the point of view of the mutual authentication. Obviously, the intruder can more easily disturb the registration. The only difference is that the intruder's actions will be discovered later in the protocol. Formally, there exists a safety preorder between the corrected version of the protocol and this simplified version regarding the six special events only. Hence the former satisfies all safety properties verified by the latter.

6   CONCLUSION

This paper presents a formal description of a security protocol. We have chosen a protocol that achieves the registration of a user to a trusted third party. We have shown how complex cryptographic operations can be abstracted away from mathematical details and specified by abstract data types. Our model of the Guillou-Quisquater algorithm is particularly simple while still capturing the essence of it.

We have shown how intrusions can be taken into account by adding an intruder process. Our model of this intruder is very simple and powerful. He can mimic very easily all reasonable real-world attacks, that is all non cryptographic and non repetitive attacks.

We have shown how to model the security properties, and in particular authentication properties as simple safety properties that can be checked automatically. The verification is based on the safety preorder which should hold between the system and the property.

Finally, we have shown on a concrete protocol how helpful formal description techniques and model-checkers can be to design security protocols. Many subtle attacks were indeed found (such as those provided in this paper) during the design that could probably not have been discovered, at least so early, by a human-being.

The computer aided design aspect of this work has been pushed further in Germeau *et al.* (1997) where we have made an improvement of the protocol. We show how to give the entities the ability to know exactly why a registration does not complete. We want to make a distinction between registration failures due to intruder's actions or due to a genuine user with bad credentials. A new version of the protocol have been designed with the verification tools to meet this additional requirement.

The results of the verification are obviously based on our set of safety properties and on some assumptions on our model. In particular, we do not prove formally the correctness of our abstract finite model with respect a more realistic model composed of more users and more TTPs. To strengthen our verification, it would be interesting to add such a proof, as in Lowe (1996), but our case-study is more complex. Another possible approach, proposed recently in Bolignano (1997), is based on an abstraction function and automates the computation of a correct abstract model. Finally, we do not prove any sort of completeness of our set of safety properties. Methods to automate the definition of security properties would be desirable. Some work in this direction is proposed in Abadi *et al.* (1997).

REFERENCES

Abadi, M. and Gordon, A.D. (1997) *A Calculus for Cryptographic Protocols The Spi Calculus*, Proceedings of the 4th ACM Conference on Computer and Communications Security.

Bolignano, D. (1997) *Towards a Mechanization of Cryptographic Protocol Verification*, Proceedings of CAV 97, LNCS 1254, Springer-Verlag.

Bolognesi, T. and Brinksma E. (1987) *Introduction to the ISO Specification Language LOTOS*, Computer Networks and ISDN Systems 14.

Bouajjani, A. Fernandez, J.C. Graf, S. Rodriguez, C. and Sifakis, J. (1991) *Safety for Branching Time Semantics*, 18th ICALP, Springer-Verlag.

Fernandez, J.C. Garavel, H. Kerbat, A. Mateescu, R. Mounier, L. and Sighireanu, M. (1996) *CAESAR/ALDEBARAN Development Package: A Protocol Validation and Verification Toolbox,* Proceedings of the 8th Conference on Computer-Aided Verification, Alur & Henzinger Eds.

Garabel, H. (1996) *An overview of the Eucalyptus Toolbox,* Proceedings of COST247 workshop.

Germeau, F. and Leduc, G. (1997) *Model-based Design and Verification of Security Protocols using LOTOS,* Proceedings of the DIMACS Workshop on Design and Formal Verification of Security Protocols.

Guillou, L. and Quiquater, J.J. (1988) *A Practical Zero-knowledge Protocol Fitted to Security Microprocessor Minimizing both Transmission and Memory,* Proceedings of Eurocrypt 88, Springer-Verlag.

Guimaraes, J. Boucqueau, J.M. Macq, B. (1996) *OKAPI: a Kernel for Access Control to Multimedia Services based on Trusted Third Parties,* Proceedings of ECMAST 96, pp. 783-798.

ISO (1989) *LOTOS, a Formal Description Technique Based on the Temporal Ordering of Observational Behaviour,* Information Processing Systems - Open Systems Interconnection: IS 8807.

ITU-T (1993) *The Directory : Authentication Framework,* Information Technology - Open Systems Interconnection: ITU-T Recommendation X.509.

Lacroix, S. Boucqueau, J.M. Quisquater, J.J. and Macq, B. (1996) *Providing Equitable Conditional Access by Use of Trusted Third Parties,* Proceedings of ECMAST 96, pp. 763-782.

Leduc, G. Bonaventure, O. Koerner, E. Léonard, L. Pecheur, C. and Zanetti, D. (1996) *Specification and Verification of a TTP Protocol for the Conditional Access to Services.,* Proceedings of 12th J. Cartier Workshop on Formal Methods and their Applications: Telecommunications, VLSI and Real-Time Computerized Control System, Canada.

Lowe, G. (1996) *Breaking and Fixing the Needham-Schroeder Public-Key Authentication Protocol using FDR,* T. Margaria and B. Steffen Eds., Tools and Algorithms for the Construction and Analysis of Systems, LNCS 1055, Springer-Verlag.

Pecheur, C. (1996) *Improving the Specification of Data Types in LOTOS,* Doctoral dissertation, University of Liège.

Schneier, B. (1996) *Applied Cryptography,* Second Edition, J. Wiley & Sons.

7   BIOGRAPHY

François Germeau has joined the Research Unit in Networking in 1996 and is studying the conception and verification of security protocols with formal description techniques.

Guy Leduc is professor at the University of Liège, his main research field is on formal languages and methods applicable to the software engineering of computer networks and distributed systems.

This work has been partially supported by the Commission of the European Union (DG XIII) under the ACTS AC051 project OKAPI: "Open Kernel for Access to Protected Interoperable Interactive Services".

10

Implementation of Distributed Systems described with LOTOS Multi-rendezvous on Bus Topology Networks

Keiichi Yasumoto[†], Kazuhiro Gotoh[††], Hiroki Tatsumoto[††], Teruo Higashino[††] and Kenichi Taniguchi[††]*

[†] *Dept. of Information Processing and Management, Shiga Univ., 1-1-1 Banba, Hikone 522, JAPAN, E-mail: yasumoto@biwako.shiga-u.ac.jp*
[††] *Dept. of Information and Computer Sciences, Osaka Univ., Toyonaka, Osaka 560, JAPAN, +81-6-850-6607 (Tel.), -6609 (Fax.) E-mail: {tatumoto,higashino,taniguchi}@ics.es.osaka-u.ac.jp*
[*] *Currently SONY Ltd., E-mail: gotoh@sm.sony.co.jp*

Abstract

In this paper, we propose an implementation method for specifications of distributed systems described in a subclass of LOTOS where operators such as choice and disabling can be used in combination with multi-rendezvous among remote processes. A LOTOS specification with the assignment of each process to a node is implemented as a set of executable codes which run on the corresponding nodes cooperating with each other by exchanging messages. The processes assigned to a node are transformed into a multi-threaded C code by our existing LOTOS compiler. Here, we focus on bus topology networks where broadcasted messages are received at all nodes in the same order, and propose a technique to implement multi-rendezvous among remote processes located on different nodes. We have also extended our LOTOS compiler to generate a set of C codes running on Ethernet. Some experimental results show that typical distributed systems can be described and implemented efficiently.

Keywords

LOTOS, multi-rendezvous, implementation, compiler, distributed systems

1 INTRODUCTION

The formal specification language LOTOS (ISO 1989) has advanced communication primitives such as multi-rendezvous as well as choice and interruption among concurrent processes. With those primitives, we can easily describe specifications of distributed systems such as mutual exclusion systems con-

Formal Description Techniques and Protocol Specification, Testing and Verification
T. Mizuno, N. Shiratori, T. Higashino & A. Togashi (Eds.) © 1997 IFIP. Published by Chapman & Hall

sisting of cooperating multiple remote workstations. For the rapid prototyping and the performance evaluation, it is desirable to describe the distributed systems in abstract way using FDT, and to automatically generate efficient executable codes (object codes). For these purposes, it is desirable (1) that system specifications can be simply described (it depends on the class of description languages) and (2) that efficient object codes can be generated.

In general LOTOS specifications, we can use choice, interleaving and disabling operators in combination with synchronization operators among multiple processes. In those specifications, several mutually exclusive rendezvous may become executable among several combinations of processes simultaneously (ISO 1989). This causes a number of message exchanges for making a consensus among remote processes when we implement multi-rendezvous in distributed environments.

Cheng *et al.* (1994) and Naik (1995) have proposed algorithms for a subclass where the combination of synchronizing processes at each gate is always same but each process may participate in either of several rendezvous at different gates. However, since they assume general networks where broadcasted messages may be received in the different order at each node, they require many message exchanges dependent upon n in the worst case to make a consensus among processes (where n is the number of all nodes in a distributed system). It is not efficient to implement those algorithms on bus topology networks in spite of their property that the broadcast messages are received in the same order at all nodes. On the other hand, for describing system specifications simply, we should be able to use a general multi-rendezvous where the combination of synchronizing processes may dynamically change and each process does not know its synchronization peers in advance. Sisto *et al.* (1991) proposed the algorithm to deal with a general multi-rendezvous, which allocates a behavior expression of LOTOS to the subprocesses on a binary tree. However, this technique requires extra control processes between any two processes and the complex hierarchical communication structure among the processes.

In this paper, we focus on bus topology networks, and propose an efficient implementation method of a general case of LOTOS multi-rendezvous. In a subclass of LOTOS implemented in our method, we can specify multiple concurrent processes where each process should be assigned to a node. In a behavior expression of each process, I/O events including internal events, events synchronizing with other remote processes and process instantiations to other nodes can be specified. A distributed system is described as a set of those processes and the relation among them with LOTOS operators. Such a LOTOS specification is implemented as a set of object codes which run on the corresponding nodes cooperating with each other by exchanging messages. The processes assigned to a node are implemented as a multi-threaded object code by our LOTOS compiler proposed in (Yasumoto *et al.* 1995). In this paper, we propose a technique to implement operators specified among remote processes executed at different nodes. In our technique, each node selects and executes the first enabled rendezvous independently from the contents and temporal ordering of the received messages under the assumption that broadcast messages are received in the same order at all nodes without failure.

In a general class of multi-rendezvous, the *executional dependence relation* which denotes the relation among processes specified with LOTOS operators, may dynamically change due to process instantiations and terminations. To deal with such a case, each object code needs to know the current relation among remote processes. Therefore, we implement each object code so that it keeps information about a syntax tree of operators, specified among the processes as their executional dependence relation. To keep the latest relation, when each process is invoked or terminated at a certain node, a message for invocation/termination is broadcasted with the location of the process in the current syntax tree so that all other object codes can update the information.

Although several LOTOS compilers have been developed (Dubuis 1990, Manas and Salvachia 1991, Nomura *et al.* 1990, Yasumoto *et al.* 1995), they can only generate object codes for a single node. So, we have extended our LOTOS compiler to generate a set of object codes running on Ethernet.

Some specifications of typical distributed systems such as mutual exclusion systems are provided for showing the expressiveness power of our subclass of LOTOS. In our experiments on Ethernet, it is shown that some example specifications can be implemented efficiently by our new compiler.

2 TARGET MODEL: CLASS AND ASSUMPTIONS

2.1 Class and assignment of processes to nodes

In our method, we describe a specification of a distributed system in a subclass of LOTOS with choice, interleaving, synchronization, disabling and enabling operators among multiple processes.

Let $M = \{M_1, ..., M_n\}$ be the set of all nodes in a distributed system. Let L_j be a set of local events (including internal events) not synchronizing with processes in other nodes, S_j be a set of events synchronizing with processes in other nodes, and $A_j = L_j \cup S_j$ be a set of all events executed in M_j.

Let $P = \{p_1, p_2, ..., p_m\}$ be a set of all processes in a LOTOS specification. We assign each process $p_i \in P$ to a node. Let $P(M_j) = \{p_{j,1}, \cdots, p_{j,k}\}(\subseteq P)$ be a set of processes assigned to node M_j. A behavior expression of each $p_{j,i}$ is described with events in A_j, process instantiations of P and operators. The syntax of the LOTOS specifications is restricted to the class of Table 1.

In Table 1, we can specify one of the following three types of behavior expressions for each process $p_{j,i}$: (1) *global behavior expression (GB)* consisting of process instantiations of P and operators, (2) *local behavior expression (LB)* consisting of events in A_j and local process instantiations of $P(M_j)$, and (3) *hybrid behavior expression (HB)* which is an expression composed of two types

Table 1 Class of description

HB ::= GB | LB | $HB >> HB$
GB ::= $GB[]GB$ | $GB\|\|\|GB$ | $GB\|[G]\|GB$ | $GB\|\|GB$ | $GB[> GB$
| $GB >> GB$ | $p\ (*\|node\_list\|*)$ | p'
LB ::= $stop$ | $exit$ | $Action; LB$ | $LB[]LB$ | $LB\|\|\|LB$ | $LB\|[G]\|LB$
| $LB\|\|LB$ | $LB[> LB$ | $LB >> LB$ | p'
$Action$ ::= $gate$ | $gate\ IOList$ | $gate\ IOList\ [guard]$ | i
$IOList$ ::= IO | $IO\ IOList$
IO ::= $?var : sort$ | $!exp$

(Here, G denotes a list of gates whose events must synchronize between both operands of $\|[G]\|$. i is an internal event. p' is a process instantiation for the same node, and $p(*\|node\_list\|*)$ for other nodes. Infinite behavior like $p := B \|\|\| p$ is prohibited)

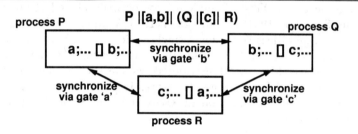

Figure 1 Mutually exclusive multi-rendezvous

of expressions GB and LB. In HB, note that different types of expressions cannot be combined with operators except '$>>$' (e.g. $a!3; exit\|\|P(*\|node_1\|*)$ cannot be described, but $Q\|\|P(*\|node_1\|*)\ where\ Q := a!3; exit$ can do).

By specifying two executional dependence relations among remote processes and among local processes separately by GB and LB, respectively, the mechanism to control remote processes with calculating the events to be executed among them can be separated from the control mechanism of local processes executed within a node.

In each process instantiation in GB, *node_list* denotes the nodes in which the specified process p may be invoked. For the purpose of load sharing, the process p can be executed in any node listed in *node_list*.

2.2 Problems to implement multi-rendezvous and assumptions for implementation

Even if only synchronization operators are specified among processes, several mutually exclusive rendezvous may be executable simultaneously when each process includes choices of different events (see Fig.1).

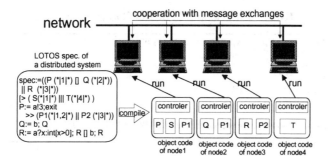

network ────── cooperation with message exchanges

LOTOS spec. of
a distributed system

spec:=((P (*|1|*) [] Q (*|2|*))
|| R (*|3|*))
[> (S(*|1|*) ||| T(*|4|*))
P:= a!3;exit
>> (P1(*|1,2|*) || P2 (*|3|*))
Q:= b; Q
R:= a?x:int[x>0]; R [] b; R

compile

controler | controler | controler | controler

P S P1 | Q P1 | R P2 | T

object code | object code | object code | object code
of node1 | of node2 | of node3 | of node4

run run run run

Figure 2 Basic policy to implement a distributed system

In order to implement the multi-rendezvous mechanism correctly, remote processes on different nodes need to exchange messages with each other for making a consensus to select a rendezvous. If we implement the mechanism on a general network where the broadcasted messages may be received in the different order at each node, we cannot take the strategy of selecting first enabled rendezvous from the temporal order of received messages.

So, on general networks, each process must decide a rendezvous after receiving the messages from all its synchronization peers. It results in $O(n^2)$ (Bochmann *et al.* 1989), $O(n\sqrt{n})$ (Cheng *et al.* 1994) or $O(n \log n)$ (Naik 1995) of message exchanges for each selection of a rendezvous where n is the number of nodes in the distributed system. Fig.1 is among the worst cases where any two exclusive rendezvous share one process.

To avoid the above problems and to implement a general case of multi-rendezvous efficiently, we focus on bus topology networks where the broadcasted messages are received in the same order at every node.

Assumptions for the target network

We assume the following properties for bus topology networks.

(1) each message is received at all nodes in a distributed system by a broadcast
(2) all messages are received at each node without failure (i.e without losses and duplications) and stored in its message buffer
(3) broadcasted messages are received in the same order at all nodes

3 ALGORITHM

3.1 Basic policy of implementation

In our implementation method, we generate a set of object codes for the corresponding nodes in a distributed system from a given LOTOS specification as described in the class of Table 1. The generated object codes implement the given LOTOS specification, cooperating with each other (Fig.2).

Table 2 An example of LOTOS specification

---

spec.:= (P (*|$node_1$|*) [] Q (*|$node_2$|*) || R (*|$node_3$|*))
    [> (S (*|$node_1$|*) ||| T (*|$node_4$|*))
where P:= (a!3;exit) >> (P1 (*|$node_1, node_2$|*) || P2 (*|$node_3$|*))
    Q:= b;Q
    R:= a?x:int[x>0]; R [] b;R
    S:= (S1 || S2) [] S3
    T:= d;stop
(Here, the declarations of processes P1, P2, S1, S2 and S3 are omitted)

For example, four object codes for $node_1$, $node_2$, $node_3$ and $node_4$ are generated from the LOTOS specification in Table 2, and those object codes can execute the processes {P,S,P1}, {Q,P1}, {R,P2} and {T}, respectively. Since P1 may be executed on either $node_1$ or $node_2$, both nodes have the code for executing P1. First, those object codes execute the processes in the initial behavior expression. In the example of Table 2, when the object code of $node_1$ is executed, processes P and S are invoked.

The initial executional dependence relation among processes corresponds to the area within the dotted rectangle in Fig.3(i). In the relation, two exclusive rendezvous a!3 and b can be executed with the combinations (P, R) and (Q, R), respectively. Several disabling events may be also executed in S and T.

The executional dependence relation among processes changes dynamically when new processes are invoked and/or processes are terminated. In the example of Table 2, when a rendezvous a!3 of (P, R) is executed, process Q must be terminated and new processes P1 and P2 will be invoked. After that, the dependence changes from (i) to (ii) in Fig.3.

Therefore, in the proposed technique, we make each node hold the newest information about the executional dependence relation among all running processes of all nodes so that the temporal ordering of events among the processes are implemented based on the relation. In each object code, we allocate a working space called *control area* which has the same structure as the syntax tree of the operators specified among processes in order to memorize the initial executional dependence relation. When each process is invoked or terminated, the process broadcasts the information to all nodes so that the control area can be updated at each node to keep the newest relation.

Each process calculates its *synchronization events* which need to synchronize with other processes, independently based on the technique in (Yasumoto *et al.* 1995) and broadcasts a request message for the events. Then the process selects and executes a rendezvous which has satisfied the synchronization conditions* earlier than other rendezvous from the current executional dependence relation and the temporal order of the received messages in its buffer.

---

*In each rendezvous, I/O values of all of its participants must be unified (ISO 1989).

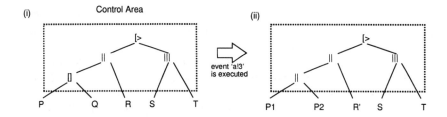

Figure 3 Executional dependence relation and its dynamic change

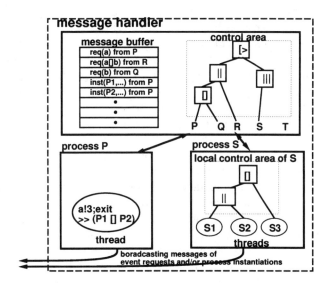

Figure 4 Environment of each node

3.2 Management of control among processes

We assume that the broadcasted messages from all nodes are received and stored in the message buffer of each node automatically and that each node has a sufficient buffer. Each process calculates a set of its synchronization events $\{a_1, \cdots, a_k\}$, and broadcasts a request message for the set (denoted by $req(\{a_1, \cdots, a_k\})$). In an object code of each node, we implement an additional process called *message handler* to select a rendezvous from the current executional dependence relation and both the contents and temporal ordering of the messages stored in the buffer. Each process executes the selected event after the message handler permits its execution. Fig.4 shows the environment of $node_1$ in the example of Table 2.

Here, it is important (1) that choice and disabling operators among processes should be implemented correctly, (2) that one of the exclusive rendezvous should be selected under a consensus of all nodes, and (3) that the dynamic change of the executional dependence relation should be managed correctly among all nodes.

The above criterion (1) can be implemented fairly with the following policy: if a process p wants to execute its events, p broadcasts a request message for those events with the process name 'p'. When the message handler of each node processes the message, it examines the path (called *control path*) from p's position (leaf) to the root node in the control area and checks whether different sides of behavior expressions have been selected and/or an interruption has occurred. Only when the handler detects that every different side of expressions has not been selected at each node, it accepts the message, and updates the control path so that p's side has been selected. After that, p executes one of those events at random. In this policy, the process which broadcasted a request message earlier than others is selected and executed.

In the example of Table 2, suppose that the request messages of events $req(\{d\})$ and $req(\{a!3\})$ from processes T and P are broadcasted in this order. The message handler of each node stores the information that interruption occurred in the sub-area corresponding to [> operator when it processes $req(\{d\})$. When the later message $req(\{a!3\})$ is processed, the message handler detects the occurrence of interruption on the control path of P. Then, $req(\{a!3\})$ is rejected and process P is terminated.

The criterion (2) is implemented as follows: when each message handler processes a request message m for synchronization events, the handler stores the information about the synchronization events in the area corresponding to $\|[G]\|$ operator which is the nearest to the leaf in the control path of m's sender. If the handler finds appropriate synchronization peers at the area, it applies the same operation to the upper areas of $\|[G]\|$ operators with new event values after matching synchronization conditions. If there is no upper $\|[G]\|$ in the control path, then the handler permits the events to be executed.

We implement the criterion (3) as follows: when a process instantiation becomes executable in a process at a node, the message handler of the node calculates the control path of the process instantiation and broadcasts a request message for the process instantiation with its control path. When the message handler at each node processes the message, it extends its control area with the control path in the message and it also invokes the process if it is requested to be invoked at the node. Once a process is selected to execute an event, the subareas corresponding to choice and disabling operators in the control area may become unnecessary. Such unnecessary areas are removed when the messages of all processes that were not selected have been processed.

We will explain more about the criteria (2) and (3). Suppose that the request messages $req(\{a!3\})$, $req(\{a?x : int[x > 0], b\})$ and $req(\{b\})$ are broadcasted from P, R and Q in this order (see Fig.4). When each message handler processes $req(\{a!3\})$ from P, it registers the information $a!3$ to the area corresponding to $\|$ on the control path of P. When the next message

$req(\{a?x : int[x > 0], b\})$ from R is processed, a rendezvous between P and R satisfies the synchronization conditions with $a!3$ and both P and R are not disabled on their control paths. So $a!3$ is selected and executed between P and R after updating the control paths of P and R (the side of P is stored to the area corresponding to []). After that, when $req(\{b\})$ from Q is processed, Q detects that the different side has been selected from the information stored at the area of [] on its control path. So, Q is terminated at $node_2$. On the other hand, if the request messages from P, Q and R are broadcasted in this order, R's request message $req(\{a?x : int[x > 0], b\})$ enables two exclusive rendezvous $a!3$ and b simultaneously. To cope with this problem, we adopt the technique based on the random number introduced in (Cheng *et al.* 1994, Naik 1995). Here, we attach a random integer to each event in the request messages for calculating the maximum value of each rendezvous so that all nodes can uniquely select the rendezvous with the largest number.

When $a!3$ is executed in P, the behavior expression of P changes to P1 || P2 and the request messages for process instantiations $inst(P1)$ and $inst(P2)$ are broadcasted. Each message includes the process name, node names and the control path. If several nodes are specified like $P1(*|node_1, node_2|*)$, the process is invoked at the node whose load is the lowest. The information about the load of each node is attached to each request message so that all nodes can keep the information to find the lowest node.

Treatment of local processing in each process

In the class of Table 1, since each process may include multiple concurrent sub-processes executed in its node. It may cause the messages for several exclusive events to be broadcasted separately. For example, in process P:= (a;exit [] b;exit) || (...;a;... ||| ...;b;...), two messages for exclusive events a and b may be broadcasted at the different time. To deal with such a case correctly, we use a message for replacing an issued request message to a new one. If we need to broadcast a request message of alternative event b after broadcasting the request message $req(\{a\})$, we can broadcast $replace(req(\{a\}), req(\{a, b\}))$ to change the issued request message. Since a message handler processes messages sequentially according to their temporal ordering, the replaced message may become obsolete when a rendezvous has been selected before the message is processed. In that case, the message will be ignored.

3.3   Correctness and efficiency of the algorithm

Here, we briefly explain the correctness and the efficiency of our algorithm.

Correctness

The executable events among processes depend on the current executional dependence relation among the processes and *active events* which want to be executed in each process. If rendezvous, choice and/or disabling operators are specified among processes, the executional dependence relation may change after an event (or a rendezvous) has been executed. So, if all nodes can keep the latest information about the relation, and if each node can calculate the same event (rendezvous) of a process (a combination of processes) from the relation and active events in each process, the algorithm will work correctly.

In the algorithm, if choice and/or disabling operators are specified among processes, the process which has sent the earliest request message among the processes is selected. For synchronization operators, the combination of the processes whose messages have been sent earlier than others is selected. The algorithm assumes that the order and the contents of the received messages are the same among all nodes as explained in Sec.2.2 and all nodes work based on the same algorithm. This fact shows that each node is capable of knowing which processes are selected/invoked, and therefore of keeping the current executional dependence relation at any time, with sequentially processing messages in its local buffer. Therefore, the algorithm works correctly.

Here, we mention a bit about the fairness of our algorithm. If the performances of nodes differ greatly, some rendezvous may be always selected among the nodes with good performances and other alternative rendezvous containing 'slow' nodes may not be selected infinitely. It is because the algorithm always selects the first enabled rendezvous from the temporal ordering of received messages. In that case, the fairness in the nondeterministic selection may not be preserved in the algorithm. However, if a message makes several exclusive rendezvous executable at the same time (see Sec.3.2), each of them can be selected based on random numbers.

Efficiency compared to other algorithms

Since each process broadcasts a request message for its active events at each stage, at most n broadcast messages are sent as a total with our algorithm (here, n is the number of all processes in the distributed system). Therefore, if our algorithm is implemented with the unicast mechanism, its message complexity will be $O(n^2)$, which may seem to be worse than other algorithms.

In the previously proposed algorithms (Bochmann *et al.* 1989, Cheng *et al.* 1994, Naik 1995), the temporal ordering of the received messages may be different because they assume general networks. This fact forces all processes to make a consensus for a rendezvous after receiving all messages from related processes if several exclusive rendezvous may exist. On the other hand, with our algorithm we can select the first enabled rendezvous owing to the property of bus topology networks. Accordingly, for selecting a rendezvous consisting of n_k participants (usually n_k is much less than n), our algorithm requires only n_k broadcast messages even in the worst case as shown in Fig.1. Although we have to consider there may be the messages broadcasted for other unselected rendezvous, most of them can be used for calculating the subsequent rendezvous. So, the average number of the required messages for a rendezvous approximately results in n_k. In addition, since the cost of a broadcast is almost same as that of a unicast on bus topology networks, the proposed algorithm may work much faster than the previous ones which assume general networks.

4 APPLICATION

This section shows some typical distributed systems which can be described simply within the class defined in Table 1.

Example 1. Duplication system

Let us suppose that a service is duplicated for high reliability. In the case that each service process fails rarely, it is popular to make several processes carry out the same task and to select their majority by getting the results from those processes.

Here, we describe a request for each task as $req?task$, and an output of the result after processing the task as $line!ProcessTask(task)$ in the following process $S[req, line]$ ($ProcessTask(task)$ is an abstract data type function which calculates the result from $task$).

$S[req, line] := req?task; line!ProcessTask(task); S[req, line]$

Next, we describe another process called *Checker* which gets the results from duplicated processes and outputs the correct result based on the majority or detects an error. *Checker* receives three results as x, y and z from three processes, respectively with multi-rendezvous. If more than two of x, y and z are the same ($Correct(x, y, z)$), *Checker* outputs their majority ($out!Major(x, y, z)$) and repeats the procedure. Otherwise, it executes the event er to stop all processes. *Checker* can be described as follows:

$Checker[ln1, ln2, ln3, out, er] :=$
$\quad (ln1?x; exit \mathbin{|||} ln2?y; exit \mathbin{|||} ln3?z; exit)$
$\quad >> ([Correct(x, y, z)]- > out!Major(x, y, z); Checker[ln1, ln2, ln3, out, er]$
$\quad [] \ [not \ Correct(x, y, z)]- > er; stop)$

We invoke three duplicated processes S in parallel on node1, 2 and 3, which send the results to *Checker* via three gates $ln1$, $ln2$ and $ln3$, respectively. The whole system can be described as follows:

$(\ (S[req, ln1](*|node_1|*) \ |[req]| \ S[req, ln2](*|node_2|*) \ |[req]| \ S[req, ln3](*|node_3|*))$
$[> Error[er](*|node_4|*) \ )$
$|[ln1, ln2, ln3, er]| \ Checker[ln1, ln2, ln3, out, er](*|node_4|*)$
$where \ Error[er] := \ er; stop$

Example 2. Client server system

In client server systems such as VOD (video on demand) systems, a series of data streams is transferred successively between client and server systems. With the class in Table 1, we can describe such a system.

If both video and audio data must be transmitted synchronously from server machines (SV and SA) to a client machine (C) and we would like to make several nodes share their load to execute a server process, such a client server system can be described as follows:

$SV(*|node_1, node_2, node_3|*) \; |[m]| \; SA(*|node_1, node_2|*) \; |[m]| \; C(*|node_4|*)$
(Here, m denotes a gate for exchanging video and audio data)

Since some nodes can execute server processes SV and SA in our model, each server process is invoked at the node with the lowest load.

Example 3. Mutual exclusion system

In designing a distributed system, frequently we treat the mutual exclusion such that a resource placed at a node can be used by at most k user processes simultaneously.

If each user process uses the resource **A** after $a!lock$ and releases **A** before $a!unlock$, the restriction for **A** can be described as follows:

$R[a](usr, max_A) :=$
$([usr < max_A]- > a!lock; R[a](usr + 1, max_A)) \; [] \; a!unlock; R[a](usr - 1, max_A)$
(max_A denotes the number of user processes which can use the resource simultaneously)

If processes $P_1, \cdots, P_n$ using resource **A** can run independently of each other except the access to **A**, the whole system can be described as follows:

$(P_1[a](*node_1|*) \; ||| \; P_2[a](*|node_2|*) \; ||| \; ... \; ||| \; P_n[a](*|node_n|*))$
$|[a]| \; R[a](0, max_A)(*|node_1|*)$

In the above specification, events $a!lock$ and $a!unlock$ in process R synchronize to either of $P_1, \cdots, P_n$ and increase and decrease the variable usr, respectively. If $usr + 1$ exceeds max_A, only $a!unlock$ can be executed. In that case, processes which want to use resource **A**, will wait until **A** is released with $a!unlock$.

We can easily extend the above system for treating several resources, say, **A**, **B** and **C**, only by adding processes for restricting those resources as follows:

$(P_1[a, b, c] \; ||| \; P_2[a, b, c] \; ||| \; ... \; ||| \; P_n[a, b, c])$
$|[a, b, c]| \; (R[a](0, max_A) \; ||| \; R[b](0, max_B) \; ||| \; R[c](0, max_C))$
(Here, the node names for processes are omitted)

5 IMPLEMENTATION AND ITS EVALUATION

5.1 LOTOS compiler and implementation on Ethernet

Each local process whose behavior expression is described as an LB in Table 1 is implemented as a multi-threaded object code by our LOTOS compiler. The compiler implements each process containing concurrent subprocesses with our portable multi-thread mechanism PTL (Abe *et al.* 1995) as follows: (1) decomposing each behavior expression into some sequentially executable

subbehavior expressions called *runtime units*, (2) mapping each runtime unit
to a thread, and (3) constructing a shared memory (called *local control area*).

To implement a wide class of behavior expressions a local control area is
used to implement the executional dependence relation among threads with a
similar technique explained in Sec.3.2. Since each area in the local control area
may be accessed by multiple concurrent threads simultaneously, the mutual
exclusion mechanism in PTL is used. The above techniques enabled us to
obtain a higher performance than other compilers in (Dubuis 1990, Manas
and Salvachia 1991). The details are in (Yasumoto *et al.* 1995).

We have extended the compiler to generate a set of object codes which run
on the corresponding nodes on Ethernet based on the algorithm in Sec.3.

Implementation of the algorithm on Ethernet

In popular LAN environments, Ethernet is widely used. Unfortunately, Ethernet does not satisfy the assumptions in Sec.2.2 partly. So, first we have
examined to what extent Ethernet satisfies the assumptions. Here, we used
UDP/IP on several UNIX workstations (BSD-OS 2.1 on Pentium 90 to Pentium Pro 200MHz) connected to fast ethernet network (100BASE-TX).

Through the experiments, we recognized (i) that a few percent of messages
are lost when the frequency of message exchanges are high (e.g. when over
3000 messages a second), and (ii) that some messages are reordered on some
nodes when several messages are broadcasted in a short time period.

The reason for (i) is that the message buffer on a slow machine overflows
due to the shortage of the processing time for messages. (ii) is caused by the
loop back mechanism of the operating system by which the messages sent
from a node is stored to its message buffer directly.

For the above problems, we can use the techniques proposed in (Melliar *et
al.* 1990, Luan and Gligor 1990) so that the assumptions hold on Ethernet.
However, for efficiency, here we adopt to place a proxy server which broadcasts
messages forwarded from all nodes. Each node does not broadcast its messages
directly but asks the server to broadcast its messages. By placing the server
at an independent node, all nodes in a distributed system can receive the
messages in the same order. By adding sequential numbers to subsequent
messages, each node can easily detect message loss and ask for retransmission
from the server. Our current implementation includes such a proxy server.

5.2　Evaluation

To evaluate our implementation method, we have examined the efficiency of
the object codes generated from several specifications of distributed systems.

In the experiments, we have measured the number of events synchronizing

Table 4 Performance of a mutual exclusion system

n=1	n=2	n=3	n=4	n=5
680	688	586	592	598

Table 3 Performance of simple rendezvous

$k \backslash n$	n=2	n=3	n=4	n=5
k=1	993	818	647	571
k=2	663	534	426	367
k=3	481	370	319	257
k=4	519	290	237	190
k=5	428	225	184	139

Table 5 Efficiency in abracadabra protocol (KBytes/sec.)

packet size	n=1	n=2	n=4
256 bytes	8.7	7.7	6.0
512 bytes	18.7	16.1	12.0
1024 bytes	35.0	31.5	23.2

among remote processes executed in a time unit (events/second). The object codes were executed on the same environment in Sec.5.1.

Basic performance

First, we have executed object codes generated from the following specification where each event synchronizes among n remote processes repeatedly. The experimental result is shown in Table 3.

$P(*|node_1|*)||P(*|node_2|*)|| \cdots ||P(*|node_n|*)$ where $P := a_1; P[]a_2; P[] \cdots []a_k; P$

According to Table 3, the performance depends on both the numbers of processes and alternative events (n and k). The latter is due to the cost of selecting a rendezvous of several exclusive ones.

Performance of more complicated distributed systems

We have examined the efficiency for more complicated distributed systems.

First, we have measured the performance in a mutual exclusion system where n remote concurrent processes use a resource exclusively from each other as explained in Sec.4. We show in Table 4 the experimental results with several numbers for n. Table 4 shows that the performance does not depend on the number of user's processes. This is owing to the property of our algorithm that the request messages for the unselected rendezvous are used for selecting the subsequent rendezvous unless their sender processes are disabled.

Finally, we have examined the efficiency of a file transfer system based on a specification of abracadabra protocol (ISO/IEC/TR 10167 1991). The specification describes the behavior of a node where a file is transferred (received) to (from) other node depending on the request of the service user on the node. In the protocol, the file is divided into multiple data packets and the packets are transmitted via unreliable medium. The specification includes several concurrent subprocesses synchronizing with each other. Here, we extended the specification so that a sender node (*Sndr*) distributes a file (the size is

1 Mbytes) to n receiver nodes (*Recv*) synchronizing with a process *Medium* which emulates the communication medium.

$Medium[m_1, m_2](*|node_1|*)$ $|[m_1, m_2]|$
$(Sndr[a_1, m_1](*|node_2|*)$ $|||$ $Recv[a_2, m_2](*|node_3|*))...$ $|||$ $Recv[a_2, m_2](*|node_{n+2}|*))$

In Table 5, we show the result with the size of a data packet from 256 to 1024 bytes and n from 1 to 4. In Table 5, if the packet size is small (256 bytes), the transmission rate is lowering loosely as the number of recipients is growing. In contrast, the rate is lowering drastically if the size is large (1024 bytes). It may be caused by the overhead of local processing for messages.

6 CONCLUDING REMARKS

In this paper, we have proposed a technique to implement a specification with 'general' multi-rendezvous among multiple remote processes as a set of object codes which can be executed at the multiple remote nodes on bus topology networks. In our implementation method, a distributed system is described as a set of processes representing the individual behaviors of nodes. It might be useful to integrate the technique in (Kant *et al.* 1996) which automatically derives protocol entities' specifications from a service specification (i.e. the whole behavior of a distributed system).

Although there is an opinion that the efficient implementations are not necessary for specification languages like LOTOS, we believe that the efficient implementations are important for rapid prototyping and other purposes. An application is the visualization of the dynamic behavior of distributed systems. With the visualization technique in (Yasumoto, Higashino *et al.* 1995), we can monitor the dynamic behavior of remote processes in a distributed system as the graphical animation in realtime. It may be useful for, say, the education of distributed systems in LOTOS. One of future work is to extend the compiler to deal with a time extended LOTOS (e.g. E-LOTOS).

REFERENCES

Abe, K., Matsuura, T. and Taniguchi, K.: "An Implementation of Portable Lightweight Process Mechanism under BSD UNIX", *J. of Information Processing Society of Japan*, Vol. 36, No. 2, pp. 296 – 303 (1995) (in Japanese).

Bochmann, G. v., Gao, Q. and Wu, C.: "On the Distributed Implementation of LOTOS", *Proc. 2nd Int. Conf. on Formal Description Techniques*, pp. 133 – 146 (1989).

Cheng, Z., Huang, T., and Shiratori, N.: "A New Distributed Algorithm for Implementation of LOTOS Multi-Rendezvous", *Proc. 7th Int. Conf. on Formal Description Techniques*, pp. 493 – 504 (1994).

Dubuis, E.: "An Algorithm for Translating LOTOS Behavior Expressions into Automata and Ports", *Proc. 2nd Int. Conf. on Formal Description Techniques*, pp. 163 – 177 (1990).

ISO : "Information Processing System, Open Systems Interconnection, LOTOS - A Formal Description Technique Based on the Temporal Ordering of Observational Behaviour", *IS 8807* (1989).

ISO/IEC/TR 10167: "Information Technology – Open Systems Interconnection – Guidelines for the application of Estelle, LOTOS and SDL" (1991).

Kant, C., Higashino, T. and Bochmann, G.v.: "Deriving protocol specifications from service specifications written in LOTOS", *Distributed Computing*, Vol. 10, No.1, pp. 29 – 47 (1996).

Manas, J. A., Salvachia, J.: "Λβ: a Virtual LOTOS Machine", *Proc. 4th Int. Conf. on Formal Description Techniques*, pp.445 – 460(1991).

Melliar, P.M., Moser, L.E. and Agrawara, V.: "Broadcast Protocols for Distributed Systems", *IEEE Trans. on Parallel and Distributed Systems*, Vol. 1, No. 1, pp. 17 – 25 (1990).

Naik, K.: "Distributed Implementation of Multi-rendezvous in LOTOS Using the Orthogonal Communication Structure in Linda", *Proc. 15th Int. Conf. on Distributed Computing Systems*, pp. 518 – 525 (1995).

Luan, S. and Gligor, V.D. : "A Fault Tolerant Protocol for Atomic Broadcast", *IEEE Trans. on Parallel and Distributed Systems*, vol. 1, No. 3, pp. 271 – 285 (1990).

Nomura, S., Hasegawa, T. and Takizuka, T. : "A LOTOS Compiler and Process Synchronization Manager", *Proc. 10th Int. Symp. on Protocol Specification, Testing, and Verification*, pp. 169 – 182 (1990).

Sisto, R., Ciminiera, L. and Valenzano, A.: "A Protocol for Multi-rendezvous of LOTOS Processes", *IEEE Trans. on Computer*, Vol. 40, No. 1, pp. 437 – 447 (1991).

Yasumoto, K., Higashino, T., Abe, K., Matsuura, T. and Taniguchi, K.: "A LOTOS Compiler Generating Multi-threaded Object Codes", *Proc. 8th Int. Conf. on Formal Description Techniques*, pp. 271 – 286 (1995).

Yasumoto, K., Higashino, T., Matsuura, T. and Taniguchi, K.: "Protocol Visualization using LOTOS Multi-Rendezvous Mechanism", *Proc. IEEE 1995 Int. Conf. on Network Protocols*, pp. 118 – 125 (1995).

7 BIOGRAPHY

Keiichi Yasumoto is an Assistant Professor in the Department of Information Management and Processing, Shiga University, Japan. Since April 1997, he is working as a visiting researcher at the University of Montreal, Canada. Kazuhiro Gotoh finished his Master course at the Department of Information and Computer Sciences, Osaka University, Japan in 1997 and is currently working in SONY Ltd. Hiroki Tatsumoto is a Master student in the same department. Teruo Higashino and Kenichi Taniguchi are an Associate Professor and a Professor in the same department, respectively. Their research interests include formal description techniques for communication protocols and distributed systems, and software engineering.

11
Disjunction of LOTOS specifications

M.W.A. Steen, H. Bowman, J. Derrick and E.A. Boiten
University of Kent at Canterbury
Computing Lab., The University, Canterbury, Kent CT2 7NF, UK.
Email: mwas@ukc.ac.uk

Abstract
LOTOS is a formal specification language, designed for the precise *description* of open distributed systems and protocols. The definition of, so called, implementation relations has made it possible also to use LOTOS as a *specification technique* for the design of such systems. These LOTOS based specification techniques usually (ab)use non-determinism to achieve *implementation freedom*. Unfortunately, this is unsatisfactory when specifying non-deterministic processes. We, therefore, propose to extend LOTOS with a disjunction operator in order to achieve more implementation freedom while maintaining the possibility to describe non-deterministic processes. In contrast with similar proposals we maintain the operational semantics.

Keywords
LOTOS, process algebra, specification, disjunction, operational semantics

1 INTRODUCTION

In this paper we investigate the extension of the formal specification language LOTOS with a disjunction operator. Such a specification construct could play a role in achieving a more expressive specification technique. As in logic, disjunction can be used to specify a choice between implementation options. If p_1 is an implementation of s_1, and p_2 is an implementation of s_2, then the specification $s_1 \bigvee s_2$ can be implemented by either p_1 or p_2. Thus, disjunction in specifications leads to greater implementation freedom. This is useful both in the specification of standards, which often describe a number of implementation classes, and in the development of distributed systems, where we do not want to tie the hands of the implementors in the initial specification.

1.1 Interpreting LOTOS specifications

LOTOS is a process algebraic language influenced by the earlier process calculi CCS [Mil89] and CSP [Hoa85]. For example, it has inherited the powerful idea of multi-way synchronisation, enabling constraint-oriented specification, from CSP. On the other hand, the language has been given an operational semantics much in the style of CCS.

Formal Description Techniques and Protocol Specification, Testing and Verification
T. Mizuno, N. Shiratori, T. Higashino & A. Togashi (Eds.) © 1997 IFIP. Published by Chapman & Hall

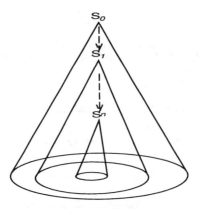

Figure 1 Design process

The operational semantics associates a labelled transition system (LTS) with each process description. The usual interpretation of a LOTOS "specification" is the set of processes that are indistinguishable (w.r.t. some notion of equivalence) from the LTS associated with it. However, this view requires that the observational behaviour of implementations is completely determined already by their initial specifications.

An alternative, more relaxed, view is, that implementations are related to specifications by a, so called, *implementation relation*[*]. Implementation relations are usually not equivalences and, therefore, allow the behaviour of implementations to be somehow more determined than the behaviour described by their specifications. Moreover, they induce a *refinement* ordering between specifications, which enables an incremental design process as depicted in figure 1. An initial abstract specification, allowing many possible implementations, goes through a series of consecutive refinement steps, each restricting the implementation space, until a final implementation is reached.

1.2   The problem

Several researchers have investigated the use of implementation relations with LOTOS to obtain a *specification technique* for concurrent processes (see section 3). Most of these approaches are inspired by CSP's failures/divergences semantics, or have been derived from testing theory. *Non-determinism* is usually (ab)used to achieve implementation freedom. We argue that this is not always satisfactory. In particular, we show that it becomes impossible to specify inherently non-deterministic processes adequately, and the wide-spread use of internal actions as an abstraction mechanism can lead to counter-intuitive implementations.

---

[*]Implementation relations are sometimes also referred to as *conformance relations* or *satisfaction relations*.

2   LOTOS: SYNTAX AND SEMANTICS

In order not to clutter the presentation of our main ideas, we will only consider a small subset of the operators that LOTOS offers for the structuring of process descriptions. The subset we use is inductively defined by the following grammar:

$$P ::= \textbf{stop} \mid a; P \mid i; P \mid P \,[]\, P \mid P \,|[G]|\, P \mid X$$

Here we assume that a set of action labels $\mathcal{L}$ is given. Then, $a \in \mathcal{L}$; i is the unobservable, or internal, action; $G \subseteq \mathcal{L}$; and X is a process name. We will assume that a definition exists for each process name used. Process definitions are written $X := P$, where P is a behaviour expression that can again contain process names, including possibly X itself, thus making the definition recursive. The set of all processes is denoted by $\mathcal{P}$, elements of $\mathcal{L}$ by $a, b, c \ldots$, and elements of $\mathcal{L} \cup \{i\}$ by μ.

The operational semantics for LOTOS associates a labelled transition system with each behaviour description through the axioms and inference rules given in table 1.

Table 1 Inference rules

	$\vdash \quad a; P \xrightarrow{a} P$
	$\vdash \quad i; P \xrightarrow{i} P$
$P \xrightarrow{\mu} P'$	$\vdash \quad P \,[]\, Q \xrightarrow{\mu} P'$
$Q \xrightarrow{\mu} Q'$	$\vdash \quad P \,[]\, Q \xrightarrow{\mu} Q'$
$P \xrightarrow{\mu} P', \mu \notin G$	$\vdash \quad P \,|[G]|\, Q \xrightarrow{\mu} P' \,|[G]|\, Q$
$Q \xrightarrow{\mu} Q', \mu \notin G$	$\vdash \quad P \,|[G]|\, Q \xrightarrow{\mu} P \,|[G]|\, Q'$
$P \xrightarrow{a} P', Q \xrightarrow{a} Q', a \in G$	$\vdash \quad P \,|[G]|\, Q \xrightarrow{a} P' \,|[G]|\, Q'$
$P \xrightarrow{\mu} P', X := P$	$\vdash \quad X \xrightarrow{\mu} P'$

The LTS for a process p is $\langle D_p, L_p, T_p, p \rangle$. Here T_p is the smallest set of transitions that can be inferred from p under the given inference rules; D_p is the set of processes derivable from p under the transitions in T_p; $L_p = \{a \mid (s, a, s') \in T_p\}$, the set of action labels.

2.1   Further notation

For the rest of the paper we need some more derived notation. Let $\mathcal{L}^*$ denote strings over $\mathcal{L}$. The constant $\epsilon \in \mathcal{L}^*$ denotes the empty string, and the variables σ, σ_i are used to range over $\mathcal{L}^*$. Elements of $\mathcal{L}^*$ are also called traces. In table 2 the notion of transition is generalised to traces. We further define $Tr(p)$, the set of traces of p, and $Ref(p, \sigma)$, the sets of actions refused by p after the trace σ:

$$Tr(p) = \{\sigma \in L^* \mid p \overset{\sigma}{\Longrightarrow} \}$$

$$Ref(p, \sigma) = \{X \subseteq L \mid \exists p' : p \overset{\sigma}{\Longrightarrow} p' \text{ and } \forall a \in X : p' \overset{a}{\nRightarrow} \}.$$

Table 2 Derived transition denotations

Notation	Meaning
$p \xrightarrow{\mu}$	$\exists p' : p \xrightarrow{\mu} p'$
$p \xrightarrow{\mu} \!\!\!\!/$	$\nexists p' : p \xrightarrow{\mu} p'$
$\overset{\epsilon}{\Longrightarrow}$	reflexive and transitive closure of $\xrightarrow{i}$
$p \overset{a\sigma}{\Longrightarrow} p'$	$\exists q, q' : p \overset{\epsilon}{\Longrightarrow} q \xrightarrow{a} q' \overset{\sigma}{\Longrightarrow} p'$
$p \overset{\sigma}{\Longrightarrow}$	$\exists p' : p \overset{\sigma}{\Longrightarrow} p'$
$p \overset{\sigma}{\Longrightarrow} \!\!\!\!/$	$\nexists p' : p \overset{\sigma}{\Longrightarrow} p'$

2.2   Equivalence

Often transition systems are considered to be too discriminating in the sense that processes that are intuitively considered to be equivalent may have different represen-
tations. The processes $a; b;$ **stop** and $a; (b;$ **stop** $[]\ b;$ **stop**$)\ []\ a; b;$ **stop**, for example, have different transition systems, but can both only perform the sequence of ac-
tions ab and then deadlock. For this reason several abstracting equivalences have been
defined over the LTS model. In this paper, we consider only the strongest of the behavioural equivalences: *strong bisimulation equivalence.* Processes are equivalent iff
they can simulate each other. This is indeed the case for the two processes above.

Definition 1 (bisimulation equivalence)
Bisimulation equivalence, $\sim \subseteq \mathcal{P} \times \mathcal{P}$, *is the largest relation such that,* $p \sim q$ *implies*
   (i)   *Whenever* $p \xrightarrow{\mu} p'$ *then, for some* q', $q \xrightarrow{\mu} q'$ *and* $p' \sim q'$; *and*
   (ii) *Whenever* $q \xrightarrow{\mu} q'$ *then, for some* p', $p \xrightarrow{\mu} p'$ *and* $p' \sim q'$.

   The choice of equivalence is fairly arbitrary. We could just as well have chosen
weak bisimulation equivalence or *testing equivalence.* We are, however, interested in
creating a specification technique that is as expressive as possible. Since bisimulation
equivalence is the strongest behavioural equivalence on processes, and by defining
satisfaction (see section 4) as an extension of it, we achieve precisely this.

3   LOTOS AS A SPECIFICATION TECHNIQUE

The "meaning" of a specification, i.e. the set of implementations that it describes, depends on the chosen *satisfaction relation.* Following [Lar90a] and [Led92], we define
a *specification technique* to be a pair $\langle \Sigma, sat \rangle$, where Σ is the set of all specifications,
and *sat* is some satisfaction relation. Using the notion of bisimulation from the pre-
vious section, we could instantiate *sat* with $\sim$. However, as argued in the introduction, this would leave very little room for manoeuvring during the implementation
phase, because the behaviour of implementations would have to be equivalent to the
behaviour of their specifications.

Several asymmetric instantiations for *sat* have been investigated for LOTOS [BSS87, Led91]. These, so called, *implementation relations* were either derived from CSP's denotational semantics [Hoa85], or from testing theory [NH84].

One of the simplest implementation relations, is the *trace preorder*. It only verifies that the implementation cannot perform sequences of observable actions (traces) that are not allowed by the specification.

Definition 2 (trace preorder) *Let* $p, s \in \mathcal{P}$. $p \leq_{tr} s$ *iff* $Tr(p) \subseteq Tr(s)$.

Example 1 *Let* $s := a; b; \text{stop}[]a; c; \text{stop}$, *then* $p_1 := a; b; \text{stop}$, $p_2 := a; c; \text{stop}$ *and* $p_3 := a; (b; \text{stop}[]c; \text{stop})$ *are all implementations of* s *according to* $\leq_{tr}$. *But, also* stop *and* $a; \text{stop}$ *are correct, since* $\leq_{tr}$ *does not* require *any behaviour to be implemented.*

The trace preorder is a very weak implementation relation. We cannot use it to specify that anything *must* happen. Another notion of validity is, that for each trace of the specification, the implementation can only refuse whatever the specification refuses after that trace. This is captured by the **conf**-relation, which was derived from testing theory. Here we give an intensional definition in terms of traces and refusals.

Definition 3 (conf) *Let* $p, s \in \mathcal{P}$. p **conf** s *iff* $\forall \sigma \in Tr(s) : Ref(p, \sigma) \subseteq Ref(s, \sigma)$.

Example 2 *For the specifications and processes given in example 1,* p_1, p_2 *and* p_3 *are all correct implementations of* s *according to* **conf**. *However,* stop *and* $a; \text{stop}$ *are not, because* s *requires either* b *or* c *to happen after* a.

The relation **red** (sometimes referred to as testing preorder, or failure preorder), which is the intersection of $\leq_{tr}$ and **conf**, gives rise to a specification technique with which we can specify both that certain actions must happen and that certain traces are not allowed. This seems to give a suitable specification technique for concurrent processes.

Example 3 *Suppose we want to specify a class of drinks machines. All machines should initially accept a coin. After that, the implementations should give the user either coffee or tea, or a choice between both. With* $\langle \mathcal{P}, \textbf{red} \rangle$ *we can capture this class of behaviours with the following specification:*

   s := coin; (i; coffee; **stop** [] i; tea; **stop**)

In the example above, note that s also allows the implementation that non-deterministically offers either coffee or tea, after accepting a coin. Since the non-determinism is solely used for achieving implementation freedom in the specification, we could require that implementations are fully deterministic. In that case we have a specification technique that is suitable for specifying deterministic processes.

Unfortunately, non-determinism is not only used to specify implementation freedom. There are some inherently non-deterministic systems, such as gambling machines. More importantly, non-determinism is needed to model non-deterministic aspects of the environment that we do not control. Examples are lossy, or erroneous communication media. In addition, the LOTOS internal action is sometimes used to model certain implementation details that cannot be modelled in LOTOS. In the example below, we show how reduction of non-determinism can lead to intuitively incorrect implementations in these cases.

Example 4 *In the following specification of a transmission protocol, the internal action is used to abstract from the occurrence of a timeout, which is currently not explicitly expressible in* LOTOS.

$$\text{TP}_{spec} := \text{send; (receive\_ack; } \textbf{stop} \text{ [] i (* timeout *); error; } \textbf{stop} \text{)}$$

This protocol sends a packet and then waits for an acknowledgement. If the acknowledgement is not received within a certain time, the protocol gives an error signal.

According to **red**, *this specification can be implemented by a process that gives an error straight away, which is counter-intuitive.*

$$\text{TP}_{error} := \text{send; error; } \textbf{stop}$$

Many more implementation relations exist, but most of them are also based on the assumption that implementations may be more deterministic than specifications. Implementation relations that require implementations to be as deterministic as their specifications are usually equivalence relations, which we have rejected for other reasons.

The solution we pursue in the next section separates the use of non-determinism to achieve implementation freedom from its other uses. A new specification construct is introduced for the specification of implementation options. An implementation is then a (possibly non-deterministic) specification in which all the implementation options have been resolved.

4   DISJUNCTION

In this section, we propose to extend LOTOS with a specification construct for explicitly specifying alternative implementation options. The construct we envisage has similarities to CSP's internal choice, but is closer to logical disjunction. In CSP the specification $P \sqcap Q$ could be implemented by $P[]Q$, but in logic, either P or Q would satisfy $P \vee Q$ (the choice is exclusive). The operator will be called disjunction, and denoted by $\vee$, because its properties are very much like those of logical disjunction. In the following S denotes the set of all specifications satisfying this extended syntax.

Disjunction is an operation on specifications that can be used to compose requirements that do not have to be satisfied simultaneously. In order to satisfy the specification $s \vee t$ it is enough to implement either s or t. Disjunction is a specification construct. Disjunctions cannot occur in implementations. Therefore disjunctions should

Table 3 Inference rules for unlabelled transitions

$s \rightarrowtail s'$	$\vdash \quad s[]t \rightarrowtail s'[]t$
$t \rightarrowtail t'$	$\vdash \quad s[]t \rightarrowtail s[]t'$
$s \rightarrowtail s'$	$\vdash \quad s\,|[G]|\,t \rightarrowtail s'\,|[G]|\,t$
$t \rightarrowtail t'$	$\vdash \quad s\,|[G]|\,t \rightarrowtail s\,|[G]|\,t'$
$s \rightarrowtail s', x := s$	$\vdash \quad x \rightarrowtail s'$

gradually be eliminated from the specification during consecutive refinement steps. Refinement should not reduce non-determinism though.

In order to define the semantics for disjunction operationally, we augment labelled transition systems with a new, unlabelled, transition: $\rightarrowtail$. These unlabelled transitions can only be introduced by the disjunction operator through the following axioms:

$$\frac{-}{s \bigvee t \rightarrowtail s} \qquad \frac{-}{s \bigvee t \rightarrowtail t}$$

i.e., a disjunction can be resolved through an unlabelled transition. The operational semantics of a specification is now given by an augmented labelled transition system.

Definition 4 (Augmented Labelled Transition System)
An augmented labelled transition system (ALTS) *is a structure* $\langle S, L, \rightarrow, \rightarrowtail, s_0 \rangle$, *with S a set of states, L a set of action labels,* $\rightarrow\, \subseteq S \times L \cup \{i\} \times S$ *a set of labelled transitions,* $\rightarrowtail\, \subseteq S \times S$ *a set of unlabelled transitions, and* $s_0 \in S$ *the initial state.*

The ALTS for a specification is determined in the usual fashion by the axioms for disjunction given above, and a set of inference rules. The inference rules that determine the normal transition relation, $\rightarrow$, are the same as the normal transition rules for LOTOS given in table 1. The rules for unlabelled transitions are given in table 3. Note that unlabelled transitions are just passed through by all binary operators and recursion. The reason for this is that we do not want a choice, for example, to be resolved by the presence of a disjunction in one of its arguments.

Example 5 *Below we have depicted the transition systems for the specifications* $S_1 := a; \mathbf{stop} \bigvee b; \mathbf{stop}$ *and* $S_2 := (a; \mathbf{stop}[]b; \mathbf{stop}) \bigvee (a; \mathbf{stop}[]c; \mathbf{stop})$.

In case of nested disjunctions (see example 6) we will usually not be interested in the disjuncts that are again disjunctions themselves. Our interest will be in the "real"

disjuncts, i.e., those states that can be reached through a sequence of unlabelled transitions, but which have no outgoing unlabelled transitions themselves. In the remainder of this paper, we therefore use a derived disjunction relation, defined below.

Example 6 *Depicted below is the transition system for* $a; \textbf{stop} \bigvee (b; \textbf{stop} \bigvee c; \textbf{stop})$.

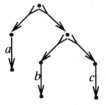

Definition 5 (derived disjunction relations)

1. *For a specification* s, *we define the following predicates:*
   $$s \longmapsto \quad \textit{iff} \quad \exists s' : s \longmapsto s' \quad (s \textit{ is a disjunction})$$
   $$s \not\longmapsto \quad \textit{iff} \quad \not\exists s' : s \longmapsto s' \quad (s \textit{ is not a disjunction})$$
2. *For specifications* s *and* t, *we define the following relations:*
   $$s \longmapsto^* t \quad \textit{iff} \quad t = s \vee \exists s' : s \longmapsto s' \wedge s' \longmapsto^* t$$
   $$\quad\quad\quad\quad (\textit{i.e., the reflexive and transitive closure of } \longmapsto)$$
   $$s \longmapsto\!\!\!\!\longmapsto^* t \quad \textit{iff} \quad s \longmapsto^* t \wedge t \not\longmapsto$$

The following lemma gives two useful properties for the $\longmapsto\!\!\!\!\longmapsto^*$-relation.

Lemma 1

1. $s \not\longmapsto \iff s \longmapsto\!\!\!\!\longmapsto^* s$;
2. $s \bigvee t \longmapsto\!\!\!\!\longmapsto^* x \iff s \longmapsto\!\!\!\!\longmapsto^* x \vee t \longmapsto\!\!\!\!\longmapsto^* x$.

Proof.

1. $\quad s \longmapsto\!\!\!\!\longmapsto^* s$
    $\Leftrightarrow \quad \{ \text{ definition of } \longmapsto\!\!\!\!\longmapsto^* \}$
    $\quad s \longmapsto^* s \wedge s \not\longmapsto$
    $\Leftrightarrow \quad \{ \text{ definition of } \longmapsto^* \}$
    $\quad (s = s \vee \exists s' : s \longmapsto s' \wedge s' \longmapsto^* s) \wedge s \not\longmapsto$
    $\Leftrightarrow \quad \{ s \not\longmapsto \}$
    $\quad s = s \wedge s \not\longmapsto$
    $\Leftrightarrow \quad \{ \text{ reflexivity of } = \}$
    $\quad s \not\longmapsto$

2. $s \lor t \rightarrowtail^* x$
   ⇔   { definition of $\rightarrowtail^*$ }
         $s \lor t \rightarrowtail^* x \land x \not\rightarrowtail$
   ⇔   { definition of $\rightarrowtail^*$ }
         $(x = s \lor t \lor \exists x' : s \lor t \rightarrowtail x' \land x' \rightarrowtail^* x) \land x \not\rightarrowtail$
   ⇔   { $s \lor t \rightarrowtail s$ and $s \lor t \rightarrowtail t$ }
         $(s \rightarrowtail^* x \lor t \rightarrowtail^* x) \land x \not\rightarrowtail$
   ⇔   { distribution of $\lor$ over $\land$ and definition of $\rightarrowtail^*$ }
         $s \rightarrowtail^* x \lor t \rightarrowtail^* x$

So far, there is nothing much new. The unlabelled transitions could just as well have been internal actions. The relation $\rightarrowtail^*$ would then correspond to the relation given by $\{(s, t) \mid s \overset{\epsilon}{\Rightarrow} t \land t \overset{i}{\nrightarrow} \}$. However, by introducing a different transition, we separate the specification of alternative implementation options from the use of internal actions and non-determinism. Note that rather than introducing an extra transition relation, we could have introduced another special action label like the i for internal actions.

In the following two sections, we define satisfaction and refinement as extensions of bisimulation equivalence. This is where we deviate from the usual approaches based on refusals.

4.1   Satisfaction

From here on we distinguish between *processes*, or *implementations*, which have no disjunctions, and *specifications*, which may have disjunctions. Processes are in the set $\mathcal{P}$, and specifications are drawn from the set $\mathcal{S}$.

A process intuitively satisfies a specification in case it is equivalent to one of its disjuncts. This intuition is reflected by the formal definition of satisfaction below. Since each disjunct can again have further disjuncts, the definition is inductive. Observe that we have used a "strong" interpretation. There is, however, no reason why this schema could not be applied to weaker interpretations of equivalence, provided they can be characterised inductively.

Definition 6 (Satisfaction)
Satisfaction, $\models \subseteq \mathcal{P} \times \mathcal{S}$, *is the largest relation such that, $p \models s$ implies $\exists s' : s \rightarrowtail^* s'$ and, for each $\mu \in L \cup \{i\}$ the following two conditions hold:*
$(\models_1)$ *Whenever $p \overset{\mu}{\longrightarrow} p'$, then $s' \overset{\mu}{\longrightarrow} s''$ for some s'' with $p' \models s''$; and*
$(\models_2)$ *Whenever $s' \overset{\mu}{\longrightarrow} s''$, then $p \overset{\mu}{\longrightarrow} p'$ for some p' with $p' \models s''$.*

Now, we can instantiate *sat* with $\models$ to obtain a powerful specification technique for both deterministic and non-deterministic processes.

Example 7 *Going back to the drinks machine specification of example 3, we can now specify the class of drinks machines that serve either coffee or tea as follows:*

    S1 := coin; (coffee; **stop** $\bigvee$ tea; **stop**)

Possible implementations, according to $\models$, *are:* coin; coffee; **stop** *and* coin; tea; **stop**. *If we also want to allow the implementation that offers a choice between coffee and tea, after a coin has been accepted, then we should add this as a disjunct to the specification:*

    S2 := coin; (coffee; **stop** $\bigvee$ tea; **stop** $\bigvee$ (coffee; **stop** [] tea; **stop**))

    Specification S2 in the example above shows that we had to trade-in some conciseness of specifications for clarity of the semantics. We believe that the semantics of logical disjunction will be better understood by most specifiers than the semantics of non-determinism.

Example 8 *In example 4 of the transmission protocol, there was no intended implementation freedom. Since the specification* TP_{spec} *does not contain disjuncts, the only possible implementation (modulo bisimulation equivalence) is the specification itself.*

The following proposition confirms that the $\bigvee$-operator behaves like logical disjunction.

Proposition 7 *Let* $s, t \in S$ *be specifications, and* $p \in \mathcal{P}$ *be a process. Then* $p \models (s \bigvee t) \Leftrightarrow (p \models s) \vee (p \models t)$.

Proof. $p \models (s \bigvee t)$
    $\Leftrightarrow$ { definition of $\models$ }
          $\exists x : (s \bigvee t) \rightarrowtail^* x \wedge$ (conditions $\models_1$ and $\models_2$ hold for(p, x))
    $\Leftrightarrow$ { lemma 1.2 }
          $\exists x : (s \rightarrowtail^* x \vee t \rightarrowtail^* x) \wedge (...)$
    $\Leftrightarrow$ { distr. of $\wedge$ over $\vee$ and distr. of $\exists$ }
          $(\exists x : s \rightarrowtail^* x \wedge (...)) \vee (\exists x : t \rightarrowtail^* x \wedge (...))$
    $\Leftrightarrow$ { definition of $\models$ }
          $p \models s \vee p \models t$

Because of this connection with logical disjunction, $\bigvee$ also enjoys the following properties.

Corollary 8 *Let* $r, s, t \in S$ *be specifications, and let* $p \in \mathcal{P}$ *be a process. Then:*

1. $p \models s \Leftrightarrow p \models (s \bigvee s)$ *(idempotency)*;
2. $p \models (s \bigvee t) \Leftrightarrow p \models (t \bigvee s)$ *(symmetry)*;

3. $p \models (r \vee (s \vee t)) \Leftrightarrow p \models ((r \vee s) \vee t)$ *(associativity)*.

It is not hard to see, that the equivalence over processes induced by the specification technique $\langle S, \models \rangle$ is precisely strong bisimulation equivalence.

Proposition 9 (process equivalence)
Let $p, q \in \mathcal{P}$ be processes, then $p \sim q \Longleftrightarrow \forall s \in S : (p \models s \Leftrightarrow q \models s)$.

Proof. (sketch) The proof for this proposition is similar to the proof that bisimulation equivalence is characterised by Hennessy-Milner logic in [Mil89, p.229]. It involves giving alternative characterisations of bisimulation and satisfaction as limits of descending chains of approximating relations. These are then used to prove the proposition by induction. □

We can also show that all other operators of the specification language distribute over disjunction. This will be a useful property when we want to establish a normal form for specifications.

Proposition 10 *Let $r, s, t \in S$ be specifications, and let $p \in \mathcal{P}$ be a process. Then the following distributivity properties hold:*

1. $p \models ((s \vee t) [] r) \Leftrightarrow p \models ((s [] r) \vee (t [] r))$;
2. $p \models ((s \vee t) |[G]| r) \Leftrightarrow p \models ((s |[G]| r) \vee (t |[G]| r))$;
3. $p \models ((s \vee t) \vee r) \Leftrightarrow p \models ((s \vee r) \vee (t \vee r))$.

Proof.

1. From left-to-right: Assume $p \models ((s \vee t) [] r)$. Then, by definition 6, there exists an x such that $((s \vee t) [] r) \rightarrowtail^* x$ and conditions ($\models_1$) and ($\models_2$) hold for p and x. Inspection of the inference rules for $\vee$ and $[]$ results in the following cases:

   $x = s' [] r'$, **where** $s \rightarrowtail^* s'$ **and** $r \rightarrowtail^* r'$: Since $((s [] r) \vee (t [] r)) \rightarrowtail (s [] r)$ and the fact that $\rightarrowtail \circ \rightarrowtail^* = \rightarrowtail^*$, we also have $((s [] r) \vee (t [] r)) \rightarrowtail^* x$, and we are done.
   $x = t' [] r'$, **where** $t \rightarrowtail^* t'$ **and** $r \rightarrowtail^* r'$: Similarly.

   From right-to-left: similar.
2. $((s \vee t) |[G]| r)$ and $((s |[G]| r) \vee (t |[G]| r))$ have isomorphic transition systems. Both specifications have the following $\rightarrowtail$-derivatives: $s|[G]|r$ and $t|[G]|r$. Neither specification has any other derivatives.
3. Follows from the idempotency, symmetry and associativity of $\vee$. □

4.2   Refinement

The definition of satisfaction above, naturally induces a refinement ordering over specifications. A specification s refines a specification t in case the set of processes satisfying s is a subset of the set of processes satisfying t, i.e. $\{p \in \mathcal{P} \mid p \models s\} \subseteq$
$\{p \in \mathcal{P} \mid p \models t\}$. However, generalising definition 6, we can also give an inductive
characterisation of refinement:

Definition 11 (Refinement)
Refinement *is the largest relation* $\sqsubseteq\, \subseteq S \times S$ *such that,* $s \sqsubseteq t$ *implies that
for each* s' *such that* $s \rightarrowtail^* s'$*, there exists a* t' *such that* $t \rightarrowtail^* t'$ *and, for each*
$\mu \in L \cup \{i\}$ *the following holds:*
  (i)   Whenever $s' \xrightarrow{\mu} s''$ *then, for some* t'', $t' \xrightarrow{\mu} t''$ *and* $s'' \sqsubseteq t''$*; and*
  (ii) Whenever $t' \xrightarrow{\mu} t''$ *then, for some* s'', $s' \xrightarrow{\mu} s''$ *and* $s'' \sqsubseteq t''$.

This definition simply states that s is a refinement of t if there is a disjunct t' in t for
each disjunct s' in s, such that s' is "bisimilar" to t'. The following theorem shows that
$\sqsubseteq$ is indeed a characterisation of refinement for the specification technique $\langle S, \models \rangle$.

Theorem 12 *Let* $s, t \in S$ *be specifications. Then*
$$s \sqsubseteq t \iff \{p \in \mathcal{P} \mid p \models s\} \subseteq \{p \in \mathcal{P} \mid p \models t\}$$

Proof. (sketch) The proof for this theorem goes very much along the lines of the proof
in [Mil89, p.229] that bisimulation is characterised by Hennessy-Milner logic. It involves giving alternative definitions for $\sqsubseteq$ and $\models$ as decreasing ω-sequences of approximating relations. We then use these to prove the given theorem by induction.
□

Proposition 13 *Let* $s, t, r \in S$ *be specifications, and let* $p \in \mathcal{P}$ *be a process. Then
the following laws for disjunction will hold:*

1. $s \sqsubseteq s \bigvee t$;
2. $t \sqsubseteq s \bigvee t$;
3. *If* $s \sqsubseteq r$ *and* $t \sqsubseteq r$, *then* $s \bigvee t \sqsubseteq r$.

In other words, $s \bigvee t$ *is the least upper bound of* s *and* t *with respect to the refinement
ordering.*

Proof. 1. and 2. follow immediately from definition 11, because $s \rightarrowtail^* s'$ implies
$(s \bigvee t) \rightarrowtail^* s'$ (using lemma 1), and similarly for t.
     3. We prove that the assumption that there is a specification r, such that $s \sqsubseteq r$ and
$t \sqsubseteq r$, but $s \bigvee t \not\sqsubseteq r$ leads to a contradiction.
     According to definition 11, $s \bigvee t \not\sqsubseteq r$ can only hold, if there exists an x, such that
$(s \bigvee t) \rightarrowtail^* x$, and for all r' such that $r \rightarrowtail^* r'$ either of the two conditions of defini-

tion 11 does not hold. However, if $(s \bigvee t) \rightarrowtail^* x$, then (by lemma 1) either $s \rightarrowtail^* x$ or $t \rightarrowtail^* x$. Since we assumed that $s \sqsubseteq_{\sim} r$ and $t \sqsubseteq_{\sim} r$, there must exist an r' such that $r \rightarrowtail^* r'$ and x and r' satisfy the two conditions, which gives us the contradiction. □

Next, we show that refinement, $\sqsubseteq_{\sim}$, is a (pre-)congruence. That is, refinement is preserved by all specification operators.

Proposition 14 *Let* $s_1, s_2, t \in S$ *be specifications, such that* $s_1 \sqsubseteq_{\sim} s_2$, *then*

1. $a; s_1 \sqsubseteq_{\sim} a; s_2$
2. $s_1 [] t \sqsubseteq_{\sim} s_2 [] t$
3. $s_1 \,||[G]||\, t \sqsubseteq_{\sim} s_2 \,||[G]||\, t$
4. $s_1 \bigvee t \sqsubseteq_{\sim} s_2 \bigvee t$

Proof. The first case is trivial. The other cases can easily be proved by constructing a relation that contains the pair (LHS,RHS) and then showing that this relation is contained in $\sqsubseteq_{\sim}$. Here, we prove just the last case.

Consider the relation $\{(s_1 \bigvee t, s_2 \bigvee t) \mid s_1 \sqsubseteq_{\sim} s_2\} \cup \sqsubseteq_{\sim}$. Whenever $s_1 \bigvee t \rightarrowtail^* x$ then either of the following two cases holds:

$s_1 \rightarrowtail^* x$: Since $s_1 \sqsubseteq_{\sim} s_2$ there exists a y such that $s_2 \rightarrowtail^* y$ and x and y satisfy the two conditions of definition 11. Since $(s_2 \bigvee t) \rightarrowtail s_2$, also $(s_2 \bigvee t) \rightarrowtail^* y$.

$t \rightarrowtail^* x$: Since $(s_2 \bigvee t) \rightarrowtail t$, also $(s_2 \bigvee t) \rightarrowtail^* x$, and we are done. □

5 APPLICATIONS

In [Hoa85], Hoare gives some examples in which the non-deterministic or, $\sqcap$, is used for loosely specifying change-giving machines in CSP. These specifications can be expressed equally well in our notation, although their interpretation is slightly different.

Example 9 *Consider the following specification of a change-giving machine, which always gives the right change in one of two combinations:*

> CH1 := in5p;
>   (out1p; out1p; out1p; out2p; CH1
> $\bigvee$
>   out2p; out1p; out2p; CH1)

This specification leaves open how the change should be given. Valid implementations are those which always return one of two possible combinations of change, but also those which return different combinations on each invocation. For example, the implementation given by CH_I1, *which alternates between the two possible combinations, satisfies* CH1.

CH_I1 := in5p;
        out1p; out1p; out1p; out2p;
        in5p;
        out2p; out1p; out2p; CH_I1

Example 10 *We saw that* CH1 *allows implementations that give different combinations of change on each invocation. The following specification allows only implementations that always give the same combination, but it leaves open which combination it will be.*

CH2 := CH2A $\bigvee$ CH2B
where
    CH2A := in5p; out1p; out1p; out1p; out2p; CH2A
    CH2B := in5p; out2p; out1p; out2p; CH2B

Although CSP's ⊓ is intended to play a similar role to logical disjunction, CSP's failures preorder allows also implementations that replace the non-deterministic choice by a deterministic one. This will then give the user a choice, at "run-time", which implementation s/he wants. For example, if the specifications CH1 and CH2 had been written with a non-deterministic choice between the alternatives, then both would have allowed the following implementation:

CH_I2 := in5p;
        (out1p; out1p; out1p; out2p; CH_I2
        []
        out2p; out1p; out2p; CH_I2)

which gives the user a chance to influence which combination of change s/he will get. However, the semantics of $\bigvee$ does not allow CH_I2 as an implementation of either CH1 or CH2, i.e. CH_I2 $\not\models$ CH1, CH2.

5.1   The most undefined specification

The disjunction operator can easily be generalised to work over a set of arguments. For S a set of specifications, $\bigvee S$ denotes the disjunction of all the specifications $s \in S$. The semantics is defined by the following family of axioms:

$$\frac{-}{\bigvee S \rightarrowtail s}(s \in S)$$

In the same fashion, choice, [], can be generalised to ΣS, with $\Sigma\{\} = \textbf{stop}$.
    Using these generalised operators, we can define the most undefined specification, i.e. the specification that allows all processes as implementations, provided the alphabet of labels is finite.

$$U := \bigvee \{ \Sigma \{ a; U \mid a \in A \} \mid A \subseteq \mathcal{L} \}$$

Example 11 *Let* $\mathcal{L} = \{a, b\}$ *be the alphabet. Then the most undefined specification* U *is given by:*

$$U := \mathbf{stop} \bigvee a; U \bigvee b; U \bigvee (a; U \; [] \; b; U)$$

This most undefined specification is very useful for partial specification. Whenever we want to leave open the behaviour at a certain point, we can just plug-in U. Later on, this can be refined to anything, thus achieving complete implementation freedom.

6   CONCLUSION

Many others before us have recognised the limited expressiveness of process algebras for the specification of non-deterministic, concurrent processes. A common approach has been to define a logic, separate from the process description language, for the specification of properties of processes (e.g. Hennessy-Milner Logic (HML) [HM85] and modal μ-calculus [Koz83]). A clear drawback is that specifications and implementations are in different notations. Step-wise refinement is not possible, and verification can only be done *a posteriori*. In order to alleviate this problem, there have been some attempts to introduce the process structuring operators into these logics. In [Hol89], HML is extended with the CCS operators, and in [BGS89], the same is done for a fragment of the μ-calculus. Unfortunately, these languages have a denotational semantics: each specification is associated with the set of processes that satisfy it. Verifying whether a process satisfies a specification amounts to checking whether it is in that set. Alternatively, the correctness of an implementation can be verified through (in-)equational reasoning.

Another way to increase the expressive power of process algebraic specifications is introduced in [Lar90b], where transitions are decorated with modalities. A distinction is made between *required* and *allowed* transitions. Bisimulation equivalence is then generalised to a refinement relation that ensures that the more concrete specification requires more and allows less. It is also possible to define the equivalent of logical *conjunction* operationally in this model [LSW95]. In fact, it has been shown that the specification technique thus obtained is as expressive as a restricted version of HML [BL92]. The restriction is caused by the inability to adequately express disjunction. However, modal transition systems can be extended with disjunction in the same way we have extended labelled transition systems with disjunction in this paper. Would this then create a specification technique with the full power of HML?

ACKNOWLEDGEMENTS

We would like to thank Rom Langerak for discussing ideas that led to this paper, and the anonymous referees for their useful comments.

REFERENCES

[BGS89] A. Bouajjani, S. Graf, and J. Sifakis. A logic for the description of behaviours and properties of concurrent systems. LNCS 354, pages 398–410, 1989.

[BL92]   G. Boudol and K.G. Larsen. Graphical versus logical specifications. *Theoretical Computer Science*, 106:3–20, 1992.

[BSS87] E. Brinksma, G. Scollo, and C. Steenbergen. LOTOS specifications, their implementations and their tests. In *PSTV VI*, pages 349–360, 1987.

[HM85]   M. Hennessy and R. Milner. Algebraic laws for nondeterminism and concurrency. *Journal of the ACM*, 32(1):137–161, January 1985.

[Hoa85] C.A.R. Hoare. *Communicating Sequential Processes*. Prentice-Hall, 1985.

[Hol89] S. Holmström. A refinement calculus for specifications in Hennessy-Milner logic with recursion. *Formal Aspects of Computing*, 1(3):242–272, 1989.

[Koz83] D. Kozen. Results on the propositional μ-calculus. *Theoretical Computer Science*, 27(3):333–354, December 1983.

[Lar90a] K.G. Larsen. Ideal specification formalism = expressivity + compositionality + decidability + testability + $\cdots$. In *CONCUR'90*, LNCS 458, pages 33–56, 1990.

[Lar90b] K.G. Larsen. Modal specifications. In J. Sifakis, editor, *Automatic Verification Methods for Finite State Systems: Proceedings*, LNCS 407, pages 232–246. Springer-Verlag, 1990.

[Led91] G. Leduc. *On the Role of Implementation Relations in the Design of Distributed Systems using LOTOS*. PhD thesis, University of Liège, Belgium, June 1991.

[Led92] G. Leduc. A framework based on implementation relations for implementing LOTOS specifications. *Computer Networks and ISDN Systems*, 25:23–41, 1992.

[LSW95] K.G. Larsen, B. Steffen, and C. Weise. A constraint oriented proof methodology based on modal transition systems. In E. Brinksma, editor, *TACAS'95*, LNCS 1019, pages 17–40, 1995.

[Mil89] R. Milner. *Communication and Concurrency*. Prentice-Hall, 1989.

[NH84]   R. de Nicola and M.C.B. Hennessy. Testing equivalences for processes. *Theoretical Computer Science*, 34:83–133, 1984.

Ir **M.W.A. Steen** obtained an MSc(Eng) in Computer Science from the University of Twente, The Netherlands, in 1993. He is currently completing a PhD degree in Computer Science at the University of Kent at Canterbury. His current research focuses on partial specification in process algebra, in particular on techniques for consistency checking and composition. Furthermore he has worked on application of these, and other, formal techniques in the area of Open Distributed Processing.

Dr. H. Bowman, Dr. J. Derrick and **Dr.Ir. E.A. Boiten** are lecturers in the Computing Laboratory at the University of Kent.

12

A timed automaton model for ET-LOTOS verification

Christian Hernalsteen
University of Brussels, Computer Science Department
Boulevard du Triomphe CP 212, 1050 Brussels, Belgium
chernals@ulb.ac.be

Abstract

We present in this paper a method to transform ET-LOTOS expressions in a subclass of timed automaton with timers, where a timer is not restarted before its reset. We show that this subclass is equivalent to timed automata and we show that this model can be used for ET-LOTOS verification. We have implemented this transformation method and interfaced our tool with the KRONOS model-checker. It is then possible to verify temporal properties, expressed in TCTL, on ET-LOTOS specifications. We illustrate our tool with a small robot controller example.

Keywords

ET-LOTOS, Timed automaton, Model-checking, Tools

1 INTRODUCTION

In the last few years, many Formal Description Techniques have been extended to support the design of real-time systems. These systems are quite complex to design due to their timing constraints. Moreover they are often safety critical; failures can have disastrous consequences which can put, for instance, human life in peril. The development of such systems can therefore be eased with the support of formal techniques.

Many process algebras have been extended to allow the description of timed dependent systems, see [10] for an overview of FDT's supporting time. ET-LOTOS [9] is a timed extension of LOTOS which allows the modeling of real-time behaviors. This extension is currently used to define the timed semantics of the future ISO standard E-LOTOS (for Extended LOTOS). Such a language can only be useful however, if tools can support it and especially its real-time characteristics. Tools which deal with time already exists, like KRONOS [4] which allows to verify that a formula of the real-time logic TCTL [1] is verified by a system described by a timed automaton. Other tools based on Hybrid automata [2] allow to describe and analyze real-time systems like HyTech [6] and SHIFT [11]. Timed automata is a subclass of hybrid automata where the

Formal Description Techniques and Protocol Specification, Testing and Verification
T. Mizuno, N. Shiratori, T. Higashino & A. Togashi (Eds.) © 1997 IFIP. Published by Chapman & Hall

automaton variables are used to represent the time passing. This model is less complex to handle than hybrid automata and is more adapted to real-time systems description.

In [5], a transformation method of an ET-LOTOS subset into timed automata is proposed but one of the timed operators of ET-LOTOS, the time capture, is not supported. We propose in this paper to extend this method in order to allow the transformation of this operator. To reach this goal, we use timed automaton with timers to represent ET-LOTOS expressions. We show that the resulting automaton model is a subclass of timed automaton with timers where a timer is never restarted before its reset and we prove that this subclass is equivalent to timed automata. Timed automata are decidable for reachability analysis [2]. We have then captured an ET-LOTOS subset including all the operators, with some restrictions on their use, which is decidable. A tool implementing the transformation has been developed and interfaced with the KRONOS model-checker. ET-LOTOS specifications can then be transformed in timed automaton and verified with KRONOS. Since timed semantics of E-LOTOS will be based on ET-LOTOS, it will be possible to extend this work to this future ISO standard.

The approach of transforming a real time process algebra in a automaton model for model-checking purposes have already bee used for RT-LOTOS [3]. In this work a specific automaton model (*Dynamic Timed Automaton*) and a transformation in two phases have been defined. With this transformation, the reachability analysis may be performed on the fly which is not possible with our compositional transformation.

2   ET-LOTOS

ET-LOTOS extends LOTOS with a global time which can be discrete or dense. We consider, in this paper, a dense time domain: $\mathbb{Q}^+$. Three timing constructs are added to the LOTOS ones. These can be used to restrict the time interval where an action can occur. With the *life reducer* (a{d}; B), the action a is limited to occur within d time units. It is not an mandatory for an observable action a to occur within d time units, but if it does not, the process behaves like **stop**. If a is the internal action (i), the action must occur within d time units (or be preempted by another alternative). The internal action is urgent. The *time capture* (a @x; B) allows to capture in an ET-LOTOS variable (x) the time at which the action a has occurred (the time elapsed between the moment the behavior has been offered and the one at which the action has occurred). The third operator is the *delay* (Δ^d B) where the behavior B is offered to the environment after a waiting time of d time units.

The formal semantics of ET-LOTOS is given in terms of labeled transition systems composed of two kinds of transitions, namely discrete and timed. This semantics can be found in found in [9].

3 TIMED AUTOMATON WITH TIMERS

We define in this section a model of automaton extended with clocks and timers (also called *integrators* or *stop watch* in the literature). Clocks value grows uniformly in all vertices proportionally with the time passing. A timer is a clock which can be started and stopped on a transition.

Let $V = \mathcal{H} \cup \mathcal{I}$ be an infinite set of variables having a value in the set of positive rationals $\mathbb{Q}^+$; $\mathcal{H}$ and $\mathcal{I}$ are respectively the infinite sets of clocks and timers with $\mathcal{H} \cap \mathcal{I} = \emptyset$. Let $\mathcal{S}$ denotes an infinite set of vertices and Lab a label set. Let $\Psi(V)$ be the set of constraints on variables defined as the smallest set satisfying: $\psi ::= true | u \prec c | u_1 - u_2 \prec c | x \prec c | \neg \psi | \psi \wedge \psi$ where $u, u_1, u_2 \in \mathcal{H}, x \in \mathcal{I}, c \in \mathbb{Z}$ and $\prec \in \{<, \leq\}$.

Definition 1 *A timed automaton with timers is defined as a tuple $< S, V, R, E, s^0, Inv >$ where*

> $S \subset \mathcal{S}$ *is a finite set of vertices,*
> $V \subset V$ *is a finite set of variables,*
> $R : S \times V \longrightarrow \{0, 1\}$ *is the rate labeling function,*
> $E \subseteq S \times Lab \times \Psi(V) \times 2^V \times S$ *is the finite set of transitions,*
> $s^0 \in S$ *is the initial vertex,*
> $Inv : S \longrightarrow \Psi(V)$ *is the vertex invariant*

where $\forall x \in V \cap \mathcal{H}, \forall s \in S, R(s)(x) = 1$.

A transition $e = (s, a, \psi, V', s')$ from the vertex s to a vertex s' is labeled by an action (a). The transition can be fired if its guard ψ is evaluated to true and the variables of V' are reset when the transition occurs. We say that a transition starts (resp. stops) a timer x if $R(s)(x) = 0$ and $R(s')(x) = 1$ (resp. $R(s)(x) = 1$ and $R(s')(x) = 0$). The vertex invariant determines for each vertex whenever the time can progress, and hence whenever the automaton can stay in the vertex.

We define two subclasses of timed automaton with timers: *timed automaton* where all the variables are clocks ($V \subset \mathcal{H}$) and *timed automaton with semi-timers* where a timer is always reseted before being restarted ($\forall e = (s, a, \psi, V', s') \in E, \forall x \in V \cap \mathcal{I} : (R(s)(x) = 0 \text{ and } R(s')(x) = 1) \implies x \in V'$).

We denote v a valuation, i.e., a function which associates to each variable $x \in V$ a value in $\mathbb{Q}^+$. The set of valuations on variables is denoted $\mathbb{V}$ and the one restricted to clocks is denoted $\mathbb{H}$. Valuation can be naturally extended to constraints on variables: for each $\psi \in \Psi(V)$ and $v \in \mathbb{V}$, $\psi(v)$ represents the value of the constraints ψ evaluated in v. The valuation $v[V := 0]$ represents the valuation v where all the variables of V have been reset to 0. For a valuation $v \in \mathbb{V}$, a rate labeling function R, and $t \in \mathbb{Q}^+$, $v + R.t$ denotes a new valuation v' such that for every variable $x \in V$, $v'(x) = v(x) + R(x).t$. For a valuation $v \in \mathbb{H}$, and $t \in \mathbb{Q}^+$, $v + t$ denotes a new valuation v' such that for every clock $u \in \mathcal{H}$, $v'(u) = v(u) + t$.

The semantics of a timed automaton with timers is given by the following definition.

Definition 2 *The model of a timed automaton with timers* $< S, V, R, E, s^0, Inv >$ *is the labeled transition system* $< \mathcal{Q}, \longrightarrow, q^0 >$ *where :*

1. $\mathcal{Q} = \{(s, v) | Inv_s(v), s \in S, v \in \mathbb{V}\}$
2. $q^0 = (s^0, v^0)$, *where* $\forall x \in V \;\; v^0(x) = 0$
3. *The transition relation* $\longrightarrow \subseteq \mathcal{Q} \times (Lab \cup \mathbb{Q}^+) \times \mathcal{Q}$ *is defined on* $\mathcal{Q}$ *by the smallest relation defined by the following rules:*

$$(R1) \;\; \frac{(s, a, \psi, V', s') \in E \;\wedge\; \psi(v)}{(s, v) \xrightarrow{a} (s', v[V' := 0])} \qquad (R2) \;\; \frac{\forall d' \le d \;\; Inv_s(v + R(s).d')}{(s, v) \xrightarrow{d} (s, v + R(s).d)}$$

where $a \in Lab, v \in \mathbb{V}, d \in \mathbb{Q}^+$.

4   TRANSFORMATION OF ET-LOTOS

We provide in this section, a method to transform an ET-LOTOS expression in a timed automaton with timers where $Lab = L \cup \{i, \Delta, \varepsilon\}$ (Δ represents the expiration of a delay and ε an empty transition). This transformation is inspired from the work presented in [5]. We restrict ourself to some of the operators but the method can be extended to all the basic operators of ET-LOTOS (see [8]).

We first define two notations to handle rate labeling functions. We define $R = R_1 \cup R_2$ the union of two rate labeling functions $R_1 : S_1 \times V_1$ and $R_2 : S_2 \times V_2$ with $S_1 \cap S_2 = V_1 \cap V_2 = \emptyset$ as $R : (S_1 \cup S_2 \cup (S_1 \times S_2)) \times (V_1 \cup V_2)$ where $\forall s_1 \in S_1, s_2 \in S_2, x_1 \in V_1, x_2 \in V_2, x \in V_1 \cup V_2$:

$$R(s_i)(x_i) = R_i(s_i)(x_i) \; i \in \{1, 2\}$$
$$R([s_1, s_2])(x) = \left\{ \begin{array}{ll} R_1(s_1)(x) & if \; x \in V_1 \\ R_2(s_2)(x) & if \; x \in V_2 \end{array} \right.$$

We define the extension a rate labeling function with a new variable and vertex. If R is a rate labeling function and $(s, x) \notin Dom(R)$ a vertex and a variable, we define $R' = R \uplus (s, x)$ the new rate labeling function resulting from the extension of R with the couple (s, x) where :

$$\forall x' \in Dom_v(R), \forall s' \in Dom_s(R) \;\; R'(s')(x') = R(s')(x')$$
$$\forall x' \in Dom_v(R) \;\; R'(s)(x') = \left\{ \begin{array}{ll} 1 & if \; x' \in \mathcal{H} \\ 0 & if \; x' \in \mathcal{I} \end{array} \right.$$

$$\forall s' \in Dom_s(R) \quad R'(s')(x) = \begin{cases} 1 & if \ x \in \mathcal{H} \\ 0 & if \ x \in \mathcal{I} \end{cases}$$

$$R'(s)(x) = 1$$

where $Dom_v(R)$ and $Dom_s(R)$ represent, respectively, the variable and vertex domain of the rate labeling function R.

The transformation method is defined in a compositional way, where the automaton corresponding to an ET-LOTOS expression depends on the automata of the operand's expressions. It results that produced automata may contain guard transitions with variables which does not belong to V since a variable is introduced when it is defined via a time capture and not when it is used via a guard. In all cases the final automaton is complete if the original ET-LOTOS specification is semantically correct.

1. The process **stop**:
   This process is transformed in $< \{s\}, \{u\}, R, \emptyset, s, Inv >$ where $u \in \mathcal{H}, Inv(s) = true$ and $R(s)(u) = 1$.
   This timed automaton with timers (automaton for short in the sequel) has no transition and has no variable, but the clock u is needed to represent the time passing as specified in the semantic rule (R2).

2. Action prefix:
   If $< S, V, R, E, s^0, Inv >$ is the automaton corresponding to the behavior B and $B' \equiv a@x\{d\}; B$ then the timed automaton corresponding to B' is given by:

$$< S \cup \{s\}, V \cup \{x\}, R \uplus (s, x), E \cup \{(s, a, x \le d, V, s^0)\}, s, Inv' >$$

   where $s \notin S, x \in \mathcal{I} \setminus V$ and for all $s' \in S, Inv'(s') = Inv(s')$.
   The activation function of the new vertex is defined by $Inv'(s) = true$ if the action a is observable, since the action is not obliged to occur before time d but only restricted to occur within this limit; if action $a = i$, the action must occur within the time limit, then the function Inv is defined as $Inv'(s) = x \le d$. The new rate labeling function states that the timer x is frozen on all the vertices but s. If the time capture is not present, we introduce a new clock $u \in \mathcal{H} \setminus V$ and use it instead of x.

3. Delay:
   If $< S, V, R, E, s^0, Inv >$ is the automaton corresponding to B and $B' \equiv \Delta^d B$ then the timed automaton corresponding to B is given by :

$$< S \cup \{s\}, V \cup \{u\}, R \uplus (s, u), E \cup \{(s, \Delta, u = d, V, s^0)\}, s, Inv' >,$$

   where $s \notin S, u \in \mathcal{H} \setminus V$.
   The label Δ represents the expiration of the delay. For all $s' \in S, Inv'(s') =$

$Inv(s')$ and $Inv(s) = u \leq d$; in this way, and due to the transition guard, the automaton is restricted and obliged to leave the vertex s after d time units.

4. Choice:

   If $< S_i, V_i, R_i, E_i, s_i^0, Inv_i >$ is the automaton of $B_i, i \in \{1, 2\}$ where $S_1 \cap S_2 = \emptyset$ and $V_1 \cap V_2 = \emptyset$ and if $B \equiv B_1 [] B_2$ then the automaton corresponding to B is given by :

   $$< S_1 \cup S_2 \cup (S_1 \times S_2), V_1 \cup V_2, R_1 \cup R_2, E \cup E_1 \cup E_2, [s_1^0, s_2^0], Inv >$$

   where E is given by:

$$
\begin{aligned}
E \;=\; & \{([s_1, s_2], a, \psi, V', s_1') \mid (s_1, a, \psi, V', s_1') \in E_1, a \in L \cup \{i\}\} \\
\cup \; & \{([s_1, s_2], a, \psi, V', s_2') \mid (s_2, a, \psi, V', s_2') \in E_2, a \in L \cup \{i\}\} \\
\cup \; & \{([s_1, s_2], \omega, \psi, V', [s_1', s_2]) \mid (s_1, \omega, \psi, V', s_1') \in E_1, \; \omega \in \{\Delta, \varepsilon\}\} \quad (1) \\
\cup \; & \{([s_1, s_2], \omega, \psi, V', [s_1, s_2']) \mid (s_2, \omega, \psi, V', s_2') \in E_2, \; \omega \in \{\Delta, \varepsilon\}\} \quad (2)
\end{aligned}
$$

For all $s_i \in S_i$, $Inv(s_i) = Inv_i(s_i)$ and $Inv([s_1, s_2]) = Inv_1(s_1) \wedge Inv_2(s_2)$ to represent that time passes at the same rate in both processes. The last two transition sets (equations 1 and 2) represent the fact that delay and empty transitions do not resolve the choice.

5. Parallel composition:

   If $< S_i, V_i, R_i, E_i, s_i^0, Inv_i >$ is the automaton of $B_i, i \in \{1, 2\}$ where $S_1 \cap S_2 = \emptyset$ and $V_1 \cap V_2 = \emptyset$ and if $B \equiv B_1 |[\Gamma]| B_2$ where $\Gamma \subseteq L$, then the automaton corresponding to B is given by:

   $$< S_1 \times S_2, V_1 \cup V_2, R_1 \cup R_2, E, [s_1^0, s_2^0], Inv >$$

   where the transition set E is given by:

$$
\begin{aligned}
E \;=\; & \{([s_1, s_2], a, \psi, V', [s_1', s_2]) \mid (s_1, a, \psi, V', s_1') \in E_1, a \in L \cup \{\Delta, i, \varepsilon\} \setminus \Gamma\} \\
\cup \; & \{([s_1, s_2], a, \psi, V', [s_1, s_2']) \mid (s_2, a, \psi, V', s_2') \in E_2, a \in L \cup \{\Delta, i, \varepsilon\} \setminus \Gamma\} \\
\cup \; & \{([s_1, s_2], a, \psi_1 \wedge \psi_2, V_1 \cup V_2, [s_1', s_2']) \mid (s_i, a, \psi_i, V_i, s_i') \in E_i, a \in \Gamma \cup \{\delta\}
\end{aligned}
$$

For all $[s_1, s_2] \in S_1 \times S_2$, $Inv([s_1, s_2]) = Inv_1(s_1) \wedge Inv_2(s_2)$ to state that the time passes at the same rate in the two processes. The first two sets represent transitions resulting from independent actions while the last one represents synchronized actions.

6. The guard:

   If $< S, V, R, E, s^0, Inv >$ is the automaton of B and $B' \equiv [G] \longrightarrow B$ where $G \in \Psi(V)$ then the timed automaton of B' is given by :

   $$< S \cup \{s\}, V \cup \{u\}, R \uplus (s, u), E \cup (s, \varepsilon, G, V, s^0), s, Inv' >$$

   where $s \notin S$ and $u \notin V$.
   For all $s' \in S$, $Inv'(s') = Inv(s')$ and $Inv'(s) = (u = 0) \vee \neg G$.

An empty transition is added to verify the value of the guard. The activation function of the initial vertex forces the transition to occur immediately whenever the guard is evaluated to true. In the other case, the function is equivalent to true since, following the ET-LOTOS semantics, the expression can be aged indefinitely when the guard is false.

It can be easily proved that the automaton model produced with this method, when the guard expressions are restricted to $\Psi(\mathcal{V})$, is a timed automaton with semi-timers. Indeed, timers are only introduced to represent time capture. These timers are never restarted once they are stopped.

5 DECIDABILITY OF TIMED AUTOMATON WITH SEMI-TIMERS

We define in this section a transformation of timed automata with semi-timers to timed automata and we show that this transformation provides equivalent automata modulo the strong bisimulation. Since reachability is decidable on timed automaton [2], we thus prove that this is also the case for timed automaton with semi timers, and for the ET-LOTOS subset considered in this paper. Moreover this result can be extended to all the basic ET-LOTOS operators [8].

The principle of the transformation is to substitute any timer x with the difference between two clocks x_u and x_l. The clock x_u represents the last time the timer has been started and x_l the last time it has been stopped. The transformation function resets both clocks when the corresponding timer is reset and x_l is reset when the transition stops the timer. The formal definition of this transformation function is given below.

Definition 3 *Let* $T_r : \mathcal{T}_{st} \longrightarrow \mathcal{T}$ *be a transformation function from timed automaton with semi-timers into timed automaton. This function is defined by* $T_r(< S_1, V_1, R_1, E_1, s_1^0, Inv_1 >) = < S_1, H_2, E_2, s_1^0, Inv_2 >$ *such that*

- $H_2 = (V_1 \cap \mathcal{H}) \cup \{x_l, x_u \mid x \in V_1 \cap \mathcal{I}\}$
- $E_2 = \{(s, a, \psi[I], H', s') \mid (s, a, \psi, V', s') \in E_1,$
  $\quad H' = (V' \cap \mathcal{H}) \cup \{x_l, x_u \mid x \in V' \cap \mathcal{I}\} \cup$
  $\quad\quad \{x_l \mid R_1(s)(x) = 1 \text{ and } R_1(s')(x) = 0, x \in V_1 \cap \mathcal{I}\},$
  $\quad I = \{x \mid x \in V_1 \cap \mathcal{I}, R_1(s)(x) = 0\}\}$
- $\forall s \in S_2, Inv_2(s) = Inv_1(s)[I], I = \{x \mid x \in V_1 \cap \mathcal{I}, R_1(s)(x) = 0\}.$

where $\forall x \in V_1 \cap \mathcal{I}, \{x_u, x_l\} \cap V_1 = \emptyset$ *and* $\forall \psi \in \Psi(\mathcal{V}), I \subseteq \mathcal{I}, \psi[I] \in \Psi(\mathcal{V})$ *represents the constraint* ψ *where all the instances of* $x \in I$ *have been replaced by* $x_u - x_l$ *and where all instances of* $x \in \mathcal{I} \setminus I$ *have been replaced by* x_u *where* $x_u, x_l \in \mathcal{H}$ *such that* $\forall x \neq y, x_l \neq y_l$ *and* $x_u \neq y_u$.

Let's remark that $\psi[I]$ belongs to $\Psi(\mathcal{V})$ since $I \subset \mathcal{I}$ and the difference between two timers $x_1, x_2 \in \mathcal{I}$ is not allowed in $\Psi(\mathcal{V})$

In [7] we proof formally that the transformation function T_r preserves the semantics of timed automaton with semi-timers modulo the strong bisimulation i.e. $\forall M \in \mathcal{T}_{st}, T_r(M)$ is strongly bisimilar to M. It results that state reachability is decidable on our timed automaton model with semi timers.

6   IMPLEMENTATION

This transformation method has been implemented in a tool which takes an ET-LOTOS specification and transform it in a timed automaton for KRONOS. The tool supports all the basic ET-LOTOS operators, including the process instantiation. Usual constraints are put on recursive behaviors: it is not allowed through a parallel and on the left hand side of the enabling and disabling operators. The implementation of the method has been optimized to reduce the number of transitions, vertices and clocks produced. For instance, a new clock is not introduced for each prefix operator and Δ transitions are, in some cases, removed and replaced by an extended guard on the subsequent transitions. Optimizations are also done on the treatment of process instantiation in order to avoid, as much as possible, the duplication for each instantiation, of an automaton representing the process instantiated.

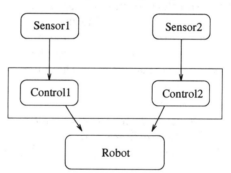

Figure 1 The robot controller structure

The KRONOS tool allows to verify TCTL formulas on timed automata. The real-time logic TCTL [1] is an extension of CTL where the two main operators have been extended with timing constraints to allow quantitative temporal reasoning. The TCTL formulas accepted by KRONOS are defined by the following grammar:

$\rho ::= init \mid enable(a) \mid after(a) \mid c \sim r \mid \neg\rho \mid \rho_1 \wedge \rho_2 \mid \rho_1 \exists \mathcal{U}_I \rho_2 \mid$
$\rho_1 \forall \mathcal{U}_I \rho_2$

where r is a positive integer, $\sim \in \{<, \leq, =, \geq, >\}$, c is a clock of the timed automaton under verification, I is a time interval, *init* represents the initial state with all the clocks set to zero, $enable(a)$ defines the states where transitions labeled with a are enabled and $after(a)$ defines the set of states reached by transitions labeled by a.

Intuitively, $\rho_1 \exists \mathcal{U}_I \rho_2$ means that it exists a run which continuously verifies ρ_1 until a state which verifies ρ_2 is reached at a time $t \in I$. In the same way, $\rho_1 \forall \mathcal{U}_I \rho_2$ means that all runs satisfy the above property. Some typical abbreviations are used such as $\forall \Diamond_I \rho$ for $true \forall \mathcal{U}_I \rho$, $\exists \Diamond_I \rho$ for $true \exists \mathcal{U}_I \rho$, $\exists \Box_I \rho$ for $\neg \forall \Diamond_I \neg \rho$ and $\forall \Box_I \rho$ for $\neg \exists \Diamond_I \neg \rho$. The unrestricted operators correspond to the operator subscripted by $[0, \infty[$. This logic allows to specify complex safety and liveness properties.

A robot controller example

Let us take a small example to illustrate the use of the tools. We consider a robot controller which provides commands, based on recent measurements of the environment, to a robot. The system consists of five components: two sensors, two controller processes and the robot itself (figure 1). The sensors probe the environment periodically; all 8 time units for the first one and 12 time units for the second. A controller process is associated to each sensor. The sensor's readings are sent to the controllers which take some amount of time to process the information and to send the new command to the robot; 1 time unit for the first one and 2 time units for the second. These processes share the same processor and are controlled by a simple scheduler. This non-preemptif scheduler gives the processor to each controller for a given amount of time. When a controller receive the processor its waits for the reading of its sensor. If this reading does not arrive in the time interval defined by the scheduler, the processor is given to the other controller. On the other hand, the scheduler waits for the end of the reading processing before giving the processor to the other controller. In all cases, the scheduler never interrupts a reading processing. The problem is to determine the time period given to each controller to insure some time limits on the processing of the sensor's readings. The processing of the first controller must be started within 6 time units after the corresponding sensor's reading. This time limit is set to 8 time units for the second controller.

This robot controller system has been specified in ET-LOTOS (figure 2). The sensors are described by two processes **Sensor** which generate periodically the action **Reading** representing a sensor reading. These actions are synchronized with the **TimeOut** processes which specify the timing requirements on the sensor's reading processing. Once a sensor's reading is captured, the process offers to its associated controller to start its processing (action **StartControl**). If the processing is not started in the time limit an exception error is raised (action **error**). The controllers are represented by two simple processes (**Controller**); they wait for the sensor reading and then execute their processing (represented by the delay) before exiting. The **Scheduler**

hide StartControl1, StartControl2, Reading1, Reading2 **in** (
    (Scheduler |[StartControl1, StartControl2]| (TimeOut1 ||| TimeOut2))
    |[Reading1, Reading2]| (Sensor1 ||| Sensor2)
where
    Scheduler ::=
        (Controller1 [] $\Delta^{\texttt{period1}}$ **exit**) >>
        (Controller2 [] $\Delta^{\texttt{period2}}$ **exit**) >> Scheduler
    Controller1 ::= StartControl1; Δ^1 **exit**
    Controller2 ::= StartControl2; Δ^2 **exit**
    TimeOut1 ::= Reading1; (StartControl1; TimeOut1 [] Δ^6 error; **stop**)
    TimeOut2 ::= Reading2; (StartControl2; TimeOut1 [] Δ^8 error; **stop**)
    Sensor1 ::= Reading1; Δ^8 Sensor1
    Sensor2 ::= Reading2; Δ^{12} Sensor2

Figure 2 The ET-LOTOS robot system

process gives the processor to the two controllers in a cyclic way. The choice expressions specify the period of processor allocation. If a controller has not begun its processing before the **period** deadline, the processor is given to the other controller. The different synchronized actions are hidden to insure their urgency; they occur as soon as all the synchronized components offered the considered action. This ET-LOTOS specification does not use the time capture operator but is, nevertheless, well adapted to illustrate our verification process.

We have used our transformation tool to produce the KRONOS timed automaton corresponding to the robot controller specification. The transformation tool has produced, in less than 2 seconds, a timed automaton of 207 states, 659 transitions and 9 clocks on a mono-processor SUN Ultra1 with 64MB. We have then used KRONOS to verify some safety requirements on the system. We first verify that two sensor's readings are not processed at the same time, which can be described in TCTL by:

$init \Rightarrow \forall\Box(after(\text{StartControl1}) \Rightarrow$
                $\forall\Box_{[0,1[} \neg(enable(\text{StartControl1}) \; or \; enable(\text{StartControl2})))$
$init \Rightarrow \forall\Box(after(\text{StartControl2}) \Rightarrow$
                $\forall\Box_{[0,2[} \neg(enable(\text{StartControl1}) \; or \; enable(\text{StartControl2})))$

These two formulas state that during the processing of one of the sensor, no other processing can start. The first formula has been verified in $7.3s$ and the second in $39.9s$.

We have then used KRONOS to define the values of **period1** and **period2** which insures that all the sensor's readings are processed in time. The system is then considered safe if the action **error** is not reachable which can be represented by the following TCTL formula:

    $init \Rightarrow \forall\Box\neg enable(\textbf{error})$

This formula states that it is not possible to reach a state where the action **error** is enabled from the initial state of the automaton.

period1	5	6	7	5	4	3	3	4	4	4	3
period2	3	3	3	4	3	3	4	4	5	6	5
time	23s	27s	30s	88s	21s	21s	28s	27s	95s	453s	39s
eval	true	true	true	false	true	true	true	true	false	false	true

Table 1 Verification results with KRONOS

The table 1 shows the results obtained for various values of the **period** parameters; the line labeled by **time** gives the running times given in seconds and the **eval** one gives the results of the TCTL formula evaluation. We have tried various configurations and have found 8 configurations which insure the safety of the system. Other kind of systems have also been verified where processing time of the controllers are non deterministic and where a premption time is considered. The ET-LOTOS specification must only be slightly changed to consider these systems.

7 CONCLUSION AND FURTHER WORKS

We have presented, in this paper, a method which allows the transformation of all the ET-LOTOS operators in a subclass of timed automaton with timers. We have shown that this subclass is equivalent to timed automaton by providing a conservative transformation between the two models. This work has allowed to capture a subset of ET-LOTOS where state reachability is decidable and to develop a tool which allows the verification of real-time logic formulas, expressed in TCTL, on an ET-LOTOS specification. This approach gives nice results, as shown with our robot controller example, but suffers of the state explosion problem. Even for quite small ET-LOTOS specifications the corresponding timed automata are large. Moreover, the KRONOS tool can only analyze, in a reasonable computation time and memory space, automata with no more than approximatively 50.000 states. The execution time depends also of the TCTL formula under verification. Further works will optimize the transformation method to obtain smaller timed automata. Nevertheless, this optimization does not really resolve the state explosion problem.

Our tool is limited to a subset of ET-LOTOS. We are extending it to full ET-LOTOS. The idea is to used the hybrid automaton model of HyTech [6] which can be used to support the data part of ET-LOTOS. This tool implements a semi-decision procedure for the reachability analysis of hybrid automata. This method does not resolve the state explosion problem and moreover, the new intermediate model will be more complex than timed automata. We are expecting from this last point, less effective results than with KRONOS.

Our approach is a first step in the analyze and tool development of timed process algebras. It can be used for other process algebras and especially for E-LOTOS whose timed semantics is based on ET-LOTOS. Another approach will be to use an intermediate representation of ET-LOTOS which avoid the explosion problem, like Timed Petri Nets, and to develop a verification technique adapted to this model. It will then be possible to develop more adapted methods to the verification of timed process algebras than the one used here.

REFERENCES

[1] R. Alur and C. Courcoubetis and D.L. Dill (1990). Model-checking for real-time systems, in Proceedings of the 5th Symposium on Logic Computer Science, pages 414-425.

[2] R. Alur, C. Courcoubetis, N. Hallbwachs, T.A. Henzinger, P.-H HO, X. Nicollin, A. Olivero, J. Sifakis and S. Yovine (1995). The algorithmic analysis of hybrid systems, in TCS 138, pages 3-34.

[3] J.-P. Courtiat and R.C. de Oliveira (1995). A Reachability Analysis of RT-LOTOS Specifications, in Formal Description techniques VIII, chapman & Hall, pages 117-124.

[4] C. Daws and A. Olivero and S. Tripakis and S. Yovine (1996). The tool KRONOS, in Hybrid Systems III, Lecture Notes in Computer Science 1066, Springer-Verlag

[5] C. Daws and A. Olivero and S. Yovine (1994). Verifying ET-LOTOS programs with KRONOS, in Seventh International Conference on Formal Description Techniques, Chapman & Hall, pages 227-242.

[6] T. A. Henzinger and P. H. Ho (1994). HyTech: the cornell HYbrid TECHnology tool, in Hybrid Systems II, Lecture Notes in Computer Science 999, Springer-Verlag, pages 265-294.

[7] C. Hernalsteen (1997). Timed automaton with semi timers for ET-LOTOS verification, Technical report TR-363, Computer science department, Free University of Brussels.

[8] C. Hernalsteen and T. Massart (1995). An extended timed automaton to model ET-LOTOS Specification, in participant's proceedings of DARTS'95, pages 95-112

[9] Luc Léonard and Guy Leduc (1994). An Enhanced Version of Timed LOTOS and Its Application to a Case Study, in FORTE-VI, pages 483-500, North-Holland.

[10] X. Nicollin and J. Sifakis (1991). An overview and synthesis on timed process algebra, in Proc. 3rd Workshop on Computer-Aided Verification.

[11] P. Varaiya (1997). SHIFT: A language for simulating Interconnected Hybrid Systems, in Hybrid and Real-Time Systems, Lecture Notes in Computer Science 1201, Springer-Verlag.

## PART FOUR

Verification Technique

13
Automatic Checking of Aggregation Abstractions Through State Enumeration

Seungjoon Park   *Satyaki Das*   *David L. Dill*
Computer Systems Laboratory, Stanford University
Gates {358, 312, 314}, Stanford University, Ca 94305, U. S. A.
{park@turnip, satyaki@turnip, dill@cs}.stanford.edu

Abstract

We present a technique for checking aggregation abstractions automatically using a finite-state enumerator. The abstraction relation between implementation and specification is checked on-the-fly and the verification requires examining no more states than checking a simple invariant property. This technique can be used alone for verification of finite-state protocols, or as preparation for a more general aggregation proof using a general-purpose theorem-prover. We illustrate the technique on the cache coherence protocol in the FLASH multiprocessor system.

Keywords

Automatic verification, cache coherence protocols, distributed systems, aggregation abstraction, formal methods

1 INTRODUCTION

Formal verification of a system design compares two different descriptions of the system: the *specification* describes the desired behavior, and the *implementation* describes the actual behavior of the system. The implementation is usually given in some (potentially) executable form. There are many specification methods, such as assertions in the implementation code, temporal logic or the other logical properties, or automata. However, the most appropriate specification for a protocol is often an abstract version of the protocol with coarser-grained atomicity. For example, most cache coherence protocols are intended to simulate atomic memory operations using non-atomic sequences of steps which execute in a distributed environment. Verification of such a protocol ultimately requires comparing the implementation protocol with the specification protocol with respect to some consistency criterion.

Previously, we developed a proof methodology called "aggregation" for relating a protocol to its abstract version by providing an abstraction function

Formal Description Techniques and Protocol Specification, Testing and Verification
T. Mizuno, N. Shiratori, T. Higashino & A. Togashi (Eds.)   © 1997 IFIP. Published by Chapman & Hall

which reassembles individual implementation steps into atomic transactions in a specification protocol [25, 24]. This method addresses the primary difficulty with using theorem proving for verification of real systems, which is the amount of human effort required to complete a proof, by making it easier to create appropriate abstraction functions.

The aggregation method is applicable when the description attempts to simulate a set of atomic *transactions*, where each transaction has a *commit step*. The user provides an *aggregation function* which maps an implementation state to a specification state by completing any committed but incomplete transactions, and correspondence between implementation steps and specification steps. Given correct correspondence, an aggregation function, and a proper invariant, a theorem prover can finish the proofs automatically (or semi-automatically depending on its level of automation).

Although the aggregation method makes verification using a theorem-prover much easier than it would otherwise be, use of theorem provers is still more labor-intensive than using algorithmic verification such as finite-state methods, especially when protocols are incorrect. In particular, finite-state methods automate the most difficult part of many verification efforts: the derivation of an adequate invariant.

In this paper, we propose a technique that checks aggregation abstractions for finite-state systems automatically using a finite-state enumerator. We reduce the problem of checking the aggregation correspondence to the simpler problem of checking an invariant (an "AG property" for those familiar with CTL [3]) by generating a set of propositional properties from the correspondence requirements of the aggregation method. This method can be used alone for verification of finite-state distributed protocols, or to debug aggregation abstractions before theorem proving. Using the finite-state method can greatly reduce the amount of human effort required to complete a proof. Of course, there is a tradeoff: finite state methods only work for small instances of a system design (e.g., a multiprocessor with three processors and one bit of memory), while a theorem prover can show the correctness of *all* instances of the system, for any number of processors and arbitrarily large memory.

Our technique is practical and much less expensive than other finite-state methods for proving abstract relations between implementation and specification (for comparison, see the section on related work). The number of states searched during verification using our technique is same as checking propositional safety properties on a finite-state system. When the aggregation correspondence holds, the method generates only the reachable states of the implementation. When the aggregation does not hold, state generation ceases when the first violating state is detected. The aggregation method is more efficient, because the abstraction is between *transition rules* of source-level descriptions, not between *state transitions* in state graphs. While it may require more human effort to define the aggregation functions than, say, finding a simulation relation automatically, this may be an appropriate tradeoff when the state explosion problem obstructs a more automatic proof.

The method is compatible with theorem proving because the same description of the specification protocol, the implementation protocol, and the aggregation function can be used for both automatic checking and theorem-proving of the aggregation abstraction. It allows the user to debug the implementation, specification, and aggregation functions quickly before invoking the theorem prover for proving the correctness of an unbounded implementation or infinite family of implementations. The state enumerator can also be used to help debug invariants and check some lemmas on examples before trying to prove them formally. Obviously, the same general technique can be used with any program capable of enumerating the reachable states of a system description, including BDD-based model checkers* [22]. Indeed, it may outperform other methods using abstraction in BDD-based model checkers, for some applications.

Background and related work

The use of abstraction functions and relations of various kinds (also called refinements [1, 19], homomorphisms [16], and simulations [23]) to compare two descriptions is a fundamental verification technique that can be applied to many different problems and representations in many different ways (e.g. [21, 18, 6]).

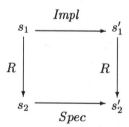

Figure 1 Abstraction relation

Since the details of these methods vary greatly, it is difficult to find a simple general principle underlying them all. However, at a very high level, many of them can be seen to involve proving a property something like (as shown in Figure 1):

$$\forall s_1, s_1', s_2 : \exists s_2' : R(s_1, s_2) \land Impl(s_1, s_1') \Rightarrow R(s_1', s_2') \land Spec(s_2, s_2'), \qquad (1)$$

---

*BDD-based model checkers use a binary decision diagram [2] to represent a Boolean function that describes a set of states symbolically.

where *Impl* is a step of the implementation, *Spec* is a corresponding step of the specification, and R is a binary relation from an implementation domain to a specification domain.

One approach using finite-state methods is to prove that a *simulation preorder* holds between the implementation and specification state graphs. For simulation preorder checking, the user provides descriptions of the implementation and specification state graphs and an initial relation which must contain the desired simulation relation. (For example, the initial relation could require that whenever an implementation state is paired with a specification state, the two states have the same label.) Given this information, the existence of a simulation relation (and the most general relation) can be computed by iterative elimination of states that cannot satisfy property (1).

Simulation preorder checking can be computed "on-the-fly" if the state graphs are given implicitly as a set of rules or finite-state programs [4, 5, 15, 14]. Unfortunately, in the worst case, this computation is linear in the size of the *product* of the implementation and specification graphs. In both theory and practice, checking simulation preorder between implementation and specification graphs is much more expensive than our technique, which costs the same as checking a simple safety property on the implementation graph alone.

Simulation preorder checking can also be performed on graphs represented by BDDs by using a symbolic fixed-point algorithm [7]. However, this requires dealing with relations containing Boolean variables of both the implementation and specification state graphs, so the cost of verification using this method is also much greater than that of checking a simple property on an implementation state graph alone.

Another approach using abstraction with BDDs is found in [20, 10, 9]. The method claims that a concrete program satisfies a property specified with CTL formulas if the abstracted program satisfies the corresponding property by an abstraction relation. To apply this method, the user must find an abstraction relation that preserves the property given in CTL formulas. There are two problems with this method from our perspective. First, we are interested in using a *protocol as the specification*. It is difficult or impossible to specify a protocol completely using CTL. Second, this method uses BDDs or some similar symbolic representation. Yet, we have found that explicit state enumeration greatly outperforms straightforward BDD-based verification for some classes of descriptions [13], such as all those described below.

A direct approach would appear to checking inclusion of the language of a finite automaton describing implementation behavior in the language of another automaton describing specification behavior, possibly using an on-the-fly algorithm. If the specification automaton is nondeterministic, this operation is generally exponential in the size of the specification automaton (some individuals have finessed this problem by requiring the specification automaton to be presented in a complemented form [16]). If the specification automaton is deterministic, the algorithm for inclusion checking is basically identical to simulation preorder checking.

Recently, there has been proposed an approach to using a model checker for comparing a specification protocol and an implementation protocol [11]. However, the technique uses a model checker simply to run the two protocols in parallel without defining a precise abstraction relation between the two protocols. Moreover, the size of each state checked by the technique is increased by that of a specification protocol.

2 THE AGGREGATION ABSTRACTION

This section describes the aggregation method in general. The verification method begins with logical descriptions of state graphs of the implementation and the specification. The implementation description contains a set of state variables; the set Q of states of the implementation is the set of assignments of values to the state variables. The specification description may contain a subset of the state variables of the implementation. Each description also specifies a transition relation between a state and its possible successors represented by a set of functions. An implementation step $Impl_i$ maps a given implementation state to its next state. Similarly, a specification step $Spec_i$ maps a given specification state to its next state. The specification contains *idle* transitions which map a state to itself.

The aggregation abstraction works when the computation can be thought of as implementing a set of transactions and each transaction has an identifiable commit step. Based on the reasoning about the commit steps, the user defines an aggregation function *aggr* which maps an implementation state to a specification state by first completing any committed but incomplete transactions, then hiding variables that do not appear in the specification. The commit steps in the implementation correspond to atomic transactions in the specification, and the other steps in the implementation correspond to an idle transition in the specification. For each pair of corresponding implementation step and specification step (for convenience, we assumed $Impl_i$ corresponds to $Spec_i$), the aggregation function should satisfy the following commutativity requirement,

$$\forall q \in Q : aggr(Impl_i(q)) = Spec_i(aggr(q)). \tag{2}$$

The number of proofs required is equal to the number of transition functions in the implementation.

The requirements (2) will generally not hold for some absurd states that cannot actually occur during a computation. Hence, it is usually necessary to provide an *invariant* predicate, which characterizes a superset of all the reachable states. If the invariant is *Inv*, the requirement can then be weakened to

$$\forall q \in Q : Inv(q) \Rightarrow aggr(Impl_i(q)) = Spec_i(aggr(q)). \tag{3}$$

In other words, *aggr* only needs to commute when q satisfies the *Inv*.

Use of an invariant incurs some additional proof obligations. First, we must find a proper invariant that makes (3) satisfied, and second, we must prove that the invariant is true in the implementation description by showing that the invariant holds at initial states and each implementation step preserves the invariant. From our experience, finding and proving an inductive invariant is the most time consuming part of many verification problems. It is especially difficult to debug faulty invariants using only a theorem prover. One of the great advantages of finite-state verification methods is that they compute the required invariant (the reachable state space) automatically.

3 CHECKING AGGREGATION ABSTRACTIONS ON-THE-FLY

To check the aggregation abstraction automatically, we use a finite-state enumerator which explores all and only the reachable states of the implementation on-the-fly. Because state enumeration generates the exact invariant of the system while searching the state space, the user can check property (2) above without proving property (3).

Given a purported aggregation function and correspondence between implementation steps and specification steps, the requirements are expressed as a Boolean condition on the implementation state which consists of a set of conjuncts corresponding to each implementation step:

$$\bigwedge_i aggr(Impl_i(q)) = Spec_i(aggr(q)). \tag{4}$$

The aggregation abstraction holds if the Boolean condition is true on all the reachable states in the implementation. Therefore, the aggregation abstraction can be automatically checked using any finite-state enumerator which is able to check such propositional properties. Although we used the Murφ verifier [8] for this purpose, the technique could be used with other model checkers, including model checkers based on BDDs [22] or other symbolic representations.

3.1 Murφ description language and verifier system

Murφ is a high-level description language for modeling finite-state asynchronous concurrent systems. The description allows the declaration of familiar data types, including subranges of integers, arrays, records, and user-defined enumerations. Additionally, procedures and functions can be declared. A Murφ program is an implicit description of a state graph, which consists of a collection of *state variables* and *transition rules*. The states of the graph are assignments to each global state variables with a value in the range of the declared type. The transition rules transform states to states by assigning to the state variables, so they define the edges of the state graph. Each rule has an enabling condition, which is a Boolean expression on the state variables, and an action, which is a statement that modifies the values of the state variables, generating a new state:

    `Rule` *condition* `==>` *action_statement* `Endrule`.

The action statement is an arbitrarily complex statement in a fairly conventional programming language with assignment, if-then-else, loops, procedure calls, and local variables.

Murφ has an automatic verifier which generates all of the reachable states of the described system. Execution of a Murφ program begins with one of a set of initial states of the graph. Then the following loop is executed forever: some rule whose condition is satisfied by the current state is chosen and its action evaluated, yielding a new current state. If there are no rules whose conditions are true, the execution halts. Although the action may be a compound statement consisting of a sequence of smaller statements, conditionals, and loops, it is executed *atomically*—no other rule can be executed before the action completes. When several rule conditions are true at the same time, a choice is made arbitrarily, resulting in several possible executions. The Murφ verifier tries them *exhaustively* by depth-first or breadth-first search.

Several types of errors can be detected while the verifier explores the state graph. An *invariant* which is a Boolean expression on the global state variables is checked in any reachable state. An *assert* statement, which is a Boolean condition specified in an action statement, can also be checked whenever the verifier gets to the specified point to execute the description. The system can detect *deadlock states*, which are states that have no other states as successors. If a problem of any type is detected, the verifier prints out a *diagnostic trace*, which is a sequence of states that leads to a state exhibiting the problem.

3.2 Checking aggregation abstractions using Murφ

Normally, a description of a single protocol implementation is specified in Murφ, and simple properties such as invariants and in-line assertions are checked. For aggregation, we need also to include a specification protocol and an aggregation function. First, we embed all of the implementation state variables in a single record type; similarly, we embed the specification protocol state variables in a second record type. The aggregation function *aggr* is written as a function in Murφ. The specification steps are also written as functions, which take a specification state (record) as an argument and which return a modified specification state.

A straightforward way of checking the Boolean condition (4) is using an *invariant* of Murφ. We specify the propositional predicate (which may be a big Boolean expression) as a single invariant and then run the verifier to check it on every reachable state. Similarly, this can be done using other model checkers: e.g., a CTL model checker by specifying the same condition as an AG property.

However, the requirements can be checked more efficiently if we exploit the property that each requirement (2) for an implementation step matters only when the step is enabled and that not all the implementation steps are enabled from a state. Using in-line *assert* commands of Murφ, each conjunct of the Boolean predicate can be checked separately on-the-fly only when the corresponding step is enabled and generates a next state by executing its action statement. To this end, we add an assert statement to each rule of the implementation description as shown in the following (the original rule was of the form `Rule CONDITION ==> Begin ACTION-STATEMENT; Endrule;`):

```
Rule "Transition relation for Impl_i"
  CONDITION
==>
Var i0, i1: ImplState;
Begin
  i0 := current_state;  ACTION-STATEMENT;  i1 := current_state;
  Assert Spec_i(aggr(i0)) = aggr(i1);
Endrule;
```

The two local variables i_0 and i_1 contain the implementation state before and after the execution of the rule respectively. The assert statement expresses the corresponding commutativity requirement using the functions defined for specification steps and the aggregation function.

The Murφ verifier will automatically check the assertions on-the-fly on all the reachable states while exploring the state space of the implementation description. Note that the specification steps are written as functions and called and computed in local variables inside the assert statements, while the implementation steps are written as statements which are executed to generate next states in the description. The specification state is not saved between rule executions—only the implementation contributes to the states. Therefore, the number of states explored by the verifier is still *the same* as that of the implementation description. Consequently, the amount of memory needed to check the abstraction is the same as that needed to check the reachable states of the implementation only.

4 EXAMPLE: FLASH CACHE COHERENCE PROTOCOL

This section illustrates our technique on the cache coherence protocol used in the Stanford FLASH multiprocessor system [17, 12].

4.1 Informal description of the protocol

The system consists of a set of nodes, each of which contains a processor, caches, and a portion of global memory of the system. The distributed nodes communicate using asynchronous messages through a point-to-point network. The state of a cached copy is in either *invalid, shared* (readable), or *exclusive* (readable and writable). The cache coherence protocol is directory-based so that it can support a large number of distributed processing nodes. Each cache line-sized block in memory is associated with *directory header* which keeps information about the line. For a memory line, the node on which that piece of memory is physically located is called *home*; the other nodes are called *remote*. The home maintains all the information about memory lines in its main memory in the corresponding directory headers.

If a read miss occurs in a processor, the corresponding node sends out a GET request to the home (this step is not necessary if the requesting processor is in the home). Receiving the GET request, the home consults the directory corresponding to the memory line to decide what action the home should take. If the line is *pending*, meaning that another request is already being processed, the home sends a NAK (negative acknowledgment) to the requesting node. If the directory indicates there is a dirty copy in a remote, then the home forwards the GET to that node. Otherwise, the home grants the request by sending a PUT to the requesting node and updates the directory properly. When the requesting node receives a PUT reply, which returns the requested memory line, the processor sets its cache state to *shared* and proceeds to read.

For a write miss, the corresponding node sends out a GETX request to the home. Receiving the GETX request, the home consults the directory. If the line is *pending*, the home sends a NAK to the requesting node. If the directory indicates there is a dirty copy in a third node, then the home forwards the GETX to that node. If the directory indicates there are shared copies of the memory line in other nodes, the home sends INVs (invalidations) to those nodes. Then the home grants the request by sending a PUTX to the requesting node*. If there are no shared copies, the home simply sends a PUTX to the requesting node and updates the directory properly. When the requesting node receives a PUTX reply which returns an exclusive copy of the requested memory line, the processor sets its cache state to *exclusive* and proceeds to write.

---

*This is the case when the multiprocessor is running in EAGER mode. In DELAYED mode, this grant is deferred until all the invalidation acknowledgments are received by the home.

During the read miss transaction, an operation called sharing write-back is necessary in the following "three hop" case. This occurs when a remote processor in node R_1 needs a shared copy of a memory line an exclusive copy of which is in another remote node R_2. When the GET request from R_1 arrives at the home H, the home consults the directory to find that the line is dirty in R_2. Then H forwards the GET to R_2 with the source of the message *faked* as R_1 instead of H. When R_2 receives the forwarded GET, the processor sets its copy to *shared* state and issues a PUT to R_1. Unfortunately, the directory in H does not have R_1 on its sharer list yet and the main memory does not have an updated copy when the cached line is in the shared state. The solution is for R_2 to issue a SWB (sharing write-back) conveying the dirty data to H with the source *faked* as R_1. When H receives this message, it writes the data back to main memory and puts R_1 on the sharer list.

When a remote receives an INV, it invalidates its copy and then sends an acknowledgment to the home. There is a subtle case with an invalidation. A processor which is waiting for a PUT reply may get an INV before it gets the shared copy of the memory line, which is to be invalidated if the PUT reply is delayed. In such a case, the requested line is marked as invalidated, and the PUT reply is ignored when it arrives.

A valid cache line may be replaced to accommodate other memory lines. A shared copy is replaced by issuing a replacement hint to the home, which removes the remote from its sharers list. An exclusive copy is written back to main memory by a WB (write-back) request to the home. Receiving the WB, the home updates the line in main memory and the directory properly.

4.2   The aggregation function

To define the aggregation function *aggr*, we first identify commit steps of each transaction in the protocol. For a transaction processing a read miss (or a write miss), the commit step occurs when the home, or a remote with an exclusive copy, sends a PUT (or PUTX) reply, granting the request. A write-back transaction begins with invalidating an exclusive copy and sending a WB request to the home; and this is the commit step of the transaction because a part of the specification variables are already updated at this moment and the write-back request can not be denied by the home.

The aggregation function simulates completing all committed transactions in the current state. If there exists a PUT message destined to a node i, the transaction for a read miss in node i must be completed by simulating the effect of node i processing the PUT message it receives at the end of the transaction: putting the data in the message into its cache and setting the state to *shared*. The transaction for a write miss is similarly completed by processing a PUTX message. There are two more kinds of messages possibly generated at commit steps and need to be processed to complete the committed transactions: SWB and WB to the home. Note that there exists at most one message of the four types destined to a particular node at any time. This processing changes values and states of cached copies, and values in main memory. Changes to implementation variables, such as removing messages from the network, and resetting the waiting flag in the processor can be omitted from the completion function, as they do not affect the corresponding specification state.

The aggregation function processing all the messages as described can be easily written in Murφ using a "for-loop" indexed on the network queue. Figure 2 shows the definition of the function. An implementation state is declared as a record consisting of an array *PNet* for network containing reply messages, an array *QNet* for network containing request messages, and an array *Procs* modeling processors with caches. Each message is also a record containing fields for its destination *dst*, source *src*, and data *Data*. From an implementation state *ist*, the function computes a specification state using a local variable *sst* to be returned. First, the specification variables of *ist* is copied into *sst*. Then, in the second for-loop, *sst* is modified by simulating to process each message in the network in the implementation state *ist* if it is one of such types that completes a committed transaction.

4.3   Checking the aggregation abstraction using Murφ

We illustrate the details on one of the implementation step (i.e., one of the requirements) of the protocol: a commit step of the transaction processing a write miss. Figure 3 shows the Murφ function for the specification step which corresponds to the transaction. As before, a specification state is declared as a record consisting of an array of cache states and data for each processor, and main memory. The function returns a specification state which is obtained by processing a write miss transaction atomically: if *oldproc* owns an exclusive copy, the exclusive data is transferred to processor *newproc*; otherwise if there is no exclusive copy in any processors, an exclusive copy is granted to processor *newproc* by copying the data in main memory to its cache.

To check the aggregation abstraction on-the-fly, we make sure that the rules in Murφ description correspond exactly to the implementation steps *Impl$_i$* of the protocol. Figure 4 presents the detailed rule for the implementation step where a remote node having an exclusive copy grants the ownership transfer by sending a PUTX reply to the requesting node. The guard condition of the rule checks if there is a GETX request on the head of the request queue from *src* to *dst* (which is a remote) and the node *dst* contains an exclusive copy of the memory line. In the action statement, the processor in the node *dst* invalidates its own copy and sends out a PUTX reply to the requesting node.

Without changing the original description, we simply add a few lines of commands to check the commutativity requirement (additions to the original

```
Function Faggr(ist:ImplState): SpecState;
Var  sst: SpecState;       -- specification state to be returned
Begin
 For i:Proc Do                               --
   sst.State[i] := ist.Procs[i].Cache.State;  -- Copy the specification
   sst.Data[i]  := ist.Procs[i].Cache.Value;  -- variables from the current
 EndFor;                                     -- state of implementation
 sst.Memory   := ist.Memory;                 --

 For i:Queue do            -- Check each message in the network
   If i < ist.PNet.Count then   -- for the reply queue
     If ist.PNet.Message[i].Mtype=Putt then
       -- Simulate processing a 'put' reply
       If ist.PNet.Message[i].dst=Home then
         sst.Memory := ist.PNet.Message[i].Data;
       Endif;
       If ! ist.Procs[ist.PNet.Message[i].dst].Cache.InvMarked then
         sst.Data[ist.PNet.Message[i].dst] := ist.PNet.Message[i].Data;
         sst.State[ist.PNet.Message[i].dst] := Shared;
       Else
         sst.State[ist.PNet.Message[i].dst] := Invalid;
       EndIf;
     Elsif ist.PNet.Message[i].Mtype=PutX then
       -- Simulate processing a 'putx' reply
       sst.Data[ist.PNet.Message[i].dst] := ist.PNet.Message[i].Data;
       sst.State[ist.PNet.Message[i].dst] := Exclusive;
     endif;
   Endif;
   If i < ist.QNet.Count then   -- for the request queue
     -- Simulate processing a 'WB' or a 'ShWB' sent to the home
     if ist.QNet.Message[i].Mtype=WB | ist.QNet.Message[i].Mtype=ShWB
     then sst.Memory := ist.QNet.Message[i].Data;
     endif;
   Endif;
 EndFor;
 Return sst;
End;
```

Figure 2 The aggregation function written in Murφ

implementation description are marked with stars in the figure). First, local variables i_0, i_1, s_0, and s_1 are declared to be used to copy the implementation states and specification states, respectively, before and after the execution of the rule. Because this implementation step is the commit step of a write-miss transaction, the corresponding specification step is "Atomic_GetX" in Figure 3. Using the declaration of the specification step and the aggregation function, the assert statement explicitly specifies the corresponding commutativity requirement.

```
Function Atomic_GetX(st:SpecState;   oldproc,newproc:Proc): SpecState;
Begin
  If st.State[oldproc] = Exclusive & oldproc != newproc then
    st.State[oldproc] := Invalid;
    st.State[newproc] := Exclusive;
    st.Data[newproc] := st.Data[oldproc];
    Return st;
  Elsif Forall i:Proc Do st.State[i] != Exclusive EndForall then
    st.State[Home] := Invalid;
    st.State[newproc] := Exclusive;
    st.Data[newproc] := st.Memory;
    Return st;
  Else Return st;
  Endif;
End;
```

Figure 3 The specification step processing a write miss in the FLASH protocol

```
Ruleset i : Queue Do       -- Parameterized rule for each message in the queue
Alias request: QNet.Message[i];   -- Choose an arbitrary request in the queue
      dst    : request.dst;       -- the destination of the request
      SRC    : request.SRC Do     -- the node that initiated the request
Rule "NI Remote GetX (Commit)"
  TopRequestTo(i)               -- The message is the oldest among those src -> dst
  & request.Mtype = GetX        -- The message is a 'getx' request
  & request.dst != Home         -- The receiving node is a remote
  & Procs[dst].Cache.State = Exclusive   -- The node dst has an exclusive copy
  & Pspace(1) & Qspace(1)       -- There are enough spaces in the network
  ==>
* Var  i0,i1: ImplState;  s0,s1: SpecState;   -- local variables
  Begin
*    i0 := ThisImplState();
*    s0 := Faggr(i0);
     Cache.State := Invalid;
     Send_Reply(dst, SRC, PutX, Cache.Value);
     if SRC != Home then Send_Request(dst, Home, FAck, SRC, void); end;
     Consume_Request(i);
*    i1 := ThisImplState();
*    s1 := Faggr(i1);
*    s0 := Atomic_GetX(s0, i0.QNet.Message[i].dst, i0.QNet.Message[i].SRC);
*    Assert Equiv(s0,s1) "NI Remote GetX (Commit)";
Endrule;
Endalias;
Endruleset;
```

Figure 4 A sample Murφ rule with an assert statement for checking the corresponding commutativity requirement. The lines marked with a star are additions to the original implementation description.

4.4 Experiments

By the aggregation abstraction, the protocol consisting of more than a hundred different implementation steps has been reduced to a specification with only six kinds of atomic transactions. It is much easier or trivial to prove important properties of the reduced model, such as the consistency of data at the user level, than the original protocol description.

To check the abstraction automatically, we have run Parallel Murφ on 32 ULTRA SPARC processors [27]. For the protocol with 3 processing nodes and request/reply message queues of size 5, the verifier explored 457,558 states in 126 seconds; for 4 processing nodes and queues of size 3, about 19 million states in 72 minutes. As expected, the number of states explored (also the usage of memory) is exactly same as that found for exploring reachable state space of the implementation model.

5 CONCLUSION

By limiting the verification problems to finite-state systems, we proposed an efficient technique for checking aggregation abstractions without reasoning about invariants of the system. The verification requires checking only the same number of states as in the implementation model by exploiting the correspondence information provided by the user.

The technique can be used alone for verification of finite-state protocols, since abstract protocols are sometimes the best properties to check. The technique can also be applied before theorem proving of aggregation abstractions to debug a purported aggregation function in early stage.

The FLASH protocol example has been verified before by applying aggregation abstraction using a general-purpose theorem-prover, which took two months [26]. However, the proof would have been much easier had we thought of this finite-state method before completing them. Use of the automatic checking can reveal any human errors in finding aggregation functions and specification models, and helps to debug them in early stage of proofs.

Acknowledgment

This research was supported by the Advanced Research Projects Agency through NASA grant NAG-2-891. We thank Ulrich Stern for helping run Parallel Murφ and David Culler at Berkeley for letting us use their NOW.

REFERENCES

[1] Martín Abadi and Leslie Lamport. The existence of refinement mappings.

Theoretical Computer Science, 82:253–284, 1991.

[2] Randal Bryant. Graph-based algorithms for Boolean function manipulation. *IEEE Transactions on Computers*, C-35(8), 1986.

[3] E.M. Clarke, E.A. Emerson, and A.P. Sistla. Automatic verification of finite-state concurrent systems using temporal logic specifications. *ACM Transactions on Programming Languages and Systems*, 8(2), April 1986.

[4] R. Cleaveland, J. Parrow, and B. Steffen. The concurrency workbench. In *Proc. of the Workshop on Automatic Verification Methods for Finite State Systems*, June 1989.

[5] Rance Cleaveland and Steve Sims. The NCSU Concurrency Workbench. In *Computer Aided Verification, 8th International Conference, CAV'96*, pages 394–397. Springer-Verlag, 1996.

[6] J. de Bakker, W. de Roever, and G. Rozenberg, editors. *Stepwise Refinement of Distributed Systems. Models, Formalisms, Correctness: LNCS 430.* Springer-Verlag, 1990.

[7] D. Dill, A. Hu, and H. Wong-Toi. Checking for language inclusion using simulation relation. In *Computer Aided Verification, 3rd International Workshop*, pages 255–265, July 1991.

[8] David L. Dill. The Murφ verification system. In *Computer Aided Verification, 8th International Conference, CAV'96*, pages 390–393. Springer-Verlag, July 1996.

[9] Susanne Graf. Verification of a distributed cache memory by using abstractions. In *6th International Conference on Computer-Aided Verification*, pages 207–219, 1994.

[10] Susanne Graf and Claire Loiseaux. A tool for symbolic program verification and abstraction. In *5th International Conference on Computer-Aided Verification*, pages 71–84, 1993.

[11] Klaus Havelund and N. Shankar. Experiments in theorem proving and model checking for protocol verification. In *Formal Methods Europe FME '96*, pages 662–681, March 1996.

[12] Mark Heinrich. *The FLASH Protocol.* Internal document, Stanford University FLASH Group, 1993.

[13] Alan John Hu. *Techniques for Efficient Formal Verification Using Binary Decision Diagrams*, chapter 4 on 'BDD Blow-Up Representing Sets of States', pages 41–49. Stanford University, December 1995. Ph.D. Thesis.

[14] C. Norris Ip and David L. Dill. Verifying systems with replicated components in Murφ. In *8th International Conference on Computer-Aided Verification*, pages 147–158, 1996.

[15] Chung-Wah Norris Ip. *State Reduction Methods for Automatic Formal Verification.* PhD thesis, Stanford University, December 1996.

[16] Robert Kurshan. *Computer-Aided Verification of Coordinating Processes: The Automata-Theoretic Approach.* Princeton, 1994.

[17] J. Kuskin, D. Ofelt, M. Heinrich, J. Heinlein, R. Simoni, K. Gharachorloo, J. Chapin, D. Nakahira, J. Baxter, M. Horowitz, A. Gupta, M. Rosenblum, and J. Hennessy. The Stanford FLASH multiprocessor. In *Proc. 21st International Symposium on Computer Architecture*, pages 302–313, April 1994.

[18] S. Lam and A. Shankar. Protocol verification via projection. *IEEE Transactions on Software Engineering*, 10(4):325–342, July 1984.

[19] Leslie Lamport. The temporal logic of actions. *ACM TOPLAS*, 16(3):872–923, May 1994.

[20] David Long. *Model Checking, Abstraction and Compositional Verification.* PhD thesis, Carnegie Mellon University, July 1993.

[21] N. Lynch. I/O automata: A model for discrete event systems. In *22nd Annual Conference on Information Science and Systems*, March 1988. Princeton University.

[22] Ken McMillan. *Symbolic Model Checking.* Kluwer Academic Publishers, 1993. Boston.

[23] R. Milner. An algebraic definition of simulation between programs. In *Proc. of the 2nd International Joint Conference on Artificial Intellegence*, pages 481–489, 1971.

[24] Seungjoon Park. *Computer Assisted Analysis of Multiprocessor Memory Systems.* PhD thesis, Stanford University, June 1996.

[25] Seungjoon Park and David L. Dill. Protocol verification by aggregation of distributed transactions. In *Computer Aided Verification, 8th International Conference, CAV'96*, pages 300–310. Springer-Verlag, July 1996.

[26] Seungjoon Park and David L. Dill. Verification of FLASH cache coherence protocol by aggregation of distributed transactions. In *Proc. 8th ACM Symposium on Parallel Algorithms and Architectures*, pages 288–296, June 1996.

[27] Ulrich Stern and David L. Dill. Parallelizing the Murφ verifier. In *Computer Aided Verification, 9th International Conference, CAV'97*. Springer-Verlag, June 1997.

14

Concept of Quantified Abstract Quotient Automaton and its advantage

G. JUANOLE and L. GALLON
LAAS-CNRS
7, avenue du colonel Roche
31077 Toulouse cedex 4 (France)
E-mail : {juanole, gallon}@laas.fr

Abstract

At first, we define the concept of quantified abstract quotient automaton (which concerns state graphs with qualitative and quantitative labels) and present a methodology for computing it. It is an abstract view which results, on the one hand, of the quotient automaton obtained according to an equivalence relation based on qualitative labels, and, on the other hand, of reduction rules applied to the quantitative labels. This abstract view has the nice property of offering both verification and evaluation capacities (interesting point for the analysis of Qualities of Service, for example). The advantage of this concept for the bottom-up modeling and analysis of multi-layer communication architectures is then demonstrated.

Keywords

Transition systems, Petri nets, qualitative and quantitative labels, verification, quality of service, evaluation, communication networks, real-time systems

1 INTRODUCTION

- The labeled transition systems (or still labeled state graphs) are formalisms extensively used for representing the dynamic behavior of distributed systems (set of communicating entities). A labeled state graph consists of a set of global states, a set of arcs linking the states and a set of labels associated

Formal Description Techniques and Protocol Specification, Testing and Verification
T. Mizuno, N. Shiratori, T. Higashino & A. Togashi (Eds.) © 1997 IFIP. Published by Chapman & Hall

with the set of the arcs. *More often the labels are of the qualitative type* appearing in the form of Predicate/Action (a Predicate can be a message reception; an Action can be a message emission). A labeled state graph allows, in particular, the analysis of specific properties of the system being modeled (properties which depend on the system mission and which can be analyzed through the interpretation of transition sequences and the associated labels).

- In many cases the observation that we want to do on a labeled state graph only concerns a few events (events which are relevant from a point of view) and then it is important, in order to facilitate the analysis, to get a reduced labeled state graph which is equivalent to the initial labeled state graph according to an equivalence relation (the reduced labeled state graph is still called the quotient automaton). The observational equivalence relation [Mil80] is a very powerful equivalence relation which has been largely used in the field of the communication protocol verification (verification relative to the service supplied [FJV90]).

- However today, in particular, *in the context of real-time systems*, we cannot be satisfied with the representation of the dynamic behavior of the systems with state graphs which only have qualitative labels on the transitions. *We have to use state graphs with both qualitative and quantitative labels* (durations and/or probabilities). In this context, the observational equivalence relation is still relevant (we want still to observe a system from the point of view of a few events) but we want too (in order to be able to evaluate the performances related to these events) to add to the quotient automaton quantitative informations (which depend, of course, on the quantitative labels of the initial labeled state graph).

The quantification of the quotient automaton offers an object which allows to merge verification and evaluation activities. In our knowledge, there were never works on the quantification of quotient automata. That is the goal of this paper. However we want to say that this study does not give new theory. It proposes to combine (that is new) already existing theoretical results (observational equivalence [Mil80]; quantitative abstract views [AJ95]) for the definition of the concept of Quantified Abstract Quotient Automaton and to show its advantage for the modeling and analysis in communication architectures.

This paper is divided into three sections. In the first section, we present the background which is necessary. In the second section, we present a methodology for computing a quantified abstract quotient automaton and illustrate it through an example of a service provided by a protocol. In the third section, by considering a two layer communication architecture, we show the advantage of the quantified abstract quotient automaton concept for evaluating qualities of service and controlling models sizes.

2 BACKGROUND

2.1 State graph with qualitative labels

(a) Definition

A finite state graph labeled with qualitative labels is a triplet $< S, E, \Delta >$ where :

- $S = (S_0, S_1, \ldots)$ is the set of the vertices of the graph; they represent the states of the system being represented; S_0 is the initial state,
- $E = (e_0, e_1, \ldots)$ is the set of labels (events),
- Δ is the set of the labeled arcs of the graph (i.e. $\Delta \subset S \times E \times S$); an element $(m, \mu, p) \in \Delta$ is denoted $m \xrightarrow{\mu} p$. $\xrightarrow{\mu}$ is a transition relation.

An example of such graph is represented on figure 1.

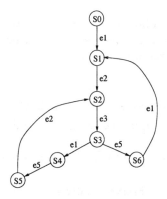

Figure 1 State graph with qualitative labels

(b) Observational equivalence relation and quotient automaton

Call E' the subset of the events of a state graph with qualitative labels that we want to observe (observable or visible events) and call τ an event which represents any event that we do not want to observe (τ is called an unobservable or invisible or internal event).

A transition relation $\{\xRightarrow{\mu}, \mu \in E' \cup \{\epsilon\}\}$ is defined in a standard way by :

- $m \xRightarrow{\epsilon} p : m = p$ or $m \xrightarrow{\tau} x_1 \xrightarrow{\tau} \ldots \xrightarrow{\tau} x_p \xrightarrow{\tau} p$
- $m \xRightarrow{\mu} p : m \xRightarrow{\epsilon} x_1 \xRightarrow{\mu} x_2 \xRightarrow{\epsilon} p$

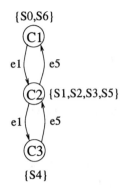

Figure 2 Quotient automaton

Figure 3 Detailed quotient automaton

Such relation represents what it is still called either ϵ-experimentation or μ-experimentation.

Now, we can define the observational equivalence relation (noted $\equiv$) which can be seen as a decreasing sequence of equivalence relations. This relation can be defined by recurrence [Ver89] :

- $\forall m, p \in S, m \equiv_0 p$
- $m \equiv_{k+1} p$ if and only if $m \equiv_k p$ and $\forall t \in E' \cup \{\epsilon\}$,
$$[m \overset{t}{\Rightarrow} n \text{ implies } \exists q \in S : p \overset{t}{\Rightarrow} q \text{ and } n \equiv_k q]$$
$$\text{and}$$
$$[p \overset{t}{\Rightarrow} q \text{ implies } \exists n \in S : m \overset{t}{\Rightarrow} n \text{ and } q \equiv_k n]$$

Informally, two states are observationally equivalent if the observed behaviors of the system being modeled from these states are similar : two states are equivalent at the order k if for any t-experimentation from one state, we have a t-experimentation from the other state which is equivalent at the order (k-1).

If we come back to the state graph of the figure 1 and if we consider the events e_1 and e_5 as the only observable events, we get the quotient automaton represented on the figure 2 (this quotient automaton has three state classes C_1, C_2 and C_3).

The decreasing sequence of the equivalence relations is :

$$\equiv_0 = \{S_0, S_1, S_2, S_3, S_4, S_5, S_6\}$$
$$\equiv_1 = \{\{S_0, S_6\}, \{S_1, S_2, S_3, S_4, S_5\}\}$$
$$\equiv_2 = \{\{S_0, S_6\}, \{S_1, S_2, S_3, S_5\}, \{S_4\}\}$$

(c) Detailled quotient automaton

In the quotient automaton, we do not worry about the links between the states in a state class (the information on these links is not relevant from the

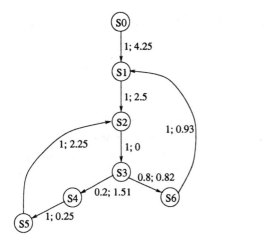

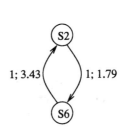

Figure 5 abstract view

Figure 4 state graph with quantitative labels

strict point of view of the observational equivalence relation). We propose to call detailed quotient automaton the quotient automaton where we have the representation of the links between the states of each state class (the detailed quotient automaton is immediately got, on the one hand, from the classes of the quotient automaton and, on the other hand, by looking at the initial state graph in order to have the links between the states in each state class).

The detailed quotient automaton of the state graph of the figure 1 is represented on the figure 3. Note that here (in the considered example) the transitions between the classes only result of the observable events (but, in a general case, event τ can also induce transitions between classes).

2.2 State graph with quantitative labels

(a) Definition

Graphs G that we consider are constituted of a set S of states and a set U of arcs connecting the states. Let an arc u_{ij} connecting a state S_i to a state S_j; the arc u_{ij} is characterized by a doublet : p_{ij} (transition probability), θ_{ij} (transition duration). An example is presented on the figure 4. Note that we only consider here graphs of the "1-graph" type i.e. graphs for which there never exists more than one arc between any two states S_i and S_j taken in this order.

In order to formally specify the quantitative informations, two types of square matrices of order equal to the cardinal of the set of the states (let Z this cardinal) are associated to the graph ([AJ95]) :

- a matrix $\mathcal{P}$ of direct transition probabilities between states : $\mathcal{P} = [p_{ij}]$ with

$$\forall i = 1, 2, \ldots, Z \text{ and } \forall j = 1, 2, \ldots, Z \text{ we have } 0 \leq p_{ij} \leq 1 \text{ and } \sum_{j=1}^{Z} p_{ij} = 1$$

- a matrix Θ of the durations of the transitions between states : $\Theta = [\theta_{ij}]$. Note that when the graph is not complete, i.e. with probabilities p_{ij} equaling zero between state pairs (S_i, S_j), it can be stated that there exists between these state pairs, fictitious transitions characterized by infinite durations.

(b)   Abstract view

An abstract view of a graph G is a reduced graph obtained from a projection on a subset of states from graph G (observed graph) : the world of observed states is equivalent to the sub-world of these states in the context of the graph G on the basis of the so-called notion of "first passage" ([AJ95]).

Let an arc u_{kl} connecting a state S_k to a state S_l of the reduced graph. The arc u_{kl} is also characterized by a doublet : F_{kl} (transition probability at the first passage between states S_k and S_l), T_{kl} (mean duration at the first passage between states S_k and S_l). F_{kl} (T_{kl}) depends on p_{kl} (θ_{kl}) but also on the probabilities and the durations between the states of the graph G which are not observed and which are on the paths between the states S_k and S_l. General formulas for obtaining F_{kl} and T_{kl} from the matrices $\mathcal{P}$ and Θ, and which are a generalization of the Beizer's reduction rules and the Chapman-Kolmogorov formulas [How71], are given in [AJ95].

For example, if we consider the graph of the figure 4, and by making a projection on the states S_2 and S_6, we get the abstract view of the figure 5.

2.3   State graph with qualitative and quantitative labels

A state graph with qualitative and quantitative labels is a state graph where each arc is labeled with a triplet (event, probability, duration). If we consider the graph of the figure 1 and we add to the arcs the labels of the graph of the figure 2, we get such a graph. Such a graph is an interesting structure, specially in the context of real-time systems, to evaluate Real-Time Computation Tree Logic [AES89] and/or Probabilistic Real-Time CTL [HJ91] formulas.

Generally, state graphs (with or without qualitative and/or quantitative labels) are not the first specification of a system, but represent the dynamic behavior of a higher level model like Petri nets, for example. We present now the Petri net based model that we use (Stochastics Timed Petri Nets (STPN) [JA91, JG95]).

2.4   Stochastic Timed Petri Nets (STPN)

(a)   Definition
An STPN is a Petri net where a probability density function is associated to each transition. A density probability function can be a continuous distribution (exponential, uniform), a deterministic distribution (duration $\neq 0$ or 0), a mixed distribution (uniform and discrete). Transitions with a duration equal to 0 are called immediate transitions (all the other transitions are called timed transitions). The selection of the transition to be fired [MBB+85] is based, in the general case, on the race model and, in some particular cases (transitions, with identical deterministic distributions, which are simultaneously enabled) on the preselection model with an equiprobable choice. The conditioning on the past history is the enabling memory [MBB+85].

The dynamic behavior is represented by a randomized state graph which is a state graph with qualitative and quantitative labels : a qualitative label is made of the name of the transition of the underlying Petri net which has induced a state change, and the name of the events associated to this transition; the quantitative label associated to a qualitative label is the doublet probability-duration of the transition firing.

(b)   Example

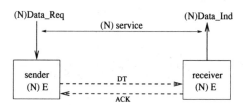

Figure 6 (N) layer

- Consider an (N) layer of a communication architecture (figure 6) which provides an unidirectional data transfer (N)service (the (N)primitives are (N)Data_ Req and (N)Data_Ind) and which is based on a "stop and wait" protocol [BSW69] (the sender (N)E sends the (N)PDU DT and waits for the (N)PDU ACK sent by the receiver (N)E; the (N)PDU losses are controlled by a time-out mechanism in the sender (N)E). Furthermore, concerning the ACK loss, in order to have not to consider a numbering scheme and then to have a simplest model, we suppose an intelligent medium in the direction receiver (N)E - sender (N)E : when there is an ACK loss, this medium signals the loss to the receiver which records this fact (then when the next DT (duplicated DT) will be received, the receiver (N)E will discard it and will send again ACK).

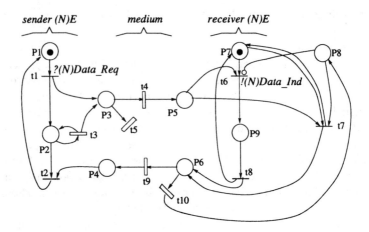

Figure 7 STPN model

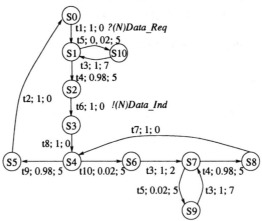

Figure 8 Randomized state graph

- The STPN model is represented on the figure 7 (immediate transitions are represented with a thin line; timed transitions are represented with a rectangle). Transitions t_1 and t_6 are respectively labeled with the (N)Data_Req reception (noted ?(N) Data_Req) and the (N)Data_Ind emission (noted !(N)Data_Ind). Transitions t_4 and t_9 represent the transmission of respectively DT and ACK (note that after an ACK loss, we put a token in place P_8 which records the fact "ACK loss"). The time-out and the retransmission are represented by the transition t_3.

  The choice of the distribution associated to the loss transitions (t_5, t_{10}) with respect to the transmission transitions (t_4, t_9) allow to simulate loss probabilities (general formulas are given in [JA91]).

- The randomized state graph (by considering the loss probability of 2.10^{-2}) is represented on the figure 8 (concerning the qualitative labels : we represent all the names of the transitions of the underlying Petri net, and only the events related to the (N) service).

3 QUANTIFIED ABSTRACT QUOTIENT AUTOMATON

3.1 Methodology for obtaining it

We have three steps.

The first step consists in obtaining the *quotient automaton according to the observational equivalence relation* (i.e. by only considering the qualitative labels and by observing a subset of events); the quotient automaton only provides the classes (with the states included in each class) and the transitions between the classes with their labels (visible or invisible event).

The second step consists in obtaining the *quantified detailed quotient automaton* i.e. we make to appear the arcs between the states in each state class (as shown in the subsection 2.1.3) and we associate to each arc (between the state classes and between the states in each state class) a probability and a duration (this quantification is immediate from the initial state graph with qualitative and quantitative labels). At this step, each class is represented by a quantified detailed view.

The third step consists in obtaining, *for each class*, a *quantified abstract view* i.e. we evaluate the quantified behavior as seen by its environment (the adjacent classes). This is the important point of the *quantified abstract quotient automaton* concept. We present it now.

3.2 Quantified abstract view of a state class

Consider a quantified detailed view of a state class C (figure 9). In the class, each transition (its qualitative labels (the invisible event τ) has not been represented here) has the quantitative labels p_x (probability) and θ_x (duration). Note that concerning the transitions coming and/or going to the others state classes, we only have represented here the qualitative labels (visible events e_i and invisible event τ).

We define two types of states : states which are at the frontier of the class with others classes and that we call frontier states (the states S_i, S_j, S_k, S_l are the frontier states of the class C), and the internal states which represent the behavior inside a class (the states S_m, S_n, S_o are the internal states of the class C).

Among the frontier states, we have input states (they represent entries in the class; the state S_i is an input state), output states (they represent exits from the class; the state S_l is an output state) and input/output states (they represent both entries and exits from the class; states S_j and S_k are input/output states).

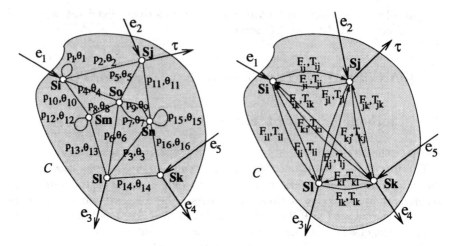

Figure 9 Quantified detailed view of a class

Figure 10 Quantified abstract view of a class

We can have loops on each type of state.

Definition :

> *A quantified abstract view of a class models the quantitative behavior of a class between the frontier states, i.e it is represented by a graph with quantitative labels which only has the frontier states as vertices and which is irreflexive (no loops).*

The quantified abstract view of the quantified detailed view of the figure 9 is represented on the figure 10. Each transition between an input frontier state S_a and an output frontier S_b is labeled with a probability F_{ab} and a duration T_{ab} which are respectively the probability and the mean duration of the transition between the states S_a and S_b on the basis of the so-called notion of "first passage" [How71] **and** knowing that since the departure instant from the frontier state S_a no other frontier state (except S_a if we have loops around it and/or circuits through internal states which come back to S_a) is traversed. The general formulas for computing F_{ab} and T_{ab} are easily obtained from the works presented in [AJ95] on the quantitative abstract views.

Remarks:

1. the reason of considering an irreflexive graph is that we want to represent the behavior of a class as strictly seen by the environment of the other classes (so loops, if any, need not to appear explicitly);
2. from a probabilistic point of view, a loop around a state S_a can always be removed by integrating its probability in the outgoing arcs probabilities;
3. if a quantified detailed view of a class has only frontier states and no loop, we have already its abstract view and we do not need obviously any computation.

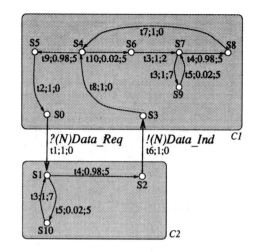

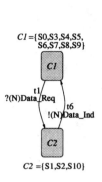

Figure 11 Quotient automaton

Figure 12 Quantified detailed quotient automaton

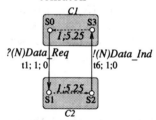

Figure 13 Quantified abstract quotient automaton

3.3 Example

Consider again the randomized state graph of the figure 8 (behavior of an (N) layer of a communication architecture) and suppose that we only are interested in the (N) service view. So we apply the quantified abstract quotient automaton concept (events ?(N)Data_Req and !(N)Data_Ind are the observed events).

Figures 11, 12 and 13 represent the three steps to obtain the quantified abstract quotient automaton. The figure 11 gives the quotient automaton (we have two state classes C_1 and C_2 (with the list of their states) which are linked with two arcs labeled with the name of the Petri net transitions and with ?(N)Data_Req and !(N)Data_Ind). The figure 12 gives the quantified detailed quotient automaton (this view combines the knowledge of the quotient automaton (figure 11) and the randomized state graph (figure 8)); the figure 13 gives the quantified abstract quotient automaton (each quantified detailed

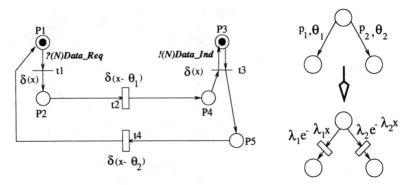

Figure 14 STPN model   **Figure 15** choice translation

class has been replaced by a quantified abstract class) i.e. we have the view which is relevant for an user of the (N) layer both in terms of qualitative and quantitative informations. Note that if we make changes in the loss probability values, the structure of the quantified abstract quotient automaton does not change (only the sojourn times in the classes C_1 and C_2 change).

3.4   Comments

We want to emphasize three points about the quantified abstract quotient automaton :

1. it allows to quantify the divergence phenomenon [Hen88] which can occur in a quotient automaton (if we have loops and/or circuits in a state class, we can then, from a qualitative point of view, stay endless in the class). The consideration of the quantitative information gives a realistic view;

2. it is a good concept to characterize both qualitatively and quantitatively Qualities of Service (QoS). For example, in the previous example, we can say :
   - from a qualitative point of view (the information was already in the quotient automaton), the (N) service is inevitably provided (sequence ?(N)Data_Req, !(N)Data_Ind);
   - from a quantitative point of view :
     - the time to provide the (N) service (i.e. from ?(N)Data_Req to !(N)-Data_Ind) is θ_1;
     - the period of the (N) service (i.e. from a ?(N)Data_Req to a new ?(N)Data_Req) is $\theta_1 + \theta_2$.

3. from the quantified abstract quotient automaton we can directly extract a stochastic timed Petri net model which can be used, for modeling the (N+1) layer (bottom-up modeling). For example, if we come back to the quantified abstract quotient automaton of the figure 13, we can translate it into the STPN model of the figure 14 (the places P_4 and P_5 and the

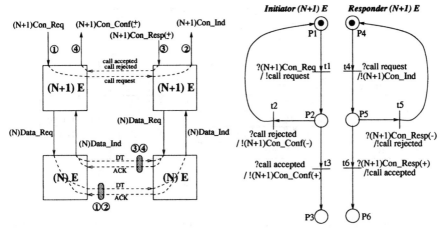

Figure 16 Two layer architecture

Figure 17 (N+1) entities

timed transition t_2 with the deterministic distribution $\delta(x - \theta_1)$ represent the class C_1; the places P_5 and P_1 and the timed transition t_4 with the deterministic distribution $\delta(x - \theta_2)$ represent the class C_2; the place P_1 represents the sender (N)E ready to receive an (N)Data_Req; the place P_2 represents the receiver (N)E ready to send (N)Data_ind; the transitions labeled with ?(N)Data_Req and !(N)Data_Ind are immediate $(\delta(x))$.

This example is simple in the sense that the automaton has no choice. But in the general case, we can have choices which will induce also Petri nets with choices (transitions in competition). Two sub-cases must be considered according to the values of the unconditional sojourn times in the states, from where there is a choice, are 0 or different from 0 :

- if 0, each transition in competition will be an immediate transition (i.e. a deterministic distribution $\delta(x)$) and will have a probability which is the probability of the corresponding transition in the automaton;
- if different from 0, each transition in competition will be a timed transition with an exponential distribution as it is represented on the figure 15. The exponential distributions are characterized by what are called equivalent rates [Flo85]: $\lambda_1 = \frac{p_1}{p_1\theta_1 + p_2\theta_2}$; $\lambda_2 = \frac{p_2}{p_1\theta_1 + p_2\theta_2}$.

4 ON THE ADVANTAGE OF THE QUANTIFIED ABSTRACT QUOTIENT AUTOMATON CONCEPT

Consider the two layer communication architecture represented on the figure 16 where the (N+1) layer must open a (N+1) connection by using a data transfer (N) service which is based on a fragmentation process ((N) SDU fragmentation) and a "stop and wait" protocol as the one presented in the

number of fragments	2	3	4	5
number of states	24	35	46	57

Table 1 unidirectional data transfer (N) service

number of fragments		2	3	4	5	
only	brute force	136	192	243	293	number of
Resp +	QAQA concept	15				states
Resp +	brute force	698	931	1139	1374	
and -	QAQA concept	26				

Table 2 (N+1) service (connection opening)

subsection 2.4.(b) We suppose that, like in the subsection 2.4.(b), we can have (N) PDU losses.

If we want to analyze and evaluate the (N+1) connection opening, we must model this communication architecture. This study can be made in two different ways :

1. either we adopt a "brute force" technique, i.e. we make a global modeling by interconnecting the STPN models of the (N+1) entities (figure 17) and the data transfer (N) service in each direction (such a model is easily obtained from the model given on the figure 7 by integrating the fragmentation (we do not represent it here for space reasons);

2. or we are more artful, i.e. we proceed in two steps :
   - we model the data transfer (N) service in each direction (the table 1 gives the number of states as a function of the number of fragments for sending a (N) SDU) and we compute the quantified abstract quotient automaton from which we extract an STPN model (their structures are independent of the fragmentation process; we have structures identical to the one presented on the figures 13 and 14, only the values of θ_1 and θ_2 change because of the fragmentation);
   - we make a global model by interconnecting the models of the (N+1) entities with the STPN models got from the quantified abstract quotient automaton (QAQA);

The table 2 summarizes the advantage of the QAQA concept to make bottom-up modeling and to master the size of the models.

Remarks :

1. When we only make a qualitative analysis, it is well known that the quotient automaton has this advantage. The quantified abstract quotient automaton extend this advantage to the quantitative area;

2. We could obviously evaluate the QoS of the (N+1) connection opening (duration) but we only wanted here to emphasize the advantage of the quantified abstract quotient automaton concept for controlling the models size;

3. We could also obviously obtain the quantified abstract quotient automaton concerning the (N+1) service and extract a STPN model in order to continue the bottom-up modeling.

5 CONCLUSION

The concept of quantified abstract quotient automaton that we have defined in this paper provides abstract views of the behavior of a system both in terms of events and quantitative informations related to these events. It is a powerful concept to study real-time distributed systems (systems growing intensively today in many areas) which require qualitative and quantitative analysis. In particular, we can characterize Qualities of Service and implement bottom-up modeling techniques which are very important with multi-layer systems in order to control the size of the models. These points have been shown on a simple example of a communication architecture (note that the sophisticated data transfer protocols with go-back n or selective reject procedures have been analysed in our laboratory works).

This concept has been implemented in the context of a formal modeling method based on the "Stochastic Timed Petri net" model which is developed in our laboratory [JA91, JG95].

REFERENCES

[AES89]   A. Emerson, A. Mok A. Sistla and J. Srinivasan. Quantitative temporal reasoning. In *Workshop on Automatic Verification Methods for Finite States Systems*, Grenoble, France, 1989.

[AJ95]   Y. Atamna and G. Juanole. Methodology for Obtaining Abstract Views of State Graphs Labeled with Probabilities and Times : an Example of Application to a Communication Protocol. In *MASCOTS'95, the Third International Workshop on Modeling, Analysis and Simulation of Computer and Telecommunication Systems*, pages 299–306, Durham, North Carolina, january 1995.

[BSW69]   K.A. Bartlett, R.A. Scantlebury, and P.T. Wilkinson. A note on realible full-duplex transmission over half-duplex links. In *Commun. Ass. Comput. Mach.*, volume 12. may 1969.

[FJV90]   C. Faure, G. Juanole, and F. Vernadat. LAPD Protocol at the ISDN Interface : Formal Modelling with Petri Nets Based Models and Verification by Abstraction (String, Observational and Behavioral Equivalences). In 10th *International Conference on Computer Communication*, New Dehli, India, november 1990.

[Flo85]   G Florin. *Réseaux de Petri Stochastiques : théorie et techniques de calcul*. PhD thesis, Université Pierre et Marie Curie (Paris

VI), 1985.

[Hen88]   M. Hennessy. *Algebric Theory of Processes*. The MIT Press, 1988.

[HJ91]   H. Hansson and B. Jonsson. A framework for reasoning about time and reliability. In *Tenth IEEE Real-Time Systems Symposium*, 1991.

[How71]   R.A. Howard. *Dynamic Probabilistic Systems, volume 2 : Semi-Markov and Decision Process*. Wiley, J and Sons, INC., 1971.

[JA91]   G. Juanole and Y. Atamna. Dealing with Arbitrary Time Distributions with the Stochastic Timed Petri Net Model. Application to Queueing Systems. In *PNPM'91, the Fourth International Workshop on Petri Nets and Performance Models*, pages 32–43, Melbourne, Australia, December 1991.

[JG95]   G. Juanole and L. Gallon. Critical Time Distributed Systems : Qualitative and Quantitative Analysis based on Stochastic Timed Petri Nets. In *FORTE'95, the 8th International IFIP Conference on Formal Description Techniques for Distributed Systems and Communication Protocols*, Montreal, Canada, october 1995.

[MBB$^+$85] A.M. Marsan, G. Balbo, A. Bobbio, G. Conte, G. Chiola, and A. Cumani. On Petri Nets with Stochastic Timing. In *TPN'85, the First International Workshop on Time Petri Nets*, Torino, Italia, 1985.

[Mil80]   R. Milner. *A Calculus of Communication Systems*, volume 92. Springer Verlag, Berlin Heidelberg, 1980.

[Ver89]   F. Vernadat. *Vérification Formelle d'Applications Réparties. Caractérisation Logique d'une Equivalence de Comportement*. PhD thesis, Université Paul Sabatier de Toulouse (FRANCE), 1989.

15

Validating Protocol Composition for Progress by Parallel Step Reachability Analysis

Gurdip Singh*
Dept. of Computing and Information Sciences, Kansas State University
234 Nichols Hall, Manhattan, KS 66506, singh@cis.ksu.edu

Hong Liu
Bellcore Applied Research, MCC 1J-222R
445 South Street, Morristown, NJ 07960, lhong@bellcore.com

Abstract

In this paper, we propose a parallel step exploration technique for protocol validation in the context of protocol composition. A protocol is modeled as a network of extended communicating finite state machines (ECFSM's). A composite protocol is defined as an interleaved execution of a set of component protocols subject to a set of constraints such as *synchronization, ordering* and *inhibition*. By encoding the constraints into the component protocols and the analysis algorithm, our method keeps each process in the component protocols as a separate entity and performs validation without constructing the composite protocol explicitly. We show that our technique not only achieves significant state reduction but also preserves the progress property of the composite protocol in the reduced state space. To our best knowledge, this is the first attempt to adapt existing state reduction techniques to the validation of protocol composition.

1 INTRODUCTION

Designing a correct protocol is a challenging task due to the complex interactions among communicating entities. One way to tackle the complexity in protocol design and analysis is through *composition*, where one divides the functionality of a protocol into subfunctions, develops component protocols for the subfunctions, and then combines them to obtain the composite protocol for the original problem. [CGL85, CM86, LT93, S94a] discuss methods for constructing a multiphase protocol,whereas [Lin88, Lin91, S93, S94b] study techniques for constructing protocols which performed multiple functions at the same time.

All these techniques impose sufficient conditions on the component protocols so that properties of the composite protocol can be inferred from those of the component protocols (which are smaller in size and therefore easier to analyze). While the analysis of the composite protocol is avoided, the sufficient conditions restrict the class of protocols that can be composed – one might still be able to construct correct protocols from a set of component protocols which do not satisfy those conditions. In this setting, the composite protocol needs to be validated for correctness.

Many techniques have been proposed to tackle the *state explosion* problem in protocol validation by eliminating redundant interleaving of independent transitions in different processes during state exploration. (Informally, two transitions are independent if they cannot enable or disable each

---

*This work was supported by NSF under grants CCR9211621 and CCR9502506.

Formal Description Techniques and Protocol Specification, Testing and Verification
T. Mizuno, N. Shiratori, T. Higashino & A. Togashi (Eds.)   © 1997 IFIP. Published by Chapman & Hall

other; otherwise they are dependent.) The *partial order* based techniques [V90, HGW92, GW93, GW94, P93, P94, LM96c] select a representative sequential execution for each set of equivalent transition sequences, while other techniques allow more than one process to make progress in a single step [YG82, RW82, II83, GH85, ZB86, CR93, OU94, LM96a, LM96b, LM96c, SU96].

In this paper, a protocol is modeled as a set of extended communicating finite state machines (ECFSM's). A composite protocol is then modeled as an interleaved execution of a set of component protocols subject to a set of constraints such as *synchronization*, *ordering* and *inhibition*. The ordering constraint was used for sequential composition in [S94a], while the synchronization constraint was used for parallel composition in [S93, S94b]. These constraints can be combined to produce a variety of composite protocols, such as serial-parallel compositions. Other than those constraints, we impose no additional restrictions on the component protocols.

Although existing techniques can potentially reduce the state space drastically, they might not be most effective if applied to the composite protocol directly. Suppose we construct R from two component protocols P and Q, and a set of constraints by composing P_i and Q_i into R_i at each site i. By definition, independent transitions can only come from *different* processes. However, in constructing R_i from P_i and Q_i, we are in fact putting many originally independent transitions between P_i and Q_i into R_i to make them artificially dependent for subsequent validation.

To achieve greater state reduction, we propose a modification to the parallel step reachability analysis [OU94] that keeps P_i and Q_i as separate entities so that both can make progress in parallel during state exploration. This provides us with a much larger set of independent actions and allows us to exploit concurrency between actions in the component protocols. By encoding the constraints into each process and the validation algorithm, we are able to enforce the composition constraints on-the-fly. We show that our technique significantly reduces the state space explored while preserves the progress property of the composite protocol.

The rest of the paper is organized as follows: Section 2 introduces the extended communicating finite state machines as the model for protocol specification; Section 3 formally specifies the composition constraints and presents a algorithm to construct a composite protocol. Our parallel step reachability analysis technique is described in Section 4. An example is given Section 5. Conclusion and future work are given in Section 6. Due to space limitations, we only outline the algorithms and omit the proofs of theorems. Please refer to the full paper [SL97] for details.

2   THE ECFSM MODEL

A communication *protocol* P is modeled as a network of $n \geq 2$ *extended communicating finite state machines* (ECFSM's), denoted as $P = P_1 \| P_2 \| \ldots \| P_n$. Each P_i is a finite state machine with local variables, denoted as $(A_i, V_i, X_i, T_i, x_i^0)$, where A_i is a finite set of *actions*, V_i is a finite set of *local variables*, X_i is a finite set of *local states*, T_i is the *transition relation*, and x_i^0 is the *initial* local state. Processes exchange messages via uni-directional FIFO *channels*. The channel from P_i to P_j is denoted as C_{ij}, bounded by a positive integer B_{ij}. The content of C_{ij} is denoted as c_{ij}. $c_{ij} = \epsilon$ if C_{ij} is empty. Define $first(c_{ij}) = m$ if $c_{ij} = m \cdot c'_{ij}$; $first(c_{ij}) = \epsilon$ if $c_{ij} = \epsilon$.

An action $a \in A_i$ is of the form $en(a) \longrightarrow a$, where $en(a)$ is a boolean function on V_i, called the *enabling condition* of a, and a is the associated computation consisting of a non-empty sequence of *statements* separated by ";". A statement is either a *local* statement involving only local variables, a *send* statement $P_j!m$ appending m at the end of channel C_{ij}, or a *receive* statement $P_j?m$ removing m from the head of channel C_{ji} if $first(c_{ji}) = m$. We assume that each action contains at most one send or receive statement. We omit the enabling condition if it is identically true. The transition relation T_i consists of tuples of the form (s_i, a, s_i'), also denoted as $s_i \xrightarrow{a} s_i'$ or $succ(s_i, a) = s_i'$. T_i is assumed to be deterministic but can be partially defined. We usually refer to transition (s_i, a, s_i') as transition a defined at s_i, or transition a if s_i is clear from the context.

Let $< v_i >$ be the tuple of values of the local variables of V_i. A *state* of P_i is defined as $s_i = (x_i, < v_i >)$. The *initial* state of P_i is denoted as $s_i^0 = (x_i^0, < v_i^0 >)$, where $< v_i^0 >$ are the initial values of the local variables. We assume that each local variable has a finite domain. So each P_i has a finite number of states. A *global state* of P is defined as $S_P = (< s_i >, < c_{ij} >)$, where s_i denotes the state of P_i and c_{ij} denotes the content of channel C_{ij}. The *initial* global state of P is denoted as $S_P^0 = (< s_i^0 >, < \epsilon_{ij} >)$, where ϵ_{ij} denotes $c_{ij} = \epsilon$.

Given a global state $S_P = (< s_i >, < c_{ij} >)$ and a transition a defined at s_i, let $stmt$ be a statement of a. $stmt$ is *enabled* in S_P iff (1) $en(a)$ is true in S_P; (2) if $stmt = P_j!m$ then $|c_{ij}| < B_{ij}$ in S_P; and (3) if $stmt = P_j?m$ then $c_{ji} = m \cdot c_{ji}'$ in S_P. Transition a is *enabled* in S_P if all the statements in a are enabled in S_P; otherwise it is *disabled* in S_P. The set of enabled and disabled transitions in S_P are denoted as $enabled(P_i, S_P)$ and $disabled(P_i, S_P)$, respectively.

The execution of a in S_P is assumed to be *atomic*. If a is executed, it will result in a global state S_P' of P such that (1) $s_i' = succ(s_i, a)$; (2) $c_{ij}' = c_{ij} \cdot m$ if a contains a send statement $P_j!m$; (3) $c_{ij} = m \cdot c_{ij}'$ if a contains a receive statement $P_j?m$; and (4) the rest of the elements in S_P' remain the same as those in S_P. We say S_P' is *directly reachable* from S_P via a, denoted as $S_P \xrightarrow{a} S_P'$ or $S_P' = succ(S_P, a)$. S_P' is *directly reachable* from S_P, denoted as $S_P \mapsto S_P'$, iff $S_P' = succ(S_P, a)$ for some action a. Denote $\mapsto^*$ as the reflexive, transitive closure of $\mapsto$. S_P' is *reachable* from S_P iff $S_P \mapsto^* S_P'$. When $S_P = S_P^0$, S_P' is a reachable global state. A reachable global state S_P is *non-progress* if there is no transition enabled in S_P. The set of reachable global states of P is denoted as $\mathbf{R}_P$.

Suppose $S_P \mapsto^* S_P'$, an *execution sequence* from S_P to S_P' is a finite sequence $ex \triangleq Z_P^0 \xrightarrow{t_1} Z_P^1 \xrightarrow{t_2} \ldots \xrightarrow{t_k} Z_P^k, k \geq 0$, such that $Z_P^0 = S_P, Z_P^k = S_P'$, and $\forall h, 1 \leq h \leq k : Z_P^h = succ(Z_P^{h-1}, t_h)$. When $S_P = S_P^0$, ex is called an execution sequence for S_P'. The length of ex is defined as the number of transitions in ex, denoted as $|ex| = k \geq 0$. The set of execution sequences from S_P^0 is denoted as *behaviors*(P).

Two transitions a and b are *independent* in a global state S_P if: (1) If a is enabled in S_P, then b is enabled in S_P iff it is also enabled in $succ(S_P, a)$; and (2) If b is enabled in S_P, then a is enabled in S_P iff it is also enabled in $succ(S_P, b)$; and (3) If both a and b are enabled in S_P, then $succ(succ(S_P, a), b) = succ(succ(S_P, b), a)$. Otherwise, a and b are *dependent*. By definition, all transitions in the same process are dependent.

Since we assume that each local variable in P_i has a finite domain, each channel has a finite capacity, and a send statement is blocked if the destination channel is full, it follows that $\mathbf{R}_P$ is finite. As a result, it is decidable whether P has the required progress property.

3 COMPOSITION OF PROTOCOLS

The composite protocol R from P and Q is defined as an interleaved execution of P and Q at each site subject to a set of constraints. Without loss of generality, we make the following assumptions: (1) P and Q have the same number of processes, with P_i and Q_i running at site i. R_i is constructed from P_i and Q_i and a set of constraints on their actions. (2) The send and receive statements from P_i and Q_i operate on the same set of channels in R_i. So each send statement $P_j!m$ $(Q_j!m')$ from P_i (Q_i) is renamed as $R_j!m$ $(R_j!m')$, and each receive statement $P_j?m$ $(Q_j?m')$ from P_i (Q_i) is renamed as $R_j?m$ $(R_j?m')$; (3) The message sets of P_i and Q_i are disjoint, and so are the local variable sets. This can be ensured through proper renaming; (4) The bound on a channel in R is the sum of the bounds on the same channel in P and Q.

3.1   Specifying the Constraints

We first define a cross product operator $\times$ for P_i and Q_i. Let $P_i = (pA_i, pV_i, pX_i, pT_i, px_i^0)^*$ and $Q_i = (qA_i, qV_i, qX_i, qT_i, qx_i^0)$ Then $G_i = P_i \times Q_i$ is an ECFSM $(gA_i, gV_i, gX_i, gT_i, gx_i^0)$ such that $gA_i = pA_i \cup qA_i, gV_i = pV_i \cup qV_i, gX_i = \{(px_i, qx_i)|(px_i \in pX_i) \wedge (qx_i \in qX_i)\}$ and $gx_i^0 = (px_i^0, qx_i^0)$. A state of G_i is denoted as $gs_i = (gx_i, < v_i >)$. The initial state of G_i is denoted as $gs_i^0 = (gx_i^0, < v_i^0 >)$. gs_i and gs_i^0 can be rewritten as (ps_i, qs_i) and (ps_i^0, qs_i^0), respectively. gT_i consists of tuples of the form (gs_i, c, gs_i'), where $gs_i = (ps_i, qs_i)$ and $gs_i' = (ps_i', qs_i')$, such that if $c \in pA_i$, then $(ps_i, c, ps_i') \in pT_i$ and $qs_i' = qs_i$; if $c \in qA_i$, then $(qs_i, c, qs_i') \in qT_i$ and $ps_i' = ps_i$.

Let $G = G_1\|G_2\|\ldots\|G_n$ be the resulting protocol, denoted as $G = P \times Q$. The set of constraints on P_i and Q_i are imposed on the set of behaviors of G. We have identified three types of constraints: *synchronization*, *ordering* and *inhibition*. These constraints are specified as pairs of actions (a, b) or (b, a), where a and b are actions of P_i and Q_i, respectively. Let $ex \triangleq S_G^0 \xrightarrow{t_1} S_G^1 \xrightarrow{t_2} \ldots \xrightarrow{t_k} S_G^k$ be an execution sequence of G, where S_G^0 is the initial global state of G. Let a^x and b^x be the xth occurrence of a and b in ex, respectively; and l_a and l_b be the number of occurrences of a and b in ex, respectively.

- ex satisfies the *synchronization constraint* (a, b) between P_i and Q_i if (1) $|l_a - l_b| \leq 1$; and (2) $\forall x, 1 \leq x \leq min(l_a, l_b)$: Let $t_h = a^x$ and $t_l = b^x$, if $h < l$ then both a and b are enabled in S_G^{h-1} and $\forall j, h < j < l : t_j$ is not an action of P_i or Q_i. A similar condition must hold if $l < h$; and (3) If $l_a > l_b$ and the last occurrence of a in ex is t_h, then both a and b are enabled in S_G^{h-1} and $\forall j, h < j \leq k : t_j$ is not an action of P_i or Q_i. A similar condition must hold for $l_b > l_a$. The set of synchronization constraints of P_i and Q_i is denoted as $synch(P_i, Q_i)$.
- ex satisfies the *ordering constraint* (a, b) from P_i to Q_i if $0 \leq l_a - l_b \leq 1$ and $\forall x, 1 \leq x \leq l_b$: (1) If $b^x = t_h$ then $\exists t_l, l < h : t_l = a^x$; and (2) if $a^x = t_h$ and $x > 1$ then $\exists t_l, l < h : t_l = b^{x-1}$. The set of ordering constraints from P_i to Q_i is denoted as $order(P_i, Q_i)$. The set of ordering constraints from Q_i to P_i, denoted as $order(P_i, Q_i)$, can be defined similarly.
- ex satisfies the *inhibition constraint* (a, b) from P_i to Q_i if the following condition is satisfied: If $t_h = a^1$ then there is no t_l such that $l > h$ and $t_l = b^x$ for any $x \geq 1$. The set of inhibition constraints from P_i to Q_i is denoted as $inhibit(P_i, Q_i)$. The set of inhibition constraints from Q_i to P_i, denoted as $inhibit(Q_i, P_i)$, can be defined similarly.

Let $constraints(P_i, Q_i)$ be the set of constraints imposed on site i. Then $constraint(P, Q) = \bigcup_{i=1}^n constraints(P_i, Q_i)$ is the set of constraints for composing P and Q. Even though they are defined with respect to (w.r.t) a finite execution sequence, they also apply to infinite behaviors of G. We say an infinite execution sequence satisfies a constraint if every prefix of the sequence satisfies the constraint. Note that while the set of synchronization constraints is a symmetric relation, the sets of ordering and inhibition constraints are not. For the latter two types, we need to distinguish the cases where P_i takes precedence over Q_i from those where Q_i takes precedence

---

*We add a prefix p to all the elements of protocol P. The same convention applies to protocols G, H, Q and R.

over P_i. We impose the following four requirements for $constraints(P_i, Q_i)$ to be *well-specified* for site i: (1) The set of synchronization, ordering and inhibition constraints be mutually disjoint. (2) Each transition of P_i be synchronized with at most one transition of Q_i and vice versa. This restriction avoids cases during execution where a transition of P_i (Q_i) has to be synchronized with more than one transition of Q_i (P_i) in the same global state of the composite protocol. (3) If a and b are synchronized, they should not both contain receive statements expecting messages from the same channel, since at any global state of G, only one of the two receive statements will be enabled. (4) There be no cyclic dependency in the sets of ordering constraints and inhibition constraints[†]. $constraints(P, Q)$ is *well-specified* iff each $constraints(P_i, Q_i)$ is well-specified. In the rest of the paper, unless otherwise specified, we assume that $constraints(P, Q)$ is well-specified.

3.2 Constructing the Composite Protocol

The construction of R_i from P_i, Q_i and $constraints(P_i, Q_i)$ is composed of three steps. The first step introduces a set of new variables for each constraint. The next step adds new conjuncts and/or local statements to the transitions in P_i and Q_i. The last step computes R_i from $P_i \times Q_i$ by deleting and modifying those transitions involved in the synchronization constraints.

Step 1: we introduce a new local variable for each constraint (a, b) or (b, a) on P_i and Q_i as follows: (1) $\forall (a, b) \in synch(P_i, Q_i)$, add syn_i^{ab}; (2) $\forall (a, b) \in order(P_i, Q_i)$, add ord_i^{ab}; (3) $\forall (b, a) \in order(Q_i, P_i)$, add ord_i^{ba}; (4) $\forall (a, b) \in inhibit(P_i, Q_i)$, add inh_i^{ab}; (5) $\forall (b, a) \in inhibit(Q_i, P_i)$, add inh_i^{ba}. Except for syn_i^{ab}, which is a three value $\{0, 1, 2\}$ variable with initial value 0, all other variables are boolean variables with initial value *false*.

Let $synV_i$, $ordV_i$ and $inhV_i$ be the sets of synchronization, ordering and inhibition variables introduced, respectively. Let $gsyn_i$ be the conjuction of the propositions $syn_i^{ab} = 0$ for all synchronization variables syn_i^{ab} created above. For each $(a, b) \in synch(a, b)$, let $gsyn_i^{ab}$ be the proposition that is formed by deleting conjunct $syn_i^{ab} = 0$ from $gsyn_i$.

Step 2: We modify each action in P_i and Q_i by adding conjunct(s) and/or local statement(s) to its enabling condition and computation. Specifically, for each $a \in pA_i$ and $b \in qA_i$, we modify a and b as follows:

(1) a (b) is not involved in any constraint in $constraints(P_i, Q_i)$. Then add $gsyn_i$ to $en(a)$ ($en(b)$) as a conjunct if it is not already there.
(2) $(a, b) \in synch(P_i, Q_i)$. Then for a, add $gsyn_i^{ab}$ and $en(b)$ as conjunts to $en(a)$ and add statement "if $syn_i^{ab} = 0$ then $syn_i^{ab} := 1$ else $syn_i^{ab} := 0$" to its computation. For b, add $gsyn_i^{ab}$ and $en(a)$ as conjuncts to $en(b)$ and add statement "if $syn_i^{ab} = 0$ then $syn_i^{ab} := 2$ else $syn_i^{ab} := 0$" to its computation.
(3) $(a, b) \in order(P_i, Q_i)$. Then for a, add conjunct $\neg ord_i^{ab}$ to $en(a)$ and add statement "$ord_i^{ab} := true$" to its computation. For b, add conjunct ord_i^{ab} to $en(b)$ and add statement "$ord_i^{ab} := false$" to its computation. The case for $(b, a) \in order(Q_i, P_i)$ can be carried out similarly.
(4) $(a, b) \in inhibit(P_i, Q_i)$. Then for a, add statement "$inh_i^{ab} := true$" to its computation. For b, add conjunct $\neg inh_i^{ab}$ to $en(b)$. The case for $(b, a) \in inhibit(Q_i, P_i)$ can be carried out similarly.

Intuitively, item (3) enforces the ordering constraints; and item (4) implements the inhibition constraints. However, items (1) and (2) together only partly enforce the synchronization constraints. The missing part will be filled in the next step.

[†] Please refer to the full paper for how these cyclic dependency can be statically checked.

Step 3: We compute R_i from $P_i \times Q_i$ as follows: for each $(a, b) \in synch(P_i, Q_i)$, for each transition $t_a = (px_i, a, px_i')$ in P_i and $t_b = (qx_i, b, qx_i')$ in Q_i, let $rx_i^1 = (px_i, qx_i), rx_i^2 = (px_i', qx_i), rx_i^3 = (px_i, qx_i')$, and $rx_i^4 = (px_i', qx_i')$, then except for these four states, t_a and t_b are removed from any other states in R_i as outgoing transitions. There are four cases to consider:

(1) $(px_i = px_i') \wedge (qx_i = qx_i')$. Both t_a and t_b are self loops in P_i and Q_i, respectively. The two corresponding transitions in R_i are also self loops: (rx_i^1, a, rx_i^1) and (rx_i^1, b, rx_i^1). We add a conjunct $syn_i^{ab} \neq 1$ to $en(a)$ and a conjunct $syn_i^{ab} \neq 2$ to $en(b)$.

(2) $(px_i \neq px_i') \wedge (qx_i = qx_i')$. t_a is not a self loop in P_i but t_b is a self loop in Q_i. The corresponding transition for t_a in R_i is (rx_i^1, a, rx_i^2) and $en(a)$ is enhanced with a new conjunct $syn_i^{ab} \neq 1$. The corresponding transitions of t_b in R_i are (rx_i^1, b, rx_i^1) and (rx_i^2, b, rx_i^2), both of which are self loops. We add a conjunct $syn_i^{ab} = 0$ to $en(b)$ of the first transition and a conjunct $syn_i^{ab} = 1$ to $en(b)$ of the second one.

(3) $(px_i = px_i') \wedge (qx_i \neq qx_i')$. t_a is a self loop in P_i but t_b is not a self loop in Q_i. The corresponding transitions for t_a in R_i are (rx_i^1, a, rx_i^1) (rx_i^3, a, rx_i^3), both of which are self loops. We add a conjunct $syn_i^{ab} = 0$ to $en(a)$ of the first transition and a conjunct $syn_i^{ab} = 2$ to $en(a)$ of the second one. The corresponding transition of t_b in R_i is (rx_i^1, b, rx_i^3) and $en(b)$ is enhanced with a new conjunct $syn_i^{ab} \neq 2$.

(4) $(px_i \neq px_i') \wedge (qx_i \neq qx_i')$. Neither t_a nor t_b is a self loop in P_i or Q_i, respectively. The corresponding transitions of t_a in R_i are (rx_i^1, a, rx_i^2) and (rx_i^3, a, rx_i^4). We add a new conjunct $syn_i^{ab} = 0$ to $en(a)$ in the first transition and a new conjunct $syn_i^{ab} = 2$ to $en(a)$ in the second one. The corresponding transitions of t_b in R_i are (rx_i^1, b, rx_i^3) and (rx_i^2, b, rx_i^4). We add a new conjunct $syn_i^{ab} = 0$ to $en(b)$ in the first transition and a new conjunct $syn_i^{ab} = 1$ to $en(b)$ in the second one.

End of Algorithm

Let $S_R = (< rs_i >, < c_{ij} >)$ be a global state of R. Then $rs_i = ((ps_i, qs_i), < sv_i >, < ov_i >, < iv_i >)$, where $< sv_i >, < ov_i >$, and $< iv_i >$ are local variables values of $synV_i, ordV_i$, and $inhV_i$, respectively. The initial global state of R is denoted as $S_R^0 = (< rs_i^0 >, < \epsilon_{ij} >)$. Here $rs_i^0 = ((ps_i^0, qs_i^0), < sv_i^0 >, < ov_i^0 >, < iv_i^0 >)$, where $< sv_i^0 >, < ov_i^0 >$, and $< iv_i^0 >$ are initial values of variables of $synV_i$ (all 0), $ordV_i$ (all *false*) and $inhV_i$ (all *false*), respectively. Let ex be an execution sequence of R and $ex|_{R_i}$ be the projection of ex on rA_i. Clearly, ex satisfies $constraints(P_i, Q_i)$ iff $ex|_{R_i}$ satisfies each constraint specified in $constraints(P_i, Q_i)$. It can be shown that each R_i thus constructed does ensure that R satisfies $constraints(P, Q)$.

Theorem 1 $\forall ex \in behaviors(R) : \forall i : ex|_{R_i}$ satisfies $constraints(P_i, Q_i)$.

4   PARALLEL STEP REACHABILITY ANALYSIS

There are two major sources that cause dependency between a transition a in P_i and a transition b in Q_i: (1) If (a, b) or (b, a) belongs to $constraints(P_i, Q_i)$, a and b cannot be executed in arbitrary order except for synchronization; (2) If both a and b involve sending a message to the same channel, the channel content will differ by the order in which a and b are executed. To take into account these dependency, we encode $constraints(P_i, Q_i)$ into P_i and Q_i, as was done Step 1 and 2 in Section 3, except that for each $(a, b) \in synch(P_i, Q_i)$, we add a conjunct $syn_i^{ab} \neq 1$ to $en(a)$ and $syn_i^{ab} \neq 2$ to $en(b)$. Then instead of constructing R_i explicitly in Step 3, we view P_i

and Q_i as two *threads* of process H_i in a hypothetical protocol $H = H_1\|H_2\|\cdots\|H_n$ that share the same set of channels and local variables. We then apply parallel step state exploration to validate H, where both P_i and Q_i can make progress from a global state. Finally, we show that the hypothetical protocol H and the composite protocol R have the same progress property.

4.1 The Hypothetical Protocol H

A global state of H is denoted as $S_H = (< hs_i >, < c_{ij} >)$, where hs_i is of the form (ps_i, qs_i). Since hs_i has the same component structure as rs_i, so are S_H and S_R. Define $S_H = \alpha(S_R)$ (or $S_R = \alpha^{-1}(S_H)$) iff S_H and S_R have the same component values. α is a one-to-one mapping from the set of global states in R and the set of global states in H. Since a transition is enabled in $\alpha(S_R)$ if it is enabled in S_R, α is a homomorphism from $\mathbf{R}_R$ to $\mathbf{R}_H$ w.r.t the reachability relation $\mapsto^*$. Let $\alpha(\mathbf{R}_R)$ be the image of $\mathbf{R}_R$ in $\mathbf{R}_H$. Then $\alpha(\mathbf{R}_R) \subseteq \mathbf{R}_H$.

Suppose $S_H = \alpha(S_R)$. It can be shown that S_H is reachable via an execution sequence ex in H if S_R is reachable via ex in R, by induction on $|ex|$. However, the converse is not always true. The main reason is that the additional conjuncts we put on actions a and b for $(a, b) \in synch(P_i, Q_i)$ are not sufficient to ensure that a (or b) occurs first iff b (or a) occurs next. For example, consider case (4) in Step 3. At rx_i^2, the conjunct added to $en(b)$ is $syn^{ab} \neq 2$ instead of $syn_i^{ab} = 1$. When $syn_i^{ab} = 0$, b can be enabled at rx_i^2 in H without executing at rx_i^1.

To avoid the above situation, we select a subset of $enabled(P_i, S_H)$, denoted as $enabled_p(P_i, S_H)$. A transition $a \in enabled_p(P_i, S_H)$ iff (1) a is not involved in any synchronization constraint, or (2) $(a, b) \in synch(P_i, Q_i)$ and $b \in enabled(Q_i, S_H)$, or (3) $(a, b) \in synch(P_i, Q_i)$, $enabled(P_i, S_H) = \{a\}$, $enabled(Q_i, S_H) = \emptyset$, and $syn_i^{ab} = 2$. a is called a *valid* transition of P_i in S_H. Similarly, a transition $b \in enabled_p(Q_i, S_H)$ iff (1) b is not involved in any synchronization constraint, or (2) $(a, b) \in synch(P_i, Q_i)$ and $a \in enabled(P_i, S_H)$, or (3) $(a, b) \in synch(P_i, Q_i)$, $enabled(P_i, S_H) = \emptyset$, $enabled(Q_i, S_H) = \{b\}$, and $syn_i^{ab} = 1$. b is called a *valid* transition of Q_i in S_H.

Suppose $S_H = \alpha(S_R)$. It can be shown that $enabled_p(P_i, S_H) \cup enabled_p(Q_i, S_H) = enabled(R_i, S_R)$. Define $\mathbf{R}_H^p$ as the set of global states reachable from S_H^0 via only valid transitions. If $S_H \in \mathbf{R}_H^p$ and ex is an execution sequence for S_H composed of only valid transitions, then ex is also an execution sequence for S_R. On the other hand, if ex is an execution sequence for S_R, then it is also an execution sequence for S_H with only valid transitions. Hence $\mathbf{R}_H^p = \alpha(\mathbf{R}_R)$, i.e., α is an isomorphism from $\mathbf{R}_R$ to $\mathbf{R}_H^p$ w.r.t $\mapsto^*$. In particular, a non-progress global state S_R is reachable in R iff $S_H = \alpha(S_R)$ is reachable in H by valid transitions only. So to study the progress property of R, it is sufficient to only generate $\mathbf{R}_H^p$. In the following, we are going to show that it actually suffices to generate only a subset of $\mathbf{R}_H^p$ via parallel step state exploration.

4.2 Parallel Step State Exploration

We partition the set of transitions for P_i defined in a global state S_H as follows. First, the set of enabled transitions is divided into two sets: $local\_enabled(P_i, S_H)$ and $global\_enabled(P_i, S_H)$. Transition $a \in enabled(P_i, S_H)$ is *locally* enabled if a does not contain a send statement that sends a message to the same channel as another transition $b \in enabled(Q_i, S_H)$; otherwise it is *globally* enabled. Then the set of disabled transitions is also partitioned into two sets: $local\_disabled(P_i, S_H)$ and $global\_disabled(P_i, S_H)$. A transition $a \in disabled(P_i, S_H)$ is *locally* disabled if $en(a) = false$, or $en(a) = true$, a contains a receive statement $P_j?m$, and $(first(c_{ji}) = m' \neq m) \wedge (m' \in pM_{ji})$; otherwise it is *globally* disabled. Let $local\_enabled_p(P_i, S_H)$ and $global\_enabled_p(P_i, S_H)$ be the set of locally and globally enabled valid transitions of P_i in S_H, respectively. The transitions for Q_i in S_H can be partitioned similarly into these four sets.

To define parallel progress, we first compute U_i, the set of *valid transition pairs* for P_i and Q_i in S_H from transitions in $enabled_p(P_i, S_H) \cup global\_disabled(P_i, S_H)$ of P_i and $enabled_p(Q_i, S_H) \cup global\_disabled(Q_i, S_H)$ of Q_i.[‡] There are four cases to consider:

(1) For each $a \in enabled_p(P_i, S_H)$ and $b \in enabled_p(Q_i, S_H)$, $(a, b) \in U_i$, i.e. P_i and Q_i can execute a and b in parallel if (i) $(a, b) \in synch(P_i, Q_i)$ and a and b do not send messages to the same channel, or (ii) neither a nor b is involved in any synchronization or inhibition constraint. Otherwise, $(a, \lambda), (\lambda, b) \in U_i$, where λ is a *null* transition, indicating no progress.

(2) For each $a \in enabled_p(P_i, S_H)$ and $b \in global\_disabled(Q_i, S_H)$, $(a, \lambda) \in U_i$ if a is not involved in any synchronization constraint or a is the only enabled transition.

(3) For each $a \in global\_disabled(P_i, S_H)$ and $b \in enabled_p(Q_i, S_H)$, $(\lambda, b) \in U_i$ if b is not involved in any synchronization constraint or b is the only enabled transition.

(4) If $global\_disabled(P_i, S_H) \neq \emptyset$ and $global\_disabled(Q_i, S_H) \neq \emptyset$, then $(\lambda, \lambda) \in U_i$.

Note that when $(a, b) \in synch(P_i, Q_i)$ but both a and b send messages to the same channel, only one of them can be executed. Suppose a (or b) is chosen, then in the following global state, $syn_i^{ab} = 1$ (or $syn_i^{ab} = 2$). Hence $gsyn_i = false$, which implies that only b (or a) can be enabled. In this case, even b (or a) is not paired with a (or b), it should still be chosen to execute. Also, when $(a, b) \in inhibit(P_i, Q_i)$, even though both are enabled, only one of them can be executed.

Now let $\vec{u}$ be a $2n$-tuple ($< pu_i, qu_i >$) such that $(pu_i, qu_i) \in U_i$. $\vec{u}$ is a *parallel progress vector* in S_H iff $\exists i : (pu_i, qu_i) \neq (\lambda, \lambda)$. Hence in a parallel progress vector, at least one thread must make progress. Since all the non-null transitions in $\vec{u}$ are independent of each other, the resulting global state S_H' is the same irrespective of the order of execution. In this case, we say that S_H' is *directly parallel* reachable from S_H via $\vec{u}$, denoted as $S_H \xrightarrow{\vec{u}} S_H'$ or $S_H' = succ(S_H, \vec{u})$. With this, we can define parallel reachability relations $\mapsto_p$ and $\mapsto_p^*$, and parallel execution sequence accordingly. Denote $\mathbf{PR}_H$ as the set of parallel reachable states in H and $behaviors_p(H)$ as the set of parallel execution sequences from S_H^0 in H.

Let $linear(\vec{u})$ be the set of permutations on the non-null transitions in $\vec{u}$. Let $\vec{u}_1, \vec{u}_2, \ldots, \vec{u}_k$, $k \geq 0$, be the sequence of progress vectors in a parallel execution sequence *pex*. Then $linear(pex)$ is defined as $\{\epsilon\}$ if $k = 0$; and as $linear(\vec{u}_1) \cdot linear(\vec{u}_2) \cdots linear(\vec{u}_k)$ if $k \geq 1$, where $\cdot$ is generalized to handle two sets of sequences, i.e., $A \cdot B = \{a \cdot b | (a \in A) \wedge (b \in B)\}$. Since each $ex \in linear(pex)$ is an execution sequence from S_H to S_H' composed of only valid transitions, $\mathbf{PR}_H \subseteq \mathbf{R}_H^p$. Moreover, *pex* satisfies $constraints(P, Q)$ iff $\forall ex \in linear(pex) : ex$ satisfies $constraints(P, Q)$. Since each ex is also an execution sequence for $S_R = \alpha^{-1}(S_H)$ in R, by Theorem 1, ex satisfies $constraints(P, Q)$ in R.

Theorem 2 $\forall pex \in behaviors_p(H) : pex$ satisfies $constraints(P, Q)$.

Now suppose $S_H \in \mathbf{R}_H^p$, S_H is a *pseudo* non-progress global state iff it has no valid transitions. From the above discussion, we know that S_H is pseudo non-progress global state in H iff $\alpha^{-1}(S_H)$ is a non-progress global state in R. Hence we only need to focus on $\mathbf{SNR}_H$, the set of pseudo non-progress global states in H. On the other hand, suppose $S_H \in \mathbf{PR}_H$, S_H is a *parallel* non-progress global state in H iff it has no parallel progress vectors. By construction, S_H has no parallel progress vectors iff it has no valid transitions. Hence S_H is also a pseudo non-progress global state in H. Let $\mathbf{PNR}_H$ be the set of parallel non-progress global state in H. Then $\mathbf{PNR}_H \subseteq \mathbf{SNR}_H$.

[‡]Due to space limitation, we only highlight the key points here, please refer to the full paper [SL97] for details.

To show the converse, let ex be an execution sequence for a pseudo non-progress global state S_H. Each transition in ex is a valid transition. Denote $pe_i = ex|_{P_i}$ and $qe_i = ex|_{Q_i}$. Then $(< pe_i, qe_i >)$ is a local execution sequence set for S_H. From $(< pe_i, qe_i >)$, we construct a parallel execution sequence pex for S_H as follows. Starting from S_H^0, in Step $k \geq 0$, for each i, we compute (pu_i^{k+1}, qu_i^{k+1}) for P_i and Q_i in global state S_H^k based on the transitions from pe_i and qe_i in S_H^k. Since ex has only a finite number of transitions, the algorithm must terminate in a finite number of steps. Moreover, since each intermediate global state contains at least one valid enabled transition, at the end of the algorithm, the final global state must be S_H. So pex thus constructed is a parallel execution sequence for S_H. (Please refer to the full paper [SL97] for details.) Hence we have $\textbf{SNR}_H \subseteq \textbf{PNR}_H$, and thus $\textbf{PNR}_H = \textbf{SNR}_H$. Since $\textbf{SNR}_H$ is exactly the set of non-progress states in R, we have the following result on fault coverage of $\textbf{PR}_H$.

Theorem 3 Given $S_R \in \textbf{R}_R$, S_R is non-progress global state in R iff $S_H = \alpha(S_R)$ is a parallel non-progress global state in H.

4.3   Discussion

The parallel step technique described in this section was adapted from the simultaneous reachability analysis method in [OU94] to fit the context of protocol composition. Similar to [SU96], we can use the "sleep-set" concept [GW93, GW94] to further eliminate redundant transitions in computing the set of parallel progress vectors. We can also correlate transitions from different processes, as was done in fair reachability analysis [RW82, GH85, LM96a, LM96b]. Doing so might result in fewer global states, but the computation in each global state becomes more elaborate.

In this paper, we assume that the component protocols have the required progress property. If the composite protocol has non-progress global states, then it is most likely that the composition constraints are not consistent with each other. Hence in analyzing error scenarios, we should focus on the set of constraints involved. Note that not all the non-progress global states are semantically incorrect. For example, if one action inhibits the other and that action corresponds to an exception in the protocol, the protocol may halt in response to that exception. So it is up to the designer to decide whether a non-progress global state is acceptable or not. However, the imposed constraints are not the only cause for non-progress in the composite protocol. This point is more subtle. Recall that in our model, a send statement is blocked if the destination channel is full. In the composite protocol, the bound on a channel might be enlarged. So it is possible that a send action that is not enabled in the original component protocol becomes enabled in the composite protocol. So a process may exhibit new behaviors after the composition. These new behaviors, together with their interactions may also cause non-progress in the composite protocol.

Last but not least, even though we presented our technique in the context of two component protocols P and Q, it can be easily extended to handle cases with more than two component protocols. Furthermore, our technique does not require that all component protocols have the same number of processes, nor does it require that the composition of processes be fixed w.r.t the indices of the processes in each component protocol. All is required is that at most one process from each component protocol can participate in each site in the composite protocol. However, not every component protocol is required to participate in the composition for a site. What we need is a composition schema that describes which process from which component protocol is needed to participate at each site. Once the schema is given, we can define composition constraints for each site with more than one process, and the rest of the work can proceed as described above. In the next section, we will give an example in this general setting.

5 EXAMPLE

Consider a network of four sites shown in Figure 1(a). We want to design a data transfer protocol in which site 1 first establishes connection with sites 2, 3 and 4 and then transfers a sequence of data items to them. Site 1 send the items directly to 2 and 4 and site 2 forwards the data items to 3. We want a stop&wait protocol in which 1 sends the next data item only after all sites have received the previous data item. Finally, site 1 may send a disconnect message at any time after the connection establishment to break the connection.

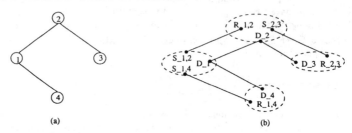

(a)   (b)

Figure 1 Topology for the data transfer protocol.

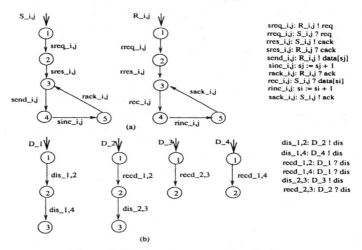

Figure 2 Data transfer component protocols.

Figure 2(a) gives a stop&wait protocol $(S_{i,j}, R_{i,j})$ with $S_{i,j}$ at site i as the sender and $R_{i,j}$ at site j as the receiver. Figure 2(b) gives a disconnect protocol in which 1 simply sends a disconnect message to 2 and 4, and 2 forwards it to 3. We will design protocols using four component protocols: $(S_{1,2}, R_{1,2})$, $(S_{1,4}, R_{1,4})$, $(S_{2,3}, R_{2,3})$ and (D_1, D_2, D_3, D_4) (see Figure 1(b)).

As a simpler case, we first compose $(S_{1,2}, R_{1,2})$ and $(S_{1,4}, R_{1,4})$ with a synchronization constraint $(send_{1,2}, send_{1,4})$ at site 1 to ensure that the first data item is sent after connection with both 2 and 4 has been established, and subsequent data items are sent only after acknowledgements from both 2 and 4 are received for the previous data item. For the composite protocol built by the algorithm in section 3, the standard reachability analysis explores 126 states, the partial order method in [HGW92] finds 81 states, whereas our method has only 9 reachable states (here we view $inc_{i,j}$ as an internal action; otherwise, the number of states are unbounded). In fact, our method explores 9 states irrespective of the number of receivers.

The next protocol with all four sites is obtained by combining all four protocols with seven more constraints: (1) An ordering constraint $(rec_{1,2}, send_{2,3})$ on 2 to ensure that a data item is forwarded to 3 only after it has been received from 1; (2) An order constraint $(rack_{2,3}, sack_{1,2})$ on 2 to ensure the stop&wait discipline w.r.t 1 and 3; (3) Two ordering constraints $(sres_{1,2}, dis_{1,2})$ and $(sres_{1,4}, dis_{1,4})$ on site 1 to ensure 1 can send a disconnect message only after connection setup; (4) Three inhibition constraints $(dis_{1,2}, send_{1,2})$, $(dis_{1,4}, send_{1,4})$ and $(dis_{2,3}, send_{2,3})$ to ensure no more data items are to be sent after the disconnect message is sent. These constraints allow the messages that have already been sent to be received and acknowledged. Although the final composite protocol is a complex one, our method explores only 47 reachable states.

6 CONCLUSION AND FUTURE WORK

In this paper, we studied the problem of validating protocol composition for progress based on the set of component protocols and a set of composition constraints. By encoding the constraints into the processes of component protocols and the analysis algorithm, we are able to perform parallel step state exploration for the composite protocol without constructing it explicitly. As a result, we are able to perform validation for the composite protocol in a significantly reduced global state space. As far as we know, this is the first attempt to adapt existing state reduction techniques to protocol composition.

However, we have just scratched the surface in this direction. First, the composite protocol construction algorithm given in Section 3 may not be the most efficient one, and the R_i constructed may not be the minimum state machine for the composite process. How to build a minimum state composite process is an interesting problem that requires further study. Second, It would be interesting to investigate other encoding schemes to fit the partial order techniques so that more general properties can be validated. We also want to include more constraint types to allow more flexible compositions. Finally, we plan to implement the parallel step method and experiment it with complex examples.

Acknowledgement

The authors would like to thank Raymond E. Miller and Jun-Cheol Park for their constructive comments on the earlier drafts of this paper.

REFERENCES

[CR93] L. Cacciari and O. Rafiq, "On Improving Reduced Reachability Analysis," Proc. FORTE'92, Perros-Guirec, France, October 13-16, 1992, pp. 137–152.

[CGL85] C.H. Chow, M.G. Gouda and S.S. Lam, "A Discipline for Constructing Multi-Phase Communicating Protocols," ACM Trans. Comput. Syst., 3(4), 1985, pp. 315–343.

[CM86] T.Y. Choi and R.E. Miller, "Protocol Analysis and Synthesis by Structured Partitions", Computer Networks and ISDN Systems, 11, 1986, pp. 367–381.

[GH85] M. Gouda and J.Y. Han, "Protocol Validation by Fair Progress State Exploration," Computer Networks and ISDN Systems, 9, 1985, pp. 353–361.

[GW93] P. Godefroid and P. Wolper, "Using Partial Orders for the Efficient Verification of Deadlock Freedom and Safety Properties," Formal Methods in System Design, 2(2), 1993.

[GW94] P. Godefroid and P. Wolper, "A Partial Approach to Model Checking," Information and Computation, 110(2), 1994, pp. 305–326.

[HGW92] G. Holzmann, P. Godefroid and P. Wolper, "Coverage Preserving Reduction Strategies for Reachability Analysis," PSTV'92.

[II83] M. Itoh and H. Ichikawa, "Protocol Verification Algorithm Using Reduced Reachability Analysis," Trans. IECE of Japan, E66(2), 1983, pp. 88–93.

[Lin88] H.A. Lin, "A Methodology for Constructing Communication Protocols with Multiple Concurrent Functions," Distributed Computing, 3(1), 1988, pp. 23–40.

[Lin91] H.A. Lin, "Constructing Protocols with Alternative Functions," IEEE Transactions on Computers, 40(4), 1991, pp. 376–386.

[LT93] H.A. Lin and C.L. Tarng, "An Improved Method for Constructing Multiphase Communications Protocols," IEEE Transactions on Computers, 42(1), 1993, pp. 15–26.

[LM96a] H. Liu and R. Miller, "Generalized Fair Reachability for Cyclic Protocols," IEEE/ACM Transactions on Networking, 4(2), April 1996, pp. 192–204.

[LM96b] H. Liu and R.E. Miller, "An Approach to Cyclic Protocol Validation," Computer Communications, 19(14), 1996, pp. 1175-1187.

[LM96c] H. Liu and R.E. Miller, "Partial-Order Validation for Multi-Process Protocols Modeled as Communicating Finite State Machines," Proc. ICNP'96, Oct. 29 – Nov. 1, 1996, pp. 76–83.

[OU94] K. Özdemir and H. Ural, "Deadlock Detection in CFSM Models via Simultaneously Executable Sets," ICCI'94, Peterborough, Ontario, Canada, May 1994, pp. 673–688.

[P93] D. Peled, "All from One, One for All: On Model Checking Using Representatives," CAV'93.

[P94] D. Peled, "Combining Partial Order Reduction with On-the-fly Model-Checking," CAV'94.

[RW82] J. Rubin and C.H. West, "An Improved Protocol Validation Technique," Computer Networks, 6, 1982, pp. 65–73.

[S93] G. Singh, "A Compositional Approach for Designing Protocols," Proc. ICNP'93, San Francisco, CA, October 19–22, 1993, pp. 98–105.

[S94a] G. Singh and M. Sammeta, "On the Construction of Multiphase Protocols," Proc. ICNP'94, Boston, MA, October 25–28, 1994, pp. 151–158.

[S94b] G. Singh, "A Methodology for Constructing Communication Protocols," Proc. ACM SIGCOMM'94, August 31 – September 2, 1994, London, U.K., pp. 245–255.

[SL97] G. Singh and H. Liu, "Validating Protocol Composition for Progress by Parallel Step Reachability Analysis," in preparation.

[SU96] H.v.d. Schoot and H. Ural, "Protocol Verification by Leaping Reachability Analysis," Proc. IC3N'96, Rockville, MD, USA, October 16-19, 1996, pp. 334-339.

[V90] A. Valmari, "A Stubborn Attack on State Explosion," Proc. CAV'90.

[YG82] Y.T. Yu and M.G. Gouda, "Deadlock Detection for a Class of Communicating Finite State Machines," IEEE Transactions on Communications, 30(12), 1982.

[ZB86] J.R. Zhao and G.v. Bochmann, "Reduced Reachability Analysis of Communication Protocols: a New Approach," Proc. PSTV'86, pp. 243-254.

16

An Improved Search Strategy for Lossy Channel Systems

Parosh Aziz Abdulla
Uppsala University, Dept. of Computer Systems
P.O. Box 325, S-751 05 Uppsala, Sweden, `parosh@docs.uu.se.`

Mats Kindahl
Uppsala University, Dept. of Computer Systems
P.O. Box 325, S-751 05 Uppsala, Sweden, `parosh@docs.uu.se.`

Doron Peled
Bell Laboratories
700 Mountain Avenue, Murray Hill, NJ 07974,
`doron@research.bell-labs.com.`

Abstract

In [1] we considered *lossy channel systems* which are a particular class of infinite state systems consisting of finite state processes communicating through channels that are unbounded and unreliable. We presented a backward reachability algorithm which, starting from a set of "bad" states, checks whether there is a backward path to the initial state of the system. Using standard techniques, the reachability algorithm can be used to check safety properties for lossy channel systems.

In this paper we adopt partial order techniques to improve the algorithm in [1]. We define a preorder, which we call the *better than relation*, among the set of transitions of the system. Intuitively a transition is better than another if choosing the first transition instead of the second preserves the reachability of the initial state during the analysis. This relation is weaker than the *independence relation*, which is an equivalence relation, used in traditional partial order methods, in the sense that two transitions are independent if and only if each of them is better than the other. Consequently, our method gives a better reduction in the number of states considered during the analysis. We demonstrate the efficiency of the approach by a number of experimental results.

1 INTRODUCTION

In the last few years, there has been a considerable interest in algorithmic verification of distributed and parallel systems. The research has led to the discovery of numerous efficient methods for the verification of *finite-state* systems ([7], [9], [14], [20], etc.). An obvious limitation of these methods is that systems with infinitely many states fall beyond their capabilities. Recently, algorithmic verification methods have been developed for some classes of infinite-state systems.

A particular class of infinite-state systems which has been important in the analy-

Formal Description Techniques and Protocol Specification, Testing and Verification
T. Mizuno, N. Shiratori, T. Higashino & A. Togashi (Eds.) © 1997 IFIP. Published by Chapman & Hall

sis of, e.g. , communication protocols consists of finite-state processes that communicate via unbounded FIFO channels [6, 5]. Such systems are infinite-state due to the unboundedness of the channels, and it is well-known that these systems can simulate Turing machines, and hence most interesting verification problems are undecidable for them [6]. In an earlier work [1], we considered a variant of this class, called *lossy channel systems*, where the FIFO channels are unreliable, in the sense that they may nondeterministically lose messages. In spite of this restriction, lossy channel systems can be used to model and analyze many interesting systems, e.g. link protocols such as the Alternating Bit Protocol [4] and HDLC [15], which are designed to operate correctly even in the case that the FIFO channels are faulty and may lose messages.

In [1, 2] we present a reachability algorithm for lossy channel systems which, for any state, checks whether the state is reachable from the initial state of the system. The idea of the algorithm is to define a partial order $\preceq$ on the set of states, where one state is smaller than another if the two states differ only in that the content of each channel in the first state is a (not necessarily contiguous) substring of the content of the same channel in the second state. An important property of lossy channel systems is that their behaviour is monotone with respect to $\preceq$, i.e. larger states have transitions to larger states. The algorithm operates on *ideals*, where the *ideal generated by a state* is the set of states larger than or equal to that state. The reachability of a state γ is obviously equivalent to the reachability of the ideal generated by γ, since if there is a path from the initial state to any state γ', which is larger than γ, then we can continue from γ' losing messages until we obtain γ. To check the reachability of an ideal I, we perform a reachability analysis starting from I and going backwards. At each step we pick an ideal I' which has already been generated during the analysis, and compute the set $pre(I)$ of states from which I' is reachable through the application of a single step of the transition relation. It is shown in [1, 2] that the monotonicity of lossy channel systems implies that $pre(I)$ is also an ideal and that it is in fact computable.

Partial order reductions [18, 12, 19] are a family of techniques which can be used to perform more efficient verification of systems consisting of asynchronously communicating concurrent processes, e.g. lossy channel systems. These methods are based on the observation that concurrent actions of the processes are often independent and hence their interleavings are equivalent in that they all lead to the same states. Most existing partial order methods work with a forward search of the state space, starting from the initial state and trying to find a path forwards to a final state. Action sequences are grouped into equivalence classes, for each of which at least one representative must to be analyzed. The algorithm then searches a reduced state-space, where at each step only a subset (an *ample set* [18]) of the enabled transitions is considered. The requirement is that an ample set should contain at least one representative of each equivalence class.

In this paper, we employ ideas based on partial order reductions to derive a more efficient version of the algorithm in [1, 2]. Instead of working with an equivalence relation (the *independence* relation), we define a *preorder* (the *"better than"* relation) among actions. An event α is *better than* an event β in a state γ if α followed by β leads to a smaller state (with respect to $\preceq$) from γ than β followed by α. We extend the preorder to sequences of actions, where a sequence is "better than" another if the first sequence can be obtained from the second by permuting adjacent actions so that "better" actions occur earlier in the first sequence than in the second sequence. We show that it is sufficient, for each set of sequences, to consider only the "best" elements of the set. The correctness of our approach can be shown again to follow from monotonicity of lossy channel systems.

Our method differs from traditional partial order techniques in the following aspects.

- We work with a preorder (the "better than" relation) which is weaker than the independence relation, since two sequences are equivalent if and only if each of them is better than the other. Instead of choosing representatives of each equivalent class, it is sufficient to choose the best elements of a certain set of sequences. This leads to a more efficient search algorithm, since two sequences may not be equivalent, but one of them may still be discarded from the analysis, if it is "worse" than the other.
- We apply our methods in the context of backward reachability analysis. This is necessary since the reachability algorithm for lossy channel systems operates backwards. In fact it is shown in [10] that forward reachability analysis is infeasible for lossy channel systems.
- The reachability algorithm works on ideals, so our method can be considered to operate on sets of states (ideals) rather than individual states. Since ideals are infinite sets, each ideal is represented symbolically by its set of generators. This set of generators can be shown always to be finite.

Using standard techniques [20], we can check safety properties for lossy channel systems through a reduction to the reachability problem.

In fact our method is quite general. In [10, 3] general theories are presented for the verification of systems with monotone behaviour. Examples of such systems include besides lossy channel systems, Petri nets, real-time automata, relational automata, Basic Parallel Processes (BPPs), and certain classes of parametrized systems. The general reachability algorithm presented in [3] exploits the monotonicity of the system in the same manner as the algorithm for lossy channel systems. Consequently our techniques are applicable even for these classes of systems.

In the next section we introduce lossy channel systems and some properties of these. In section 3 we introduce the reachability algorithm. In section 4 we introduce the better-than relation and some properties of action sequences. In section 5 we show the modifications needed to improve the reachability algorithm. In section 6 we give an algorithm to compute ample sets. In section 7 we show some experimental results. In section 8 we give some conclusions and directions for future research.

2   PRELIMINARIES

A *lossy channel system* consists of a *control part* and a *channel part*. The control part is modelled as a number of finite-state processes communicating via the channels, while the channel part consists of a finite set of channels. Each channel behaves as a FIFO buffer which is unbounded and unreliable in the sense that it can lose messages. A channel is used to perform asynchronous communication between a pair of processes, so for each channel there is unique process sending messages to the channel, and a unique process receiving messages from the channel.

For a tuple $\vec{x} = \langle x_1, x_2, \ldots, x_n \rangle$, we use $\vec{x}(i)$ to denote x_i and $\vec{x}[i \mapsto e]$ to denote the tuple $\langle x_1, x_2, \ldots, x_{i-1}, e, x_{i+1}, \ldots, x_n \rangle$. The *domain* $\text{dom}(\vec{x})$ of $\vec{x}$ is the set $\{1, 2, \ldots, n\}$ of indices, while the *range* $\text{rng}(\vec{x})$ of $\vec{x}$ is the set $\{x_1, x_2, \ldots, x_n\}$ of values. For strings x and y, we use $x \cdot y$ to denote the concatenation of x and y.

Formally, lossy channel systems are defined as follows.

Definition 1 A *lossy channel system* is a tuple $\mathcal{L} = \langle \vec{S}, \vec{s}_0, A, C, M, \delta \rangle$, where

$\vec{S} = S_1 \times \cdots \times S_n$ is a tuple of finite sets of *control states*, where n is the number of processes in the control part, and S_i denotes the set of control states of the ith process.

$\vec{s}_0 \in \vec{S}$ is a tuple of *initial control states*,

A is a finite set of *actions*,

C is a finite set of *channels*,

M is a finite set of *messages*,

δ is a finite set of *transitions*, each of which is a triple of the form $\langle s_1^i, l, s_2^i \rangle$, where $s_1^i, s_2^i \in \vec{S}(i)$ for some process i, and l is of the form

$c!m$ where $c \in C$ and $m \in M$, representing sending a message m to channel c (m is appended to the end of c);

$c?m$ where $c \in C$ and $m \in M$, representing receiving a message m from channel c (m is removed from the head of c); or

a where $a \in A$, representing an observable interaction with the environment without changing the contents of the channels.

A *global state* (or just *state*) $(\vec{s}, \vec{w})$ consists of a tuple $\vec{s} \in \vec{S}$ of control states and a tuple $\vec{w}$ of strings over M. The string $\vec{w}$ is indexed by the elements of C, with $\vec{w}(c)$ representing the content of c. We use γ to denote a global state and use Γ to denote the set of all global states.

We formalise the intuitive behaviour of lossy channel systems by introducing a labelled transition system on global states.

Definition 2 Given a lossy channel system $\mathcal{L} = \langle \vec{S}, \vec{s}_0, A, C, M, \delta \rangle$ we construct the labelled transition system $\mathcal{G} = \langle \Gamma, \longrightarrow, A \cup \{\tau\} \rangle$ where Γ is the set of global states, $A \cup \{\tau\}$ is a set of labels, and $\longrightarrow \subseteq \Gamma \times (A \cup \{\tau\}) \times \Gamma$ is a transition relation among global states. We will use $\gamma_1 \xrightarrow{a} \gamma_2$ to denote $(\gamma_1, a, \gamma_2) \in \longrightarrow$. The set $\longrightarrow$ is the smallest set such that

- if $\langle s_1^i, c!m, s_2^i \rangle \in \delta$ and $(\vec{s}, \vec{w})$ is a global state such that $\vec{s}(i) = s_1^i$, then $(\vec{s}, \vec{w}) \xrightarrow{\tau} (\vec{s}[i \mapsto s_2^i], \vec{w}[c \mapsto \vec{w}(c) \cdot m])$. This corresponds to sending message m to channel c.

- If $\langle s_1^i, c?m, s_2^i \rangle \in \delta$ and $(\vec{s}, \vec{w})$ is a global state such that $\vec{s}(i) = s_1^i$ and $\vec{w}(c) = m \cdot v$, then $(\vec{s}, \vec{w}) \xrightarrow{\tau} (\vec{s}[i \mapsto s_2^i], \vec{w}[c \mapsto v])$. This corresponds to receiving message

m from channel c. Notice that this transition may be performed only if the element in the head of channel c is equal to m.

- If $\langle s_1^i, a, s_2^i \rangle \in \delta$ and $(\vec{s}, \vec{w})$ is a global state such that $\vec{s}(i) = s_1^i$, then $(\vec{s}, \vec{w}) \xrightarrow{a} (\vec{s}[i \mapsto s_2^i], \vec{w})$. This corresponds to performing an observable interaction a with the environment.
- If $(\vec{s}, \vec{w})$ is a global state such that $\vec{w}(c) = x \cdot m \cdot y$, then $(\vec{s}, \vec{w}) \xrightarrow{\tau} (\vec{s}, \vec{w}[c \mapsto x \cdot y])$. This corresponds to losing message m from channel c.

We use $(\vec{s}_1, \vec{w}_1) \longrightarrow (\vec{s}_2, \vec{w}_2)$ to denote that $(\vec{s}_1, \vec{w}_1) \xrightarrow{a} (\vec{s}_2, \vec{w}_2)$ for some $a \in A \cup \{\tau\}$, and use $\xrightarrow{*}$ to denote the transitive closure of $\longrightarrow$.

We define a partial order $\preceq$ on elements of M^* by letting $w_1 \preceq w_2$ iff w_1 is a (not necessarily contiguous) substring of w_2, e.g. given $M = \{a, b, c, x, y, z\}$, $ab \preceq abc \preceq axbyc \npreceq axyc$. We use ε to denote the empty string. We extend the partial order $\preceq$ to states by letting $(\vec{s}_1, \vec{w}_1) \preceq (\vec{s}_2, \vec{w}_2)$ iff $\vec{s}_1 = \vec{s}_2$ and $\vec{w}_1(c) \preceq \vec{w}_2(c)$ for each $c \in C$. An interesting property of this partial order is that it is a well quasi-order, i.e. there is no infinite set of strings such that all strings are pairwise incomparable.

Lemma 1 (Higman's Lemma) Let M be a finite set. If $S \subseteq M^*$ is a subset such that all strings in S are pairwise incomparable with respect to $\preceq$, then S is finite.

The proof of this lemma can be found in [17], where it is attributed to Higman [13].

The well quasi-orderedness of our partial order on global states follows directly from the lemma.

A set I of states is said to be an *ideal* if it is the case that $\gamma \in I$ and $\gamma \preceq \gamma'$ implies $\gamma' \in I$. For a state γ, the *ideal generated by* γ is the set $\{\gamma'; \gamma \preceq \gamma'\}$.

3 A REACHABILITY ALGORITHM

In this section we describe a reachability algorithm [1, 2] for lossy channel systems.

The *initial state* of a lossy channel system (denoted ι) is given by $(\vec{s}_0, \vec{w}_0)$, where $\vec{s}_0$ is the initial control state and $\vec{w}_0(c) = \varepsilon$ for each $c \in C$. We say that a state γ is *reachable* if $\iota \xrightarrow{*} \gamma$. We define the reachability problem formally as follows.

Definition 3 [Reachability Problem]

Instance: A lossy channel system $\mathcal{L} = \langle \vec{S}, \vec{s}_0, A, C, M, \delta \rangle$ and a set $F \subseteq \Gamma$ of *final states*.

Question: Is there a state $\gamma \in F$ such that $(\vec{s}_0, \vec{w}_0) \xrightarrow{*} \gamma$.

The set F is usually used to represent a set of "bad" states which we do not want to occur during the execution of the system. It can be shown [1, 2] that, using standard techniques [20], all safety properties, formulated as regular sets of allowed finite traces, can be reduced to the reachability problem.

To decide reachability, we perform a backward reachability analysis, starting with the states in F and try to find a path back to ι. Since the state space is infinite, the analysis is not *a priori* bounded. It turns out that it is inconvenient to choose the inverse of the forward transition relation (i.e. $\longrightarrow^{-1}$) as our basic backward transition step, since this would add messages to the channels in an uncontrolled manner. Instead we define a "backward" semantics for lossy channel systems as follows.

Let an *operation* be a partial function from states to states. An operation α is *enabled* at a state γ if $\alpha(\gamma)$ is defined for γ. The set of all enabled operations at state γ is denoted *enabled*(γ).

Definition 4 Let $(\vec{s}, \vec{w})$ be a state of a lossy channel system. We define three different operations, each of which corresponds to going backwards in the transition relation. Notice that we execute the operations in reverse, e.g. receiving corresponds to adding a message to the beginning of a channel, etc.

rcv c, m A backwards *receive* operation. Enabled from $(\vec{s}, \vec{w})$ if $\langle s_2^i, c?m, s_1^i \rangle \in \delta$
and $\vec{s}(i) = s_1^i$. When executed, the state $(\vec{s}[i \mapsto s_2^i], \vec{w}[c \mapsto m \cdot \vec{w}(c)])$ is obtained.
snd c, m A backwards *send* operation. Enabled from $(\vec{s}, \vec{w})$ if $\langle s_2^i, c!m, s_1^i \rangle \in \delta$ and
$\vec{s}(i) = s_1^i$. When executed, the state $(\vec{s}[i \mapsto s_2^i], \vec{w}[c \mapsto v])$ is obtained if $\vec{w}(c) = v \cdot m$ for some $v \in M^*$, otherwise $(\vec{s}[i \mapsto s_2^i], \vec{w})$ is obtained, corresponding to a send with a subsequent message loss.
lcl A backwards *local* operation. Enabled from $(\vec{s}, \vec{w})$ if $\langle s_2^i, a, s_1^i \rangle \in \delta$ and $\vec{s}(i) = s_1^i$. When executed, the state $(\vec{s}[i \mapsto s_2^i], \vec{w})$ is obtained.

Intuitively, for a state γ and an operation α, if I and I' are the ideals generated by γ and $\alpha(\gamma)$ respectively, then I' is the set of states from which a state in I is reachable through a single application of α.

We are now ready to introduce the algorithm to decide reachability for lossy channel systems [2]. The algorithm uses a set V of "visited" states, and a "working" set W of states which are yet to be investigated. The algorithm proceeds by selecting and removing a state γ from the working set W. From γ the set of all enabled operations *enabled*(γ) is computed. Each operation is used to compute a predecessor state $\alpha(\gamma)$, which is in turn added to the working set W. The state γ is then added to the set V. Observe that, during our search, if we find a state larger than some state in V, we know that the state is redundant since it gives no new information, hence we can throw it away. A detailed description of the algorithm can be seen in Figure 1.

Theorem 2 The reachability algorithm is correct in the sense that it always terminates, and it returns the value *true* if and only if a state in F is reachable.

The proof can be found in [2], where termination of the algorithm is shown to follow from Lemma 1.

4   DEPENDENCY AMONG OPERATIONS

In this section we define a dependency relation among operations.

For a state γ and a sequence of operations $\rho = \alpha_1 \alpha_2 \cdots \alpha_n$, we say that ρ is *enabled* from γ if, for each $1 \leq i < n$, α_i is enabled from $\alpha_{i-1}(\alpha_{i-2}(\ldots \alpha_2(\alpha_1(\gamma))))$. If ρ is enabled from γ then we use $\rho(\gamma)$ to denote the state $\gamma' = \alpha_n(\cdots \alpha_2(\alpha_1(\gamma)))$.

Algorithm 1 *(Reachability Algorithm)*
Input: A lossy channel system $\mathcal{L}$ and a set F of states.
Output: *true* if any state in F is reachable, *false* otherwise.
Local: A set V containing the already visited states and a set W containing states to be investigated.

1) Add all states in F to the set W
2) If W is empty, exit with the result *false*
3) Select and remove a state $\gamma \in W$
4) If $\gamma = \iota$, exit with the result *true*
5) If there is $\gamma' \in V$ such that $\gamma' \preceq \gamma$, goto 2
6) For each operation $\alpha \in enabled(\gamma)$
7)   Add $\alpha(\gamma)$ to W
8) Add γ to V
9) Goto 2

Figure 1 Reachability Algorithm

An operation α is *monotone* iff $\gamma \preceq \gamma'$ implies $\alpha(\gamma) \preceq \alpha(\gamma')$. Observe that this means that α is enabled in both γ and γ'.

Proposition 3 The operations snd c, m, rcv c, m and lcl are monotone.

Proof. Follows directly from the definitions. $\square$

For a state γ and operations α and β, we say that α *is better than* β *in* γ when executing α before β results in a smaller state with respect to $\preceq$. It is defined formally as follows.

Definition 5 An operation α is better than an operation β in a state γ (denoted $\beta \sqsubseteq_b \alpha$) if and only if:

- $\alpha, \beta \in enabled(\gamma)$,
- $\alpha \in enabled(\beta(\gamma))$,
- $\beta \in enabled(\alpha(\gamma))$, and
- $\alpha\beta(\gamma) \preceq \beta\alpha(\gamma)$.

If $\beta \sqsubseteq_b \alpha$ and $\alpha \sqsubseteq_b \beta$ then α and β are *independent* and if $\beta \not\sqsubseteq_b \alpha$ and $\alpha \not\sqsubseteq_b \beta$ the operations are *dependent*.

Since $\alpha\beta(\gamma) \preceq \beta\alpha(\gamma)$, it is "better", from the point of view of the reachability algorithm, to perform α followed by β than the opposite. The reason is that we know that smaller states generate larger ideals. Consequently, it is safe to discard the sequence $\beta\alpha$ since the ideal generated by $\beta\alpha(\gamma)$ is a subset of the ideal generated by $\alpha\beta(\gamma)$.

Proposition 4 Given lossy channel system $\mathcal{L}$ with operations snd c, m, rcv c, m, and lcl enabled in a state $(\vec{s}, \vec{w})$.

1. Operations of the same process are always dependent.
2. rcv c, n is better than snd d, m iff $c = d$, $\vec{w}(c) = \varepsilon$, and $m = n$; otherwise they are independent.
3. all other combinations are independent.

Proof. Follows directly from the definitions.   □

Notice that the operation snd c, n has two possible behaviours, depending on the state from which it is taken. The fact that rcv c, n is better than snd c, n makes use of this, replacing one type of execution by another.

Definition 6 Let γ be a state, ρ be a finite sequence such that $\rho \in enabled(\gamma)$ and α an operation. An operation α is *contributory* to ρ from state γ iff for any partition of ρ into $\sigma\beta\sigma'$, where β is an operation, $\beta \sqsubseteq_b \alpha$ holds in $\sigma(\gamma)$.

Intuitively, a contributory operation has the property that when it is moved forward in a sequence, we reach "smaller" states (according to $\preceq$).

Lemma 5 Let γ be a state, ρ be a finite sequence of monotone operations such that $\rho \in enabled(\gamma)$, and α a single operation. If α is contributory to ρ from γ then $\alpha\rho(\gamma) \preceq \rho\alpha(\gamma)$.

Proof. Proof by induction on the length of ρ. For the base case $|\rho| = 0$ the lemma holds trivially. For the induction case we let ρ be a sequence of operations such that $|\rho| > 0$ and α is contributory to ρ from γ. Let $\rho = \rho'\beta$ for some sequence of operations ρ' and some operation β. The operation α is better than every operation in ρ and in particular the operation β. Hence $\alpha\beta\rho'(\gamma) \preceq \beta\alpha\rho'(\gamma)$, i.e.

$$\rho'\alpha\beta(\gamma) \preceq \rho'\beta\alpha(\gamma). \tag{1}$$

From the induction hypothesis we know that $\alpha\rho'(\gamma) \preceq \rho'\alpha(\gamma)$. Since β is monotone and enabled at $\alpha\rho'(\gamma)$ it follows that

$$\alpha\rho'\beta(\gamma) \preceq \rho'\alpha\beta(\gamma). \tag{2}$$

By combining (1) and (2) we see that $\alpha\rho'\beta(\gamma) \preceq \rho'\beta\alpha(\gamma)$.   □

In Lemma 5, we define a preorder $\sqsubseteq_B$ on sequences of operations, such that $\rho_1 \sqsubseteq_B \rho_2$ whenever $\rho_1 = \rho'\beta\alpha\rho'$ and $\rho_2 = \rho'\alpha\beta\rho'$, with $\beta \sqsubseteq_b \alpha$ in $\rho'(\gamma)$. We define the "better than" relation among sequences to be the reflexive transitive closure $\sqsubseteq_B^*$ of $\sqsubseteq_B$. Intuitively a sequence ρ_2 is better than a sequence ρ_1, if ρ_2 can be obtained from ρ_1 by permuting adjacent actions so that "better" actions occur earlier in ρ_2 than in ρ_1. It follows from Lemma 5 that $\rho_2(\gamma) \preceq \rho_1(\gamma)$.

This approach is more general than that used in traditional partial order reduction methods, where $\sqsubseteq_b$ is taken to be an equivalence relation and hence $\sqsubseteq_B^*$ becomes an equivalence (rather than a preorder) on operation sequences.

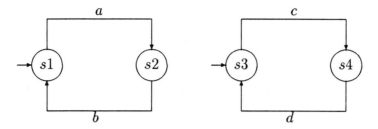

Figure 2 A system of two concurrent processes.

5   IMPROVING THE REACHABILITY ALGORITHM

In this section we introduce two strategies to improve the search conducted in Algorithm 1. The general idea is to consider only a subset of the enabled operations for each state. We replace *enabled*(γ) on line 6 in Algorithm 1 with *ample*(γ) [18], where *ample*(γ) $\subseteq$ *enabled*(γ).

We now proceed by investigating how to select a subset *ample*(γ) of *enabled*(γ). For Algorithm 1 to be correct when replacing *enabled*(γ) with *ample*(γ), the selection of a subset of *enabled*(γ) is guided by the following rules.

B0   The set *ample*(γ) is empty if and only if *enabled*(γ) is empty.
B1   Each of the operations in *ample*(γ) is contributory to every finite sequence of operations ρ enabled from γ that does not contain an operation from *ample*(γ).

It can be shown that selecting a set *ample*(γ) satisfying conditions **B0** and **B1** is not guaranteed to preserve reachability of the initial state. The reason is that because of independence between operations, we may reach a local state in one process before doing so in a second process; we may then continue going backwards in the first process, before reaching the control initial state in the second process and thus we miss the configuration where all processes are in their initial control states. This is a problem inherent in partial order reductions and is not particular for lossy channel systems.

As an example, consider the system in Figure 2. It consists of two processes, executing concurrently. All the transitions (and hence the operations) are local. When trying to search for the initial state (s_1, s_3) from the state (s_2, s_3), the righthand process has already reached its initial control state. However, according to condition **B0**, we may select the operation which is the reverse of transition d. We then reach the state (s_2, s_4). In fact, we may then continue to take the reverse of transition c and reach (s_2, s_3) again, terminating the search with the wrong conclusion that the initial state is not reachable.

We have several alternative ways to remedy the problem. One of the simplest ways is to ensure that it is not possible to leave an inital control states once the search has reached it. This can be accomplished by making the system *rooted*.

Definition 7 A system is *rooted* if none of its operation is enabled from its initial state.

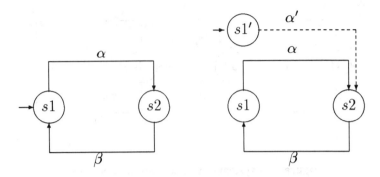

Figure 3 A process transformed to become rooted.

Thus we have our first search strategy:

> STRAT1 A backward search in a rooted system (or a system transformed to become rooted), where each ample set satisfies conditions **B0** and **B1**.

Theorem 6 The correctness of the backward reachability problem is preserved under strategy STRAT1.

Proof. It is easy to see that the reduced state space generates only states that can be generated by the full search. Thus, if an initial state is not reachable, the STRAT1 search will terminate (with the same termination argument of the original algorithm [2]) with a negative answer.

For the other direction, assume that an initial state is reachable. Let γ be a state of the system, with $\alpha \in enabled(\gamma) \setminus ample(\gamma)$, such that there is a path $\alpha\rho$ from γ to an initial state ι. Consider first the case where we partition $\alpha\rho$ into a sequence $\sigma\beta\sigma'$ such that the sequence σ does not contain any operation from $ample(\gamma)$. Then according to condition **B1** and Lemma 5, $\beta\sigma\sigma'(\gamma) \preceq \sigma\beta\sigma'(\gamma)$. Hence $\beta\sigma\sigma'$ is a sequence leading to the initial state ι.

Consider now the case that $\alpha\rho$ does not contain any operation from $ample(\gamma)$. Then, according to condition **B0** and **B1**, $ample(\gamma)$ contains some operation β that is contributory to $\alpha\rho$. But then according to Lemma 5, β is enabled from the initial state ι, contradicting the fact that the system is rooted.   □

It is easy to see that one can convert any system to a rooted one (if it is not rooted already). The disadvantage of having a rooted system is that it may introduce additional non-determinism. Figure 3 demonstrates a system that is not rooted (on the left) that is converted into a rooted system.

Another strategy, trying to avoid enlarging the state space in order not to miss the initial state, is the following.

Definition 8 A *delayed* operation is one that can lead out of the initial state. More precisely, α is said to be *delayed* if $\alpha(\iota)$ is defined.

In fact, in our case an operation is delayed if and only if it executes from an initial control state of a process. We add the following condition on ample sets:

B2 Either *ample*(γ) contains no delayed operations or *ample*(γ) = *enabled*(γ).

For example, in the left process in Figure 3, the operation β is delayed, since one can take a β operation from the initial state. Thus, we have the following search strategy:

STRAT2— A backward search where each ample set satisfies conditions **B0**, **B1**, and **B2**.

Theorem 7 The correctness of the backward reachability problem is preserved under strategy STRAT2.

Proof. Similar to the proof of Theorem 6. The only difference is that in the contradiction case, if β is contributory to the sequence $\alpha\rho$ that transforms γ into ι, β can still be enabled from ι (there is no requirement that the system is rooted). However, this means that β is a delayed operation. But then according to **B2**, we must have selected *ample*(γ) = *enabled*(γ), contradicting the fact that we did not select α. $\square$

6 ALGORITHMS FOR AMPLE SETS

Having defined conditions for ample sets, we need to provide an algorithm that guarantees selection of a subset of the enabled operations satisfying **B0** and **B1** and, depending upon the strategy, also **B2**. Conditions **B0** and **B2** require only simple checks. However, to guarantee **B1**, we need to 'predict' what can happen along sequences of operations, starting from the current state. Deciding condition **B1** for an arbitrary subset of the enabled transitions can be easily shown to be as difficult as reachability. Instead, some heuristics have to be used.

One simple algorithm is the following.

Algorithm 2 *(Simple Algorithm)*
Input: A set of processes $P_1, \ldots, P_n$ and a state γ.
Output: A set *ample*(γ) being a subset of *enabled*(γ).

1) For each process P_i
2)   If each operation is either lcl, rcv d, m, or snd c, m with channel c non-empty
3)   select *ample*(γ) as the enabled operations of P_i, exit
4) If there were no process satisfying the above conditions
5)   select *ample*(γ) = *enabled*(γ)

Lemma 8 The rules **B0** and **B1** are satisfied when using Algorithm 2 to compute *ample*(γ).

Proof sketch. It is trivial to check that **B0** is satisfied. Observe that **B1** is satisfied since all operations belonging to a process are in *ample*(γ) and that no two processes may receive from the same channel. Hence, it cannot happen that an alternative operation of the same process is disabled but will become enabled by the execution of some snd or rcv event in another process.

To use strategy STRAT2 we alter steps 3 and 6 in Algorithm 2: **B2** is applied by selecting the transitions of a certain process only if they also satisfy the condition that they are not delayed.

Theorem 9 Algorithm 1 correctly decides reachability under STRAT1 when using Algorithm 2 to compute *ample*(γ).

Proof. Follows from Theorem 6 and Lemma 8.   □

7   EXPERIMENTS

In this section we consider two examples and show the reductions obtained when using the improved algorithm for reachability.

First we consider the *go back n* protocol of the well known Sliding Window protocol family. We model the protocol as a lossy channel system with two unbounded and unreliable channels.

In the protocol, n corresponds to the size of the sender window. The receiver window is always of size 1. We have added a distinguished initial state to make the system rooted.

In Table 1 we compare the performances of Algorithm 1 with and without partial order reduction, for different sizes of the sender window.

Window Size	No Partial Order Memory	Time	With Partial Order Memory	Time	Reduction (%) Memory	Time
1	16152	0.02	14744	0.02	9%	0%
2	92416	0.10	67024	0.08	27%	20%
3	431096	0.42	297016	0.34	31%	19%
4	1446040	1.67	968184	1.26	33%	25%
5	4022080	4.94	2718224	4.24	32%	14%
6	9656504	13.34	6435560	11.22	33%	16%
7	21013480	32.71	13840888	29.81	34%	9%

Table 1 Verification of *Go Back n*

In the second example, we consider a token ring to implement mutual exclusion among a number of processes. A number of processes communicate through lossy channels. In Table 2 we compare the performances of Algorithm 1 with and without partial order reduction, for different numbers of processes in the ring. The high reduction achieved in this example compared to the first example, is due to the fact that here we have more parallel processes instead of only two processes in the case of the "go back n" protocol.

	Algorithm 1		Algorithm 2		Reduction (%)	
Processes	Memory	Time	Memory	Time	Memory	Time
4	66072	0.08	48264	0.04	27%	50%
5	733200	4.18	238416	0.40	67%	90%
6	5154440	161.16	1201856	5.63	77%	97%
7	34531204	6913.97	6555268	222.07	81%	97%

Table 2 Verification of the Mutex Protocol

8 CONCLUSIONS AND FUTURE WORK

We have considered the application of partial order reduction techniques for lossy channel systems.

In contrast to most existing partial order methods, our approach is applied to a backward reachability algorithm. The choice of transitions selected during the analysis is guided by a preorder which we call the *better than relation*. This is a weaker relation than the *independence relation* used in traditional partial order methods, and leads consequently to a more effective analysis. The idea of using the "better than" relation is general as it is applicable to all systems which have "monotone" behaviour, e.g. Petri nets, real-time automata, relational automata, etc. As part of our future research, we shall try to extend our theory to these classes of systems. We also intend to carry out experiments to study the performance of our algorithms on more advanced examples.

REFERENCES

[1] Parosh Aziz Abdulla and Bengt Jonsson. Verifying programs with unreliable channels. In *Proc. 8th IEEE Int. Symp. on Logic in Computer Science*, pages 160–170, 1993.

[2] Parosh Aziz Abdulla and Bengt Jonsson. Verifying programs with unreliable channels. *Information and Computation*, 127(2):91–101, 1996.

[3] Parosh Aziz Abdulla, Karlis Čerāns, Bengt Jonsson, and Tsay Yih-Kuen. General decidability theorems for infinite-state systems. In *Proc. 11th IEEE Int. Symp. on Logic in Computer Science*, pages 313–321, 1996.

[4] K. Bartlett, R. Scantlebury, and P. Wilkinson. A note on reliable full-duplex transmissions over half duplex lines. *Communications of the ACM*, 2(5):260–261, 1969.

[5] G. V. Bochmann. Finite state description of communicating protocols. *Computer Networks*, 2:361–371, 1978.

[6] D. Brand and P. Zafiropulo. On communicating finite-state machines. *Journal of the ACM*, 2(5):323–342, April 1983.

[7] J.R. Burch, E.M. Clarke, K.L. McMillan, D.L. Dill, and L.J. Hwang. Symbolic model checking: 10^{20} states and beyond. In *Proc. 5th IEEE Int. Symp. on Logic in Computer Science*, 1990.

[8] S. Christensen, Y. Hirshfeld, and F. Moller. Bisimulation equivalence is decidable for basic parallel processes. In *Proc. CONCUR '93, Theories of Concurrency: Unification and Extension*, pages 143–157, 1993.

[9] E.M. Clarke, E.A. Emerson, and A.P. Sistla. Automatic verification of finite-state concurrent systems using temporal logic specification. *ACM Trans. on Programming Languages and Systems*, 8(2):244–263, April 1986.

[10] A. Finkel. Reduction and covering of infinite reachability trees. *Information and Computation*, (89):144–179, 1990.

[11] S. M. German and A. P. Sistla. Reasoning about systems with many processes. *Journal of the ACM*, 39(3):675–735, 1992.

[12] P. Godefroid and P. Wolper. Using partial orders to improve automatic verification methods. In *Proc. Workshop on Computer Aided Verification*, 1990.

[13] G. Higman. Ordering by divisibility in abstract algebras. *Proc. London Math. Soc.*, 2:326–336, 1952.

[14] G.J. Holzmann. *Design and Validation of Computer Protocols*. Prentice Hall, 1991.

[15] ISO. Data communications – HDLC procedures – elements of procedures. Technical Report ISO 4335, International Standards Organization, Geneva, Switzerland, 1979.

[16] B. Jonsson and J. Parrow. Deciding bisimulation equivalences for a class of non-finite-state programs. *Information and Computation*, 107(2):272–302, Dec. 1993.

[17] M. Lothaire. *Combinatorics on Words*, volume 17 of *Encyclopedia of Mathematics and its Applications*. Addison-Wesley, 1983.

[18] D. Peled. Combining partial order reductions with on-the-fly model checking. *Formal Methods in System Design*, 8:39–64, 1996.

[19] A. Valmari. On-the-fly verification with stubborn sets. In Courcoubetis, editor, *Proc. 5th Int. Conf. on Computer Aided Verification*, number 697 in Lecture Notes in Computer Science, pages 59–70, 1993.

[20] M. Y. Vardi and P. Wolper. An automata-theoretic approach to automatic program verification. In *Proc. 1st IEEE Int. Symp. on Logic in Computer Science*, pages 332–344, June 1986.

Conformance Testing

17
A weighted random walk approach for conformance testing of a system specified as communicating finite state machines

Deukyoon Kang, Sungwon Kang, Myungchul Kim and Sangjo Yoo
Protocol Engineering Team, Technical Standards & Requirements Laboratory
Korea Telecom Research & Development Group
17 Woomyun-dong Sucho-ku, Seoul, Korea
Tel: +82-2-526-5156, Fax: +82-2-526-5567
{dykang, kangsw, mckim, sjyoo}@sava.kotel.co.kr

ABSTRACT

It is very important to test protocol implementations to verify conformance to their specifications (standards) in order to promote interoperability between them. This kind of testing is referred to as conformance testing. For that purpose, a kind of test scenario need prepared in advance and the involved work is called test generation. On the other hand, often a protocol can be specified succinctly and in an understandable way as a collection of communicating finite state machines. In this paper, we propose a test generation scheme called weighted random walk that can be applied to the test generation of communicating finite state machines. The proposed scheme

Formal Description Techniques and Protocol Specification, Testing and Verification
T. Mizuno, N. Shiratori, T. Higashino & A. Togashi (Eds.) © 1997 IFIP. Published by Chapman & Hall

is applied to an example protocol and some results of comparison with existing schemes such as pure random walk and guided random walk are presented. Our scheme is superior to the existing schemes in that it tends to test communicating finite state machines with fewer external test inputs. In an illustrated example in the paper, our scheme shows about 48% improvement over the existing schemes in terms of the number of necessary external test inputs.

Key words

Conformance testing, protocol testing, communicating finite state machines, random walk

1   INTRODUCTION

Conformance testing checks if an implementation conforms to its specification. Especially, in protocol engineering area, it is considered important in order to achieve interoperable networks since they consist of communication equipment from many different vendors implementing the same specification like Q.2931, the ATM user-network signalling protocol standardized by ITU-T.

In order to perform conformance testing on an implementation, test generation should be done first. It is well known that manual test generation is error-prone and very time consuming. Thus, a large amount of research work has been done for automatic test generation from various formal specifications including Finite State Machine (FSM) models. In fact, much research effort was focused on test generation from a single FSM model and produced concrete results [3,4]. However, often communication protocols can be specified more succinctly and in an understandable way as a collection of communicating FSM's [1,2].

We can classify conformance testing into structured testing and non-structured testing. In the case of structured testing, a test is generated based on the structure of a single FSM which is composed from a set of communicating FSM's in terms of I/O behaviour. Otherwise the test is non-structured. Furthermore, conformance testing can be classified into static testing or adaptive testing. In the static testing, a test campaign is carried out based on a pre-determined test sequence. For example, the test generation methods such as TT, UIO and W are for static testing. In the case of adaptive testing, there is no such pre-determined test sequence. Only when the current state of an IUT is known, the next external input to be applied is selected. Adaptive testing is very useful to cope with the difficulties arising from non-deterministic behaviour of the IUT since for a non-deterministic FSM we cannot know in advance which transition would be exercised at a given state.

A naive approach to test generation for a system specified as a collection of communicating FSM's would be to first compose them into a single FSM, then apply to it existing test generation methods such as TT, UIO,

W and so on. But the well-known state explosion problem [1,2,4] would be encountered in the process of composing the communicating FSM's into a single FSM. Thus, structured testing would be difficult to apply in practice. In fact, some test generation approaches were proposed to avoid the state explosion problem. They attempt to test in respect of each of communicating FSM's rather than a single FSM. In other words, they attempt to test each of communicating FSM's separately instead of composing them into a single FSM and then testing the one complex FSM. However, even in this approach, static testing would be impractical due to inherent difficulties to be described in next section.

In this paper, we propose a heuristic test generation scheme for a system specified as a collection of communicating FSM's, which is based on random walk and enhanced with weight information. A closely related idea is the guided random walk approach [2]. But its drawback is to converge to random walk rapidly resulting in inefficient test generation as a test campaign goes on. Meanwhile, our approach keeps working effectively with the help of the weight information.

The paper is organized as follows: in Section 2, background concepts and existing test generation approaches based on random walk are explained briefly. Also fundamental difficulties in conformance testing of communicating FSM's are described. In Section 3, we present in detail our test generation approach called weighted random walk. In Section 4, we show some results of comparison of our approach with existing ones. Finally, the paper concludes with Section 5.

2   BACKGROUND AND PREVIOUS WORK

In this paper, a protocol is specified as a set of communicating FSM's. For convenience, we refer to the protocol simply as a CFSM. Each FSM constituting the CFSM is called a component FSM F_i and defined as follows:

Definition 1. $F_i = (Q_i, M_i, \delta_i, s_0^i)$
- Q_i: *a finite set of states*
- M_i: *a finite set of I/O messages*
- δ_i: *non-deterministic transition function defined as $Q_i X M_i \rightarrow 2^{Qi}$*
- s_0^i: *initial state of F_i.*

Hence, we represent a component FSM F_i as a directed graph (V, E) where V is a set of states (Q_i) and E is a set of edges connecting a state s_j^i to another state s_m^i. Each edge is labeled with an I/O (M_i).

A CFSM P consisting of k component FSM's is denoted as follows:
Definition 2. $P=(F_0, F_1, F_2, ..., F_{k-1})$

We assume that communication between component FSM's occurs in synchronous manner, that is, a sender is blocked until the message sent is received by a receiver and a receiver is blocked until an expected message is received. The following notation is used to represent the message exchange

between FSM's:
- To send a message M to Process (or FSM) B: B!M
- To receive a message M from Process A: A?M

For a CFSM consisting of k component FSM's, we can build a single FSM equivalent to the CFSM in respect of I/O behavior. The single FSM is referred to as a composite FSM.

In this paper, for conformance testing of a CFSM, we deal with each component FSM instead of a composite FSM derived from the CFSM to avoid the state explosion in the stage of converting the CFSM to the composite FSM. The approach is justified by the following proposition [1]:

> *Proposition 1: Given a test sequence, if it is a conformance test of each component FSM, then it is also a conformance test of the composite FSM.*

However, there is a difficulty in generating a test sequence systematically for each component FSM due to the following proposition [1]:

> *Proposition 2: Given two states in a component FSM, it is a PSPACE problem to calculate an exact test sequence leading one state to another considering side effects of other component FSM's.*

Thus, random walk based approaches are viable solutions in the situation. In past years random walk was used for the validation of an FSM [5] and some other validation purposes. However, it is not appropriate to apply to conformance test generation directly since the odds are that it produces a very long test sequence for an even very small FSM. Therefore, P-method and guided random walk were proposed in [2] and [1] respectively.

Both of them are unstructured testing in terms of a composite FSM and are based on random walk. However, the assumptions on component FSM's and the communication between them are different. The P-method assumes deterministic component FSM's and asynchronous communication between them. Whereas, the guided random walk assumes non-deterministic component FSM's and synchronous communication.

It was claimed in [2] that for given a real world protocol, it is very difficult or practically impossible to make a complete state space exploration provided that the protocol is specified as a collection of asynchronous communicating FSM's. Its proposed solution is to test   more probable part of a protocol first with the aid of pre-information as to which transitions are more likely to happen. Hence, the P-method generates a set of test sequence prior to a test campaign. Thus, its approach can be said to be based on static testing.

The other scheme [1] is based on adaptive testing to deal with non-determinism. Its goal is different from that of the P-method in that it attempts to cover all transitions of each component FSM and does not generate a fixed test sequence. In this approach, when reaching a state of a CFSM during a test campaign, the next external input is dynamically selected depending on the state. More specifically, given a state, it divides possible external input transitions into two classes, unvisited transitions and visited transitions. Then it gives higher priority to the transitions in the unvisited class than those in the visited class. Actually, it would not try to traverse transitions belonging to visited class if there exists unvisited transition(s). Notwithstanding it is an improvement on the pure random walk approach, the guided random walk loses its advantage over the pure random walk fast as the test campaign goes on. Because if many transitions are traversed, there are seldom chances to exercise the guided selection of transitions.

3   WEIGHTHED RANDOM WALK

We propose the weighted random walk approach for test generation for a CFSM. The proposed approach performs well even in the situation where the guided random walk loses its advantage over the pure random walk. It is made possible by incorporating pre-knowledge from a specification in the form of weights. In this paper, it is assumed that the minimal requirement of test generation for a CFSM is to traverse at least once all transitions of each component FSM of the CFSM.

3.1   Protocol model and assumptions

The protocol model to be used throughout this paper is basically the same as the one in [1]. It was already described in Definitions 1 and 2 in Section 2. That is, synchronous communicating non-deterministic FSM's. Now we refine the notions of transition and state in those definitions and make some additional assumptions on them.

   Transitions are either external or internal. A transition is classified as an external one if it is associated with the environment. Otherwise it is classified as an internal transition. An external transition is denoted as ?M (external input transition) or !M (external output transition). Especially, only external output transitions can be observed and we can exercise limited control to external input transitions. It is limited because the external input transitions may be non-deterministic. For example, for an external input, there can be more than one corresponding transitions and it cannot be known in advance which transition would be exercised. Hence, each external input transition has an integer value, *weight*. The usage of the weight and how to determine its value will be discussed in next subsection.

   There are two kinds of states a component state and a global state. A component state is defined as a state of a component FSM. A global state is defined as tuples of component states and represents a unique state of a CFSM. For example, given a CFSM consisting of k component FSM's, a global state s is defined as *k-tuples $(s^0, s^1,...,s^{k-1})$*, where s^i is a component state corresponding to each component FSM of the CFSM. Hence, global states are classified into stable global states and transient global states. Stable global states are states in which the CFSM are waiting for an external input. All other global states are transient. Transitions can proceed without external inputs in transient global states. We assume that only stable global states can be observed. In other words, when an IUT reaches a stable global state, it would stay there as long as no external inputs are applied to. When an external input is applied to the IUT, it would make some internal input and/or output transitions and/or external output transitions while going through transient global states. Then, it would eventually reach a stable global state. And, we assume that the initial global state can be reached from any global state.

3.2   Overall architecture of weighted random walk

Our proposed testing approach is depicted in Figure 1. Note that there are three procedures such as the selection procedure, the conformity decision procedure and the test manager. The selection procedure takes weight information list as input in order to choose the next external input. The conformity decision procedure determines if observed stable global states and external outputs conform to the specification. The test manger checks if the test termination criteria is satisfied and manages some data structures to keep various information regarding a test campaign. They will be explained in detail through next subsections.

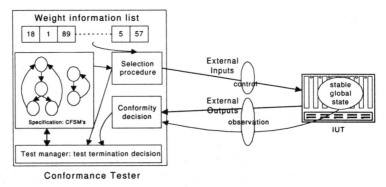

Figure 1 Overall test architecture of weighted random walk.

3.3   Selection of external inputs

When a stable global state is reached, it is very important to select an appropriate external input if more than one external inputs are applicable since it may affect the number of necessary external test inputs significantly. There are three different existing schemes for the purpose such as 1) the pure random walk, 2) the P-method and 3) the guided random walk as mentioned in the previous section.

    In our scheme, we introduce an integer value, *weight,* for the decision on which external input transitions would be applied next. We represent it in a directed graph by augmenting labels on corresponding edges as follows:

?external message (weight value)

We classify external input transitions into *unvisited* and *visited* depending on whether they were already traversed or not. However, we cannot decide for certain that an external input transition was traversed due to non-determinism. This problem will be addressed in next subsection. For a while we assume that we can classify states into *unvisited* and *visited* for certain. Given a state, a transition among the ones in unvisited class would be chosen at random as the next transition before any one in the visited class is chosen. In case that the unvisited class is empty, a transition among ones with the highest weight would be chosen at random from visited class.

The following example shows the basic idea of our approach. In the FSM given in Figure 2, the external input transition *(0,?A,1)* is on a path leading to a bigger behaviour space than the one resulting from the external input transition *(0,?B,4)*. Provided that the FSM is in the state 0 and *(0,?A,1)* and *(0,?B,4)* were already traversed, the guided random walk would choose one of them at random. However, even if we traversed both *(0,?A,1)* and *(0,B,4)*, one of States 2 and 3 may remain untraversed. Thus, it is more reasonable to apply an external input *A* than to apply *B* at State 0 if both *(0,?A,1)* and *(0,?B,4)* were traversed. That is what our approach chooses. However, as we select *(0,?A,1)* repeatedly, its advantage over *(0,?B,4)* decreases. For example, if we traversed *(0,?A,1)* more than a certain number of times, it would be better to give a chance to *(0,?B,4)* next time. Actually, when an external input transition is selected by the associated weight, the weight will be decreased by one if it is greater than *0*. Hence, the mechanism prevents the traversal of a CFSM from being caught in livelock, for example, bouncing back and forth between two global states.

3.4   Weight and weight calculation

In our weighted random walk, weight values should be prepared prior to a test campaign as shown in Figure 1. The weight values should guide selection of the next external input so that the following requirement is satisfied:

> **Requirement:** *When we meet with a stable global state with more than one candidate external input transitions, it should guide us to the next global state from which untraversed transitions are likely to be discovered.*

In order to satisfy the above requirement, we define *weight* as follows. Suppose that we have a composite FSM. For an external input transition e_i at the state s_i in the FSM, its weight is defined as the number of different reachable transitions from s_j to s_0, where s_j is a state reached immediately after applying e_i and s_0 is the initial state of the composite FSM. The weight w_{ei} for e_i can be obtained using the following formula:

$$w_{ei} = \sum_{s_k \in rs_j} \rho_k$$

where rs_j is *{x/x ∈ reachable states from s_j without passing through the initial state and x ≠ the initial state}* and ρ_k is the number of outgoing transitions of s_k. Table 1 shows weights obtained using the formula.

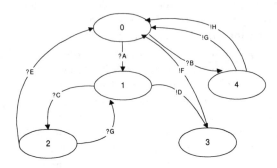

Figure 2 An example FSM.

However, in order to apply the above formula, communicating FSM's should be converted into a composite FSM in advance and we may run into the state-explosion problem again.

Table 1 Weights for the example FSM

Ext. Input transition	Weight	Ext. Input transition	Weight
(0, ?A, 1)	5	(0, ?B, 4)	2
(1, ?C, 2)	5	(2, ?E, 0)	0
(1, ?D, 3)	1	(2, ?G, 1)	5

Thus, instead of using the formula directly, we propose an algorithm that can satisfy the requirement. Let us assume that all transitions in a CFSM have associated weights. From the initial state $(s_0^0, s_0^1,..., s_0^{k-1})$, we can construct a path by concatenating a transition at random iteratively until the path reaches the initial state. In the concatenating process, if we meet with a new transition, weights associated with transitions on the path are increased and the transition is concatenated at the end of the path. Otherwise, the transition is concatenated at the end of the path without an increment of weights. On the other hand, when reaching the initial state, we discard the path and preserves associated weights. We repeat the path construction process until all transitions are traversed. Then, it is obvious that given a weight W_t associated with a transition t, there must be at least W_t transitions reachable via the transition. Thus if a weight associated with an external input transition greater than the weights of other external input transitions, we can claim that the external input transition leads to a bigger behaviour space than others. Therefore, the weight values obtained as described satisfy *Requirement*.

ALGORITHM 1 /* Weights calculation algorithm */
BEGIN
/* *the CFSM consist of k component FSM's* */
　　　　　initialize W; /* *W is an integer array of size equal to the number of external*
　　　　　　　　　　input transitions and indexed by a transition. It keeps track of
　　　　　　　　　　weights of external inputs */
　　　　　initialize Q of size MAX_Q; /* *Q is a queue to keep external inputs leading the*
　　　　　　　　　　CFSM from the initial state to the initial state
　　　　　　　　　　again */
　　　　cur_state := $(s_0^0, s_0^1,..., s_0^{k-1})$; / initial state */*

SUBPROCEDURE *transition(tr: a transition)*
BEGIN
　　　　IF *tr is an external input transition* **THEN**
　　　　　　put tr in Q;
　　　　IF *tr is a newly traversed transition* **THEN BEGIN**
　　　　　　FOR *all t $\in$ a set of transitions in Q* **DO**
　　　　　　　　W[t] := W[t]+1;
　　　　　　mark tr as 'traversed';
　　　　ENDIF;
　　　　update cur_state according to tr;
　　　　IF *cur_state = $(s_0^0, s_0^1,..., s_0^{k-1})$ or the size of Q $\geq$ MAX_Q* **THEN**
　　　　　　discard all elements in Q and reset Q;
　　　　RETURN;
ENDSUBPROCEDURE

WHILE *(there is an untraversed transition)* **DO BEGIN**
　　　　T_o := {t|t $\in$ possible internal and external output transitions at
　　　　　　cur_state};
　　　　WHILE($T_o \neq \varnothing$) DO BEGIN
　　　　　　t_o := choose one at random from T_o;
　　　　　　transition(t_o);
　　　　　　IF *t_o is an internal output transition* **THEN BEGIN**
　　　　　　　　t_i := input transition matching t_o;
　　　　　　　　transition(t_i);
　　　　　　ENDIF;
　　　　　　T_o := {t|t $\in$ possible internal and external output transitions at
　　　　　　　cur_state};
　　　　ENDWHILE;
　　　　/* *stable global state* */
　　　　T_e:= {t|t $\in$ possible external input transitions at cur_state};
　　　　t_e := choose one at random from T_e;
　　　　transition(t_e);
ENDWHILE
RETURN(W);
ENDALGORITHM

The prescribed sketch is refined in Algorithm 1. Note that the length of a path can be arbitrarily long due to the possible cycles existing in a CFSM. Thus, in the algorithm the length of a path is limited by MAX_Q. Hence, it is not necessary to keep information about internal transitions or about external output transitions. All we need is the information of how many transitions are possible from each external input transition. Thus, we construct and keep a path consisting only of external input transitions. Hence, the path is kept in a queue.

3.5   A conformity decision algorithm

It is essential in testing to decide whether observed states and output transitions from an IUT conform to its specification. Before addressing this problem, let us define spontaneous transitions as follows.

> **Definition 3:** *A spontaneous transition is a transition having an external output or internal input/output.*

For any pair of two states, $(s_{(i-1)}{}^j, s_{(i)}{}^j)$ in a component *FSM j*, we can determine if $s_{(i)}{}^j$ is reachable from $s_{(i-1)}{}^j$ through only spontaneous transitions at the cost $O(l_j)$, where l_j is the number of spontaneous transitions in the component *FSM j*. The proof is straightforward and is omitted here. For convenience, in order to denote that $s_{(i)}{}^j$ is reachable from $s_{(i-1)}{}^j$ through only spontaneous transitions, let's use the following notation:

$$s_{(i-1)}{}^j \rightarrow^* s_{(i)}{}^j$$

In order to denote that $s_{(i)}{}^j$ is reachable immediately from $s_{(i-1)}{}^j$ after an input or output message t, the following notation is used:

$$s_{(i-1)}{}^j \rightarrow^t s_{(i)}{}^j$$

Suppose that we observed a global state $S_{(i-1)}$ after applying an external input *Ext* to a component *FSM m* at a global state $S_{(i)}$, where $S_{(i)} = <s_{(i-1)}{}^0, s_{(i-1)}{}^1, ..., s_{(i-1)}{}^{k-1}>$ and $S_{(i-1)} = <s_{(i)}{}^0, s_{(i)}{}^1, ..., s_{(i)}{}^{k-1}>$ respectively. For convenience, we denote it as $(S_{(i-1)}, ?Ext^{(m)}, S_{(i)})$ and refer to it as a stable external input transition. The observed $(S_{(i-1)}, ?Ext^{(m)}, S_{(i)})$ is considered conforming if the following condition is satisfied:

$s_{(i-1)}{}^j \rightarrow^* s_{(i)}{}^j$ *for all j, $0 \leq j < k, j \neq m$ and* $s_{(i-1)}{}^m \rightarrow^{Ext} s_{(t)}{}^m \rightarrow^* s_{(i)}{}^m$····**Condition 1**

For instance, if we observed $(0, ?B^{(0)}, 0)$ in an IUT implementing the FSM in Figure 2, it is considered valid because it satisfies the above condition as follows:

$$0^{(0)} \rightarrow^{?B} 4^{(0)} \rightarrow^{!G} 0^{(0)}$$

In this example, note that $k = 1, m = 0$.

On the other hand, for the outputs observed during the stable external input transition $(S_{(i-1)}, ?Ext^{(m)}, S_{(i)})$, we verify if they belong to a set of expected external output transitions. The expected output transitions can be obtained in the following way. First, obtain the set A as follows:

$A = \{x/x \in s_{(t)}{}^j, \text{ where } s_{(i-1)}{}^j \rightarrow^* s_{(t)}{}^j \text{ for all } j, 0 \leq j < k, j \neq m\}$

Then, obtain the set B as follows:

$B = \{x/x \in s_{(t)}{}^m \text{ or } s_{(t)}{}^{m'}, \text{ where } s_{(i-1)}{}^m \rightarrow^{Ext} s_{(t)}{}^m, s_{(t)}{}^m \rightarrow^* s_{(t)}{}^{m'}\}$

Finally, we obtain O, a set of expected outputs, defined as follows:

$O = \{x/x \in \text{External outputs possible at } s_i{}^j, s_i{}^j \in A \cup B\}$··················**Condition 2**

For instance, we can calculate A, B and O as follows for $(0, ?B^{(0)}, 0)$ in the example FSM. $A = \{\}, B = \{0, 4\}, O = \{G, H\}$.

For a set of observed outputs O', if $O' \subseteq O$, then O' is conisided valid.

Otherwise, invalid. Note that we do not concern with the order or the number of occurrences of external outputs.

3.6 Termination criterion

As mentioned in Section 3, given a CFSM, the minimal requirement of conformance testing is to traverse all transitions of the CFSM at least once. So, in our scheme, the termination criterion is to check if all transitions are exercised at least once. However, for all kinds of transitions such as external input transitions, external output transitions, internal input transition and internal output transition, we could not determine for certain that they are exercised since we assumed non-deterministic FSM model. Thus, we can say only with certain confidence level that they are exercised. For the purpose, a real array T is introduced. Each entry of it indicates the possibility that a corresponding transition is exercised. For the proper manipulation of the real array, we define outgoing degree as follows:

> **Definition 3:** *For a component state s_i^j of a component FSM j, outgoing degree ρ_i^j is defined as the number of possible spontaneous transitions at the state.*

When observing $(S_{(i-1)}, ?Ext^{(m)}, S_{(i)})$, for each transition at $s_{(i)}^j$, where $s_{(i)}^j \in A \cup B$, we increase the corresponding entry of T by $1/\rho_i^j$. If all entries of T is greater than or equal to a certain thresh-hold value, α, then the termination criteria is considered satisfied and the test campaign ends; where α is given prior to the test campaign by a test operator regarding the characteristics of a system under test.

3.7 The conformance testing procedure of weighted random walk

So far we have described the pieces constituting the weighted random walk approach one by one. The algorithm for a conformance test campaign using those pieces is elaborated in Algorithm 2.

The algorithm works in the following way. External input transitions possible at the current stable global state are classified into *class_0, class_1,* and *class_1_high*. Untraversed external input transitions belong to *class_0* and traversed ones to *class_1*. The ones with the highest weight among external input transitions in *class_1* belongs to *class_1_high*. As we noted already, we cannot determine if external input transitions were traversed for certain due to non-determinism. Nevertheless for the external input transitions with corresponding entries of T are equal to zero, we can assert that they have not been traversed for certain. Thus, we assign those external input transitions to *class_0* and others to *class_1*. When the classification is done, one of external input transitions from *class_0* is selected at random if it is not empty. Otherwise, an external input transition that belongs to *class_1_high* is selected at random. Then, the corresponding external input is applied to the IUT. Note that it is not guaranteed that the applied external input results in the execution of the selected external input transition. It is this reason that we do not update the *class_0* and *class_1* at the time when the selection procedure occurs in Algorithm 2. The classification occurs immediately after updating the entries of *T*. The whole process is iterated until the termination criterion is satisfied.

ALGORITHM 2 /* *the CFSM are assumed to consist of k component FSM's */*
BEGIN
    /* *W is a integer array to keep weights and indexed by an external input transition id; for e.g, W[t]. T is a real array to keep confidence level of traversal of a transition and indexed by a transition id; for e.g, T[t] */*
    W := **call** *ALGORITHM 1*;
    initialize T;
    /* *class_0, class_1, and class_1_high are sets. See Section 3.7 */*
    class_0 := $\varnothing$; class_1 := $\varnothing$; class_1_high := $\varnothing$;
    observed_outputs := $\varnothing$;
    class_0:={t|t $\in$ possible external input transitions at the initial state};
    cur_state := $(s_0^0, s_0^1, ..., s_0^{k-1})$;,
    WHILE *termination criteria is not satisfied* **DO BEGIN** /* *see Section 3.6 */*
        IF *class_0 $\neq \varnothing$* **THEN**
            sel_t := select one at random from class_0;
        ELSE BEGIN
            class_1_high := {t|t $\in$ class_1 and $^\forall t' \in$ class_1 $\leq t$};
            sel_t := select one at random from class_1_high;
            IF *W[sel_t] > 0* **THEN**
        /* *note that the selected external input transition may not be exercised in the IUT due to non-determinism. but we consider it exercised */*
                W[sel_t] := W[sel_t] - 1;
        ENDELSE;
        apply to the IUT the corresponding external input for sel_t;
        observe the next stable global state and outputs;
        pre_state := current_state;
        cur_state := the observed global state;
        observed_outputs:= the observed outputs;
        expected_outputs:= calculate expected outputs regarding pre_state,
            cur_state and the selected external input;
        FOR   *regarding pre_state, cur_state and sel_t* **DO BEGIN**
            IF *cur_state is not reachable from pre_state* **THEN**
                RETURN*(fault);* /* *see Condition 1*/*
            IF *observed_outputs $\not\subset$ expected_outputs* **THEN**
                RETURN*(fault);* /* *see Condition 2*/*
            update involved entries of T properly;
            class_0 := {t|T[t]=0 and t is a valid external input transition
                at cur_state};
            class_1 := {t|T[t]>0 and t is a valid external input transition
                at cur_state};
        ENDFOR
    ENDWHILE;
    RETURN*(ok);*
ENDALGORITHM;

There is the possibility that the termination criteria cannot be satisfied due to faults existing in an IUT. To solve this problem, one may incorporate the progress observers proposed in [1]. Their role is to check if the test campaign makes a progress in a sense that new transitions are being traversed. For example, if some entries of T are zero and external inputs have been applied more than a certain number of times, we can declare that a fault may exist in the IUT.

4   Experiments and comparison with the previous approaches

Including ours, there are now four random-walk based testing schemes. They are (1) pure random walk, (2) P-method, (3) guided random walk and (4) weighted random walk. Among them, P-method is quite different from the others in that it is based on asynchronous deterministic communicating FSM's model while the others are based on synchronous non-deterministic communicating FSM's model. Hence, P-method can be classified as static testing while the others are adaptive. Thus, we only consider the schemes (1), (3) and (4) for comparison.

For comparison, we implemented the simulators for (1), (3) and (4) with the C language on UNIX based workstation. For the purpose, we use the example protocol in Figure 2 as a specification as well as an IUT. The weight values for each of external input transitions were obtained by iterating Algorithm 1 ten times and taking the mean values. The results were calculated in terms of the number of external inputs required to traverse all transitions of each component FSM.

In case of a real IUT, the odds is very low that we can observe internal transitions as assumed in Section 3. However, for our comparison purpose, without losing generality all internal transitions can be assumed to be observable.

For an example protocol in Figure 3, external inputs were applied until all transitions of each component FSM are covered. This was done for each of the three schemes. Furthermore, each scheme was iterated 500 times with different random number seeds to get fair results. The comparison results are shown in Table 2 and Figure 4.

Table 2 shows that our scheme is superior to the two existing schemes by about 48%. Weighted random walk covers all transitions of each component FSM with the average of 29 external inputs while the pure random walk and the guided random walk cover them with the average of 57 and 56 external inputs, respectively.

Table 2 Results of the three schemes

Scheme	Avg. Number of external inputs
Pure random walk	57
Guided random walk	56
Weighted random walk	29

From Figure 4, we can observe that when the number of external inputs is between 1 and 4, all three schemes show similar performance with respect to the number of untraversed transitions and the number of applied external inputs since most of the transitions remain untraversed.

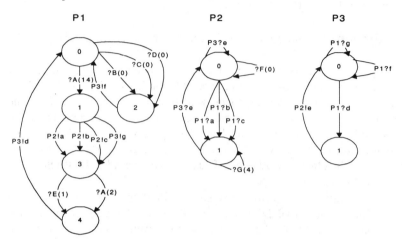

Figure 3 An example protocol.

Where the number of applied external inputs is between 4 and 8, the guided random walk and the weighted random walk are superior to the pure random walk because the former two schemes pay attention to untraversed transitions and their advantage begins to show up. Hence, the two schemes show similar performance in the range.

When the number of applied external inputs is more than 16, the guided random walk and the pure random walk show similar performance. Now the advantage of the guided random walk diminishes since that it has already achieved high traversal ratio. However, we can see that the weighted random walk still performs well by exploiting weight information.

To accomplish higher than 95% coverage in terms of the number of traversed transitions, 24 external inputs were necessary in the case of weighted random walk while 48 and 46 external inputs were necessary in the case of the pure random walk and the guided random walk, respectively. In order to achieve high coverage by random walk based schemes in conformance testing of communicating FSM's, we believe that the pre-knowledge about a CFSM such as the weight information plays an essential role.

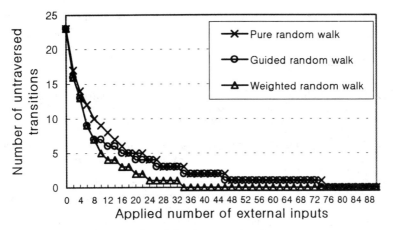

Figure 4 Comparison results of the three approaches.

5 Conclusion

In this paper, we developed a weighted random walk testing scheme for the conformance testing of a system specified as a collection of communicating FSM's. Also, we developed a heuristic method to obtain weight information which is again based on random walk. For the example protocol in Figure 3, it was shown that the proposed scheme shows about 48% improvement over the existing schemes in terms of the number of external inputs required to cover transitions of each component FSM constituting the protocol. Especially, it was addressed that our weighted random walk is expected to achieve high coverage.

The adaptive random walk approach like the guided random walk [1] and ours may at first sight look unrealistic when compared with the methodology ISO 9646 [6]. However, the conformance testing based on the ISO 9646 has the following problems:

- It is very tedious and time consuming to derive test cases from specifications.
- The size of test suites tends to be so big that it is very difficult to validate them to satisfaction and hence impractical to make them standard documents

We believe that our approach is very promising to overcome such problems. For it is straightforward to derive a specification in a set of communicating FSM's from the original protocol specification, for example, from a specification written in SDL (Specification and Description Language). Hence, we can validate the specification with various formal methods. In fact, the example protocol in this paper was validated through random walk.

For further work, we plan to apply our weighted random walk scheme to a real world protocol and to conduct a fault coverage analysis of the scheme.

Also, we look forward to implementing a test software for it on a commercial protocol tester.

6   REFERENCES

[1] D. Lee, K. Sabnani, D. M. Kristol and S. Paul, "Conformance testing of protocols   specified as communicating finite state machines - A guided random walk based approach", *IEEE Trans. on Communication*, Vol.44, No. 5, May 1996.

[2] A. Chung and D. Sidhu,   "Fault coverage of probabilistic test sequence", In Proc. 3rd *International workshop on protocol test systems*, November 1990.

[3] D. Sidhu and T. Leung, "Formal methods for protocol testing: A detailed study", *IEEE Trans. on Software engineering*, Vol.15, No.4, April 1989.

[4] D. Lee and M. Yannakakis, "Principles and methods of testing finite state machines - A survey", Technical report, AT&T Bell Labs., September, 1995.

[5] C. West, "Protocol validation by random state exploration",   In Proc. 6th *International Workshop Protocol specification, testing, and verification*, 1986.

[6] ISO DIS 9646, Conformance Testing Methodology and Framework, part 1, December, 1989.

7   BIOGRAPHY

Deukyoon Kang received B.A. in electronics engineering from Kumoh Nat'l Institute of Technology in 1993 and M.S. in computer science from Pohang Institute of Science and Technology in 1995. Currently he is with Korea Telecom R&D Group as a member of technical staff. He is involved in the design and implementation of a protocol test system for the ATM/B-ISDN protocol family.

Sungwon Kang received a B.A. from Seoul National University in Korea in 1982 and received M.S. and Ph.D. in computer science from the University of Iowa in U.S.A in 1989 and 1992. Since 1993, he has been a senior researcher at Korea Telecom R&D Group. In 1995-1996, he was a guest researcher at National Institute of Standards and Technology of U.S.A. In 1997, he was co-chair of the 10th International Workshop on Testing of Communicating Systems. Currently he is the head of the Protocol Engineering Team at Korea Telecom R&D Group. His research interests include communication protocol testing, program optimization and programming languages.

Myungchul Kim received B.A. in electronics engineering from Ajou Univ. in 1982, M.S. in computer science from the Korea Advanced Institute of Science and Technology in 1984, and Ph.D. in computer science from the Univ. of British Columbia in 1992. Since 1984, he has been working for Korea Telecom. In 1997, he was co-chair of the 10th International Workshop on Testing of Communicating Systems. Currently he is the managing director of Testing Technology Research Section at Korea Telecom R&D Group and chairman of Profile Test Specifications - Special Interest Group of Asia-Oceania Workshop. His interests include protocol engineering on multimedia and telecommunications.

Sangjo Yoo received B.A. in electric communication engineering from Hanyang Univ. in 1988 and M.S. in electrical engineering from the Korea Advanced Institute of Science and Technology in 1990. Currently he is with the Korea Telecom R&D Group as a member of technical staff.

18
Friendly Testing as a Conformance Relation

David de Frutos-Escrig, Luis Llana-Díaz and Manuel Núñez
Dept. de Sistemas Informáticos y Programación
Universidad Complutense de Madrid. E-28040 Madrid. Spain.
e-mail:{defrutos,llana,manuelnu}@dia.ucm.es

Abstract

In this paper we present a new kind of testing, namely *friendly testing*, which has been developed to obtain a satisfactory conformance relation sharing the good properties of the more popular conformance relations, that is *must-testing* and conf, while avoiding their respective problems. In particular, our friendly tests cannot punish a process when it is able to execute some action, while classical testing did it. This was a clear drawback of must testing when considered as a conformance relation. Finally, We prove that the preorder induced by friendly testing is just the transitive closure of conf. As a consequence we obtain an interesting characterization of this closure, from which we derive several its properties.

Keywords

Semantical foundations, Conformance testing, Formal methods.

1 INTRODUCTION AND RELATED WORK

Conformance is the term used by system analyzers to describe the situation in which an implementation is adequate with respect to a given specification. In order to properly define this notion, and thus to have the formal basis for the process of testing, there has been a considerable effort, that in particular has been the seed for the joint ISO/ITU–T working group on "Formal Methods in Conformance Testing". In a recent paper [Cavalli, Favreau & Phalippou 1996], some members of this group have presented a short, but nice, summary of the work carried out by the group, that was included in their working

Research supported in part by the CICYT project TIC 97-0669-C03-01.

Formal Description Techniques and Protocol Specification, Testing and Verification
T. Mizuno, N. Shiratori, T. Higashino & A. Togashi (Eds.) © 1997 IFIP. Published by Chapman & Hall

documents [JTC1/SC21/WG1/54.1 1995*b*, JTC1/SC21/WG1/54.1 1995*a*]. In the same special issue of "Computer Networks and ISDN Systems" devoted to testing where [Cavalli et al. 1996] appeared, there is a longer paper on conformance testing which will be our main reference for definitions and main results [Tretmans 1996]. There, the reader is pointed out for a clear and rather intuitive presentation of the subject. In addition, a complete list of references is provided and therefore we will omit in this paper most of them.

In order to formalize the notion of conformance, two are the most extended methods: by means of an *implementation relation* or by *requirements*. We will concentrate ourselves on the first approach, that is the one in which more work has been developed. An implementation relation relates implementations from a given set Imp with specifications from another set Spec. We are interested in the case in which both sets are somehow formalized, and more specifically, in the case in which both sets are the same. Thus, we will explore relations imp $\subseteq$ Proc $\times$ Proc, for some classes of processes Proc. The most known implementation relation is conf [Brinksma, Scollo & Steenbergen 1986, Brinksma 1988]. This relation is defined from traces and refusals of processes in the following way:

$$i \text{ conf } s \text{ iff } \forall t \in \text{Tr}(s) : \text{Ref}(i, t) \subseteq \text{Ref}(s, t)$$

This relation is derived from the plain refusal ordering [Hoare 1985], which is obtained by removing the constraint $\text{Tr}(s)$ in the universal quantification above. As it is well known, when restricted to non-divergent processes, the refusal ordering is an alternative characterization of the must testing preorder [Hennessy 1988]. Once the (must) passing of tests is defined (a detailed definition can be found at the beginning of Section 2), we can define the preorder $\sqsubseteq_{\text{must}}$ as follows: *

$$s \sqsubseteq_{\text{must}} i \text{ iff } \forall T \ (s \text{ must } T \Rightarrow i \text{ must } T)$$

Since conformance relations are defined to establish the framework in which to formally define testing, it seems that to define one of them by means of passing of tests is a very natural choice. Unfortunately, and even if the (must) testing relation has many pleasant properties, it proves to be too strong to adequately formalize the implementation process. For instance, if we use the notation in [Hennessy 1988] (where there are two choice operators: external, denoted by $+$, and internal, denoted by $\oplus$.), we have $a \not\sqsubseteq_{\text{must}} a + b$. This does not seem very reasonable, because if we are restricted to execute only the

---

*There is some notation disagreement between the testing [Hennessy 1988] and the conformance communities. We have adopted the conventions of the first one to define $\sqsubseteq_{\text{must}}$, since in a testing scenario it seems natural to consider that a process is *better* than another one when it passes more tests, and it is usual to read greater than relations as *better than*. On the contrary, in [Tretmans 1996] testing ordering $\leq_{\text{te}}$ is represented by $i \leq_{\text{te}} s$, probably to maintain the left to right convention between the implementation and the specification in the conformance relation. Finally, it is easy to check that the reduction relation red [Brinksma et al. 1986, Leduc 1992], which can be defined by i red s iff i conf s and $\text{Tr}(i) \subseteq \text{Tr}(s)$, is equal to $\leq_{\text{te}}$ above. In fact, this could be seen as a more convincing justification of the use of the left to right notation for the ordering, since in this alternative definition of the relation there is no reference to tests, and instead the stress is put on the conformance relation conf .

action a, it should not matter if we are also able to execute the action b. It is the case that the relation conf solves this problem; actually, $a + b$ conf a. But this relation does not possess good formal properties. Probably, its most important weakness is that conf is not transitive, and thus neither an order relation. For example, it is easy to check that we also have a conf $a \oplus (b \, ; c)$ but not $a + b$ conf $a \oplus (b \, ; c)$, since after the possible execution of b by the specification, the implementation cannot execute the expected c, while this b plays no role when comparing a and $a \oplus (b \, ; c)$.

G. Leduc has thoroughly worked on the theoretical study of conformance relations [Leduc 1991, Leduc 1992]. He has studied the equivalence induced by an implementation relation, which is defined by:

$$s_1 \; \text{imp-eq} \; s_2 \;\; \text{iff} \;\; \forall i : \; (i \; \text{imp} \; s_1 \Longleftrightarrow i \; \text{imp} \; s_2)$$

Whenever imp is an order relation, it is immediate to prove that imp-eq is the usual equivalence relation induced by it, thus we have imp-eq $= \text{imp} \cap \text{imp}^{-1}$. But if imp is not an order relation, we only have imp-eq $\subseteq \text{imp} \cap \text{imp}^{-1}$. This is the case for conf, for which we have

$$s_1 \; \text{conf-eq} \; s_2 \;\; \text{iff} \;\; s_1 \; \text{conf} \; s_2 \wedge s_2 \; \text{conf} \; s_1 \;\; \wedge \; \forall t \in \text{Tr}(s_1) - \text{Tr}(s_2) : L \in \text{Ref}(s_1, t)$$
$$\wedge \; \forall t \in \text{Tr}(s_2) - \text{Tr}(s_1) : L \in \text{Ref}(s_2, t)$$

where L denotes the full alphabet of observable actions. The last two conditions above are necessary indeed, as the following example shows: Let $s_1 = a$ and $s_2 = a \, ; (\text{STOP} \oplus (b \, ; c))$; we have s_1 conf s_2 and s_2 conf s_1, but not s_1 conf-eq s_2. As a matter of fact, conf $\cap$ conf^{-1} is not an equivalence relation, as the following example shows: Let $s_3 = a \, ; (\text{STOP} \oplus (b \, ; d))$; we have s_1 conf s_3 and s_3 conf s_1, but neither s_2 conf s_3 nor s_3 conf s_2. Finally, conf-eq is weaker than must-equivalence, since for $s_4 = a \, ; (\text{STOP} \oplus b)$ we have s_1 conf-eq s_4, but $s_1 \not\sqsubseteq_{\text{must}} s_4$.

Since conf is not an order relation, we need a stronger relation if we want to follow a refinement process to obtain implementations from specifications. Thus, confrestr is introduced, which is the strongest order relation weaker than conf which preserves that relation, that is, conf $\circ$ confrestr $=$ conf. The relation confrestr can be defined in any of the following alternative ways:

- s_1 confrestr s_2 iff $\forall i : (i \; \text{conf} \; s_1 \Rightarrow i \; \text{conf} \; s_2)$.
- s_1 confrestr s_2 iff s_1 conf $s_2 \wedge \forall t \in \text{Tr}(s_2) - \text{Tr}(s_1) : L \in \text{Ref}(s_2, t)$.

It is easy to check that conf-eq $=$ confrestr $\cap$ confrestr^{-1}.

Based on these somehow negative facts about the two most extended relations, that is $\sqsubseteq_{\text{must}}$ and conf, we have looked for a compromise between them which could inherit the good properties of both, while avoiding their problems. As we already said, we think that to maintain a testing interpretation for a relation that will be the basis for the testing framework seems to be very desirable. So, we have tried to find a new notion of test, and of the passing of tests mechanism, by means of which the desired ordering could be defined following the usual testing way: an implementation is better than (or adequate with respect to) a specification, relatively to our desired conformance relation, if it passes more tests than this last one.

It is clear that any relation defined in this way is an ordering. As an immediate consequence we have that the conformance relation cannot be characterized in this way. Thus we concluded that the adequate starting point for our new notion of testing was not the conformance relation conf, but the classical testing scenario defining $\sqsubseteq_{must}$. So we concentrated ourselves on how tests, and the passing of tests, are defined.

As we will define in detail in the following section, tests are just processes over the alphabet of actions **Act** extended with a special action ω to express successful passing of tests. We apply a test T to a process P by considering the system $P \parallel T$; then, a computation succeeds whenever it reaches a point where the action ω can be executed. If we consider must passing of tests as defined in [de Nicola & Hennessy 1984, Hennessy 1988], we have found that tests have the power to *punish* processes being able to execute actions. So, $a \not\sqsubseteq a + b$, since the former process passes the test $(1 ; \omega) + (b ; \text{STOP})$, while the latter does not.* We are interested on a testing framework in which tests cannot punish processes when they are able to execute some action. This is why we call *friendly testing* to our new testing scenario, and we denote by $\sqsubseteq_{fr}$ the induced preorder. Intuitively, friendly tests can just *reward* with success when the desired traces are executed, but not to *punish* with a failure when some other traces are executable by the tested process. A possible interpretation of this fact leads to the conclusion that the problem comes because we allow both successful (ω) and unsuccessful (STOP) terminations in tests, but this is not the case. In fact, if we restrict the set of tests to *always successful* tests, i.e. tests whose *leaves* are always labeled by ω, nothing is gained, since we could always assume the existence of a new action *reject*, to be read as failure, such that any STOP (failure) termination in the original tests could be simulated by a *reject* ; ω termination.

We could think that if we are working with a language including an internal choice operator, as it is $\oplus$ in [Hennessy 1988], then internal actions are not needed in order to have nondeterministic choices in tests. But, even in the presence of internal choices, the possibility of having internal actions increases the discriminatory power of tests, which does not seem to have a clear intuitive justification. Then it could be thought that all our problems would be solved just by considering tests without internal actions, and indeed this was our first attempt, but this is far from being true. We would get that STOP is the minimum element (if we do not allow divergent processes) since STOP only passes trivial tests. Moreover, $a ; \text{STOP} \sqsubseteq_{fr} a ; P$ and so on; but unfortunately when there are choices among several observable actions, the problem still remains. For instance, $a + b$ would not be *better* than a because the test $(a ; \omega) + (b ; reject ; \omega)$ is passed by the latter process but not by the former. This means that we cannot just restrict the family of tests to reach our goal,

---

*In [Hennessy 1988] the symbol 1 is used to denote internal actions. Other alternative notations are τ and i.

but also the definition of test passing must change, if we desire to obtain $a \sqsubseteq_{\text{fr}} a + b$.

In the following section we will show the adequate changes leading to our new notion of testing. Moreover, we will show that $\sqsubseteq_{\text{fr}}$ is indeed related with conf, as it was our intention. Actually, we have proved that $\sqsubseteq_{\text{fr}}$ is just the transitive closure of conf, namely conf*, which is the strongest order relation weaker than conf where, as usual, we say that $\sqsubseteq_1$ is stronger that $\sqsubseteq_2$ iff $S_1 \sqsubseteq_1 S_2$ implies $S_1 \sqsubseteq_2 S_2$. This is, in our opinion, a nice alternative to the original conformance relation conf, which solves most of its problems, and thus represents the searched compromise between conf and $\sqsubseteq_{\text{must}}$, even if it is not somewhere between them, but instead it is weaker than both. Thus, we have followed the opposite direction that led to confrestr. The reason is very simple: when we studied the examples showing that conf is not an order relation we found no problem on taking conf* instead of conf, thus having, for instance, $a + b$ conf* $a \oplus (b;c)$. We think that the only reason because this is not allowed under conf is that conf was defined looking for an elegant solution in terms of traces and refusals, even if the obtained relation did not posses some good properties (for instance, being an order relation).

The relation conf* was introduced in [Leduc 1992] where also some of its properties were studied. In particular, two interesting results are: conf* = conf ∘ conf and conf* = ext ∘ red. Both results will be somehow used in our proof of the fact that $\sqsubseteq_{\text{fr}}$ is equal to conf*. From this characterization one can find many interesting properties of this relation, which would be more difficult to obtain directly using its definition. For example, by means of this characterization we have defined a complete axiomatization which is obtained by adding a single axiom to that for must testing. Moreover, we also have obtained an explicit characterization based on acceptance sets (or equivalently on refusals). All these results contribute to get a justification and support of friendly testing (equivalently conf*) as a satisfactory conformance relation.

The rest of the paper is structured as follows. In Section 2 we present our new notion of testing. We first introduce *friendly testing* for a particular class of processes that we call *normal forms*. Next, we define friendly testing for arbitrary processes in our language. In Section 3, an alternative characterization of the friendly testing relation, based on a modification of acceptance sets, is defined. Section 4 is devoted to prove that $\sqsubseteq_{\text{fr}}$ is equal to conf*. Finally, in Section 5 we present our conclusions and sketch some of the results on friendly testing that we have obtained, including the complete axiomatization announced above.

2 FRIENDLY TESTING: BASIC DEFINITIONS

Since its introduction in [de Nicola & Hennessy 1984, Hennessy 1988] *Testing Semantics* has been broadly studied and used as a natural way to define an observational semantics with a reasonable power to distinguish semantically different processes. It is defined by observing the operational semantics of processes by means of *tests*. Tests are just processes which may execute a new action ω reporting *success* of the test application. To define the application of a test to a process, we consider the different computations of the experimental system which is obtained by composing in parallel the test and the tested process. We say that a computation is *successful* if there exists a step in the computation such that the associated test can execute the action ω. Since it is possible that some, but not all of the computations may succeed, we can distinguish three families of tests for each process: those whose computations are all unsuccessful, those for which some computations are successful, and those whose computations are all successful. From the last two classes of tests we define two different semantics which are called *may* and *must* semantics. A process P *may* pass a test T (in short P *may* T) if the composition of P and T has at least a successful computation, while P *must* pass T (in short P *must* T) if every computation is successful. By combining these two semantics we can obtain a third one: the *may-must* semantics. Two processes are (*may*, *must*) equivalent iff they pass (in the corresponding sense) the same tests. In addition to these equivalences, we obtain respective partial orderings between processes: Q is *better* than P if any test passed by P is also passed by Q. As a matter of fact, the previous equivalence notions are just the ones induced by the preorders, which could also be studied by themselves.

It is well known that, for divergence-free processes, the different testing preorders and equivalences are related in the following way:

- $P \sqsubseteq_{\text{must}} Q \implies Q \sqsubseteq_{\text{may}} P$
- $P \approx_{\text{may-must}} Q \iff P \approx_{\text{must}} Q$ and $P \approx_{\text{must}} Q \implies P \approx_{\text{may}} Q$

As a consequence, the axiom $P \sqsubseteq_{\text{must}} P + Q$ is not fulfilled at all, and so the testing preorder does not capture the notion of conformance.

In order to present our proposal, we will concentrate ourselves on a syntactic definition of processes, considering a process algebra, instead of using arbitrary transition systems.

Besides we will follow a step by step approach, considering incrementally more general languages, because we think that this process contributes to a better understanding of the definition itself, and also of its properties. To make the comparison with classical testing semantics easier, we will work with the same signature considered in [Hennessy 1988], which is defined by:

Definition 2.1 The set of *finite processes*, denoted by *Proc*, is defined as the set of expressions given by the following BNF-expression:

$$P ::= \text{STOP} \mid a\,;P \mid P + P \mid P \oplus P$$

where $a \in \textbf{Act}$. For the sake of clarity we will omit trailing occurrences of STOP.   □

Operator $+$ corresponds with the external choice operator □ in CSP, and also with the same operator + in CCS when internal actions are not involved in the choice. Besides, $\oplus$ corresponds with the internal choice operator $\sqcap$ in CSP. All the actions in **Act** are assumed to be visible.

In order to introduce friendly testing, we will first consider finite processes in *normal form*. They are defined by the following BNF expression:

$$NFP ::= \bigoplus_{A \in \mathcal{A}} P_A, \qquad \text{where } P_A \in DP ::= \sum_{a \in A}(a\,;P_a), \text{ and } P_a \in NFP$$

where $\mathcal{A} \subseteq \mathcal{P}_f(\mathbf{Act})$, $\mathcal{A}$ is non-empty, and $\bigoplus$ and $\sum$ are the obvious generalizations of $\oplus$ and $+$ to an arbitrary (but finite) number of arguments. By convention $\bigoplus_{A \in \{\emptyset\}}$ represents the process STOP.

As usually, tests will be just processes over the alphabet $\mathbf{Act} \cup \{\omega\}$. For the same reasons that for processes, we will consider a restricted version of tests: those finite deterministic tests with acceptance actions at the end of each trace. We will show that this family of deterministic tests is a set of *essential* tests, in the sense that whenever two processes are not friendly equivalent then there exists a deterministic test distinguishing them. *Deterministic tests* are defined by the BNF expression:

$$DT ::= \omega \mid \sum_{a \in A \subseteq \mathbf{Act}} a\,;DT$$

In Figure 1 we give a graphical representation of **(a)** normal forms and **(b)** deterministic tests. Note that we could see deterministic tests as a particular case of normal forms for which $|\mathcal{A}| = 1$.

Definition 2.2 Given a normal form process P and a deterministic test T, we say that P *friendly passes* T iff

1. $T = \omega$, or
2. $P = \bigoplus_{A \in \mathcal{A}} P_A$, and for each $A \in \mathcal{A}$, P_A *friendly passes* T, or
3. $P = \sum_{a \in A}(a\,;P_a)$, $T = \sum_{b \in B}(b\,;T_b)$, and there exists some $a \in A \cap B$ such that P_a *friendly passes* T_a.

$\square$

Note that this definition, although recursive, is sensible since it is well founded as far as we only consider finite tests. Let us note that the first two cases of this definition are equivalent to those for classical must testing. The differences appear in the last case. If we are testing a generalized external choice, and the test offers several of the actions in the choice, we do not impose that *all* the possible computations must succeed; on the contrary, we only impose that the computations starting with one of the (common) offered actions succeed. In Figure 1 **(c)** we illustrate this definition. In order to friendly pass the test, it is enough that all the computations that are obtained by following the arrows succeed. Next, we compare our definition with the plain must testing by means of an illustrative example.

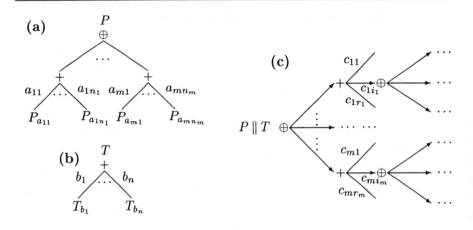

Figure 1 Normal Forms, Deterministic Tests, and P *friendly passes* T.

Example 2.3 Let us consider the following processes $P_1 = a$; P_a, $P' = (a$; $P_a) + (b$; $P_b)$, $P_2 = P_1 \oplus P'$, and $P'' = (a$; $P_a) \oplus (b$; $P_b)$. Let $T = (a$; $\omega) + (b$; STOP). It is easy to check that under the classic notion of testing we have P_1 *must* T but not P_2 *must* T. The reason for this is that in order to get P_2 *must* T all the computations of $P_2 \| T$ must be successful. In particular, this must be true for the computations of $P' \| T$. But when we apply a test like T, offering several actions that could be executed by the tested process, it does not matter if the involved choices in this process are either internal or external. So, for this kind of tests we have P' *must* T iff P'' *must* T.* Such a behavior could be justified by the assumption of testing being the only way to observe the behavior of the tested process. As a matter of fact, and even if that would have no effect in its definition of passing tests, [Hennessy 1988] does not label the transitions of experimental systems of the form $P \| T$. As a consequence, the computations tree corresponding to both $P' \| T$ and $P'' \| T$ are equivalent. On the contrary, we consider that the test is not the final way to observe the behavior of the process. Thus, we do not hide the synchronization actions, and so we maintain some information which allows us to distinguish $P' \| T$ and $P'' \| T$. This is indeed the case, because if we apply the classic (expansion) axioms for the parallel operator we obtain on the one hand $P' \|_{\textbf{Act}} T \approx (a$; $(P_a \|_{\textbf{Act}} \omega)) + (b$; $(P_b \|_{\textbf{Act}}$ STOP$))$, while on the other hand $P'' \|_{\textbf{Act}} T \approx (a$; $(P_a \|_{\textbf{Act}} \omega)) \oplus (b$; $(P_b \|_{\textbf{Act}}$ STOP$))$. So, under our notion of friendly testing we have that P' and P'' can be distinguished by the

---

*It is clear that P' and P'' can be distinguished under plain must testing by a test like a ; ω. In fact, if this would not be the case, they could neither be distinguished under friendly testing. However, it is interesting to observe that P' and P'' cannot be distinguished under must testing by a test like T that offers both a and b; on the contrary, under friendly testing we are able to distinguish P' and P'' by such a test.

test T. Thus we have P_2 *friendly passes* T, and in fact it is the case that for any test T' we have P_1 *friendly passes* T' iff P_2 *friendly passes* T'. □

Then, our justification of the way friendly test passing is defined is that the observer maintains the control, even after a test is applied, as far as external choices remains, as it is the case for process P' in the example above. In such a case the observer can select the action to be executed taking into account when a success (or more exactly, when a set of successful computations) will be reached. The existence of such an action is enough to pass the test. In this way the computations leading to a failure could possibly be avoided, and a test that is not passed in the classic way could be friendly passed.

The reader could think this new notion of passing tests is much more involved than the classic one, but we advocate that this is not the case. Actually, if we consider a recursive definition of the classical notion of must test passing for normal form processes, we see that it can be obtained from our definition of friendly test passing just by changing the existential quantification in the third condition of Definition 2.2 by a universal quantification. Anyway, one could insist on the fact that to impose that all the computations have to be successful is simpler than to check our (apparently) more complicated condition, but this is not the case. In order to check any of these notions we must (in the worst case) explore the full tree of computations; sometimes to check must testing will be faster (when the test fails), and sometimes it is faster to check friendly testing (when the test is successfully passed). Next we present a collection of examples showing the strength and properties of our new notion of testing.

Example 2.4

1. $P \oplus Q \sqsubseteq_{fr} P$. This is because we already had $P \oplus Q \sqsubseteq_{must} P$, and in general we have $P \sqsubseteq_{must} Q \implies P \sqsubseteq_{fr} Q$. As a particular case we have $a \oplus (a+b) \sqsubseteq_{fr} a+b$. On the contrary, we have $(a;c) \oplus (b;c) \not\sqsubseteq_{fr} a+b$, since the test $(a;c;\omega)+(b;c;\omega)$ is friendly passed by the former process but not by the latter.

2. $a \sqsubseteq_{fr} a + b$. Note that under our notion of testing we cannot *punish* the second process when applying a test like $(a \, ; \, \omega) + (b \, ; \, c \, ; \, \omega)$. Even if the computation executing b will not succeed, we can select instead the computation executing a, which immediately succeeds (note that this test is not passed by the second process in the must sense). Actually, we have $P \sqsubseteq_{fr} P + Q$ whenever the sets of actions that can be executed by P and Q in their first steps are disjoint.

3. $a \oplus (a+b) \approx_{fr} a$, because on the one hand we have $a \oplus (a+b) \sqsubseteq_{fr} a$, again as a particular case of the property asserted in 1. On the other hand, note that $a \oplus a \approx_{fr} a$ and then we can apply the fact that all the operators of the language are monotonic with respect to the friendly testing relation. □

2.1   Friendly Testing for arbitrary finite processes and tests

In this section we will consider arbitrary finite processes and tests generated by the syntax given in Definition 2.1. The operational semantics of the language is defined as in [Hennessy 1988]:

$$\overline{a;P \xrightarrow{a} P} \qquad \overline{P \oplus Q \succ\!\!\!\longrightarrow P} \qquad \overline{P \oplus Q \succ\!\!\!\longrightarrow Q}$$

$$\frac{P \xrightarrow{a} P'}{P+Q \xrightarrow{a} P'} \qquad \frac{Q \xrightarrow{a} Q'}{P+Q \xrightarrow{a} Q'} \qquad \frac{P \succ\!\!\!\longrightarrow P'}{P+Q \succ\!\!\!\longrightarrow P'+Q} \qquad \frac{Q \succ\!\!\!\longrightarrow Q'}{P+Q \succ\!\!\!\longrightarrow P+Q'}$$

The following conventions will be used:

$P \xrightarrow{a}$ stands for $\exists P' : P \xrightarrow{a} P'$, $\quad P \xrightarrow{a}\!\!\!\!/\,$ for $\not\exists P' : P \xrightarrow{a} P'$,

$P \not\longrightarrow$ for $\not\exists P', a : P \xrightarrow{a} P'$,

$P \succ\!\!\!\longrightarrow$ for $\exists P' : P \succ\!\!\!\longrightarrow P'$, $\qquad P \succ\!\!\!\longrightarrow\!\!\!\!/\,$ for $\not\exists P' : P \succ\!\!\!\longrightarrow P'$, and

$\succ\!\!\!\longrightarrow^*$ for the transitive and reflexive closure of $\succ\!\!\!\longrightarrow$.

Moreover, for $s = a_1, \ldots, a_n$ we write $P \xRightarrow{s} P'$ if there exist $P_1, \ldots, P_n$, $P'_1, \ldots, P'_n$ such that $P \succ\!\!\!\longrightarrow^* P_1 \xrightarrow{a_1} P'_1 \succ\!\!\!\longrightarrow^* P_2 \cdots P_n \xrightarrow{a_n} P'_n \succ\!\!\!\longrightarrow^* P'$.

Tests are just finite processes over the alphabet **Act** $\cup \{\omega\}$, and the previous operational semantics is also valid for tests. We define the operational semantics of *experimental systems*, $P \parallel T$, by

$$\frac{P \xrightarrow{a} P' \wedge T \xrightarrow{a} T'}{P \parallel T \xrightarrow{a} P' \parallel T'} \qquad \frac{P \succ\!\!\!\longrightarrow P'}{P \parallel T \succ\!\!\!\longrightarrow P' \parallel T} \qquad \frac{T \succ\!\!\!\longrightarrow T'}{P \parallel T \succ\!\!\!\longrightarrow P \parallel T'}$$

Let us remark that, in contrast with the classical testing semantics, we do not hide the actions that experimental systems execute. Now, we introduce some auxiliary concepts for the definition of *friendly* testing.

Definition 2.5 Let P be a process. We say that P is *stable* if $P \succ\!\!\!\longrightarrow\!\!\!\!/\,$. Moreover, given a test T we say that a configuration $P \parallel T$ is *stable* if $P \parallel T \succ\!\!\!\longrightarrow\!\!\!\!/\,$.

Given a process P and $a \in$ **Act**, we define the process P *after the execution of* the action a, denoted by P/a, as $P/a = \bigoplus \{P' \mid P \xRightarrow{a} P'\}$. □

Definition 2.6 (*Friendly Test Passing*). Given a process P and a test T, we say that P *friendly passes* T if the following conditions hold:

- If $P \parallel T$ is stable, then either $T \xrightarrow{\omega}$, or there exists some $a \in$ **Act** such that $P \parallel T \xrightarrow{a}$ and (P/a) *friendly passes* (T/a).
- If $P \parallel T$ is not stable, then for each P', T' such that $P \parallel T \succ\!\!\!\longrightarrow P' \parallel T'$ we have P' *friendly passes* T'.

□

Let us remark that the *first condition* in the previous definition is equivalent to the following one: *If $P \parallel T$ is stable, then either $T \xrightarrow{\omega}$, or there exist $P', T', a \in \mathbf{Act}$ such that $P \parallel T \xrightarrow{a} P' \parallel T'$, and for all P'', T'' such that $P \parallel T \xrightarrow{a} P'' \parallel T''$, we have P'' friendly passes T''.* Thus it is easy to check that the definition above is an extension of the one for normal forms.

Next we present some properties of the general definition of friendly testing. The proofs, by structural induction, are easy.

Proposition 2.7 Let P, P_1, P_2 be processes, and T, T_1, T_2 tests. We have

1. P *friendly passes* ω.
2. P *friendly passes* $T_1 \oplus T_2$ iff P *friendly passes* both T_1 and T_2.
3. $P_1 \oplus P_2$ *friendly passes* T iff both P_1 and P_2 *friendly pass* T.
4. If P_1, P_2 are stable, and $\{a \mid P_1 \xrightarrow{a}\} \cap \{b \mid P_2 \xrightarrow{b}\} = \emptyset$ then for any test T we have $P_1 + P_2$ *friendly passes* T iff P_1 *friendly passes* T or P_2 *friendly passes* T.
5. If P *must* T then P *friendly passes* T.

Definition 2.8 Let P, Q be processes. We write $P \sqsubseteq_{\mathrm{fr}} Q$ iff for all test T we have P *friendly passes* T implies Q *friendly passes* T. Besides, we write $P \approx_{\mathrm{fr}} Q$ iff $P \sqsubseteq_{\mathrm{fr}} Q$ and $Q \sqsubseteq_{\mathrm{fr}} P$. □

Concluding this section we state a result showing that deterministic tests constitute indeed a set of *essential* tests.

Proposition 2.9 Let P, Q be processes. Then we have $P \sqsubseteq_{\mathrm{fr}} Q$ iff for any deterministic test T whenever P *friendly passes* T we also have that Q *friendly passes* T.

3 ALTERNATIVE CHARACTERIZATION OF $\sqsubseteq_{\mathrm{fr}}$

In this section we provide an alternative characterization of the friendly testing preorder given in Definition 2.8. This characterization is based on a modification of acceptance sets [Hennessy 1988]. These adapted acceptance sets are called *friendly acceptance sets*. The last result of the previous section will be very helpful in order to prove that the preorder induced by the alternative characterization is equivalent to $\sqsubseteq_{\mathrm{fr}}$.

Definition 3.1 Let P be a process, and $s = a_1, \ldots, a_n$ a (possibly empty, denoted by ϵ) sequence of actions. We define the following concepts:

- *Initial actions* of P: $S(P) = \{a \mid P \xrightarrow{a}\}$.
- *Acceptance sets* of P after s: $\mathcal{A}(P, s) = \{S(P') \mid P \xRightarrow{s} P'\}$.

- *friendly acceptance sets* of P: $\mathcal{F}(P) = \{A \in \mathcal{A}(P, \epsilon) \,|\, \nexists A' \in \mathcal{A}(P, \epsilon) : A' \subsetneq A\}$
  □

Note that we have defined friendly acceptance sets of a process only for the empty trace. Anyway, friendly acceptance sets for each trace $s = a_1, \ldots, a_n$ could be defined as the friendly acceptance sets of the process $((P/a_1) \cdots)/a_n$. By comparing the friendly acceptance sets of processes we can obtain a new preorder. This preorder is obtained by adapting the preorder for acceptance sets to the new setting.

Definition 3.2 Let P, P' be processes. We write $P \ll_{\text{fr}} P'$ if for all $A' \in \mathcal{F}(P')$ there exists $A \in \mathcal{F}(P)$ such that $A \subseteq A'$, and for all $a \in A$, $P/a \ll_{\text{fr}} P'/a$.   □

Now we will prove that the preorders $\sqsubseteq_{\text{fr}}$ and $\ll_{\text{fr}}$ coincide. We split the proof in two parts.

Theorem 3.3 Given P and P' be processes, we have $P \sqsubseteq_{\text{fr}} P'$ implies $P \ll_{\text{fr}} P'$. *Proof:* The proof will be done by the contrapositive, and structural induction. Let us suppose $P \not\ll_{\text{fr}} P'$, then there exists some $A' \in \mathcal{F}(P')$ such that one of the following conditions hold:

- $\forall A \in \mathcal{F}(P): \ A \not\subseteq A'$, or
- $\forall A \in \mathcal{F}(P): \left(A \subseteq A' \implies \exists a_A \in A : \ P/a_A \not\ll_{\text{fr}} P'/a_A \right)$.

As a matter of fact the first case is just a particular case of the second, but we think that by considering first this particular case we contribute to make the proof more understandable.

In the first case we construct a set S including for each $A \in \mathcal{F}(P)$ one action in $A - A'$. Then, if we consider the deterministic test $T = \sum_{a \in S} a \,;\, \omega$, we get P *friendly passes* T but P' does not.

In the second case, by induction hypothesis we can assume that for each $A \subseteq A'$ there exists T_{a_A} such that P/a_A *friendly passes* T_{a_A}, but P'/a_A does not. Besides, for each $A'' \in \mathcal{F}(P)$ such that $A'' \not\subseteq A'$ we take $a_{A''} \in A'' - A'$, and we consider the deterministic test

$$ T = \sum_{\substack{A \subseteq A' \\ A \in \mathcal{F}(P)}} a_A \,;\, T_{a_A} + \sum_{\substack{A'' \not\subseteq A' \\ A'' \in \mathcal{F}(P)}} a_{A''} \,;\, \omega $$

It is easy to check that P *friendly passes* T but P' does not, since each P/a_A does not *friendly pass* the test T_{a_A}.   □

Theorem 3.4 Given P and P' processes, we have $P \ll_{\text{fr}} P'$ implies $P \sqsubseteq_{\text{fr}} P'$. *Proof:* Let T be a deterministic test such that P *friendly passes* T. We will prove, by induction on the depth of T, that P' also *friendly passes* T.

If depth$(T) = 1$ then $T = \omega$ and the result is trivial. Otherwise we have $T = \sum_{i \in I} a_i ; T_i$. Then, in order to check that P' *friendly passes* T we have to show that for each $A' \in \mathcal{A}(P', \epsilon)$ there exists some $a' \in A'$ with $a' = a_i$, for some i, and such that P/a' *friendly passes* T_i. Since for any $A' \in \mathcal{A}(P', \epsilon)$ there exists $A'' \in \mathcal{F}(P')$ such that $A'' \subseteq A'$, it is enough to prove the previous property for the sets in $\mathcal{F}(P')$.

Given that $P \ll_{\mathrm{fr}} P'$, we have that for any $A' \in \mathcal{F}(P')$ there exists $A \in \mathcal{F}(P)$ with $A \subseteq A'$ such that for all $a \in A : P/a \ll_{\mathrm{fr}} P'/a$. By hypothesis P *friendly passes* T, and thus there exists $a \in A$, with $a = a_i$ for some i, such that P/a *friendly passes* T_i. Therefore we can take $a' = a = a_i$, and by applying the induction hypothesis we obtain P'/a' *friendly passes* T_i, and thus we conclude P' *friendly passes* T. □

Corollary 3.5 Let P, P' be processes. Then $P \ll_{\mathrm{fr}} P' \iff P \sqsubseteq_{\mathrm{fr}} P'$.

4 RELATION BETWEEN conf* AND $\sqsubseteq_{\mathrm{fr}}$

In this section we will prove that the relations conf* and $\sqsubseteq_{\mathrm{fr}}$ are the same.

First, to make easier the comparison with the conformance relation conf, we give a characterization of $\sqsubseteq_{\mathrm{fr}}$ in terms of refusals. We have obtained an explicit non–recursive characterization by introducing the notion of *friendly admissible sets of traces* which gathers the information about the traces that must be taken into account to friendly compare two given processes.

Definition 4.1 Given two processes P, P' we define the family of *friendly admissible sets of traces* for them, denoted by $\mathcal{F}at(P, P')$, as the class of sets S verifying the following conditions:

- $\epsilon \in S$
- $t \in S \Longrightarrow \forall R' \in \mathrm{Ref}(P', t) \ \exists R \in \mathrm{Ref}(P, t) : (R' \subseteq R \land \forall a \notin R : ta \in S)$

□

Theorem 4.2 Given two processes P, P' we have:

$$P \sqsubseteq_{\mathrm{fr}} P' \text{ iff } \exists S \in \mathcal{F}at(P, P') \ \forall t \in S : \ \mathrm{Ref}(P', t) \subseteq \mathrm{Ref}(P, t)$$

Let us remark that the condition on the traces of S in the formula above is already coded in the definition of friendly admissible sets of traces and thus could be removed here, but we include it in order to make easier the comparison with conf.

Corollary 4.3 P' conf $P \Rightarrow P \sqsubseteq_{\mathrm{fr}} P'$.
Proof: We only have to notice that for any $S \in \mathcal{F}at(P, P')$ whenever we have $t \in S$ we also have $t \in \mathrm{Tr}(P')$. □

Let us note that $t \in \mathrm{Tr}(P)$, too. This means that only common traces have to be explored. This makes possible P' being friendly better than P when the former has either more or less traces than the latter.

The following two theorems prove the desired equivalence between conf* and $\sqsubseteq_{\mathrm{fr}}$.

Theorem 4.4 Let P, P' be processes. We have P conf* $P' \implies P' \sqsubseteq_{\mathrm{fr}} P$.
Proof: Trivial, just noticing that conf* is the transitive closure of the relation conf, that P conf $P' \implies P' \sqsubseteq_{\mathrm{fr}} P$ (Corollary 4.3), and that $\sqsubseteq_{\mathrm{fr}}$ is an order relation.   □

Theorem 4.5 Let P, P' be processes. We have $P \sqsubseteq_{\mathrm{fr}} P' \implies P'$ conf* P.
Proof: We will present the proof for normal form processes. In order to extend it to arbitrary processes, we would use the result in [Hennessy 1988] saying that any finite process can be transformed into normal form up to must–testing equivalence, and the fact that $\sqsubseteq_{\mathrm{must}}$ is stronger than $\sqsubseteq_{\mathrm{fr}}$.

Let P_1, P_2 be normal form processes such that $P_1 \sqsubseteq_{\mathrm{fr}} P_2$. We will prove by induction on the depth of P_1 that we also have P_2 conf* P_1.

If $\mathrm{depth}(P_1) = 0$ we have $P_1 = \mathrm{STOP}$, and so we trivially get P_2 conf* P_1.

Let $\mathrm{depth}(P_1) = n + 1$ with $P_1 = \bigoplus_{A \in \mathcal{A}} P_A^1$ and $P_2 = \bigoplus_{B \in \mathcal{B}} P_B^2$ such that $P_1 \sqsubseteq_{\mathrm{fr}} P_2$. Then we have that for any $B \in \mathcal{B}$ there exists some $A_B \in \mathcal{A}$ such that $A_B \subseteq B$ and for all $a \in A_B$ we have $P_1/a \sqsubseteq_{\mathrm{fr}} P_2/a$. Then, if we take $\mathcal{A}' = \{A_B \mid B \in \mathcal{B}\}$ we have that for any $a \in A'$ with $A' \in \mathcal{A}'$, we also have $P_1/a \sqsubseteq_{\mathrm{fr}} P_2/a$. This means that if we consider $P_1' = \bigoplus_{A' \in \mathcal{A}'} P_{A'}^1$, and we define $P_2' = \bigoplus_{B \in \mathcal{B}} P_B'$, where $P_B' = \sum_{b \in B} P_b'^2$ and

$$P_b'^2 = \begin{cases} P_b^1 & \text{if } \exists A' \in \mathcal{A}' : b \in A' \\ P_b^2 & \text{otherwise} \end{cases}$$

we have P_2' conf P_1'. Besides, by applying induction hypothesis, we have that $\forall A' \in \mathcal{A}', a \in A' : P_2/a$ conf* P_1/a. Given that conf is substitutive in the context of the arguments of normal forms, if we recover the original continuations of P_2, by substituting those from P_1 in P_2' by those from P_2, we conclude P_2 conf* P_1', and since obviously we have P_1' conf P_1, we finally obtain P_2 conf* P_1.   □

It is interesting to observe that it is just this final step of the proof which makes (in general) not possible to conclude P_2 conf P_1, since when relating P_2 and P_1 by using an intermediate process P_1', we have that P_1' is a restriction of P_1 (i.e. P_1' red P_1) while P_2 could extend P_1' (i.e. P_2 ext P_1'), and if we eliminate this intermediate process we could obtain some common traces that conf must explore, what $\sqsubseteq_{\mathrm{fr}}$ only partially does. Let us note that we could make a more detailed proof to directly conclude $\sqsubseteq_{\mathrm{fr}} \equiv$ ext ∘ red, but given that conf* $\equiv$ ext ∘ red [Leduc 1992], it is enough to prove $\sqsubseteq_{\mathrm{fr}} \equiv$ conf* even if we were interested in the final characterization $\sqsubseteq_{\mathrm{fr}} \equiv$ ext ∘ red.

Corollary 4.6 Let P, P' be processes. We have $P \sqsubseteq_{\text{fr}} P' \iff P' \text{ conf}^* P$.

5 CONCLUSIONS AND FURTHER WORK

We have presented a new kind of testing, *friendly testing*, which proves to behave as a conformance relation better than the classical must testing does. This is because we reduce the power of tests in such a way that processes cannot be punished when they are able to execute more actions than others. More exactly, we have proved that the order relation induced by friendly testing is just the transitive closure of the conformance relation, **conf**. As a consequence we have obtained an interesting characterization of this relation, from which many properties of it can be derived.

In [Frutos-Escrig, Llana-Díaz & Núñez 1997] we have developed a full theory of friendly testing similar to that for classical testing [Hennessy 1988]. First, we have adapted the results in this paper to deal with general labeled transitions systems which in particular cover the case of recursive processes. Moreover, we have provided both a denotational model and a complete axiomatization. This axiomatization is obtained by adding to the set of axioms for must testing in [Hennessy 1988] the following one

$$\sum_{a \in A} a \,; P_a \leq_{\text{fr}} \sum_{a \in A'} a \,; P_a \text{ whenever } A \subseteq A'$$

In order to obtain both, the denotational model and the complete axiomatization we have found an important technical problem: as it was the case for **conf**, conf^* is not substitutive for arbitrary contexts. More exactly, we have that $\sqsubseteq_{\text{fr}}$ is not a pre–congruence with respect to the external choice operator, as the following example shows:

$$\text{STOP} \oplus b \,; P \approx_{\text{fr}} \text{STOP}$$
$$(\text{STOP} \oplus b \,; P) + b \,; Q \approx_{\text{fr}} b \,; (P \oplus Q) \not\approx_{\text{fr}} b \,; Q \approx_{\text{fr}} \text{STOP} + b \,; Q$$

The problem disappears if there are no interferences between the offerings of the two involved processes.

In the axiomatization we only have to substitute the external choice substitutivity axiom for a more restrictive version covering the case where the involved processes do not offer any common action, to obtain a sound system for friendly testing which can be proved to be also complete by adequating the concept of normal form to the new framework, by means of the characterization by friendly acceptance sets.

Concerning the denotational semantics, it is obvious that we cannot obtain a fully abstract model, since the friendly testing equivalence is not substitutive. This leads to study the weaker pre–congruence $\sqsubseteq_{\text{frext}}$ stronger than $\sqsubseteq_{\text{fr}}$.

We have seen that the relation $\sqsubseteq_{\text{frext}}$, which is the pre–congruence induced by $\sqsubseteq_{\text{fr}}$, is somewhere between $\sqsubseteq_{\text{fr}}$ and $\sqsubseteq_{\text{must}}$, but closer to the first than to the last. In fact, we still have that $\sqsubseteq_{\text{frext}}$ is not stronger than **conf** (it is not weaker either). Thus if we work under $\sqsubseteq_{\text{frext}}$ we still have a rather satisfactory behavior as expected for a conformance relation. Besides $\sqsubseteq_{fr}$ is *almost* a pre–congruence and so we can, in most of the contexts, substitute a process by another related by that relation, with the guarantee that the relation will be preserved.

REFERENCES

Brinksma, E. [1988], A theory for the derivation of tests, *in* 'Protocol Specification, Testing and Verification VIII', pp. 63–74.

Brinksma, E., Scollo, G. & Steenbergen, C. [1986], LOTOS specifications, their implementations and their tests, *in* 'Protocol Specification, Testing and Verification VI', pp. 349–360.

Cavalli, A., Favreau, J. & Phalippou, M. [1996], 'Standardization of formal methods in conformance testing of communication protocols', *Computer Networks and ISDN Systems* **29**, 3–14.

Frutos-Escrig, D., Llana-Díaz, L. & Núñez, M. [1997], Introducing friendly testing, Technical Report DIA 53/97, Dept. Informática y Automática. Universidad Complutense de Madrid.

de Nicola, R. & Hennessy, M. [1984], 'Testing equivalences for processes', *Theoretical Computer Science* **34**, 83–133.

Hennessy, M. [1988], *Algebraic Theory of Processes*, MIT Press.

Hoare, C. [1985], *Communicating Sequential Processes*, Prentice Hall.

JTC1/SC21/WG1/Project 54.1 [1995a], 'FMCT guidelines on Test Generation Methods from Formal Descriptions'.

JTC1/SC21/WG1/Project 54.1 [1995b], 'Working Draft on "Framework: Formal Methods in Conformance Testing"'.

Leduc, G. [1991], Conformance relation, associated equivalence, and minimum canonical tester in LOTOS, *in* 'Protocol Specification, Testing and Verification XI', pp. 249–264.

Leduc, G. [1992], 'A framework based on implementation relations for implementing LOTOS specifications', *Computer Networks and ISDN Systems* **25**(1), 23–41.

Tretmans, J. [1996], 'Conformance testing with labelled transition systems: Implementation relations and test generation', *Computer Networks and ISDN Systems* **29**, 49–79.

19

Generalized metric based test selection and coverage measure for communication protocols

Jinsong Zhu and Son T. Vuong
Department of Computer Science
University of British Columbia
Vancouver, B.C., Canada V6T 1Z4
Email: {jzhu,vuong} @cs.ubc.ca

Abstract

This paper presents an important generalization of the metric based test selection and coverage measure, originally proposed in [14, 5]. Although the original method introduces a significant analytical solution to the problem of coverage and test selection for protocols, its applicability is limited to only the control part of protocols. We extend this method to handle a protocol behavior space where both the control sequences and the data valuations for event parameters are included. We prove that the important properties of total boundedness and completeness are preserved in the generalized metric space, thus ensuring the possible approximation of the specification (infinite number of execution sequences) with a finite test suite within arbitrary degree of precision. A generalized test selection algorithm and coverage measure are also discussed.

Keywords

protocol testing, testing distance, test generation, coverage measure

1 INTRODUCTION

The handling of both the control and data parts of protocols in the process of test generation and selection has been considered as a difficult practical prob-

Formal Description Techniques and Protocol Specification, Testing and Verification
T. Mizuno, N. Shiratori, T. Higashino & A. Togashi (Eds.)   © 1997 IFIP. Published by Chapman & Hall

lem for which no tractable analytical solution has been found. The original work on coverage measure and metric based test selection [14, 5] provides a significant analytical method in assessing the quality of a test suite in terms of its coverage of the specification. Contrary to fault targeting models, where detection of a predefined fault classes (also called fault model [1]) is the test purpose, this metric based method seeks to cover the behavior space of a protocol and to achieve trace equivalence of the test suite with the specification in the limit.

The metric based approach is interesting in that the metric definition with testing distance can be shown to lead to a compact metric space, and the test selection process is convergent where the more test sequences are selected the closer the selected set tends to the original set, *i.e.,* there are no relevant, peculiar test cases or groups of test cases that may be missed out in the selection process due to mere overlook, or due to the limitation of testing cost. Furthermore, the metric defined is made general and flexible by a number of parameters which can be tuned according to the expert knowledge of the specific protocols and potential faults.

A serious limitation of the method, however, lies in the fact that it can only handle the control space of a protocol, thereby making the analysis of a large number of real life protocols infeasible. Many real life protocols have to be extended with data storage to cope with increased complexity of the protocol behavior space. The importance of data storage in practical protocols and their testing can be seen from the numerous research work on testing data flow in a protocol [13, 12, 10, 2, 9]. It would be technically infeasible trying to expand an extended transition system into a pure transition system by unrolling data variables into discrete values, as this would cause the well known problem of state space explosion.

In order to handle the data part feasibly and effectively, we have generalized the original metric based method (hereafter also referred to as the *basic* metric based or MB method). We propose a new definition of the testing distance which incorporates distance contribution from both the control elements and data values. The resulting metric space remains totally bounded and complete, which makes it possible to apply a convergent test selection process to approximate the whole protocol behavior space. A generalized test selection algorithm is proposed for this purpose, with special consideration of data variations. The coverage of the generated test set is measured by the density of the test set by using the general coverage measure as defined in the basic MB method. The extended MB method not only nicely handles protocol properties such as recursion levels, concurrent connections, and transition patterns, as supported by the basic method, but also provides a way to deal with data variations in an execution sequence. If data part is omitted, it then gracefully degrades to the basic method.

The rest of the paper is organized as follows. After a brief overview of the basic metric based method, we describe our generalized model and metric definition in Section 3. Section 4 proves the total boundedness and completeness properties of the generalized metric space. We then proceed to describe our test selection algorithm and coverage measure definition. We conclude by discussing the ramifications of the method and further research work.

2 OVERVIEW OF THE BASIC METRIC BASED METHOD

In [14, 5], a basic coverage metric and the basic metric based test selection method were proposed, with the purpose of generating test cases that cover the control part of the specification. The control behavior of a protocol is considered as composed of execution sequences which represent the interaction between the protocol system and its environment. Within an execution sequence, a concise notation is defined for recursive events: an event a with recursion depth ρ is denoted as a pair (a, ρ). The control space so defined can be infinite: either an execution sequence can be infinite, or there are infinite number of execution sequences. Therefore, in order to cover the space within the computer system and time resources limit, approximations have to be made.

The basic metric based method solves this problem by defining a metric space over the behavior space made of execution sequences. A set of finite execution sequences (a *test suite*, or a *test set*) as approximations of infinite sequences, can be selected based on the metric. Furthermore, a finite number of test suites, which approximates the infinite behavior space, can be generated based on the restriction of test cost. The important property of this approximation process is that the series of test suites can converge to the original specification in the limit. Thus, we have a way to achieve coverage of the specification with an arbitrary degree of precision limited only by the test cost.

The metric is built on the concept of testing distance between two execution sequences. The distance satisfies the requirement that the resulting space be a metric space and be totally bounded, so that we have the nice property of finite covers of an infinite space [5]. It should also capture the intuitive requirement of testing relationships between execution sequences, so that a concept of "closeness" of sequences can be understood. This closeness actually represents a notion of testing representativeness: the closer the two sequences, the more likely they'll yield the same test result.

Formally, testing distance is defined as [5]:

Definition 1 (Basic testing distance) *Let s, t be two (finite or infinite) execution sequences in S, where $s = \{(a_k, \alpha_k)\}_{k=1}^{K}$, and $t = \{(b_k, \beta_k)\}_{k=1}^{L}$, $K, L \in \mathbf{N} \cup \{\infty\}$. The testing distance between two execution sequences s and* *t is defined as*

$$dt(s,t) = \sum_{k=1}^{\max\{K,L\}} p_k \delta_k(s,t)$$

where

$$\delta_k(s,t) = \begin{cases} |r_{\alpha_k} - r_{\beta_k}| & \textit{if } a_k = b_k \\ 1 & \textit{if } a_k \neq b_k \end{cases}$$

If s and t are of different lengths then the shorter sequence is padded to match the length of the longer sequence, so that $\delta_k = 1$ for all k in the padded tail.

In the above definition, δ_k measures the difference in recursion depths. The functions p and r satisfy the following properties:

P1 $\{p_k\}_{k=1}^{\infty}$ is a sequence of positive numbers such that $\sum_{k=1}^{\infty} p_k = p < \infty$. This sequence defines the weights given to the differences of two events within an execution sequence.

P2 $\{r_k\}_{k=0}^{\infty}$ is an increasing sequence in $[0,1]$ such that $\lim_{k\to\infty} r_k = 1$. Put $r_{\infty} = 1$. This function is a normalization function that limits the distance contribution from recursion depths to $[0,1]$.

The convergence of p_k and monotonicity of r_k guarantee that the space (S, dt) is a metric space, and more importantly it is totally bounded and complete [5, 4]. It ensures the existence of finite covers for infinite metric space (S, dt), which is the theoretical foundation for the approximation process and also the test selection algorithm.

The coverage measure for a set of test sequences T with respect to a set S is defined as

$$Cov_S(T) = 1 - m_S(T),$$

where

$$m_S(T) = \frac{\sup\{dt(s,T)|s \in S\backslash T\}}{\sum_{k=1}^{\infty} p_k}.$$

$m_S(T)$ represents the normalized maximum distance from T to S. A large value of $m_S(T)$ implies that T is "farther" from S, hence a smaller coverage.

3   GENERALIZED METRIC SPACE

The basic metric based method handles the protocol control space nicely, with particular considerations of recursion depths. However, in a general protocol behavior space where data flow plays an important role, the omission of the data part limits the applicability of the method to protocols at higher level of abstraction with data part abstracted away.

A simple-minded method to deal with data part may be to expand a protocol with data to the underlying pure finite transition system, by unrolling the data part into discrete values. However, this will cause the well-known state space explosion problem and is only feasible when the data part are small in number of variables and their value ranges. Moreover, even if this expansion is possible, events with different data values would be treated as totally different events, which makes the testing distance insensitive to data differences.

In order to solve this problem and broaden the applicability of the MB method, we have purported to generalize the metric space so as to accommodate both the control flow and data flow of protocols. The idea is to define one metric which incorporates distance contributions from both the execution sequences and data variations. The protocol space can thus be still considered as one metric space, and with proof of total boundedness and completeness of this space, the convergent test selection process and coverage measure can be naturally extended. Moreover, the test selection process can actually be performed in two steps: first for control sequences, and then for data variations. We show that within our distance definition, these two steps can indeed cover the whole protocol space.

We first define the concept of *generalized execution sequence* in our generalized protocol space, G, where both control flow and data flow are present. Each event now has associated data values for its parameters. A generalized execution sequence is thus a sequence of events each of which has its parameters instantiated with certain values within their respective ranges. The range for any value is assumed to be finite, which is reasonable in protocols. For recursive events, the same concise notation as in the basic method is employed. However, if the same event recurses with different data values, we do not consider as recursion depth the number of times that the event is repeated. This is because the events are essentially different when data values are different. To avoid confusion, we distinguish between an event and its name:

Definition 2 (Event) *An event in an executions sequence consists of an event name, and a set of associated data values for its parameters. Formally, an event can be represented as $e(v_1, v_2, ..., v_n)$ where e is the event name and $v_1, ..., v_n$ are a particular data valuation.*

The type of the data valuations can be either numeric or alphabetic. In communication protocols, data are typically integers (*e.g.*, sequence number) or strings (*e.g.*, identifiers).

Definition 3 (Generalized execution sequence) *A generalized execution sequence can be represented as a set of pairs: $\{(a_k(v_1, v_2, ..., v_n), \rho)\}$, where $a_k(v_1, v_2, ..., v_n)$ is an event with name a_k and data values $v_1, v_2, ..., v_n$, and ρ is the recursion depth of the event. Where ambiguity is not possible, we simply use the term "execution sequence".*

All generalized execution sequences of a protocol constitutes the protocol space G. A generalized execution sequence in G is also called a *point* in the space G. The testing distance between two generalized execution sequences is defined as follows:

Definition 4 (Generalized testing distance) *Let* s, t *be two (finite or infinite) generalized execution sequences in* G, *where* $s = \{(a_k(u_{k1}, ..., u_{km}), \alpha_k)\}_{k=1}^{K}$, *and* $t = \{(b_k(v_{k1}, ..., v_{kn}), \beta_k)\}_{k=1}^{L}$, $K, L \in \mathbb{N} \cup \{\infty\}$. *The testing distance between* s *and* t *is defined as*

$$dt(s,t) = \sum_{k=1}^{\max\{K,L\}} p_k \delta_k(s,t)$$

where

$$\delta_k(s,t) = \begin{cases} \alpha \cdot r_{d_k} + \beta \cdot |r_{\alpha_k} - r_{\beta_k}| & \text{if } a_k = b_k \\ \gamma & \text{if } a_k \neq b_k \end{cases}$$

where α, β, γ *are constants satisfying:*

$$\alpha, \beta \geq 0, \gamma > 0, \gamma \geq (\alpha + \beta)/2,$$

and r_{d_k} *is the vector distance between two events of the same name:*

$$r_{d_k} = r(\sum_{i=1}^{m} |u_{ki} - v_{ki}|).$$

The definition of r_{d_k} deserves a special mention in the case where the data values are alphabetic or structured. Suppose the ith data value is of a general string type, *i.e.*,

$$u_{ki} = p_1 p_2 \cdots p_k, \quad v_{ki} = q_1 q_2 \cdots q_{k'}.$$

where $p_k, q_{k'}$ are string elements and each has an integer value (say ASCII code for an alphanumerical element, and 0 or 1 for a bit element). Their distance can be defined as:

$$|u_{ki} - v_{ki}| = \sum_{j=1}^{\min\{k,k'\}} |p_j - q_j| + |k - k'|,$$

which is further used in the definition of r_{d_k}. The distance for structured data values, such as a record, a set, or a sequence, can be recursively computed as the sum of each component element.

In the definition of $dt(s,t)$, the functions p and r satisfy the same properties as in the basic metric based method (Section 2). If s, t differ in length, the shorter one will be padded with a null event and $\delta_k = \gamma$ for the padded part.

We now prove the following theorem.

Theorem 1 *The pair* (G, dt) *is a metric space.*

Proof. See appendix.

Interpretation of the distance

We now give an intuitive interpretation of the testing distance. Same as in the basic MB method, the sequence $\{p_k\}_{k=1}^{\infty}$ gives the weight to the difference of two events at each position k. The requirement of p_k being convergent implies that the weight should be decreasing with the increase of the length of a sequence. This is consistent with our intuition that the farther a sequence goes, the less significant the difference should be, *i.e.*, that difference contributes less to the testing distance. Of course, one can choose to have different weights for finite number of leading events, but after that point, the weights must converge in a decreasing order. For example, a typical function p_k is $p_k = 1/a^k (a > 1)$. One may also define a p_k as

$$p_k = \begin{cases} 1 & \text{if } 0 < k < 10 \\ 1/a^k & \text{if } k \geq 10 \end{cases}$$

The function δ_k measures the difference in recursion depths and data variations. The purpose of using r_k in δ_k is to normalize the distance contributions. r_k is restricted in $[0, 1]$ and must be increasing so that dt is a distance in G. For example, we can choose

$$r_k = \frac{ak}{ak + 1} \quad \text{where} \quad a > 0.$$

We also use the constants α, β and γ as "fine-tuning" parameters in determining the relative importance of the difference in recursion depths, data variations, and events. As a rule of thumb, we can fine-tune the parameters as follows.

1. If difference in data variations (r_{d_k}) is considered as more (or same, less) important than that in recursion depths ($|r_{\alpha_k} - r_{\beta_k}|$), choose α, β such that $\alpha > \beta$ (or $\alpha = \beta$, $\alpha < \beta$, respectively).
2. If difference in events is considered more (or same, less) important than that in recursion depths and data variations, choose α, β, γ such that $\gamma > \alpha + \beta$ (or $\gamma = \alpha + \beta$, $(\alpha + \beta)/2 \leq \gamma < \alpha + \beta$, respectively).

In the simplest case, we can choose $\alpha = \beta = \gamma = 1$, indicating the same importance of differences in recursion depths, data variations, and actions. If we choose to neglect the data part, *i.e.*, $\alpha = 0$, and choose $\beta = \gamma = 1$, we get

$$\delta_k(s, t) = \begin{cases} |r_{\alpha_k} - r_{\beta_k}| & \text{if } a_k = b_k \\ 1 & \text{if } a_k \neq b_k \end{cases}$$

which is exactly the basic metric space.

Example

Let $a(x, y), b(z, w), c(x, w), d(y, z)$ be events, where x, y, z are integers, and w a character string. Consider the following three test cases:

1. $A = \{(a(x, y), 3), (d(y, z), 2), (b(z, w), 1), \}$.
2. $B = \{(a(x, y), 2), (c(x, w), 2), (b(z, w), 2), (c(x, w), 1)\}$.
3. $C = \{(b(z, w), 2), (c(x, w), 3), (d(y, z), 1), (c(x, w), 2)\}$.

Using the generalized testing distance in Definition 4 with the following parameters:

$$p_k = \begin{cases} 2 & \text{if } 1 \leq k < 2 \\ 1/2^{k-2} & \text{if } k \geq 2 \end{cases}$$

and

$$r_k = \frac{k}{k+1}, \quad \alpha = \beta = 1, \ \gamma = 2.$$

Suppose in test case A, $x = 1, y = 2, z = 3, w =$ "DataString1", in B, $x = 4, y = 3, z = 9, w =$ "AnotherData", and in C, $x = 8, y = 6, z = 5, w =$ "GoodTests!", we have

$$
\begin{aligned}
\delta_1(A, B) &= r_{d_1} + |r_{\alpha_1} - r_{\beta_1}| \\
&= r(3 + 1) + |r(3) - r(2)| \\
&= 0.883, \\
\delta_2(A, B) &= 2, \\
\delta_3(A, B) &= r_{d_3} + |r_{\alpha_3} - r_{\beta_3}| \\
&= r(6 + 173) + |r(1) - r(2)| \\
&= 1.161, \\
\delta_4(A, B) &= 2.
\end{aligned}
$$

Therefore the testing distance between A and B is:

$$dt(A, B) = \sum_{k=1}^{4} p_k \delta_k = 7.927.$$

Similarly, we can calculate

$$dt(A, C) = 11, \quad dt(B, C) = 7.119.$$

What this means is that, among the three test cases, B and C are the closest while A and C are the farthest. If we draw a picture with A, B and C as points in the protocol behavior space, they would appear as a triangle as in Figure 1.

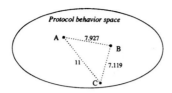

Figure 1 Testing distance between test cases A, B, and C

4 TOTAL BOUNDEDNESS AND COMPLETENESS

In order to be able to use finite test suite to approximate the specification (infinite number of execution sequences) with arbitrary degree of precision, we must show that the generalized metric space (G, dt) is totally bounded and complete, as is the case in the basic MB method.

Theorem 2 *The generalized metric space (G, dt) is totally bounded.*

Proof. We need to show that for every $\epsilon > 0$ it is possible to cover G by a finite number of spheres with radius ϵ. Let $\sum_{k=1}^{\infty} p_k = p$ and k_ϵ be such that $\sum_{k=k_\epsilon+1}^{\infty} p_k < \epsilon/2\gamma$. Let n_ϵ be such that for all $k \geq n_\epsilon$, $r_k > 1 - \eta$, where $\eta = \epsilon/2p\beta$. E is the set of all events with data valuations. Since we assume finite value range of data parameters, E is finite. Therefore, the set

$$F = \{\{(a_i(v_{i1}, ..., v_{in}), \alpha_i)\}_{i=1}^{k_\epsilon} | a_i(v_{i1}, ..., v_{in}) \in E, \alpha_i \in \{1, ..., n_\epsilon\}\}$$

is finite.

We claim that the set of all spheres of radius ϵ centered at elements of F covers the entire space. To see this, let $s = \{(b_i(u_{i1}, ..., u_{in}), \beta_i)\}_{i=1}^{L}$ be arbitrary. Then there exists $t = \{a_i(v_{i1}, ..., v_{in}), \alpha_i)\}_{i=1}^{k_\epsilon}$ in F such that $a_i = b_i$ and $u_{ij} = v_{ij}$ for $i = 1, ..., k_\epsilon$ and $j = 1, ..., n$. Make $\alpha_i = \beta_i$ if $\beta_i \leq n_\epsilon$ or $\alpha_i = n_\epsilon$ if $\beta_i > n_\epsilon$. Note that with this choice of t, $r_{d_k}(s, t) = 0$ for $i = 1, ..., k_\epsilon$. We have

$$dt(s, t) = \sum_{k=1}^{\max\{k_\epsilon, L\}} p_k \delta_k(s, t)$$

$$\leq \sum_{k=1}^{k_\epsilon} p_k(\alpha \cdot r_{d_k} + \beta \cdot |r_{\alpha_k} - r_{\beta_k}|) + \gamma \cdot \sum_{k=k_\epsilon+1}^{\infty} p_k$$

$$\leq \sum_{k=1}^{k_\epsilon} p_k(\beta \cdot |r_{n_\epsilon} - 1|) + \gamma \cdot \sum_{k=k_\epsilon+1}^{\infty} p_k$$

$$\leq p \cdot \beta\eta + \gamma \cdot \epsilon/2\gamma$$

$$< \epsilon/2 + \epsilon/2 = \epsilon.$$

This completes our proof of the theorem.□

Total boundedness is very important for the subsequent test selection because it implies the existence of finite number of balls of radius ϵ that covers the metric space. This makes it possible to choose finite number of test sequences (points in G), each belonging to one ball, to cover the protocol behavior space. The points that fall into a particular ball are all represented by the chosen point (a test sequence) in the sense that their testing distances are small enough.

The total boundedness property itself is not enough for test selection, since we want to choose test sequences incrementally, and eventually, the protocol space will be fully covered. This relies on the completeness property that we now proceed to prove. In general, we want the metric space to be compact, *i.e.*, both totally bounded and complete.

To prove the compactness of the metric space (G, dt), we need to use the following definition and preposition:

Definition 5 (Completeness) *A metric space is* complete *if every Cauchy sequence in it is convergent to a point of the space.*

Proposition 1 *[11]: A metric space is compact if and only if it is complete and totally bounded.*

As we have proved the boundedness of (G, dt), we now only need to prove the completeness of (G, dt), *i.e.*, every Cauchy sequence in G converges to a point in G.

Theorem 3 *The generalized metric space (G, dt) is complete.*

Proof. Let $\{g_n\}_{n=1}^{\infty}$ be a Cauchy sequence in G, where $g_n = \{(a_i^{(n)}(v_{i1}^{(n)}, ..., v_{ik}^{(n)}),$ $a_i^{(n)})\}_{i=1}^{K^{(n)}}$, $K^{(n)} \in \mathbf{N} \cup \{\infty\}$. By definition,

$$\forall \epsilon > 0 \ \exists N_{\epsilon}[\forall m, n > N_{\epsilon} \Rightarrow dt(g_m, g_n) < \epsilon]$$

where $N_{\epsilon}, m, n \in \mathbf{N}$.

Let k be arbitrary. We first show that $\lim_{m,n\to\infty} \delta_k(g_m, g_n) = 0$. For an arbitrary $\eta > 0$, choose ϵ such that $\epsilon < \eta p_k$. By definition, there exists N_{ϵ} such that $dt(g_m, g_n) < \epsilon$ for all $m, n > N_{\epsilon}$. Therefore

$$p_k \delta_k(g_m, g_n) \leq dt(g_m, g_n) < \epsilon < \eta p_k.$$

Consequently, $\delta_k(g_m, g_n) < \eta$ for all $m, n > N_{\epsilon}$. This proves $\lim_{m,n\to\infty} \delta_k(g_m, g_n) = 0$.

By the definition of δ_k, we can conclude that the sequence $\{K^{(n)}\}_{n=1}^{\infty}$ must converge to K, $K \in \mathbf{N} \cup \{\infty\}$. To see this, consider two cases: a) $K^{(n)}$ are all finite, then $K^{(n)}$ must be eventually constant, *i.e.*, equals $K, K \in \mathbf{N}$, since otherwise $\delta_k(g_m, g_n) \geq \gamma$ and cannot converge to 0; b) $K^{(n)} \to \infty$ when $n \to \infty$, in which case we also conveniently say $K^{(n)}$ converges to ∞.

Similarly, we can conclude that

$$\lim_{m,n\to\infty} r_{d_k} = 0 \quad \text{and} \quad \lim_{m,n\to\infty} |r_{\alpha_k} - r_{\beta_k}| = 0.$$

This implies that the sequences $\{a_i^{(n)}\}_{n=1}^\infty$, $\{v_{ik}^{(n)}\}_{n=1}^\infty$'s, and $\{\alpha_i^{(n)}\}_{n=1}^\infty$ are all convergent. Let $a_i(v_{i1}, ..., v_{ik}) = \lim_{n\to\infty} a_i^{(n)}(v_{i1}^{(n)}, ..., v_{ik}^{(n)})$ and $\alpha_i = \lim_{n\to\infty} \alpha_i^{(n)}$. Since i is arbitrary, we get the sequence $g = \{(a_i(v_{i1}, ..., v_{ik}), \alpha_i)\}_{i=1}^K$, $K \in \mathbf{N} \cup \{\infty\}$, which is a point in space G.

We claim that the sequence $\{g_n\}_{n=1}^\infty$ converges to g. This is obvious when K is finite since eventually $g_n = g$. To prove the case where $K = \infty$, let $\epsilon > 0$ be arbitrary. Choose k_ϵ such that $\sum_{k_\epsilon+1}^\infty p_k < \epsilon/2$. Since $\lim_{n\to\infty} \delta_k(g_n, g) = 0$, we can choose n_ϵ such that $\delta_k(g_n, g) < \epsilon/2p$ (where $p = \sum_{k=1}^\infty p_k$) for all $1 \leq k \leq k_\epsilon$ and for all $n \geq n_\epsilon$. Therefore, for all $n > n_\epsilon$ we have

$$
\begin{aligned}
dt(g_n, g) &= \sum_{k=1}^\infty p_k \delta_k(g_n, g) \\
&\leq \sum_{k=1}^{k_\epsilon} p_k \delta_k(g_n, g) + \sum_{k=k_\epsilon+1}^\infty p_k \\
&< \frac{\epsilon}{2p} \sum_{k=1}^{k_\epsilon} p_k + \frac{\epsilon}{2} \\
&\leq \frac{\epsilon}{2p} p + \frac{\epsilon}{2} = \epsilon.
\end{aligned}
$$

This proves that the Cauchy sequence $\{g_n\}_{n=1}^\infty$ in G converges to a point in G with respect to dt, therefore (G, dt) is complete.□

From Theorem 2 and 3 and the Preposition, we obtain the following compactness theorem.

Theorem 4 *The generalized metric space (G, dt) is compact.*□

The compactness (*i.e.*, total boundedness plus completeness) lays the foundation of test selection and coverage measure in the next section. We will modify the test selection process in the basic MB method to make it applicable to the generalized space.

5   TEST SELECTION AND COVERAGE MEASURE

The basic idea in test selection with the basic MB method is to generate ϵ-dense* set of test sequences with respect to the protocol specification or some original test suite. By choosing ϵ arbitrarily small and incrementally adding more test sequences, we can achieve arbitrary accuracy in covering the protocol specification in a convergent manner.

We adopt the same idea in our generalized metric space, with extensions to handle the data valuations. For simplicity, we omit the cost factor [14] in our algorithm, and focus on the selection of ϵ-dense test sets with respect to a set of execution sequences which can be finite or infinite.

Algorithm: Selection of an ϵ-dense set of test sequences

Input: G: original set of execution sequences, ϵ: density requirement.
Output: T: selected ϵ-dense set of test sequences.

Step 1: Initially, $T = \phi$ (empty). Also let $X = G$.
Step 2: If $X = \phi$ return T and exit. Otherwise randomly remove a test sequence t from X (*i.e.*, $X \leftarrow X - \{t\}$), apply appropriate data valuations where data are involved, and calculate $dt(t, T)$.
Step 3: If $dt(t, T) \geq \epsilon$, $T \leftarrow T \cup \{t\}$. Go to Step 2.

Using the same idea of multi-pass algorithm with decreasing ϵ density, we obtain a set of Cauchy sequences over the successive passes. These Cauchy sequences converge to infinite execution sequences in the specification when G is infinite. Since the space (G, dt) is totally bounded, we can eventually find the finite covering of G with density ϵ, at which point the algorithm terminates.

It is also worthwhile to note that in Step 2 when data valuations are generated, we can either simply generate random data values, or intentionally select fault sensitive data values such as boundary values and typical values. In this way, some test data generation criteria that have been proved effective in software testing can be incorporated. In practice, we need also consider the executability of an execution sequence. This executability checking can be performed on-the-fly as data values are chosen, using the method in [3] or [9].

In the implementation of the algorithm, we may also consider all data variations for an execution sequence before proceeding to a new execution sequence, or consider all execution sequences with distinct control part before embarking on data variations. Either way we can have a more clear-cut two pass algorithm that clearly distinguishes between the control and data part,

---

*A set T is said to be ϵ-dense ($\epsilon > 0$) in a set S if for every $s \in S$ there is $t \in T$ such that $dt(s, t) < \epsilon$.

and yet they together cover the whole protocol space with a desired test set density.

From the test selection process, we can actually measure the coverage of a test set with density ϵ. This can be seen by observing that

$$\sup\{dt(g,T) : g \in G\backslash T\} \leq \epsilon$$

Using the same definition of the basic MB method (Section 2), we have

$$m_G(T) = \frac{\sup\{dt(g,T) : g \in G\backslash T\}}{\sum_{k=1}^{\infty} p_k} \leq \frac{\epsilon}{p},$$

where $p = \sum_{k=1}^{\infty} p_k$. Consequently,

$$Cov_G(T) = 1 - m_G(T) \geq 1 - \frac{\epsilon}{p}.$$

Using $1 - \epsilon/p$ as the coverage of T with respect to G would be a fair measure since it is the lower bound of the exact coverage. The property $\lim_{\epsilon \to 0} Cov_G(T) = 1$ also indicates that with arbitrarily small ϵ density, the test set can ultimately cover the whole space.

6 CONCLUDING REMARKS

The problem of test generation and selection, which can handle both the control and data parts of protocols has been considered as a difficult practical problem for which no tractable analytical solution has been found. In this paper, we have presented a generalization of the basic metric based test selection method where both control and data parts of protocols can be handled. The major contributions include a generalized metric space definition, the proof of total boundedness and completeness (*i.e.*, compactness) properties of the space, and a generalized test selection algorithm. This lays out a metric based testing framework where extended protocol space can be tackled. In our generalized test selection algorithm, it is also possible to incorporate test data generation criteria that have been proved effective in discovering program faults, such as boundary values, typical values, and values that cover data dependencies, etc. This could be a point where our metric based method is joined by the traditional software testing techniques, and therefore its effectiveness and applicability are boosted. We are currently doing experiments with practical protocols in order to show this effectiveness.

It is true that the MB method critically hinges on the definition of testing distance, and with different definitions, the effectiveness may be different. However, the testing distance is intended to provide a way to measure the "closeness" or "representativeness" of test sequences. The exact meaning of the "closeness" concept relies on our experience and expertise in protocol testing. With our definition of the testing distance, we believe many important protocol properties, such as recursion, concurrency, and data variations, are already captured in the definition. With appropriate parameterization of the functions r and p, and the fine-tuning constants α, β and γ, a "good" testing distance definition which reflects our experiences in protocol testing is possible. It is of course a non-trivial task to incorporate expert knowledge into the parameters. This is actually a further research area where heuristics need to be developed to provide guidance to the choice of metric functions and parameterizations. An initial research work that evaluates the sensitivity of metric functions to recursion depths, concurrent connections, and transition patterns is reported in [6].

It is worth noting that our method is not intended as a replacement of fault targeting methods; instead we view them as complementary methods for testing protocols and software in general. In a typical test campaign, functional testing is first carried out to make sure an implementation indeed performs the functions it is supposed to perform. Structural testing follows by checking that the protocol or program space is sufficiently exercised. Our method can be very useful in the second stage in that it can systematically generate test sets with increased coverage, and the test sets converge to the whole space eventually, with the accuracy only limited by cost.

As many software testing researchers have noticed and showed with experiments that code coverage does imply fault coverage [7], we conjecture that our method should also have good fault coverage. How good it is should still depend on the definition of testing distance and possibly the structure of a protocol. We plan to explore this further with experiments, by observing and analyzing the sensitivity of metric functions to the fault detection capability of the test sets. As a final remark, we observe and conjecture that there is link between our method and software reliability which can be measured with test set density in some way. Further work will be necessary to validate (or invalidate) the conjecture.

7   ACKNOWLEDGMENTS

The authors wish to thank Dr. Jadranka Alilovic-Curgus for many fruitful discussions during the course of the work reported in this paper, and the anonymous reviewers for their helpful comments.

8   APPENDIX

Proof of Theorem 1: It suffices to show that dt is a distance in G.

1. $dt(s,t)$ is a non-negative real number for any s, t.

This is straightforward from the fact that δ_k is up-bounded by $\max\{\alpha + \beta, \gamma\}$, and the sequence p_k converges.

2. $dt(s,t) = 0$ iff $s = t$.

This can be easily seen from the fact that $dt(s,t) = 0$ iff $\delta_k(s,t) = 0$ for all

k, which is equivalent to $K = L, a_k = b_k, u_{ki} = v_{ki}$ for all i, and $\alpha_k = \beta_k$, for all $k = 1, 2, ..., K = L$.

3. $dt(s,t) = dt(t,s)$.

This directly comes from the fact that $\delta_k(s,t)$ is symmetric, *i.e.*, $\delta(s,t) = \delta(t,s)$.

4. $dt(s,t)$ satisfies the triangular inequality.

Let s, t, u be three execution sequences in G, $u = \{(c_k(w_{k1}, ..., w_{kl}), \gamma_k)\}_{k=1}^{M}$. Without loss of generality, consider the following cases at δ_k:

Case I: s, t, u all have the same event, *i.e.*, $a_k = b_k = c_k$. We have

$$
\begin{aligned}
&\delta_k(s,u) + \delta_k(u,t) \\
=\ & (\alpha \cdot r_{d_k}(s,u) + \beta \cdot |r_{\alpha_k} - r_{\gamma_k}|) + (\alpha \cdot r_{d_k}(u,t) + \beta \cdot |r_{\gamma_k} - r_{\beta_k}|) \\
=\ & \alpha \cdot (r_{d_k}(s,u) + r_{d_k}(u,t)) + \beta \cdot (|r_{\alpha_k} - r_{\gamma_k}| + |r_{\gamma_k} - r_{\beta_k}|) \\
\geq\ & \alpha \cdot r_{d_k}(s,t) + \beta \cdot |r_{\alpha_k} - r_{\beta_k}| \\
=\ & \delta_k(s,t).
\end{aligned}
$$

The inequality $r_{d_k}(s,u) + r_{d_k}(u,t) \geq r_{d_k}(s,t)$ holds due to a special case of the Minkowski's inequality [8] and also the fact that the function r is increasing.

Case II: s, t have the same event, but u's event is different, *i.e.*, $a_k = b_k \neq c_k$. Since $\gamma \geq (\alpha + \beta)/2$, it follows that

$$
\begin{aligned}
\delta_k(s,u) + \delta_k(u,t) \ =\ & 2\gamma \\
\geq\ & \alpha + \beta \\
\geq\ & \alpha \cdot r_{d_k} + \beta \cdot |r_{\alpha_k} - r_{\beta_k}| \\
=\ & \delta_k(s,t).
\end{aligned}
$$

Case III: s, t have different event, but u has the same event as s, *i.e.*, $a_k = c_k \neq b_k$. We have

$$
\delta_k(s,u) + \delta_k(u,t) = (\alpha \cdot r_{d_k}(s,u) + \beta \cdot |r_{\alpha_k} - r_{\gamma_k}|) + \gamma \geq \gamma = \delta_k(s,t).
$$

Case IV: s, t, u all have different event, *i.e.*, a_k, b_k, c_k are all different. Obviously,

$$
\delta_k(s,u) + \delta_k(u,t) = \gamma + \gamma > \gamma = \delta_k(s,t).
$$

Therefore, in all cases, we have $\delta_k(s,u) + \delta_k(u,t) \geq \delta_k(s,t)$, which leads to:

$$
\begin{aligned}
dt(s,u) + dt(u,t) \ =\ & \sum_{k=1}^{\max\{K,L\}} p_k \delta_k(s,u) + \sum_{k=1}^{\max\{L,M\}} p_k \delta_k(u,t) \\
\geq\ & \sum_{k=1}^{\max\{K,L\}} p_k \delta_k(s,t) = dt(s,t). \qquad \square
\end{aligned}
$$

REFERENCES

[1] G.v. Bochmann and et al. Fault models in testing. In J. Kroon, R.J. Heijink, and E. Brinksma, editors, *Protocol Test Systems, V*, Leidschendam, The Netherlands, October 1991.

[2] S.T. Chanson and J. Zhu. A unified approach to protocol test sequence generation. In *Proc. IEEE INFOCOM*, San Francisco, March 1993.

[3] S.T. Chanson and J. Zhu. Automatic protocol test suite derivation. In *Proc. IEEE INFOCOM*, Toronto, Canada, June 1994.

[4] J.A. Curgus. *A metric based theory of test selection and coverage for communication protocols.* PhD thesis, Dept. of Computer Science, Univ. of British Columbia, June 1993.

[5] J.A. Curgus and S.T. Vuong. A metric based theory of test selection and coverage. In *Proc. IFIP 13th Symp. Protocol Specification, Testing, and Verification*, May 1993.

[6] J.A. Curgus, S.T. Vuong, and J. Zhu. Sensitivity analysis of the metric based test selection. In *IFIP 10th Int. Workshop on Testing of Communicating Systems*, Cheju Island, Korea, September 1997.

[7] P.G. Frankl and S.N. Weiss. An experimental comparison of the effectiveness of branch testing and data flow testing. *IEEE Transactions on Software Engineering*, 19(8):774–787, August 1993.

[8] E. Hewitt and K. Stromberg. *Real and Abstract Analysis*. Springer-Verlag, 1965.

[9] T. Higashino and G.v. Bochmann. Automatic analysis and test case derivation for a restricted class of LOTOS expressions with data parameters. *IEEE Transactions on Software Engineering*, 20(1), January 1994.

[10] R.E. Miller and S. Paul. On generating test sequences for combined control and data flow for conformance testing of communication protocols. In *Proc. IFIP 12th Int. Symp. on Protocol Specification, Testing, and Verification*, June 1992.

[11] H.L. Royden. *Real Analysis*. Macmillan Publishing Company, New York, 1988.

[12] B. Sarikaya, G.v. Bochmann, and E. Cerny. A test methodology for protocol testing. *IEEE Transactions on Software Engineering*, May 1987.

[13] H. Ural. Test sequence selection based on static data flow analysis. *Computer Communication*, 10(5), 1987.

[14] S.T. Vuong and J.A. Curgus. On test coverage metrics for communication protocols. In *Proc. 4th Int. Workshop on Protocol Testing System*, 1991.

Future Information Technology and its Impact on Society

Future Information Technology and Its Impact on Society

Shoichi Noguchi, President, The University of Aizu

Ikki-machi, Aizu-Wakamatsu 965-80, Japan
phone: +81-242-37-2525   fax: +81-242-37-2528
noguchi@u-aizu.ac.jp

In the near future, the most important infrastructure for information communication and processing will be the Highly Intelligent Information Network (HIIN). HIIN will be supported by terabit technology for data communication and processing and a hoard of intelligent softwares. Most of the social activities will depend on this HIIN platform. We will discuss two basic issues here, (1) the technology of HIIN, and (2) the impact of this omnipresent information technology on the society.

The Model of HIIN could be represented by a three layer hierarchical structure. The first layer (**L-I**) supports the low level functions (optical communication, high speed switching etc.). The second layer (**L-II**) supports network operation and management (network resource allocation and management, distributed information processing, implementing agent oriented design paradigm et.). The highest layer (**L-III**) is the middleware to support all social activities (highly intelligent application softwares for banking & financial business, industries, education etc.). As the HIIN will pervade the whole World, standardization at every layer is of utmost importance for seamless communication and interaction.

The impact of HIIN on society has basically two aspects, (1) the impact on industries, and (2) the impact on culture.

Industrial aspect: Different industries could mainly be categorized as (1) Primary industries **IN-I** (e.g. agriculture, mining, fishing etc. i.e. collection of raw materials from nature), (2) Secondary industries **IN-II** (mainly manufacturing industries e.g. car, computer, food processing etc.), (3) Tertiary industries **IN-II** (e.g. financial market, banking etc.). Four important aspects for every industries are *production, logistic, marketing* and *finance*. Their relative importance are different as shown in Table 1, where **O**, **Y** and **X** stand for *Very Important, Moderately Important* and *Not So Important*, respectively.

	Production	Logistic	Marketing	Finance
IN-I	X	Y	O	Y
IN-II	O	O	O	Y
IN-III	O	X	O	O

Table 1: Industries and corresponding important functionalities

If the volume of business for the three are denoted by **M-I**, **M-II** and **M-III** respectively, then we can approximately say that

10 × **M-I** < **M-II** and 10 × **M-II** < **M-III**.

Excepting the production aspects, the other three important functions will depend heavily on **HIIN**. From table.1 it is thus evident that **IN-II** and **IN-III** will be entirely dependant on **HIIN** in near future. CALS (Commerce At Light Speed) based on **HIIN** will facilitate improved production, logistic and marketing. There will be EC (Electronic Commerce) and internet shopping based on **HIIN**. The ensuing deregulation (e.g. Big Bang as named in Japan), a consensus of G-7 participants, will create a borderless business World. It will at the same time bring the toughest possible competition never experienced before by the industries. It may create a chaos in the World economy, but there is no way to stop the course.

Cultural Aspects: The origin of networking and global information network is in US, and is still by far the most advanced country with its information superhighway project and all. Naturally in the global network its influence is supreme and will continue for years to come. One important result is that, a unique culture is forming around this global information network, which is so to say pro-American and is grasping local cultures by its shear omnipresence. This is not healthy for a global harmonious World, and it is very important that the local cultures also should be nurtured on this global information structure. If this aspect is neglected in the very beginning, it may be too late and we will lose many of the colorful cultures and heritage of this World.

Formal Description Techniques and Protocol Specification, Testing and Verification
T. Mizuno, N. Shiratori, T. Higashino & A. Togashi (Eds.)   © 1997 IFIP. Published by Chapman & Hall

## PART SIX

Real Time Systems

20
Dynamic priorities for modeling real-time

Girish Bhat, Rance Cleaveland
Department of Computer Science, North Carolina State University
Raleigh, NC 27695-8206, USA, e-mail: {gsbhat1,rance}@eos.ncsu.edu

Gerald Lüttgen
Fakultät für Mathematik und Informatik, Universität Passau
D-94030 Passau, Germany, e-mail: luettgen@fmi.uni-passau.de

Abstract
This paper describes an approach for modeling real-time systems using *dynamic priorities*. The advantage of the technique is that it drastically reduces the state space sizes of the systems in question while preserving properties of their functional behavior. We demonstrate the utility of our approach by formally modeling and verifying aspects of the widely-used *SCSI-2 bus-protocol*. It turns out that the state space of this model is about an order of magnitude smaller than the one resulting from traditional real-time semantics.

Keywords
model checking, modeling, process algebra with priority and real-time, SCSI-2 bus-protocol, verification

1 INTRODUCTION

A variety of formal approaches have been introduced for modeling and verifying distributed systems including *process-algebraic* frameworks (Milner 1989) and *model checking* (Clarke et al. 1986, Kozen 1983). However, only with the advent of verification tools (Bengtsson et al. 1995, Cleaveland & Sims 1996, Holzmann 1991) in the last decade they have emerged as practical aids for system designers (Baeten 1990, Cleaveland et al. 1996, Elseaidy et al. 1996). This paper addresses the problem of modeling and verifying concurrent systems where *real-time* plays an important role for functional behavior. On the one hand, real-time is used to implement abstract *synchronization constraints* in distributed environments. As an example of a synchronization constraint, consider a communication protocol where the next protocol phase may be entered only if some or all components agree. On the other hand, electric phenomena like *wire glitches*, that may lead to malfunction, can be avoided using *deskew delays*. Thus, for accurately modeling such systems it is necessary to

Formal Description Techniques and Protocol Specification, Testing and Verification
T. Mizuno, N. Shiratori, T. Higashino & A. Togashi (Eds.) © 1997 IFIP. Published by Chapman & Hall

capture their real-time aspects, thereby motivating the need for implementing real-time process algebras (Moller & Tofts 1990, Yi 1991) efficiently.

Traditional implementations of real-time process algebras typically cause state spaces to explode, the reason for this being that time is considered as part of the state, i.e. a new state is generated for every clock tick. We tackle this problem by using *dynamic priorities* to model real-time. We introduce a new process algebra, called CCS^{dp} (CCS with dynamic priorities), which essentially extends CCS (Milner 1989) by assigning priorities to actions. Unlike traditional process algebras with priorities (e.g. Cleaveland & Hennessy 1990), actions in our algebra do not have fixed or *static* priorities; priorities may change as systems evolve. It is in this sense that we refer to CCS^{dp} as a process algebra with *dynamic* priorities. In contrast to traditional real-time algebras, e.g. a version of Temporal CCS (Moller & Tofts 1990), which we refer to as CCS^{rt} (CCS with real-time), the CCS^{dp} semantics interprets delays preceding actions as priority values attached to these actions. In other words, the longer the delay preceding an action, the lower is its priority. The semantics of CCS^{dp} avoids the unfolding of delay values into sequences of elementary steps, each consuming one time unit, thereby providing a formal foundation for *efficiently* modeling real-time. The soundness and completeness of this approach is proven by establishing a one-to-one correspondence between the CCS^{rt} and the CCS^{dp} semantics of arbitrary systems. It is important to note that our approach does not abstract away aspects of real-time. Thus, all quantitative timing constraints explicit in CCS^{rt} semantics can still be analyzed within CCS^{dp} semantics.

The utility of our technique is shown by means of a practical example, namely modeling and verifying several aspects of the SCSI-2 bus-protocol, a protocol used in many of today's computers. The protocol's model is derived from the official standard (ANSI 1994) where real-time delays are recommended for implementing synchronization constraints as well as for ensuring correct behavior in the presence of signal glitches. Accurate modeling of the SCSI-2 bus-protocol thus requires considering discrete quantitative real-time. To this end, we model our protocol in the syntax common to both CCS^{rt} and CCS^{dp}. We then generate the state space according to both semantics. We show that the size of our model is an order of magnitude smaller in the CCS^{dp} semantics than in the CCS^{rt} semantics. The modeling of the protocol was carried out in the *Concurrency Workbench of North Carolina* (Cleaveland & Sims 1996), CWB-NC, a tool for analyzing and verifying concurrent systems. In order to verify and to prove the accuracy of our model, we extract several mandatory properties from the ANSI document and validate them for our model. We use the well-known *modal μ-calculus* as our specification language, and automatically check the formalized properties by using the *local model checker* (Bhat & Cleaveland 1996) integrated in the CWB-NC. Due to space constraints all proofs and part of the formalization of the SCSI-2 case study are left out and can be found in a technical report (Bhat et al. 1997).

2 PROCESS-ALGEBRAIC FRAMEWORK

In this section we introduce the process algebra CCS^{rt} and develop the new process algebra CCS^{dp}, which has the same syntax but different semantics. Whereas CCS^{rt} is an extension of CCS (Milner 1989) in order to capture *discrete quantitative timing aspects* with respect to a single, global clock, CCS^{dp} extends CCS by a concept of *dynamic priorities*. Our syntax differs from CCS by associating delay and priority values with actions, respectively, and by including the *disabling* operator known from LOTOS (Bolognesi & Brinksma 1987).

Formally, let Λ be a countable set of *action labels* or *ports*, not including the so-called *silent* or *internal* action τ. With every $a \in \Lambda$ we associate a *complementary action* $\bar{a}$. Intuitively, an action $a \in \Lambda$ may be thought of as representing the receipt of an input on port a, while $\bar{a}$ constitutes the deposit of an output on a. We define $\overline{\Lambda} =_{df} \{\bar{a} \mid a \in \Lambda\}$ and take $\mathcal{A}$ to denote the set of all actions $\Lambda \cup \overline{\Lambda} \cup \{\tau\}$. In what follows, we let $a, b, \ldots$ range over $\Lambda \cup \overline{\Lambda}$ and $\alpha, \beta, \ldots$ over $\mathcal{A}$. Complementation is lifted to actions in $\Lambda \cup \overline{\Lambda}$, also called *visible actions*, by defining $\bar{\bar{a}} =_{df} a$. As in CCS an action a communicates with its complement $\bar{a}$ to produce the internal action τ.

In our syntax actions are associated with *delay values*, or *priority values*, taken from the natural numbers, respectively. More precisely, the notation $\alpha : k$, where $\alpha \in \mathcal{A}$ and $k \in \mathbb{N}$, specifies that action α is ready for execution after a minimum delay of k time units or, respectively, that action α possesses priority k. In the priority interpretation, smaller numbers encode higher priority values; so 0 represents the highest priority. The syntax of our language is defined by the following BNF:

$$P \quad ::= \quad nil \quad \mid \quad \alpha : k.P \quad \mid \quad P + P \quad \mid \quad P \wr P \quad \mid$$
$$P \mid P \quad \mid \quad P[f] \quad \mid \quad P \backslash L \quad \mid \quad C$$

where $k \in \mathbb{N}$, the mapping $f : \mathcal{A} \to \mathcal{A}$ is a *relabeling*, $L \subseteq \mathcal{A} \backslash \{\tau\}$ is a *restriction set*, and C is a *process constant* whose meaning is given by a defining equation $C \overset{\text{def}}{=} P$. A relabeling f satisfies the properties $f(\tau) = \tau$ and $f(\bar{a}) = \overline{f(a)}$. We adopt the usual definitions for *closed* terms and *guarded* recursion, and refer to the closed guarded *terms* as *processes*. Let $\mathcal{P}$ represent the set of all processes, ranged over by $P, Q, R, \ldots$.

Regarding the semantics of processes we first introduce a real-time semantics, referred to as CCS^{rt} semantics, which explicitly represents timing behavior. We concentrate here on the operational semantics for our notion of prefixing since the semantics of the other operators is standard (Moller & Tofts 1990). Formally, the semantics of a process is defined by a labeled transition system which contains explicit time transitions – each representing a delay of one time unit – as well as action transitions. With respect to time transitions, the operational semantics is set up such that processes willing to communicate with some process running in parallel are able to wait until the communication partner is ready. However, as soon as it is available the communication has to take place, i.e. further idling is prohibited. This assumption is usually referred to as *maximal progress assumption* (Yi 1991). Accordingly, the process $\alpha : k.P$, where $k > 0$, may delay one time unit and then behave like $\alpha : (k-1).P$. The process $\alpha : 0.P$ performs an α transition leading to P. Moreover, if $\alpha \neq \tau$, it may also idle by performing a time transition to itself. Unfortunately, CCS^{rt} semantics unfolds every delay value into a sequence of elementary time units, thereby creating many additional states. For example, the process $\alpha : k.nil$ has $k + 2$ states, namely *nil* and $\alpha : l.nil$ where $0 \leq l \leq k$. It would be much more efficient if we could represent $\alpha : k.nil$ by a single transition labeled by $\alpha : k$ leading to the state *nil*. This idea of compacting the state space of real-time systems can be realized by viewing k as a *priority value* assigned to action α. In other words, one may consider the delay value k as the time stamp of action α.

To this end, we present a new semantics for our language that uses a notion of priority taken from Cleaveland & Hennessy (1990), generalized to a *multi-level* priority scheme. We refer to our process algebra as CCS^{dp} when interpreted with respect to the new semantics which, in contrast to the priority approach mentioned above, dynamically adjusts priorities along transitions. Intuitively, visible actions represent potential synchronizations that a process may be willing to engage in with its environment. Given a choice between a synchronization on a high priority and one on a low priority, a process should choose the former. Thus, high-priority τ-actions have pre-emptive power over low-priority actions. The reason that high-priority visible actions do *not* have pre-emptive power over low-priority actions is that visible actions only indicate the potential of a synchronization, i.e. the potential of progress, whereas τ-actions describe complete synchronizations, i.e. *real* progress, in our model. Note that this notion of pre-emption naturally mimics the maximal progress assumption employed in CCS^{rt} semantics.

Formally, the CCS^{dp} semantics of a process $P \in \mathcal{P}$ is given by a labeled transition system $\langle \mathcal{P}, \mathcal{A} \times \mathrm{N}, \longrightarrow, P \rangle$ where $\mathcal{P}$ is the set of states, $\mathcal{A} \times \mathrm{N}$ the set of labels, $\longrightarrow$ the transition relation which is defined in Table 1 via structural operational rules, and P is the initial state. For the sake of simplicity we write $P \xrightarrow{\alpha k} P'$ for $\langle P, \alpha : k, P' \rangle \in \longrightarrow$ and say that P engages in action α with priority k and thereafter behaves like process P'. The presentation of the operational rules requires two auxiliary definitions which are formally given in Appendix 1. First, we introduce *initial action sets* which are inductively defined on the syntax of processes as usual. More precisely, $\mathcal{I}^k(P)$ denotes the set of all potential initial actions of P with priority greater than k, where $\mathcal{I}^0(P)$ is defined to be the empty set. Second, we define a *priority adjustment function*. Intuitively, our semantics is set up in a way such that if one parallel component of a process engages in an action with priority k, then the priority values of all initial actions at every other parallel component have

Table 1 Operational semantics for CCS^{dp}

$$\text{Act1} \quad \frac{-}{a:k.P \xrightarrow{a:l} P} \ l \geq k \qquad\qquad \text{Act2} \quad \frac{-}{\tau:k.P \xrightarrow{\tau:k} P}$$

$$\text{Sum1} \quad \frac{P \xrightarrow{\alpha:k} P'}{P+Q \xrightarrow{\alpha:k} P'} \ \tau \notin \mathcal{I}^k(Q) \qquad \text{Sum2} \quad \frac{Q \xrightarrow{\alpha:k} Q'}{P+Q \xrightarrow{\alpha:k} Q'} \ \tau \notin \mathcal{I}^k(P)$$

$$\text{Dis1} \quad \frac{P \xrightarrow{\alpha:k} P'}{P \wr Q \xrightarrow{\alpha:k} P' \wr [Q]^k} \ \tau \notin \mathcal{I}^k(Q) \qquad \text{Dis2} \quad \frac{Q \xrightarrow{\alpha:k} Q'}{P \wr Q \xrightarrow{\alpha:k} Q'} \ \tau \notin \mathcal{I}^k(P)$$

$$\text{Com1} \quad \frac{P \xrightarrow{\alpha:k} P'}{P|Q \xrightarrow{\alpha:k} P'|[Q]^k} \ \tau \notin \mathcal{I}^k(P|Q) \qquad \text{Rel} \quad \frac{P \xrightarrow{\alpha:k} P'}{P[f] \xrightarrow{f(\alpha):k} P'[f]}$$

$$\text{Com2} \quad \frac{Q \xrightarrow{\alpha:k} Q'}{P|Q \xrightarrow{\alpha:k} [P]^k|Q'} \ \tau \notin \mathcal{I}^k(P|Q) \qquad \text{Res} \quad \frac{P \xrightarrow{\alpha:k} P'}{P\backslash L \xrightarrow{\alpha:k} P'\backslash L} \ \alpha \notin L \cup \overline{L}$$

$$\text{Com3} \quad \frac{P \xrightarrow{a:k} P' \quad Q \xrightarrow{\overline{a}:k} Q'}{P|Q \xrightarrow{\tau:k} P'|Q'} \ \tau \notin \mathcal{I}^k(P|Q) \qquad \text{Con} \quad \frac{P \xrightarrow{\alpha:k} P'}{C \xrightarrow{\alpha:k} P'} \ C \stackrel{\text{def}}{=} P$$

to be decreased by k, i.e. those actions become 'more important.' Thus, the semantics of parallel composition deploys a kind of *fairness assumption*, and priorities have a *dynamic* character. More precisely, the priority adjustment function applied to a process $P \in \mathcal{P}$ and a natural number $k \in \mathbb{N}$, denoted as $[P]^k$, returns a process term which is 'identical' to P except that the priorities of the initial, top-level actions are decreased by k. Note that a priority value cannot become smaller than 0.

Intuitively, $a:k.P$ may engage in action a with priority $l \geq k$ yielding process P. The side condition $l \geq k$ reflects that k does not specify an exact priority but the *maximal* priority of the initial transition of $a:k.P$. It may also be interpreted as *lower-bound* timing constraint. Due to the notion of pre-emption incorporated in CCS^{dp}, $\tau:k.P$ may not perform the τ-transition with a lower priority than k. The summation operator $+$ denotes *non-deterministic* choice, i.e. the process $P+Q$ may behave like P (Q) if Q (P) does not pre-empt it by being able to engage in a higher prioritized internal transition. Thus, our notion of pre-emption reflects implicit *upper-bound* timing constraints. Also the process $P \wr Q$ behaves like P and, additionally, it is capable of *disabling* P by engaging in Q. The *restriction* operator $\backslash L$ prohibits the execution of actions in $L \cup \overline{L}$ and thus permits the scoping of actions. $P[f]$ behaves exactly as P where actions are renamed by the *relabeling* f. The process $P|Q$

stands for the *parallel composition* of P and Q according to an interleaving semantics with synchronized communication on complementary actions of P and Q both having some priority k which results in the internal action $\tau : k$. The side conditions of the interleaving rules implement pre-emption. Finally, $C \stackrel{\text{def}}{=} P$ denotes a *constant definition*, i.e. C is a recursively defined process that is a distinguished solution of the equation $C = P$.

For our framework we obtain the following important results, which are formally stated and proved in a technical report (Bhat et al. 1997). Given an arbitrary process in our language there exists a one-to-one semantic correspondence between the associated transition systems according to CCS^{rt} and to CCS^{dp} semantics. Moreover, the standard strong bisimulations (Milner 1989), which can be defined straightforwardly for CCS^{rt} and CCS^{dp}, coincide.

We conclude this section by discussing a related approach by Jeffrey (1992) who has established a formal relationship between a quantitative real-time process algebra and a process algebra with (*static*) priorities. He also translates real-time into priorities based on the idea of time stamping. In contrast to Temporal CCS semantics, a process modeled in Jeffrey's framework may either *immediately* engage in an action or idle forever. However, this semantics does not reflect our intuition about the semantic behavior of *reactive* systems, i.e. a process should wait until a desired communication partner becomes available instead of engaging in a 'livelock.' Only because of these counter-intuitive assumptions, Jeffrey does not need to choose a *dynamic* priority framework.

3   SCSI-2 – AN OVERVIEW

We demonstrate the utility of the process algebra CCS^{dp} by a case study dealing with the bus-protocol of the widely-used *Small Computer System Interface* (ANSI 1994), or *SCSI* for short. The *SCSI bus* is designed to provide an efficient peer-to-peer I/O connection for peripheral devices such as disks, tapes, printers, processors, etc. It usually connects several of these devices with one *host adapter* which often resides on a computer's motherboard. In contrast to the host adapter, peripherals are not attached directly to the bus but via *SCSI controllers*, also called *logical units* (LUNs). Thus, LUNs provide the physical and logical interface between the bus and the peripherals. Conceptually, up to seven LUNs can be connected to one bus, and one LUN can support up to seven peripherals. However, in practice most peripherals contain their own SCSI controller (cf. Figure 1).

The *SCSI-2 bus-protocol* implements the logical mechanism regulating how peripherals and the host adapter communicate with each other on the bus. Communication on the SCSI bus is point-to-point, i.e. at any time either none or exactly two LUNs may communicate with each other. In order to allow easy addressing each LUN is assigned a fixed SCSI id in form of a number ranging from one to seven. Id 0 is reserved for the host adapter which is also,

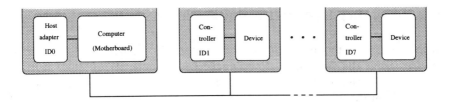

Figure 1 Typical SCSI configuration

conceptually, a LUN. Communication on the bus is organized by the use of eight signal lines whereas the actual information, like *messages*, *commands*, *data*, and *status information*, are transferred over a data bus.

Figure 2 Usual progression of the SCSI-2 bus-phases

The SCSI-2 bus-protocol is organized in eight distinct phases, called `Bus Free`, `Arbitration`, `Selection`, `Reselection`, `Command`, `Data`, `Status`, and `Message` Phase. At any given time, the SCSI bus is exactly in one phase. The usual progression of phases is shown in Figure 2. During the `Bus Free` Phase no device is in possession of the bus, i.e. LUNs may request access. If more than one device competes for the bus in order to initiate a communication, the one with the highest SCSI id is granted access. In the `Arbitration` Phase, every LUN that has posed a request determines if it has been granted access. All LUNs which lose may compete for the bus again later, whereas the winner, also referred to as *initiator*, proceeds to the `Selection` Phase. In this phase the initiator tries to connect to the desired destination, called *target*. When the link between initiator and target has been established, the so-called *information transfer phases*, including the `Command`, the `Data`, the `Status`, and the `Message` Phase are entered. In the `Command` Phase the target may request a command from the initiator. During a `Message` Phase information is exchanged between the initiator and the target concerning the bus-protocol itself. Finally, the `Status` Phase is used to transfer status information to the initiator upon completion of a command executed by the target. The key idea for accelerating communication on the bus, which has significantly contributed to the success of SCSI, is that the target can free the bus whenever it receives a time-intensive command from the initiator. As soon as the execution of such a command is finished, the target competes for the bus in order to transmit the result to the former initiator.

4   MODELING THE SCSI-2 BUS-PROTOCOL

In this section we model the SCSI-2 bus-protocol in our language. The syntax we use here is the one implemented in the *Concurrency Workbench of North Carolina* in which CCS^{rt} and CCS^{dp} are integrated as front ends. It slightly departs from the syntax introduced in Section 2 in that output actions $\overline{a} \in \overline{\Lambda}$ are notated as 'a, the internal action τ as t, and process definitions $C \overset{\text{def}}{=} P$ as proc C = P. Moreover, we use the notation $\alpha(\text{obs}):k$ which, for the purposes of this section, may be interpreted as $\alpha:k$. Actions obs come into play in the next section where they serve as 'probes' for verification purposes.

Before we present the actual modeling of the SCSI-2 bus-protocol, we comment on some assumptions we imposed. First, we restrict ourselves to modeling two LUNs, called LUN0 and LUN1, having id 0 and id 1, respectively. This is sufficient for dealing with the aspects of the SCSI-2 bus-protocol we are interested in. Note that even in the situation of two LUNs there exists competition for the bus. Moreover, we abstract away from time-out procedures and from the contents of most messages, commands, and data. These abstractions are justified since they do not affect the conceptual parts of the bus-protocol's behavior. For example, the sole purpose of a timeout is to determine if a target is alive or not. The contents of information sent over the bus, except from messages presenting the completion of some transmission, are only relevant for the device-specific part of LUNs but not for the bus-protocol itself. Additionally, the bus signals BSY (*busy*) and SEL (*select*) are *wired-or* signals in reality. However, we need not model this 'or'-behavior, since our model only deals with two LUNs, and just one LUN at a time can assert the BSY or SEL signal. Finally, all quantitative timing information occurring in the model is measured relative to a time unit of 5 ns, including *arbitration delays* (480 time units), *bus clear delays* (160 time units), *bus settle delays* (80 time units), *deskew delays* (9 time units), and *cable skew delays* (9 time units).

The underlying structure of the bus-protocol is explicitly reflected in our model. Each LUN connected to the bus is modeled as a separate parallel component containing models of the different bus phases as discussed in the previous section. The logical behavior of the bus control is implemented by bus signals. Each signal physically consists of a wire which we model as a separate process similar to a global Boolean variable. Note that signal delays are not modeled in the wires but in the operations used for transmitting information over the SCSI bus. Since we abstract away from the content of most information, we do not need to model each bit of the data bus. Hence, arbitration is modeled via a global variable which stores the highest id of all LUNs requesting access to the bus. Accordingly, the structure of our model, called SCSIBus, consists of the parallel composition of both LUNs, and the BusSignals, including the regular signals and the data path. Formally,

```
proc SCSIBus = (LUN0 | LUN1 | BusSignals) \ Restriction
```

Table 2 Modeling the bus signals and the data bus

```
proc BusSignals =   DataBus
                  | Arbitrator
                  | Off[setBSY/sset,relBSY/rel,isBSY/on,noBSY/off]
                  | Off[setSEL/sset,relSEL/rel,isSEL/on,noSEL/off]
                  | ...

proc Off  =    'off:0.Off + sset:0.On + rel:0.Off
proc On   =    'on:0.On   + sset:0.On + rel:0.Off

proc DataBus  = DataBus' [> release(obsrelease):0.DataBus
proc DataBus' =    placemsgIn(obsplace):0.'readmsgIn(obsread):0.DataBus'
                 + placemsgOut(obsplace):0.'readmsgOut(obsread):0.DataBus'
                 + placefinished(obsplace):0.'readfinished(obsread):0.DataBus'
                 + placedata(obsplace):0.'readdata(obsread):0.DataBus'
                 + placecmd(obsplace):0.'readcmd(obsread):0.DataBus'
                 + placestatus(obsplace):0.'readstatus(obsread):0.DataBus'
                 + sentdisconnect(obssentdiscon):0.
                   'readdisconnect(obsreaddiscon):0.DataBus'
                 + sentcomplete(obssentcompl):0.
                   'readcomplete(obsreadcompl):0.DataBus'
                 + writetarget0(obswritet0):0.'readtarget0(obsreadt0):0.DataBus'
                 + writetarget1(obswritet1):0.'readtarget1(obsreadt1):0.DataBus'
```

where Restriction contains all actions that are internal to the protocol, i.e.
those concerned with setting/releasing signals, requesting signal status, and
placing/reading messages, commands, and data on/from the data bus.

4.1   Modeling the bus signals and the data bus

Conceptually, each bus signal is modeled as a Boolean variable which is either
true (signal on) or false (signal off). Thus, the processes representing the
signals BSY (*busy*), SEL (*select*), C/D (*command/data*), I/O (*input/output*),
MSG (*message*), ATN (*attention*), REQ (*request*), and ACK (*acknowledgment*) are
generically created by relabeling the actions of the process Off (see Table 2).
Using the ports sset and rel one can set or release the signal and, hereby,
switching the state to On or Off, respectively. Actions 'off ('on) indicate that
the signal is currently in state Off (On). Note that the atomicity of actions
in process algebras guarantees that conflicts, arising by setting several signals
simultaneously, are avoided.

   In the following, we abstract away the contents of most messages. Only the
distinguished messages disconnect and complete are explicitly considered
since they require to exit the information transfer phases and to switch to the
initial state of the LUN. Accordingly, we may model the data bus, as seen
in Table 2, as a variable which can store and read out information (actions
placeXXX and readXXX, respectively). The labels obsXXX are used to record
the events of placing and reading messages on the bus.

Table 3 Bus Free and Command Phase

```
proc LUN0   =   t(start0):9.'relI0:0.(BusFree0 + GetSelected0)
              + t(start0):9.'setI0(obs_setI0):0.(BusFree0 + GetSelected0)
              + t:9.LUN0
              + GetSelected0

proc BusFree0 =   t(busfree):80.'setBSY(obs_setBSY):80.'setid0:0.Arbitrate0
                + isSEL(obs_isSEL):0.LUN0
                + isBSY(obs_isBSY):0.LUN0

proc CommandI0  = isREQ:0.( 'placecmd:0.'setACK:9.noREQ:0.'release:0.
                              'relACK:0.CommandI0
                          + 'placefinished:0.'setACK:9.noREQ:0.'release:0.
                              'relACK:0.Initiator0'
                          )
proc CommandT0  = 'relMSG:0.'setCD:0.'relI0(begin_Command):0.t(begin_Phase):0.
                  CommandT0'
proc CommandT0' = 'setREQ:0.isACK(obs_isACK):0.
                  ( readcmd:0.'relREQ(obs_relREQ):0.noACK:0.CommandT0'
                  + readfinished:0.'relREQ(obs_relREQ):0.noACK:0.
                    t(end_Phase):0.
                    (MsgOutT0 + MsgInT0 + DataOutT0 + DataInT0 + StatusT0)
                  )
```

4.2   Modeling the bus-phases

Now we focus on modeling the logical characteristics of the SCSI-2 bus-protocol (see Section 6 of ANSI 1994). Due to space constraints we only provide models of the Bus Free Phase and the Command Phase for LUN0 here. For the complete model we refer the reader to a technical report (Bhat et al. 1997).

In the Bus Free Phase, no device is in possession of the bus, hence it is available for arbitration. The SCSI bus is defined to be in the Bus Free Phase as soon as the signals SEL and BSY have been false for at least a bus settle delay. Accordingly, the process BusFree0 detects the Bus Free Phase when the actions isBSY and isSEL are absent for 80 time units (cf. Table 3). If one of the actions isBSY or isSEL is observed, the bus is occupied and LUN0 returns to the start state. If the bus is free, the logical unit asserts the BSY signal (action 'setBSY) and sets the arbitration variable accordingly (action 'setid0) before it performs an arbitration delay and switches to the Arbitration Phase.

The processes Target0 and Initiator0 initiate the Information Transfer Phases (ITP) which include the Command, Data, Status, and Message Phases. In those phases, information is exchanged between the initiator and the target. The Data and the Message Phases are further divided in DataIn, DataOut, MessageIn, and MessageOut Phases according to the direction of information flow. The 'In' phases are concerned with transferring information from the target to the initiator whereas the 'Out' phases are concerned with transferring information in the other direction. The information transfer takes place by byte-wise *handshakes*. The phase of the SCSI-2 bus-protocol, in which it is currently in, is encoded via the MSG, C/D, and I/O signals. In the following, we explain the Command Phase and its modeling in detail, especially the underlying handshake mechanism (cf. Table 3).

The Command Phase is entered if the target, the master of the bus-protocol, intends to request a command from the initiator. The target indicates the Command Phase by deasserting the MSG and I/O signals and asserting the C/D signal. After waiting for a deskew delay the target requests a command from the initiator by setting the REQ signal (action 'setREQ). In the meantime, the initiator detects that the target has switched to the Command Phase by observing the status of the MSG, C/D, and I/O signals. Upon detection of the asserted REQ signal (action isREQ) the initiator places the first byte of the command on the data bus (action 'placecmd), waits for a deskew delay, and asserts the ACK signal (action 'setACK). After the target detects the asserted ACK signal (action isACK) it reads the command from the data bus (action readcmd) and releases the REQ signal (action 'relREQ). At this point the handshake procedure for receiving (the first byte of) the command is completed. Now, the initiator may release the data bus (action 'release) and the ACK signal (action 'relACK). Alternatively, since a command may consist of more than one byte, the bus may remain in the Command Phase, and the handshake mechanism may be repeated, until the message *finished* (action readfinished) has been transferred. Note that in practice the length of a command can always be determined from its first byte.

4.3 State spaces of our model

We have created front-ends for both process algebras, CCS^{rt} and CCS^{dp}, for the *Concurrency Workbench of North Carolina* (Cleaveland & Sims 1996), CWB-NC, by using the *Process Algebra Compiler* (Cleaveland et al. 1995) which is a generic tool for integrating new interfaces in the CWB-NC. Whereas the integration of CCS^{rt} has been straightforward, we have needed some more effort regarding CCS^{dp}. The reason is that Rule Act1 gives rise to an infinite branching transition system. However, for practical purposes infinite branching can be eliminated by providing an upper bound upper which reflects the maximal priority value of any initial action of the considered process. The validity of this approach stems from the fact that a delay of more than upper time units does not change the system state but results in global idling.

We have run the CWB-NC on a SUN SPARC 20 workstation to construct and minimize the state spaces of our models. Whereas the CCS^{rt} version of our model has 62 400 states and 65 624 transitions, the CCS^{dp} possesses only 8 391 states and 14 356 transitions. This drastic saving in state space emphasizes the utility of using dynamic priorities in order to encode discrete quantitative real-time.

5   VERIFYING THE SCSI-2 BUS-PROTOCOL

In this section we specify and verify several safety and liveness properties which our CCS$^\text{dp}$ model of the SCSI-2 bus-protocol is expected to satisfy. The one-to-one correspondence between CCS$^\text{rt}$ and CCS$^\text{dp}$ semantics ensures that the properties hold with respect to CCS$^\text{rt}$ semantics, too. As specification language for the properties we use the *modal μ-calculus* (Kozen 1983), and for verification we employ the *model-checker* (Bhat & Cleaveland 1996) integrated in the CWB-NC. The following desired requirements of the SCSI-2 bus-protocol have been extracted from the official standard (ANSI 1994).

- *Property 1:* All bus phases are always reachable. This implies that the model is free of deadlocks.
- *Property 2:* Whenever a bus phase is entered, it is eventually exited.
- *Property 3:* The signals REQ and ACK do not change between two information transfer phases.
- *Property 4:* The signal BSY is on and the signal SEL off during information transfer phases.
- *Property 5:* Whenever a device sends a message on the bus, the message is eventually received by the intended LUN.
- *Property 6:* Whenever the initiator sets the ATN signal, eventually the bus enters the MessageOut Phase.

We formalize these properties within the modal μ-calculus which is a simple but expressive language for specifying temporal properties. Its syntax and semantics has been given e.g. by Kozen (1983). For our purposes it is sufficient to introduce the intuitive meaning of the following meta-formulas, where $\alpha, \beta \in \mathcal{A}$, $L \subseteq \mathcal{A}$, and Φ is a temporal formula.

$$\texttt{between}(\alpha, \beta, \Phi) =_{\text{df}} \nu X.[\alpha](\nu Y.(\Phi \wedge [\beta]X \wedge [-\beta]Y)) \wedge [-\alpha]X$$

$$\texttt{fair-follows}(\alpha, \beta, L, \Phi) =_{\text{df}}$$
$$\nu X.[\alpha](\nu Y.\mu Z.(\Phi \wedge [\beta]X \wedge [L]Y \wedge [-\beta, L]Z)) \wedge [-\alpha]X$$

The meta-formula $\texttt{between}(\alpha, \beta, \Phi)$ can be interpreted as follows: On every path it is always the case that after α, the formula Φ is true at every state until the action β is seen. Note that action β need not occur after α since β only *releases* the requirement that Φ be true at every state. The meta-formula $\texttt{fair-follows}(\alpha, \beta, L, \Phi)$ encodes that on every path it is always the case that after action α is seen, either Φ is always true until β is seen or Φ is always true, and an action from L occurs infinitely often on the path. Note that on fair-paths, i.e. paths on which actions from L do not occur infinitely often, action β has to occur eventually. Without this notion of fairness, which we use to encode e.g. that messages transferred over the SCSI bus have finite length, some properties cannot be validated.

Unfortunately, our process algebra CCS^{dp} turns any visible action a and $\overline{a}$ into the internal action τ when communicating on channel a. However, in order to prove any interesting property except deadlock, we have to observe certain actions of the system, e.g. asserting and deasserting bus signals. Therefore, we attach to each output action $\overline{a}$ a visible action or *probe* o, thus leading to a complex action $\overline{a}(o)$. Whenever a transition labeled by $\overline{a}(o)$ synchronizes with a transition labeled by a, the resulting τ is annotated by o, i.e. $\tau(o)$ is produced. Hence, a communication on port a is immediately observed by probe o, as intended. Our model includes the probes begin_Phase and end_Phase marking the beginning and end of each information transfer phase, respectively, and the probes obs_setSIG and obs_relSIG indicating the assertion and deassertion of some signal SIG, respectively.

Now, we can formalize the desired properties in the modal μ-calculus as shown for Properties 2 and 3. For Property 2 we have to check for every path that probe begin_Phase is eventually followed by probe end_Phase before another begin_Phase is observed.

$$\text{fair-follows}(\text{begin\_Phase}, \text{end\_Phase}, \{\text{obs\_setATN}\}, \langle - \rangle tt) \ .$$

The implicit fairness constraint ensures that the initiator does not forever ignore the target's wish to enter a new phase by continuously asserting the ATN signal. Regarding Property 3 we encode that on all paths the probes obs_setREQ, obs_relREQ, obs_setACK, and obs_relACK do not occur in between end_Phase and begin_Phase by the formula

$$\text{between}(\quad \text{end\_Phase}, \text{begin\_Phase}, \\ [\text{obs\_setREQ}, \text{obs\_relREQ}, \text{obs\_setACK}, \text{obs\_relACK}]\!f\!f \quad) \ .$$

We were able to validate each property in our model within at most two minutes when running the CWB-NC on a SUN SPARC 20 workstation. The model checker we used is a local model checker (Bhat & Cleaveland 1996). Applying a *local* model checker in contrast to a *global* one remarkably speeds-up the task of verification. In fact, the modeling of the SCSI-2 bus-protocol has been done in several stages, after each of which the above mentioned properties have been checked. At early modeling stages the model checker has invalidated most properties immediately. The encountered errors have ranged from missed fairness constraints to wrong timing information. However, the diagnostic information in form of failure traces provided by the model checker simplifies the task of finding bugs in models.

During the process of verification, we also realized that the timing constraints of the bus-protocol are not only imposed for avoiding wire glitches but also in order to implement necessary synchronization constraints during the initial bus-phases. Without these synchronization constraints, two LUNs may gain access to the bus for arbitration which leads to a deadlock. This emphasizes the necessity of dealing with real-time constraints in reactive systems.

6 CONCLUSIONS AND FUTURE WORK

We introduced the process algebra CCS^{dp} with dynamic priorities whose semantics corresponds one-to-one with the discrete quantitative real-time semantics of CCS^{rt}. However, the CCS^{dp} semantics yields significantly more compact models and, thus, provides a means for efficiently implementing traditional real-time process algebras. Moreover, our approach does not abstract away any aspects of real-time, i.e. all quantitative timing constraints can still be verified within CCS^{dp} semantics.

We implemented the process algebras CCS^{dp} and CCS^{rt} in the Concurrency Workbench of North Carolina, an automated verification tool, which we used to formally model and reason about the SCSI-2 bus-protocol. The size of our model is about an order of magnitude smaller when constructed with CCS^{dp} instead of CCS^{rt} semantics and could be handled easily within the Workbench. In addition, we specified several desired properties of the bus-protocol in the modal μ-calculus and validated them by using model checking. Regarding future work, the SCSI-2 bus-protocol should be modeled in more detail, thereby enabling the verification of additional interesting properties.

Acknowledgments. We would like to thank the anonymous referees, Michael Mendler, and Pranav K. Tiwari for their comments and suggestions. Research support for the first two authors has been provided by NSF/DARPA grant CCR-9014775, NSF grant CCR-9120995, ONR Young Investigator Award N00014-92-J-1582, NSF Young Investigator Award CCR-9257963, NSF grant CCR-9402807, and AFOSR grant F49620-95-1-0508. The research of the third author was partly supported by the German Academic Exchange Service under grant D/95/09026 (Doktorandenstipendium HSP II / AUFE).

REFERENCES

Albert, J. L., Monien, B. & Artalejo, M. R., eds (1991), *Automata, Languages and Programming (ICALP '91)*, Vol. 510 of *Lecture Notes in Computer Science*, Springer-Verlag, Madrid.

ANSI (1994), *ANSI X3.131–1994, Information Systems — Small Computer Systems Interface-2*, ANSI. See also http://abekas.com:8080/SCSI2/.

Baeten, J., ed. (1990), *Applications of Process Algebra*, Vol. 17 of *Cambridge Tracts in Theoretical Computer Science*, Cambridge University Press, Cambridge, England.

Bengtsson, J., Larsen, K., Larsson, F., Pettersson, P. & Yi, W. (1995), UPAAL — a tool suite for automatic verification of real-time systems, *in* 'Proccedings of the 4th DIMACS Workshop on Verification and Control of Hybrid Systems', Lecture Notes in Computer Science, Springer-Verlag.

Bhat, G. & Cleaveland, R. (1996), Efficient local model-checking for fragments of the modal μ-calculus, *in* T. Margaria & B. Steffen, eds, 'Second

International Workshop on Tools and Algorithms for the Construction and Analysis of Systems (TACAS '96)', Vol. 1055 of *Lecture Notes in Computer Science*, Springer-Verlag, Passau, Germany, pp. 107–126.

Bhat, G., Cleaveland, R. & Lüttgen, G. (1997), Dynamic priorities for modeling real-time, Technical report, North Carolina State University, Raleigh, NC, USA. To appear.

Bolognesi, T. & Brinksma, E. (1987), 'Introduction to the ISO specification language LOTOS', *Computer Networks and ISDN Systems* 14, 25–59.

Clarke, E., Emerson, E. & Sistla, A. (1986), 'Automatic verification of finite-state concurrent systems using temporal logic specifications', *ACM Transactions on Programming Languages and Systems* 8(2), 244–263.

Cleaveland, R. & Hennessy, M. (1990), 'Priorities in process algebra', *Information and Computation* 87(1/2), 58–77.

Cleaveland, R., Madelaine, E. & Sims, S. (1995), Generating front-ends for verification tools, *in* E. Brinksma, W. R. Cleaveland, K. G. Larsen, T. Margaria & B. Steffen, eds, 'First International Workshop on Tools and Algorithms for the Construction and Analysis of Systems (TACAS '95)', Vol. 1019 of *Lecture Notes in Computer Science*, Springer-Verlag, Aarhus, Denmark, pp. 153–173.

Cleaveland, R., Natarajan, V., Sims, S. & Lüttgen, G. (1996), 'Modeling and verifying distributed systems using priorities: A case study', *Software–Concepts and Tools* 17(2), 50–62.

Cleaveland, R. & Sims, S. (1996), The NCSU Concurrency Workbench, *in* R. Alur & T. Henzinger, eds, 'Computer Aided Verification (CAV '96)', Vol. 1102 of *Lecture Notes in Computer Science*, Springer-Verlag, New Brunswick, New Jersey, pp. 394–397.

Elseaidy, W., Baugh, J. & Cleaveland, R. (1996), 'Verification of an active control system using temporal process algebra', *Engineering with Computers* 12, 46–61.

Holzmann, G. (1991), *Design and Validation of Computer Protocols*, Prentice-Hall.

Jeffrey, A. (1992), Translating timed process algebra into prioritized process algebra, *in* J. Vytopil, ed., 'Proceedings of Symposium on Real-Time and Fault-Tolerant Systems (FTRTFT'92)', Vol. 571 of *Lecture Notes in Computer Science*, Springer-Verlag, Nijmegen, The Netherlands, pp. 493–506.

Kozen, D. (1983), 'Results on the propositional μ-calculus', *Theoretical Computer Science* 27, 333–354.

Milner, R. (1989), *Communication and Concurrency*, Prentice-Hall, London.

Moller, F. & Tofts, C. (1990), A temporal calculus of communicating systems, *in* J. Baeten & J. Klop, eds, 'CONCUR '90', Vol. 458 of *Lecture Notes in Computer Science*, Springer-Verlag, Amsterdam, pp. 401–415.

Yi, W. (1991), CCS + time = an interleaving model for real time systems, *in* Albert et al. (1991), pp. 217–228.

APPENDIX 1   AUXILIARY DEFINITIONS

Tables 4 and 5 formally present auxiliary relations used for defining the operational semantics of CCS^{dp}.

Table 4 Initial action sets

$\mathcal{I}^k(nil) =_{df} \emptyset$ $\qquad\qquad\qquad\qquad$ $\mathcal{I}^k(\alpha:l.P) =_{df} \{\alpha \mid l < k\}$

$\mathcal{I}^k(P + Q) =_{df} \mathcal{I}^k(P) \cup \mathcal{I}^k(Q)$ $\qquad$ $\mathcal{I}^k(P \rangle Q) =_{df} \mathcal{I}^k(P) \cup \mathcal{I}^k(Q)$

$\mathcal{I}^k(P|Q) =_{df} \mathcal{I}^k(P) \cup \mathcal{I}^k(Q) \cup \{\tau \mid \mathcal{I}^k(P) \cap \overline{\mathcal{I}^k(Q)} \neq \emptyset\}$

$\mathcal{I}^k(P[f]) =_{df} \{f(\alpha) \mid \alpha \in \mathcal{I}^k(P)\}$ $\quad$ $\mathcal{I}^k(P \backslash L) =_{df} \{\alpha \notin L \cup \overline{L} \mid \alpha \in \mathcal{I}^k(P)\}$

$\mathcal{I}^k(C) =_{df} \mathcal{I}^k(P)$ where $C \stackrel{def}{=} P$

Table 5 Priority adjustment function

$[nil]^k =_{df} nil$ $\qquad\qquad\qquad$ $[\alpha:l.P]^k =_{df} \begin{cases} \alpha:(l-k).P & \text{if } l > k \\ \alpha:0.P & \text{otherwise} \end{cases}$

$[P + Q]^k =_{df} [P]^k + [Q]^k$ $\qquad$ $[P \rangle Q]^k =_{df} [P]^k \rangle [Q]^k$

$[P|Q]^k =_{df} [P]^k | [Q]^k$ $\qquad$ $[C]^k =_{df} [P]^k$ where $C \stackrel{def}{=} P$

$[P[f]]^k =_{df} [P]^k[f]$ $\qquad\qquad$ $[P \backslash L]^k =_{df} [P]^k \backslash L$

21

On-line timed protocol trace analysis based on uncertain state descriptions*

Marek Musial
Technical University of Berlin
Department for Technical Computer Science, Real Time Systems Group

www: http://www.cs.tu-berlin.de/~musial
email: musial@cs.tu-berlin.de

Abstract

This paper presents a new approach to the task of passive protocol tracing. The method called *FollowSM* for the first time meets all requirements of practical in-field use, including the checking of time constraints, the independence of the current state when starting the analysis, the admittance of nondeterminism, and on-line real time analysis capability. This is achieved by a suitable modeling of the implementation under test and the generalization of the tracing algorithm to operate on state information with any degree of uncertainty. *FollowSM* has been implemented as a prototype system and proved capable of minimizing the time required for troubleshooting.

Keywords

Trace analysis, time constraints, EFSM model, real time, passive monitoring

1 INTRODUCTION

The area of the formal specification, verification, and testing of communication protocols and their implementations has been a fertile research field from the late seventies up to today. But with regard to the testing of protocol implementations against an underlying formal specification, researchers have ever been focusing their interest on *active testing* (see e.g. [vBP94] for an overview or [FvB91, ISO91, TKB91]). Active testing means that a special tester system *actively* confronts the implementation under test (IUT) with certain preselected input messages and awaits the IUT's reactions. These reactions are matched against expected message patterns to obtain some verdict about the implementation's conformance with the protocol specifica-

---

*This work has been made possible by a grant of Siemens AG, Germany, who have applied for a patent on the presented approach and are planning to implement it as a feature of their protocol analyzers.

Formal Description Techniques and Protocol Specification, Testing and Verification
T. Mizuno, N. Shiratori, T. Higashino & A. Togashi (Eds.) © 1997 IFIP. Published by Chapman & Hall

tion. All input messages and output message patterns are typically determined before test time, constituting *test cases* and *test suites*. Their derivation out of formal specifications is one of the key issues in the protocol engineering area (just a few examples: [A$^+$88, Bri88, LL91]).

On the other hand, relatively little work has been presented about *protocol trace analysis*. This concept means that the IUT is *passively* observed to obtain a *trace* of its observable interactions at some of its interaction points (IPs). The tester does not at all interfere with the communication progress. It has to answer the question whether there exists a sequence of internal actions permitted by the protocol specification that can accept the inputs in the observed trace, producing the observed outputs.

Apart from the well-known fact that testing can never prove the absence of faults but only sometimes their presence, testing by trace analysis provides a few specific advantages over active testing:

- It can be performed while the whole communication system is working under normal conditions. This means that it is highly probable that faults causing operational problems do occur during the test process as well.
- Sometimes PICS/PIXIT-parameters or user-settings might cause interworking problems between protocol implementations. These aspects cannot be covered by conformance tests of an isolated IUT.
- Passive testing within a largely functional communication system is possible without interrupting its operation.

This paper introduces a new non protocol-specific solution for passive trace analysis called *FollowSM* (for "follow a state machine"). To the author's knowledge, *FollowSM* is the first approach that addresses *all* practical requirements of the in-field trace analysis of modern protocols. These include the validation of timing constraints, the ability to start the analysis at any intermediate communication state, the admission of almost any kind of nondeterminism and efficient on-line operation. The method presented has been implemented in a prototype trace analyzer and applied to relevant B-ISDN protocols, i.e. Q.2110 (SSCOP) [IT94a] and Q.2931 [IT94b].

The remainder of this section lists related work from the literature. In section 2 the above-mentioned requirements on the presented method are explained in more detail. Section 3 gives a formal specification of the protocol model the approach is based on and defines the task of basic trace analysis. In section 4 trace analysis is extended to cover uncertain state information for the handling of initial states. Section 5 outlines how the formal model has been implemented in a prototype trace analyzer. In section 6 some experimental results are presented, and section 7 concludes the paper.

1.1   Related Work

Some of the first papers providing an overview of the trace analysis of communication protocols are [JvB83] and [UP86]. These papers primarily focus on the validation of refined specifications, but they already establish basic concepts and suggest solutions related to the field of the testing of protocol implementations.

An implementation of a passive monitor for the simultaneous trace analysis of several OSI layers is described in [CL91], with some emphasis put on a sample application to the X.25 protocol. This system is based on state machines coded in C and begins its analysis after a complete link reset.

The paper [KCV92] presents an approach which first enumerates *all* action paths possible due to the FSM projection of a restricted Estelle [ISO87a] specification, then deletes paths infeasible due to state variables and interaction parameters by symbolic evaluation. This method is very appealing from a theoretical point of view, especially since it may be used for both trace analysis and for test case generation. The fact that it operates on paths rather than on states might cause complexity problems with respect to on-line operation, particularly in the case of transitions that do not change the FSM state but are solely controlled by the data part of the protocol.

In [BvBDS91] trace analysis against LOTOS [ISO87b] specifications is investigated and the analysis tool *TETRA* is presented. Its Prolog implementation of the trace analysis algorithm utilizes uninstantiated Prolog variables to cope with non-deterministic choices and value generation. However, LOTOS (without extensions such as in [Sch94]) does not include time, and on-line evaluation tends to be inefficient.

The excellent work [EvB95] presents the *Tango* compiler, which translates a single-module Estelle specification without *delay* statements into a protocol-specific trace analyzer. This approach is elegant due to the direct use of Estelle, and all major issues of practical on-line trace analysis are discussed.

2 DESCRIPTION OF THE PROBLEM

This section describes the problem solved by the *FollowSM* approach to on-line timed protocol trace analysis in an informal way. The requirements stated below have been collected as a set of minimal demands for the in-field use of a trace analyzer as a troubleshooting tool.

Figure 1 depicts the context for the operation of the *FollowSM* trace analyzer. The latter gets all the PDUs exchanged between the IUT and its peer ("B") as its input, but does not interfere with communications in any way. In contrast to a protocol verification setting [Boc78], the analyzer shall not care about the validity of any inputs from B to the IUT, since it has to verify whether the IUT correctly reacts to possible protocol violations of its peer entity.

Since the approach presented here was designed for practical in-field use, there is no intention to observe additional interaction points (IPs), which would require special knowledge about the observed implementation and something like an upper tester (UT).

For practical application to modern protocols, *FollowSM* is required to observe and validate the PDU timing. Time checking in the context of trace analysis has to utilize tolerance intervals. For the apparent (observed) action timing will always deviate from the exact times specified because of signal propagation times, buffer delays and message transmission times.

Some complication arises from the fact that observations made on a bidirectional

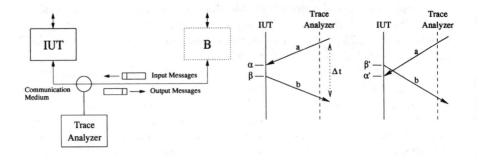

Figure 1 Operational context of *Fol-lowSM*.

Figure 2 Ambiguous ordering of internal actions.

channel cannot be considered fully sequential. Figure 2 depicts the atomic case that leads to nondeterministic action ordering: The trace analyzer first observes an input a to the IUT, and some time later an output b of the IUT. The two PDUs might, but need not, have crossed. This means both of the action sequences $\langle \alpha, \beta \rangle$ and $\langle \beta', \alpha' \rangle$ respectively could explain the observation. This is usually not an issue in active testing.

In [EvB95] this kind of problem is solved by an additional option to disable sequence checking among certain queues. In the timed approach presented here, it is necessary to be more restrictive: If and only if the time interval Δt falls below a specified upper bound, the "output first" action sequence has to be taken into consideration as well.

A trace analyzer should be able to start its analysis *at any intermediate state of the communication*, for it is practically undesirable to have to wait for some – possibly rare – synchronization conditions. The same ability is required after the detection of an error in the IUT's behavior, because in any such case the subsequent protocol state of the IUT is completely uncertain. *Ezust* and *Bochmann* [EvB95] suggest the attachment of an "undefined" attribute to all state variables in the analyzer's state machine as a possible enhancement to their approach. In this paper a more general approach is introduced which will turn out to cover the "undefined" attribute solution as a special case.

Another indispensable requirement to the *FollowSM* method is the processing of actually nondeterministic protocol specifications. Even if the target protocol is basically deterministic, uncertain initial states, ambiguous action ordering, and hidden requests at the upper service access point (SAP) will introduce nondeterminism.

In order to run a trace analyzer on-line, i.e. simultaneously with the communication, mainly the following three conditions have to be met: The analyzer's message processing speed must catch up with the message throughput of the communication, the algorithm has to be prepared for new messages to arrive during the analysis process, and its accesses of the trace data must be local. The last two conditions constitute qualitative design guidelines for the *FollowSM* algorithm, whereas the first one is just a quantitative condition.

3 PROTOCOL MODELING AND BASIC TRACE ANALYSIS

The analysis algorithm is based on a single extended finite state machine (EFSM) specification of the observed protocol. The employed EFSM model carries the necessary extensions to define timing restrictions and some specializations that guarantee the efficient computability of the trace analysis problem. This model will be referred to as *observable timed extended finite state machine*, OTEFSM, during the rest of the paper, and formally defined in this section. The restriction to a *single* state machine for protocol modeling imposes no mathematical limitation because *extended* finite state machines can simulate Turing machines by state variables with infinite range.

The approach presented here focuses on the computational aspects of practical trace analysis. How to derive the single OTEFSM specifying the permissible IUT behavior given a formal protocol specification in Estelle [ISO87a], LOTOS [ISO87b] or SDL [CCI87] is beyond the scope of this paper. The experimental results reported in section 6 are based on manual conversions of the respective SDL specifications. Since it is highly desirable to directly base the trace analysis on existing formal specifications, work is currently in progress to automate this transformation task for the language SDL.

3.1 Observable Timed Extended Finite State Machine

In the OTEFSM model, time is represented by natural numbers, but countability is not relied on in any context.

An OTEFSM M is a quadruple $M = (Q, T, \Sigma, \Delta, d)$ such that:

- Q is a set of *simple states*, denoting the part of the protocol machine's internal state information that is *not time-related*. Q may be infinite.
- T is a finite and possibly empty set of *timer labels*.
- Σ is a possibly infinite set of *input or output symbols*.
- Δ is a finite set of *transitions*. Their structure is refined below.
- $d \in \mathbb{N}$ is the maximum *output delay*, the amount of time that may pass before generated outputs become visible.

In the following, indexing by name will be used to clarify the component relation whenever necessary. For example, Q_M denotes the state space component of the OTEFSM M. Indexing by natural numbers will sometimes mean projection, e.g. $p_2 = 4$ if $p = (3, 4)$.

Additionally, let S denote the set of *complete states* of the OTEFSM, that is, both time-related and not time-related state information. Formally:

$$S \stackrel{def}{=} Q \times (T \to \mathbb{N}_\infty^2)$$

with elements $(q, \tau) \in S$, and $\mathbb{N}_\infty$ denoting the set of natural numbers plus "infinity". Practically, the two natural numbers associated with each timer label represent the start and end of the time interval within which the expiry of the timer may occur. The pair (∞, ∞) marks a disabled timer.

For two timer states $\tau, \tau' : T \to \mathbb{N}_\infty^2$ a partial order $\leq$ shall be defined expressing interval inclusion, formally $\tau \leq \tau'$ iff $\forall \psi \in T \, ([\tau(\psi)_1, \tau(\psi)_2] \subseteq [\tau'(\psi)_1, \tau'(\psi)_2])$.

Furthermore, let $\perp$ be a special value $\perp \notin T \cup \Sigma$ to indicate nonexistence. The abbreviations T_0 and Σ_0 will be used for T or Σ, respectively, plus $\perp$.

Each transition $\delta \in \Delta$ is a quintuple $\delta = (\psi, e, \alpha, \theta, \pi)$:

- $\psi \in T_0$ determines the timer that has to expire to enable the firing of δ. In case $\psi = \perp$, transition δ is not time-dependent.
- $e \in \mathbb{P}(Q \times \Sigma \times \Sigma_0) \cup \mathbb{P}(Q \times \{\perp\} \times \Sigma_0)$ is the enabling predicate of transition δ. e defines both the conditions under which δ may or must fire and its input/output behavior. The first powerset above denotes transitions triggered by inputs, while inputless transitions are covered by the second. Only inputless transitions are allowed to be time-dependent.
- $\alpha : e \to Q$ is the state transformation function describing the simple state change effected by the firing of δ.
- $\theta : T \to \mathbb{N}^2 \cup \{(\infty, \infty)\}$ is a partial function that determines the effect of δ on the timer state. $\theta(\psi) = (a, b)$ means that timer ψ is started to expire after at least a and at most b time units, whereas $a = b = \infty$ is used to indicate the cancelling of a timer.
- $\pi \in \mathbb{N}$ is the *priority* of δ. Greater π means higher priority.

3.2   Observational Semantics

To define the observational semantics of an OTEFSM $M = (Q, T, \Sigma, \Delta, d)$, it is useful to start with a function $\Psi : \Delta \times \mathbb{N} \times S \times \Sigma_0 \to \mathbb{N}^2$ expressing an *activation priority*. $\Psi(\delta, t, s, i)$ denotes the activation priority of transition δ at time t given the complete OTEFSM state s and an input i to be consumed by δ (otherwise $i = \perp$). This priority is represented by a pair of numbers because it depends both on the transition type and on π_δ, with the total order given by $(a, b) < (c, d)$ iff $a < c \vee (a = c \wedge b < d)$. The definition of Ψ distinguishes the following transition types:

spontaneous $\Psi(\delta, t, s, i) = (1, 0)$ if $\psi_\delta = \perp \wedge \pi_\delta = 0 \wedge \exists o \in \Sigma_0((s, \perp, o) \in e_\delta)$
immediate $\Psi(\delta, t, s, i) = (4, \pi_\delta)$ if $\psi_\delta = \perp \wedge \pi_\delta > 0 \wedge \exists o \in \Sigma_0((s, \perp, o) \in e_\delta)$
input $\Psi(\delta, t, s, i) = (2, \pi_\delta)$ if $\psi_\delta = \perp \wedge \exists o \in \Sigma_0((s, i, o) \in e_\delta)$
timed $\Psi(\delta, t, s, i) = (1, \pi_\delta)$ if $\psi_\delta \neq \perp \wedge \tau_s(\psi_\delta)_1 \leq t < \tau_s(\psi_\delta)_2 \wedge \exists o \in \Sigma_0((s, \perp, o) \in e_\delta)$
timeout $\Psi(\delta, t, s, i) = (3, \pi_\delta)$ if $\psi_\delta \neq \perp \wedge t \geq \tau_s(\psi_\delta)_2 \wedge \exists o \in \Sigma_0((s, \perp, o) \in e_\delta)$
not activated $\Psi(\delta, t, s, i) = (0, 0)$ otherwise.

Next, define the local state transition relation $\lambda \subseteq S \times \Delta \times \mathbb{N}^2 \times \Sigma_0^2 \times S$, which specifies the effect of firing a transition on the complete OTEFSM state. It expresses that state (q, τ) is transformed into state (q', τ') by the firing of δ at a time between t_1 and t_2, consuming symbol i as an input and producing symbol o as an output (unless i respectively o are $\perp$). Introducing an arrow notation, the definition of λ is:

$$(q, \tau) \xrightarrow[i,o]{\delta, t_1, t_2} (q', \tau') \text{ iff}$$

$$\forall t \in [t_1, t_2](\Psi(\delta, t, s, i) > (0, 0)) \wedge q' = \alpha_\delta(s, i, o)$$

$$\wedge \quad \forall x \in T : \tau'(x) = \begin{cases} (t_1 + \theta_\delta(x)_1, t_2 + \theta_\delta(x)_2) & \text{if } x \in \text{dom}(\theta_\delta) \\ (\infty, \infty) & \text{if } x = \psi_\delta \wedge x \notin \text{dom}(\theta_\delta) \\ \tau(x) & \text{otherwise} \end{cases}$$

The second case of the timer state change expresses that a timer is automatically disabled after it has triggered a transition.

Now the global state transition relation λ^* can be formulated, which extends λ to obey the prioritization imposed by Ψ. Its functionality is $\lambda^* \subseteq S \times \mathbb{N}^2 \times \Delta \times (\Sigma_0 \times \mathbb{N})^2 \times S \times \mathbb{N}^2$. It means that from state s entered at any time between t_1 and t_2 the next occurrence of a transition may be the firing of δ at any time between t'_1 and t'_2 resulting in the subsequent state s', consuming symbol i with its time stamp t_i and producing symbol o with its time stamp t_o. Again, $i = \bot$ or $o = \bot$ means no observable input respectively output, in which case the respective time stamps are irrelevant. With the abbreviation $P = \Psi(\delta, t'_2, s, i)$, the relation λ^* (in arrow notation) is defined by:

$$s, t_1, t_2 \xrightarrow[(i,t_i),(o,t_o)]{\delta} s', t'_1, t'_2 \text{ iff } s \xrightarrow[i,o]{\delta, t'_1, t'_2} s'$$

$$\wedge \quad t_1 \leq t_2 \wedge t_1 \leq t'_1 \leq t'_2 \wedge t'_2 \leq t_o \leq t'_1 + d \wedge i \neq \bot \Rightarrow t'_1 = t'_2 = t_i$$

$$\wedge \quad \forall \delta^* \in \Delta(\Psi(\delta^*, t'_2, s, i) \leq P \wedge \forall t^* \in [t_1, t'_2[$$

$$(\Psi(\delta^*, t^*, s, \bot) < (2, 0) \wedge \psi_{\delta^*} = \psi_\delta \Rightarrow \Psi(\delta^*, t^*, s, \bot) < P))$$

From the viewpoint of executing an OTEFSM specification, this means that transitions may be triggered either solely by their state-dependent enabling condition (*spontaneous* and *immediate* transitions) or by the arrival of an input PDU (*input* transitions) or by the expiry of a timer (*timed* and *timeout* transitions). *Spontaneous* transitions have "may fire" semantics, i.e., their enabling does not enforce their firing, whereas *immediate* transitions must fire as soon as they become enabled, suppressing all other transition types. Triggered *input* transitions have "must fire" semantics as well, but may be overruled by *immediate* transitions. Time-dependent transitions have to be regarded as "must fire", too, with the actual timer expiry time drawn randomly from the expiry interval $[t_1, t_2]$ of the respective timer. If the enabling predicate of such a transitions becomes true *after* timer expiry, the transition is referred to as of the *timeout* type, overruling *input*, but not *immediate* transitions. From the viewpoint of trace analysis, this behavior of time dependent transitions has to be equivalently modeled as a "may fire" condition within the timer expiry interval and a "must fire" rule after its upper bound t_2, because the actual expiry time remains unknown.

Among triggered transitions of equal type, prioritization occurs according to π_δ. If this leaves more than one transition enabled, one is selected nondeterministically. While the firing of *input* transitions is always tied to the time stamp of the respective input PDU, output observations may be delayed by at most d time units after their generation.

An observation is a permissible protocol trace iff there exist an initial state $s^0 \in S$ and a sequence of globally permissible transitions of the form

$$s^0, 0, 0 \xrightarrow[(i^1,t_i^1),(o^1,t_o^1)]{\delta^1} s^1, t_1^1, t_2^1 \xrightarrow[(i^2,t_i^2),(o^2,t_o^2)]{\delta^2} \cdots \xrightarrow[(i^l,t_i^l),(o^l,t_o^l)]{\delta^l} s^l, t_1^l, t_2^l$$

that consumes and produces the observed symbols with their proper time stamps (again, $\perp$'s ignored for inputless respectively outputless transitions). This concludes the definition of the OTEFSM model.

3.3   Computability

The OTEFSM definition contains a number of subtle features that are both necessary and sufficient to facilitate the efficient computability of trace analysis. The following lemma describes the most important of them.

Lemma 1 *Assume there are a state transition*

$$(q, \tau), t_1, t_2 \xrightarrow[(i,t_i),(o,t_o)]{\delta} (q', \tau'), t'_1, t'_2$$

permitted by λ^* and a complete state $(q, \hat{\tau}), \hat{t}_1, \hat{t}_2$ such that $\hat{\tau} \geq \tau \wedge [\hat{t}_1, \hat{t}_2] \supseteq [t_1, t_2]$. Then,

a) there exists a firing interval $[\hat{t}'_1, \hat{t}'_2] \supseteq [t'_1, t'_2]$ such that

$$(q, \hat{\tau}), \hat{t}_1, \hat{t}_2 \xrightarrow[(i,t_i),(o,t_o)]{\delta} (q', \hat{\tau}'), \hat{t}'_1, \hat{t}'_2$$

b) for any λ^* transition satisfying a, $\tau' \leq \hat{\tau}'$.

Outline of proof: By the definition of λ, the wider timer expiry intervals in $\hat{\tau}$ and the wider expansion start interval $[\hat{t}_1, \hat{t}_2]$ permit a superset of firing times compared to τ and $[t_1, t_2]$, because all transitions being λ-activated according to τ and $[t_1, t_2]$ are as well λ-activated according to $\hat{\tau}$ and $[\hat{t}_1, \hat{t}_2]$ and overruling by *timeout* transitions does not occur earlier (refer to the definition of Ψ). The firing time does not influence the simple state transformation $q \rightarrow q'$ at all. Therefore at least $[\hat{t}'_1, \hat{t}'_2] = [t'_1, t'_2]$ provides a valid transition, which proves **a**. **b** is a direct consequence of the way the resulting timer states τ' and $\hat{\tau}'$ are computed according to λ. $\square$

Theorem 1 *Trace analysis against an OTEFSM specification is computable, provided that the initial state s^0 is known and the length of possible transition sequences without input and output is bounded by the specification.*

Outline of proof: For a permissible observation (I, O) there exists an explaining λ^* action sequence A. By the definition of λ^*, the enabling times of any transition to be the next to fire in any intermediate step form a single coherent interval. Consequently there is a *largest* enabling time interval $[t^k_1, t^k_2]$ unambiguously defined for any action in A. By iteratively substituting these largest enabling intervals into A, a normalized action sequence A_n is unambiguously determined that is a valid λ^* expansion and also explains (I, O), as can be concluded by induction from lemma 1. Since the intermediate states s^k are deterministically determined* by the function α_δ, T and the number of observed symbols are finite and the number of transitions per symbol is bounded by assumption, the normalized action sequences that might explain (I, O) can be enumerated by exhaustive state space search. $\square$

---

*Note that the output message of δ need *not* be deterministically determined, as it is required in [EvB95], since e_δ is a relation.

4   UNCERTAIN TRACE ANALYSIS

In order to overcome the known-initial-state restriction, the *FollowSM* method makes the trace analysis capable of processing *uncertain state descriptions*. This means the state space search algorithm is generalized to operate on range subsets rather than on single values of state variables. This section explains the extension of basic trace analysis to cover uncertainty.

For an informal outline, consider a state space of $(a, b) \in \{0, 1, .., 255\}^2$. The initial state description is $s^0 \in (0..255, 0..255)$. Let δ be a candidate transition defined by an OTEFSM specification. Assume its enabling predicate e_δ defines an associated output message $\sigma(x)$ with $x \in \{0, .., 255\}$ and requires $a < x$, and the transformation function α_δ performs $a \leftarrow a + 10; b \leftarrow 17$. Assume the next observed output message is $\sigma(42)$. Trace analysis based on a straightforward reference implementation cannot make any use of this information. However, *FollowSM* will instantiate the enabling predicate to $a < 42$, restrict the given state description s^0 to the subset enabling δ, and apply δ speculatively:

$$s^0 \in (0..255, 0..255) \xrightarrow{a<42} s' \in (0..41, 0..255) \xrightarrow{a \leftarrow a+10;\, b \leftarrow 17} s^1 \in (10..51, 17)$$

s^1 is – after considering just a single transition – a much more certain description of a possible state than s^0 has been. Of course, all other transitions have to be considered for speculative execution as well, and quite a lot of them may be enabled by s^0. In fact, at least all transitions without associated interactions will be.

However, this kind of uncertain state expansion cannot be reasonably performed in a complete and correct way, i.e., yielding a "pass" if and only if there exists a permissible action sequence for an observation. This is due to space and time limitations, especially due to the fact that the satisfiability of general boolean expressions is not efficiently computable. Therefore, the problem is solved approximately, obeying these two requirements:

1. No permissible trace may be rejected as a "fail". That is, the algorithm shall be *correct*, but not necessarily *complete*, with respect to trace rejection.
2. As soon as state space search has reached a "certain" state description, further expansion of this state has to be both correct and complete.

Formally, first a *representation space R* has to be chosen for a state space Q such that at least the one completely uncertain and all completely certain state descriptions are representable:

$$R \subseteq \mathbb{P}Q \setminus \emptyset \quad \wedge \quad Q \in R \wedge \forall q \in Q(\{q\} \in R)$$

Let further $U = R \times (T \to \mathbb{N}^2_\infty)$ denote the set of *uncertain complete state descriptions*. Let $\mathrm{st} : U \to \mathbb{P}S$ be a function to directly refer to those certain complete states covered by the argument, i.e., $\mathrm{st}(r, \tau) \overset{def}{=} \{(q, \tau)|q \in r\}$

Uncertain trace analysis can now be specified as an extension of λ^* to the relation μ^*, substituting U for S in its signature and u for s in its arguments, in this way: With

$$v' \overset{def}{=} \left\{ s' \in S \,\middle|\, \exists s \in \mathrm{st}(u) : s, t_1, t_2 \xrightarrow[(i,t_i),(o,t_o)]{\delta} s', t_1', t_2' \right\}$$

collecting those states actually reachable from states in u,

$$u, t_1, t_2 \xrightarrow[(i,t_i),(o,t_o)]{\delta} u', t_1', t_2' \text{ iff}$$

$$t_1 \le t_2 \wedge t_1 \le t_1' \le t_2' \wedge t_2' \le t_o \le t_1' + d \wedge i \ne \perp \Rightarrow t_1' = t_2' = t_i$$
$$\wedge \quad \text{st}(u') \supseteq v' \wedge (|\text{st}(u)| = 1 \Rightarrow \text{st}(u') = v')$$

In uncertain trace analysis, an observation is considered a permissible protocol trace iff there exist a sequence of transitions permitted by μ^* of the form

$$u^0, 0, 0 \xrightarrow[(i^1,t_i^1),(o^1,t_o^1)]{\delta^1} u^1, t_1^1, t_2^1 \xrightarrow[(i^2,t_i^2),(o^2,t_o^2)]{\delta^2} \cdots \xrightarrow[(i^l,t_i^l),(o^l,t_o^l)]{\delta^l} u^l, t_1^l, t_2^l$$

that consumes and produces exactly the observed symbols with their proper time stamps, with $u^0 = (Q, T \times \{(0, \infty)\})$ being the initial state description denoting complete uncertainty.

Above requirements 1 and 2 will now be investigated by theorems 2 and 3 respectively.

Theorem 2 *Every observation permissible according to basic trace analysis is permissible according to uncertain trace analysis.*

Outline of proof: It has to be shown that for any λ^* expansion path there exists a μ^* expansion path starting from u^0 yielding the same observations. As an inductive hypothesis consider some state $s = (q, \tau) \in S$ and some corresponding state description $u = (r, \tau_u) \in U$ that fulfill $q \in r \wedge \tau \le \tau_u$. Further assume some valid λ^* state transition

$$(q, \tau), t_1, t_2 \xrightarrow[(i,t_i),(o,t_o)]{\delta} (q', \tau'), t_1', t_2'$$

It follows from lemma 1 that for some $\tau_u' \ge \tau'$

$$(q, \tau_u), t_1, t_2 \xrightarrow[(i,t_i),(o,t_o)]{\delta} (q', \tau_u'), t_1', t_2'$$

is also valid. Consequently, by the definition of μ^* there exists a valid uncertain state expansion

$$(r, \tau_u), t_1, t_2 \xrightarrow[(i,t_i),(o,t_o)]{\delta} (r', \tau_u'), t_1', t_2'$$

such that $q' \in r'$, which completes the inductive step. Since the fixed u^0 for uncertain and any s^0 for basic trace analysis fulfill the inductive hypothesis, the theorem follows by induction over the length of the expansion path. $\square$

Theorem 3 *Basic and uncertain trace analysis are equivalent provided an initial state description $s^0 = (q^0, \tau^0)$ resp. $u^0 = (\{q^0\}, \tau^0)$ without uncertainty is given.*

Outline of proof: It suffices to show that singleton states are always expanded into singleton states again by μ^*, formally that $|v'| = 1$ whenever $|\text{st}(u)| = 1$. Then, μ^* and λ^* become obviously equivalent by their definitions. $|v'| = 1$ is valid for singleton state descriptions because the resulting state s of an expansion step is deterministically defined by α_δ, ϕ_δ, and λ if the transition δ, the firing interval, and the observable interactions are given. $\square$

Theorem 4 *Uncertain trace analysis against an OTEFSM specification is computable, provided that the length of possible transition sequences without input and output is bounded by the specification.*

Outline of proof: This proof works quite similar to that of theorem 1. Here, a normalized explaining action sequence A_n is derived by setting all intermediate simple states to $Q \in R$ and, again, iteratively selecting the maximum firing-time intervals as determined by the timer states τ^k, the current time intervals $[t_1^k, t_2^k]$, and the interaction time stamps t_i^k, t_o^k according to λ. By the definition of μ^*, such A_n is permissible. It is unambiguously determined because the resulting timer states are. It also explains the observation because of part b of lemma 1. Thus, computation is possible by exhaustive search for such a normalized action sequence explaining the observation. $\square$

5 IMPLEMENTATION

This section deals with the actual implementation of OTEFSM-based uncertain trace analysis and describes some properties of the *FollowSM* prototype.

One additional issue omitted in the formalization is that the first observed output messages might have been produced before the start of the analysis, such that the analysis algorithm must independently consider to ignore any prefix of the output sequence up to the maximum output delay d. This can be done easily because the number of such alternatives is finite and mostly small.

The proof of theorem 4 shows that the specification of uncertain trace analysis is extremely weak: It only ensures the validation of the timing constraints, but not at all the checking of the intermediate simple state descriptions $r^i \in R$ (with $u^i = (r^i, \tau^i)$). This weak specification is necessary to allow an implementation to use the selected representation space R for any intermediate results as well, e.g. during the evaluation of complex enabling predicates. The convergence of the state refinement process depends on the selected representation space, implementation details and the given protocol. It cannot be deduced from the given specification.

In practice, the representation space R is constituted by the combination of representation spaces individually selected for each state variable. In the implementation of *FollowSM*, several data types exist for state variable representations, which differ in the respective tradeoff between exactness and computational cost.

5.1 Search Strategy and Complexity

The theoretical results imply that OTEFSM based trace analysis is of exponential worst-case time complexity because only the number of branches in each step is bounded. But this is not of any practical relevance provided that the search strategy for the state space search is suitably chosen, as will be explained now.

The *FollowSM* implementation uses breadth-first search, in which it is different from the approaches in [BvBDS91] and [EvB95] where depth-first search is preferred. The decisive advantage of breadth-first search in the context of *FollowSM* is that it allows the algorithm to detect and make use of the reunification of state expansion paths: Whenever a state description is derived that equals or includes or specializes a previously generated one, that expansion path with the less general state description can be abandoned without loss of correctness. Therefore, the search paths expanded in parallel are synchronized according to their current time components. Figure 3 depicts an example where half of the search tree (dotted) turns out to be redundant because of a single path unification at a and b. The practical benefit from this solution is that there is some effective bound on the number of simultaneously open nodes in the search tree. Consequently, the trace analysis can be computed in almost linear time, regardless whether the given trace is valid or not. Furthermore, the space needed to store the open search nodes does not depend on the length of the trace, as it is the case in depth-first search with backtracking. Finally, the algorithm processes the trace in a message-by-message manner without jumping back and forth in time.

In short, breadth-first search with path unification appears to be the method of choice for on-line and real time trace analysis.

5.2   Derivation of Diagnostics

The diagnostics *FollowSM* can provide for the user are based on the reconstruction of a single sequence of internal actions that explains the observed trace. This single action sequence is derived from the search graph, which is often complicated and highly branched. Whenever path unification according to section 5.1 has occurred, the selection of the explaining action sequence becomes ambiguous. If one of the unified state descriptions is more general, its path is chosen since only this path can explain *all* subsequent behaviors. If some unified state descriptions are equal, one path is arbitrarily chosen.

When *FollowSM* detects a protocol violation by failure to expand the latest search node generated so far, the error cause is automatically characterized according to the condition that made the state expansion algorithm fail. Seven types of failure causes are distinguished in this way, including *"Enabled immediate/timeout/input transition ... did not occur according to output observation"*, *"Enabled transition ... suppresses all others at expiry of timer ..."*, and *"No transition was enabled to produce/consume PDU ... at time ..."*.

A single implementation fault in the IUT can easily result in thousands of observed protocol violations that are not of any individual interest. Therefore, *FollowSM* can combine "similar" protocol violations into a single error class diagnosis and provide statistical information about the occurrence of protocol violations related to this error class. Ideally, each error class represents exactly one cause of errors, for instance a single fault in the IUT software. The decision which detected violations are similar is made depending on the failure cause as explained above and on the internal action sequence immediately preceding a protocol violation.

5.3   Prototype Architecture

The prototype has been implemented in C++. It shows the basic modular structure drawn in figure 4. The protocol specific parts, dashed in the figure, consist of the OTEFSM specification of the target protocol and the *decoder* module. All the logic

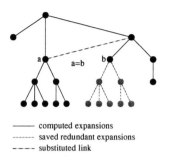

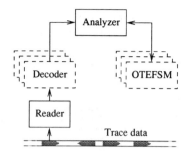

Figure 3 Reduction of complexity by path unification.

Figure 4 Basic architecture of the *FollowSM* prototype.

for the computation of the trace analysis and the handling of uncertainty is completely protocol-independent and implemented in the *analyzer* module. Only the very small *reader* module provides the interface to the trace data, so that it makes almost no difference whether the analysis is performed off-line, from a file, or on-line, connected to an actual communication channel.

Due to the breadth-first search strategy, *FollowSM* is able to run several independent state machines for different protocol layers and/or different virtual connections in parallel. A new instance of an OTEFSM is generated and initialized to an "unknown" state whenever a new connection identifier is found by the decoder.

6 EXPERIMENTAL RESULTS

The prototype implementation has been applied to two protocols of high practical relevance and considerable complexity: The protocol SSCOP according to the ITU-recommendation Q.2110 and the signaling protocol according to Q.2931, which constitute the layer 2 and 3 protocols respectively for the *Digital Subscriber Signalling System No. 2* (DSS2) on the *user-network interface* (UNI) in the broadband ISDN environment. The OTEFSM model of the SSCOP protocol consists of 138 transitions.

The most interesting experiment performed with the prototype was the on-line analysis of traffic between an active protocol tester running an executable SSCOP test suit (ETS) against a real SSCOP implementation. Overall, the results are very promising.

First, the convergence of the state descriptions to singleton states had to be confirmed by experiment (refer to section 5). It turned out that the trace analysis synchronizes with the communication after only a few messages – say between 3 and 10 depending on the currently executed protocol procedures. This does not hold for *all* state variables, because the SSCOP protocol includes state variables that are only involved in rare exceptional procedures. The values of such variables inevitably remain uncertain as long as they are not referred to. This does not prevent *FollowSM* from detecting those protocol violations that can be recognized from parts of the state description that do have converged. Figure 5 shows a small excerpt from the very beginning of an analysis report. There are three state variables included, namely the major state variable, a timer and a counter.

Time	Transition	PDU in	PDU out	State	Timer_CC	VT(CC)
0.0000				?	???	???
0.0000	SendENDAK		ENDAK	?	???	???
0.0250	U3BGAK	BGAK		3..10	???	???
0.0250..0.0252	U4TimedOut			1	off	4..-96
0.0250..0.0252	RequestConnection		BGN	2	0.9250..4.0252	1
0.0252	U2END	END		2	0.9250..4.0252	1
0.0747	U2BGREJ	BGREJ		1	off	1
0.0747..0.0749	RequestConnection		BGN	2	0.9747..4.0749	1
0.0749	ConnectionBegin	BGN		10	off	1

Figure 5 Example synchronization phase.

Second, during one experiment *FollowSM* detected an error in the ETS, which formulated an incorrect acceptance criterion for a message parameter. This resulted in a "fail" verdict from the ETS, while the *FollowSM* trace analysis correctly accepted the observation. On the other hand, *FollowSM* reported protocol violations for one of the two SSCOP implementations used in the experiments. These diagnostics could be validated manually and are consistent with the respective implementation's being known to be faulty.

Third, the experiments delivered some information about the throughput of the on-line trace analysis with *FollowSM*. On a SUN Sparc5 workstation also running the *FollowSM* user interface and a standard UNIX operating system, SSCOP trace analysis was performed at a speed ranging between 200 and 400 transitions per second. This is, in fact, an order of magnitude that makes real time operation possible in many cases.

7   CONCLUSION

In this paper a new approach to protocol trace analysis has been presented. This approach meets all requirements of practical in-field use, including the checking of action timing and the ability to synchronize with the observed communication almost instantly and in any intermediate state.

The approach is based on the new concept of uncertain trace analysis and the new OTEFSM model for the specification of the target protocol. Both the OTEFSM model and uncertain trace analysis have been defined formally. Computability, correctness, and a restricted form of completeness have been shown for the presented approach.

Experiments with a prototype trace analyzer implementing uncertain trace analysis have proven that the method can be of high practical benefit. The trace analysis can even be performed fast enough such that real time on-line operation will be feasible in many cases.

Future work on the presented approach will focus on the automatic transformation of common protocol specifications, preferably in the language SDL, into OTEFSM models for uncertain trace analysis.

7.1 Acknowledgements

This work has been funded by Siemens AG, Germany. I would like to thank Dr. Wendisch and Dr. Xu of Siemens Communications Test Equipment for their help in making this publication possible. I am grateful to Ina Schieferdecker of German National Research Center for Information Technology, Research Institute for Open Communication Systems (GMD FOKUS) for the opportunity to perform the described experiments, and to my advisor, Günter Hommel.

REFERENCES

[A+88]   A. V. Aho et al. An optimization technique for protocol conformance test generation based on UIO sequences and rural chinese postman tours. In *Protocol Specification, Testing, and Verification, VIII* [IFI88], pages 75–86.

[Boc78]   Gregor V. Bochmann. Finite state description of communication protocols. *Computer Networks*, 2:361–372, 1978.

[Bri88]   Ed Brinksma. A theory for the derivation of tests. In *Protocol Specification, Testing, and Verification, VIII* [IFI88], pages 63–74.

[BvBDS91] O. B. Bellal, G. v. Bochmann, M. Dubuc, and F. Saba. Automatic test result analysis for high-level specifications. Technical report #800, University of Montreal, Department IRO, 1991.

[CCI87]   CCITT. Recommendation Z.100: Specification and Description Language SDL. Contribution Com X-R15-E, CCITT, 1987.

[CL91]   Samuel T. Chanson and Jeffrey K. H. Lo. Open systems interconnection passive monitor OSI-PM. In *Protocol Test Systems 3*, pages 423–442. University of British Columbia, 1991.

[EvB95]   S. Alan Ezust and Gregor v. Bochmann. An automatic trace analysis tool generator for Estelle specifications. *Computer Communication Review*, 25(4):175–184, October 1995. Proceedings of the ACM SIGCOMM 95 Conference, Camebridge.

[FvB91]   S. Fujiwara and G. v. Bochmann. Testing non-deterministic state-machines with fault coverage. In *Protocol Test Systems 4* [IFI91].

[IFI88]   IFIP. *Proceedings of the IFIP WG 6.1 8th International Symposium on Protocol Specification, Testing, and Verification (1988)*, Amsterdam, 1988. North-Holland.

[IFI91]   IFIP. *Protocol Test Systems 4, Proceedings of the 4th International Workshop on Protocol Test Systems*, Amsterdam, 1991. Elsevier Science Publishers, North Holland.

[ISO87a]   ISO. ESTELLE: A formal description technique based on an extended state transition model. International Standard ISO/IS 9074, ISO, 1987.

[ISO87b]   ISO. LOTOS: Language for the temporal ordering specification of observational behaviour. International Standard ISO/IS 8807, ISO, 1987.

[ISO91]   ISO. OSI conformance testing methology and framework. International Standard ISO/IS-9646, ISO, 1991.

[IT94a]   ITU-T. B-ISDN ATM Adaption Layer – Service Specific Connection Oriented Protocol (SSCOP). Draft new Recommendation Q.2110, ITU-T, 1994.

[IT94b]   ITU-T. Digital Subscriber Signalling System No. 2 (DSS 2). User network interface (UNI) layer 3 specification for basic call/connection control. Draft new Recommendation Q.2931, ITU-T, 1994.

[JvB83]   Claude Jard and Gregor v. Bochmann. An approach to testing specifications. *The Journal of Systems and Software*, 3:315–323, 1983.

[KCV92]   M. C. Kim, Samuel T. Chanson, and Son T. Vuong. Protocol trace analysis based on formal specifications. In K. R. Parker, editor, *Formal Description Techniques (FORTE), IV*, pages 393–408. IFIP, North-Holland, 1992.

[LL91]   D. Y. Lee and J. Y. Lee. A well-defined Estelle specification for the automatic test generation. *IEEE Transactions on Computers*, 40(4):526–542, April 1991.

[Sch94]   Ina Kathrin Schieferdecker. *Performance-Oriented Specification of Communication Protocols and Verification of Deterministic Bounds of their Qos Characteristics*. PhD thesis, Technical University of Berlin, Department of Computer Science, November 1994.

[TKB91]   J. Tretmans, P. Kars, and E. Brinksma. Protocol conformance testing: A formal perspective on ISO 9646. In *Protocol Test Systems 4* [IFI91].

[UP86]   Hasan Ural and Robert L. Probert. Step-wise validation of communication protocols and services. *Computer Networks and ISDN Systems*, 11(3):183–202, March 1986.

[vBP94]   Gregor v. Bochmann and Alexandre Petrenko. Protocol testing: Review of methods and relevance for software testing. In *Proceedings of the 1994 International Symposium on Software Testing and Analysis (ISSTA)*, ACM SIGSOFT Software Engineering Notes, Special issue, pages 109–124, August 1994.

AUTHOR'S BIOGRAPHY

Marek Musial received his diploma in computer science from the Technical University of Berlin in 1995. He did some work on genetic algorithms and participated in the development of an autonomous flying robot as the TUB entry to the 1995 International Aerial Robotics Competition in Atlanta, which took second place. Now he is working on his PhD thesis on "intelligent" protocol monitoring and trace analysis in the real time systems group of Günter Hommel at the TUB Institute for Technical Computer Science.

## PART SEVEN

Languages and Applications

22
Algebraic Specification through Expression Transformation

M. J. Fernández Iglesias, M. Llamas Nistal
Área de Ingeniería Telemática
Dpt. de Tecnologías de las Comunicaciones
Campus Universitario s/n. E-36200 Vigo, Spain
Tel. +34 86 813777. Fax +34 86 812116.
E-mail: {manolo,martin}@ait.uvigo.es

Abstract

In this paper we present a framework to help the specifier to define, handle, and gather properties of ACT-ONE-based Abstract Data Types. This framework may also be used to add data support to other formal design tasks like normalization, or verification of relations like bisimulation. This framework has been implemented in the LOTOS design tool LOLA to fulfill these objectives.

Keywords

FDT-based system and protocol engineering. Tools and tool support. Algebraic Specification. LOTOS.

1 INTRODUCTION

Formal Description Techniques (FDTs) provide a model for system specification whose structure usually resembles that of programming languages. A part of the language is devoted to the specification of the system's *behaviour* or evolution, and another part is used to specify the corresponding data structures.

These two parts are complementary. Specification styles have been defined and analyzed which put more emphasis in either behaviour or data definitions(Logrippo *et al.* 1992)(Vissers 1990). It can be said that behaviour specification affects the description of data structures and vice versa.

Several of the most used FDTs support a data model based on Abstract Data Types (ADTs) (e.g. LOTOS(ISO 1988)(ISO 1997) and SDL(ITU 1993)). With respect to formal specification, ADTs have several advantages: they are fully implementation independent, have enough expressive power to support formal analysis of system properties, and have a sound underlying mathematical model.

On the other side, practice has shown that ADTs have some drawbacks, mainly for prototyping or testing (i.e. as we come close to the final implementation) (Turner 1993). Although ADT's descriptive power is very suitable for the specification of abstract (implementation independent) properties of a given system, it is very difficult to map ADT descriptions to real data structures handled by real systems. Indeed, the average system designer is not required to be a mathematician trained in Universal Algebra, and consequently lacks the needed background to fully understand the consequences of an ill-constructed Algebraic Specification. This problem is not completely solved with standard libraries, because the specifier may freely *extend* standard data types in an inconsistent way.

In this paper we present a framework to help the specifier to define, handle, and gather properties of ADTs. This framework may also be used to add data support to other formal design tasks like normalization, or verification of relations like bisimulation. This framework has been implemented in the LOTOS design tool LOLA (Quemada 1987) to fulfill the objectives described here.

2   TOOL DESCRIPTION

Usually, LOTOS ADTs are interpreted as rewrite systems(Mitchell 1996) and rewriting is used as the only tool to process data expressions. This significatively reduces the application field, and does not permit abstract analysis of general data properties. For example, we can only assess if a data expression is valid for concrete instantiations of its free variables, but we cannot perform other tasks like obtaining a solution set, or conditioning data expressions for behaviour analysis.

Sometimes (e.g. SMILE(Eertink *et al.* 1992)), a narrowing tool is used to extend the functionality offered by a rewriting tool. Typically, a narrowing tool takes an equational theorem as input and tries to produce a solution set for it. This helps to reduce the state space during symbolic execution detecting unsolvable boolean predicates, and assists the environment (i.e. the specifier) to assign values to variables.

In both cases, standard tools from equational reasoning are used basically to assess if a boolean predicate is valid in a given theory. On the other side, we propose a framework specifically designed to assist the specifier along the whole design process. We tried to design a tool which permits the specifier to study the properties of the ADTs being defined. For example, to see the (side) effects of introducing new equations, to analyze boolean predicates which affect the evolution of the system, etc.

Furthermore, this framework has been constructed to support automatic transformation of data expressions as a subtask of behaviour transformation. Behaviour transformation may be related to design tasks like normalization, test generation or verification. As a consequence, the data transformation process should be easily configured to obtain the most suitable data expressions in a given context.

The whole process is based on the transformation of an initial boolean predicate (equational theorems are particular cases) driven by a set of transformation rules. On each transformation step, a target expression is transformed into an equivalent one applying the rules defined below.

In the next paragraphs we describe this set of rules and the corresponding control strategy. Then, we provide some examples on the use of the proposed technique in contexts related to the FDT-based design process. At the end of this paper, we present some conclusions.

2.1 Basic Definitions

The boolean operators *true*, *false*, *and*, and *or* are represented respectively as $\top$, $\bot$, $(\cdot, \prod)$ and $(+, \sum)$.

A **signature** is a pair $\Sigma = (S, \Omega)$, where S is a set of *sorts* and Ω is a set of S^+-sorted *operation symbols*. X is an S-sorted set of *variables*. $Var(t) \subset X$ is the set of variables in term t. A term t with $Var(t) = \emptyset$ is called *ground*. The symbol "$\emptyset$" denotes an empty set. t is said to be *linear* if each variable in t appears only once.

For each sort $s \in S$ an (infix) boolean-sorted operation $=_s$ which represent semantic equivalence in s is assumed. b represents the sort for boolean operation symbols.

Contexts and *positions* are denoted as usual. $t|p$ represent the sub-term of t at position p. $t[p \leftarrow s]$ represents a term t whose sub-term at position p has been replaced by term s.

Substitutions are endomorphisms that extend mappings from variables to the set of terms. *(Variable) renamings* are mappings from variables to variables. If ρ is a bijective renaming and t a term, the term ρt is said to be a *variant* of t. We will also use an alternative representation for substitutions, based on the *solved forms* introduced in (Martelli *et al.* 1982). That is, the substitution $\sigma = \{t_1 \rightarrow x_1, \dots t_n \rightarrow x_n\}$ is represented as the boolean predicate $\prod_{i=1}^{n}(x_i =_{s_i} t_i)$.

Substitution ι represents the identity substitution: $\iota t = t$.

A **conditional rewrite system** is a pair $\mathcal{R} = (\Sigma, R)$, where Σ is a signature and R a set of conditional rewrite rules $R = \{\prod_{j=1}^{n} P_{ji} : l_i \rightarrow r_i\}$. The boolean terms P_{ji} are called *premises*. l_i is a *search pattern* and r_i a *substitution pattern*. A rewrite rule is said to be *left linear* if the corresponding search pattern is linear. A rewrite system $\mathcal{R}$ is said to be left linear if all rewrite rules in R are left linear. Termination, confluence and convergence are defined as usual(Mitchell 1996). We write $s \xrightarrow{r!} t$ if s rewrites to t and t is a normal form.

The set $\Omega_c \subseteq \Omega$ is the set of *constructors*. Constructors are operation symbols that do not appear in any outermost position of a search pattern.

Definition 1 *Let* $C_l = \{c_i : l \rightarrow r_i \in R \mid \rho_i l_i = l\}$ *where the* ρ_i *are bijective renamings.* C_l *is said to be* **completely defined** *if and only if for all substitutions* σ *such that* $Var(\sigma c_i) = \emptyset$ *we have that* $\sum_i^{Card(C_l)} \sigma c_i \xrightarrow{r!} \top$, *and* $\sigma c_i \cdot \sigma c_j \xrightarrow{r!} \bot$ *for all* $i \neq j$.

$R = \bigcup_i C_l^i$ *is said to be completely defined if and only if every* C_l^i *is completely defined.*
$\square$

C_l is the set of all (variants of) rewrite rules which share the same search pattern l. Note that for a rewrite rule without premises there is always a (one element) completely defined set $C_l = \{l \to r\}$.

The following relation will be used to relax the requisites for variable binding during mutation (see below). Linearity for t is required to obtain well-formed substitutions.

Definition 2 *Let t be a linear term. We define the* **conditioning substitution** *for two terms t and s ($\sigma_c(t, s)$) inductively as follows:*

- $\sigma_c(t, s) = \iota$ *if $t = op_1(t_1, \ldots, t_n)$ and $s = op_2(s_1, \ldots s_n)$, with $op_1 \neq op_2$*
- $\sigma_c(t, s) = \iota$ *if s, t are both ground.*
- $\sigma_c(t, s) = \iota$ *if $t \notin X$ and $s \in X$*
- $\sigma_c(t, s) = \{s \to x\}$ *if $t = x \in X$*

- $\sigma_c(t, s) = \bigcup_i \sigma_c(t_i, s_i)$, *if $t = op_1(t_1, \ldots, t_n)$ and $s = op_2(s_1, \ldots s_n)$, with $op_1 = op_2$.*

□

Example 1 *Let x_i, w_i be variables, and f, g, h, m, n, a, b, c operation symbols. Then $\sigma_c(t, s) = \{c \to x_1, \; m(w_3) \to x_3\}$ for*

$$
\begin{aligned}
t &= f(\quad g(h(x_1, a), b), \quad n(x_2), \quad x_3) \\
s &= f(\quad g(h(c, w_1), b), \quad w_2, \quad m(w_3))
\end{aligned}
$$

□

2.2 Expression Representation

To be transformed expressions are represented as sums of boolean products (SBP). An SBP is a boolean expression of the form $\sum_i S_i$, where each S_i is a boolean product $S_i = \prod_j P_{ij}$. The P_{ij} are boolean predicates which do not contain the operators + (boolean or) and · (boolean and). The P_{ij} are named *target* predicates.

Every boolean expression can be straightforwardly rewritten into an SBP using the usual rewrite rules for boolean operators. Equations are boolean predicates of the form $(t_1 =_s t_2)$

Our objective is to transform an SBP e into an equivalent SBP e_t. When the proposed technique is applied to equation solving, solution construction can be easily supervised. In a given stage of the process, a set of solutions can be represented as an SBP (one product representing each solution). We can also apply the properties of boolean algebra to rearrange or transform the solution set into an equivalent one satisfying some required properties for a given application context. Furthermore, the transformation process can be halted at any point with an expression equivalent to the initial one. This permits also to easily represent incomplete and parameterized solutions.

To sum up, boolean expressions will be eventually transformed into an equivalent

- ground term. If this term is $\top$, we conclude that the starting SBP is satisfied for every variable assignment. For an initial expression $(t_1 =_s t_2)$, the result is the proof of the corresponding equational theorem. If this term is $\bot$, the initial SBP is never satisfied.
- new SBP. For an initial equational theorem, each product in the transformed expression represents a *solution* to the starting equation (i.e. conditions that must be satisfied for the theorem to be valid). Unification(Hussman 1985)) can be seen as a particular case of expression transformation. If $\mathcal{R}$ is adequately constructed, a goal $(t_1 =_s t_2)$ can be transformed into an expression $\sum_{i=1}^{m} \prod_{j=1}^{n} (x_j = u_{ij})$, where the x_j are the variables of the goal, and the substitutions $\sigma_i = \prod_{j=1}^{n} (x_j = u_{ij})$ represent the corresponding unifiers.

2.3   Rule Description

The process is driven by a set of transformation rules. Associativity and commutativity of variable assignments and boolean operators $\prod$ and $\sum$ is assumed. Rules are applied to target predicates in S. The transformation process is based on a convergent, completely defined and left linear rewrite system $\mathcal{R}$.

The rules are presented in Table 1. Rule 1 (rewriting) is used to provide normalized expressions to the rest of the rules. Consequently, we will assume that the rest of the rules transform normalized expressions.

Rules 2, 3 and 4 are used to eliminate the top-level operation symbol. Let t be a non-ground, normalized term. If the top-level operation symbol is not a constructor, rule 2 is applied. If this symbol is a constructor, we apply rules 3 or 4 depending on the shape of the target.

Rule 2 (mutation) is based on the properties of narrowing and boolean algebra. An equivalent expression is obtained by mutation. Further processing is simplified due to the properties of the conditioning substitution σ_c. R_{op} is supposed to be constructed from fresh variants of rewrite rules. As a particular case, if the expression to be transformed is a variable assignment $(x =_s t)$, the target $(t =_s x)$ is transformed instead.

In a typical narrowing process, backtracking is used to recover a previous state when a sequence of rewrite rule applications leads to no solution. In our case, all available information which leads to a solution is introduced in the target expression. This allows us to halt the transformation process at any point with an equivalent expression, and consequently to support an interactive tool for expression transformation.

Rule 3 (decomposition) is based on the properties of constructors. If the outermost operation symbols in t_1 and t_2 are both constructors, the corresponding target can

Table 1 Transformation rules

Rule	Description	Comments
1	$$\frac{(t_1 =_s t_2)}{(t_1' =_s t_2')}$$	$t_1 \xrightarrow{r!} t_1'$ and $t_2 \xrightarrow{r!} t_2'$
2	$$\frac{(t_1 =_s t_2)}{\sum_{i=1}^{Card(R_{op})}(t_2 =_s \sigma_i r_i) \cdot \sigma_i c_i \cdot \prod_{j=1}^{n}(u_j =_{s_j} \sigma_i l_j^i)}$$	$R_{op} = \{c_i : op(l_1^i, \ldots, l_n^i) \to r_i\}$ $t_1 = op(u_1, \ldots, u_n)$, $op \notin \Omega_c$. t_1, t_2 normalized terms. $\sigma_i = \sigma_c(op(l_1^i, \ldots, l_n^i), t_1)$
3a	$$\frac{(t_1 =_s t_2),\ op_1 \neq op_2}{\bot}$$	$t_1 = op_1(u_1^1, \ldots u_n^1)$ $t_2 = op_2(u_1^2, \ldots u_n^2)$,
3b	$$\frac{(t_1 =_s t_2),\ op_1 = op_2}{\prod_{i=1}^{n}(u_i^1 =_{s_i} u_i^2)}$$	$op_1, op_2 \in \Omega_c$ t_1, t_2 normalized terms.
4a	$$\frac{(x =_s t),\ x \in Var(t),\ Op(t) \subseteq \Omega_c}{\bot}$$	$t = op(u_1, \ldots u_n)$, normalized, $op \in \Omega_c$
4b	$$\frac{(x =_s t),\ Op(u_i) \nsubseteq \Omega_c}{(x =_s op(x_1, \ldots x_n)) \cdot \prod_{i=1}^{n}(x_i =_{s_i} u_i)}$$	
5a	$$\frac{P_A \cdot (x =_s t) \cdot P_B,\ x \in IV}{P_A[t/x] \cdot (x =_s t) \cdot P_B[t/x]}$$	$t = op(u_1, \ldots, u_n)$, $Op(t) \subseteq \Omega_c$, $x \notin Var(t)$
5b	$$\frac{P_A \cdot (x =_s t) \cdot P_B,\ x \notin IV}{P_A[t/x] \cdot P_B[t/x]}$$	

lead to a solution only if both constructors are the same. In this case, subterms must be unified one to one.

Rule 4 (imitation) is used to transform a variable assignment with an outermost constructor symbol. Rule 4a is correct because only finite terms and well-formed constructors are considered. Using rule 4b we obtain a substitution which permits to rebuild the initial target, together with a new variable assignment which is normalized with respect to rewriting. The subterms of the obtained substitution may be further transformed. $Op(t)$ is the set of operation symbols in term t.

Rule 5 (variable simplification) is used to discard duplicate variable assignments and to detect inconsistencies (e.g $(x =_s a)(x =_s b)$ provided that $(a \neq_s b)$). *IV* is the set if variables belonging to the initial SBP.

Example 2 *The following rewrite rules provide the classical definition of the list* append *operation. List elements belong to sort e, l is the list sort.* cons *is the list constructor.* nil *represents an empty list.*

$app(nil, x) \rightarrow x$

$app(cons(a, x), y) \rightarrow cons(a, app(x, y))$

Our goal is to transform a target $(app(x_1, x_2) =_l cons(e_1, cons(e_0, nil)))$ using the proposed rules. The transformation process evolves as presented in Figure 1 (expressions being transformed in the next step are boxed, stable expressions (marked as †) are not repeated in further transformation steps, a_i, z_i are variables).

Rule	Expression
—	$(app(x_1, x_2) =_l cons(e_1, cons(e_0, nil)))$
2	$(x_2 =_l cons(e_1, cons(e_0, nil)))(x_1 =_l nil)^\dagger +$
	$\boxed{(cons(a_1, app(z_1, x_2)) =_l cons(e_1, cons(e_0, nil)))(x_1 =_l cons(a_1, z_1))}$
3	$(a_1 =_l e_1)(app(z_1, x_2) =_l cons(e_0, nil)))(x_1 =_l cons(a_1, z_1))$
5	$(app(z_1, x_2) =_l cons(e_0, nil)))(x_1 =_l cons(e_1, z_1))$
2	$(z_1 =_l nil)(x_2 =_l cons(e_0, nil)))(x_1 =_l cons(e_1, z_1)) +$
	$\boxed{(z_1 =_l cons(a_2, z_2))(cons(a_2, app(z_2, x_2)) =_l cons(e_0, nil)))(x_1 =_l cons(e_1, z_1))}$
3, 5	$(x_2 =_l cons(e_o, nil)))(x_1 =_l cons(e_1, nil))^\dagger +$
	$\boxed{(x_1 =_l cons(e_1, cons(e_0, z_2)))(app(z_2, x_2) =_l nil)}$
2, 3, 5	$(x_1 =_l cons(e_1, cons(e_0, nil)))(x_2 =_l nil)^\dagger$

Figure 1 Transformation process for $(app(x_1, x_2) =_l cons(e_1, cons(e_0, nil)))$

The final expression is

$((x_2 =_l cons(e_1, cons(e_0, nil))) \cdot (x_1 =_l nil)) +$
$((x_2 =_l cons(e_o, nil)) \cdot (x_1 =_l cons(e_1, nil))) +$
$((x_1 =_l cons(e_1, cons(e_0, nil))) \cdot (x_2 =_l nil))$

This expression, equivalent to the initial one, can be seen as the solution set of the initial expression $(app(x_1, x_2) =_l cons(e_1, cons(e_0, nil)))$

□

2.4   Control Strategy

The transformation process is performed through successive *transformation steps*. Each step starts with an SBP and terminates with an equivalent SBP.

For each step, transformation rules are selected and applied to one target predicate from the active SBP. Target predicates are marked as **stable** (no more transformation steps will be performed for the corresponding target), **unstable** (the target may be further transformed) or **loop** (the target stems from a syntactically equivalent target through previous transformation steps).

Each transformation step is performed as follows:

1. The active SBP is simplified using usual rewrite rules for booleans (Fernández 1997). Commutativity and associativity of boolean operators is applied implicitly. During this process, rule 1 (rewriting) is applied to all unstable SBPs. Rewritten ground terms are marked as stable.
2. The first unstable SBP is selected, and one of rules 2 (mutation), 3 (decomposition) and 4 (imitation) is applied depending on the shape of the target as described above. Loops in targets are detected and marked accordingly (see below).
   Rule 2 may generate new products. These products are appended to the right of the active SBP and the new targets are marked as unstable.
3. Rule 5 (variable simplification) is applied to the whole product where the target was taken.

Target predicates are selected left to right from the active SBP, and the process continues until no more targets are available, or a step limit, configured by the user, is reached. As we have an expression equivalent to the initial predicate at the beginning of every transformation step, we may adapt the target selection process to fulfill the requirements of an underlying behaviour analysis (variable separation, normalization, factorization, etc.). We may also halt the transformation process at the beginning of every transformation step to ask the specifier for the next target. This permits the implementation of an interactive transformation tool where the designer may make use of a prior knowledge of the problem to configure the best transformation strategy for a given design task.

With the transformation rules and the control above, infinite derivation sequences may be generated during mutation (application of rule 2). After the application of these rules, we check if any of the new predicates is a variant of a previously mutated ancestor. If this happens, the new expression is marked as **loop** and is not further transformed.

2.5   Correctness

The proposed set of rules in Table 1 together with the simplification process described in point 1. of section 2.4 above is sound and complete for rewrite systems which have the properties enumerated in section 2.3 (Fernández 1997). Nevertheless, theoretical correctness does not guarantee termination.

Furthermore, this framework should be applicable in environments where these properties are not fulfilled. For instance, it would be desirable that a specifier who is defining abstract data types can see the behaviour of the produced data definitions to correct or improve them. The final objective is to provide a tool for the specifier to assess if data descriptions indeed define the wanted functionality. This is the reason to introduce the above outlined control strategy and loop detection.

3 EXAMPLES

The proposed examples are based on the LOTOS(ISO 1988) ADT description appearing in Figure 2. This LOTOS ADT specifies data for the transfer of primitives between a set of client nodes and a central hub. Natural numbers and booleans are specified as usual (c.f. standard library in (ISO 1988)). Although this description is fairly simple, we think that it is adequate enough for our purpose, that is, illustrate how a transformation-oriented data tool may be useful for an FDT-based design process.

```
TYPE pdu_type IS Naturals

sorts                                   eqns
  par, prim                               forall x:nat,y:nat ofsort bool
                                            IsUserName(UserId(x))    = true;
opns                                        IsAddress(AddressId(x))  = true;
  Connect    : par, par -> prim            IsUserName(AddressId(x)) = false;
                                            IsAddress(UserId(x))     = false;
  UserId     : nat        -> par
  AddressId  : nat        -> par           IsHubAddr(AddressId(0))  = true;
                                            IsNodeAddr(AddressId(1)) = true;
  IsUserName : par        -> bool          IsNodeAddr(AddressId(2)) = true;
  IsAddress  : par        -> bool
                                        ofsort prim
  IsNodeAddr : par        -> bool          Connect(UserId(x), AddressId(y))
  IsHubAddr  : par        -> bool             = ValidPrim;
                                            ....

  ValidPrim  :            -> prim

  ....

ENDTYPE (* pdu_type *)
```

Figure 2 ADT description for the examples.

Example 3 *We have the following LOTOS behaviour expression:*

```
a?x: nat; (b!IsNodeAddr(AddressId(x)); NodeConn(x) []
          b!IsHubAddr(AddressId(x)); HubConn(x))
```

The systems takes from the environment the information needed to construct an address id. Further system evolution depends on the kind of address constructed (i.e., a node address or a hub address).

Let us suppose that along the design process a new requirement is introduced which states that the address space of the hub is included in the address space of

the rest of the nodes. That is, the equation `IsNodeAddr(AddressId(0)) = true`
is appended to the equation set.

In this case, the data transformation tool transforms the predicate

$$IsNodeAddr(AddressId(x)) =_{bool} IsHubAddr(AddressId(x))$$

into the equivalent predicate $(x =_{nat} 0)$. *The specifier sees that if* $(x =_{nat} 0)$ *the
evolution of the system is not determined because the environment may synchronize
in both branches. The specifier may use this information to analyze the introduced
requirements and test if they describe the desired functionality.*
□

In the example above we can see that it is not needed to instantiate x nor a (dynamic) testing procedure to detect the side effects of the new requirement. This analysis is performed statically. Data definitions affect system evolution and, as a consequence, abstract reasoning on data definitions may be useful even at the early stages of the design process.

Example 4 *If we transform the predicate* $(Connect(x,y) =_{prim} ValidPrim)$ *we
obtain the equivalent expression* $(x =_{par} UserId(x')) \cdot (y =_{par} AddressId(y'))$,
*This predicate specify conditions that must be fulfilled in order to satisfy the initial
expression.*

In other words, for $Connect(x,y)$ *being a valid primitive, the corresponding parameters must be respectively a well-constructed user id. and a well-constructed
address id.*
□

The specifier may use the proposed tool to see if the defined data indeed specify the desired functionality.

Example 5 *A specifier sees unexpected results during the transformation of LOTOS
specifications based on the ADT definitions in Figure 2. Specifically, observes that in
some interactions with the environment, the parameters of connection primitives are
lost.*

She/He decides to use an interactive data transformation tool to analyze the properties of Connect. To start with, she/he tries to analyze when two connection primitives are seen as equivalent. The following initial predicate is transformed interactively:

$$Connect(x,y) =_{prim} Connect(w,z)$$

After a transformation step, (mutation of $Connect(x,y)$ *applying the rewrite rule
for Connect in Figure 2 using the substitution* $\sigma = \{x \leftarrow UserId(x'), y \leftarrow
AddressId(y')\})$ *obtains:*

$$(x =_{par} UserId(x')) \cdot (y =_{par} AddressId(y')) \cdot (Connect(w,z) =_{prim} ValidPrim)$$

The specifier selects the predicate $(Connect(w, z) =_{prim} ValidPrim)$ *for transformation to obtain*

$$(x =_{par} UserId(x')) \cdot (y =_{par} AddressId(y')) \cdot$$
$$(w =_{par} UserId(w')) \cdot (z =_{par} AddressId(z')) \cdot$$
$$(ValidPrim =_{prim} ValidPrim)$$

The specifier detects that all connection primitives are equivalent if the corresponding parameters are well-constructed. Next, the specifier eliminates the rule which produced the mutation and transforms the initial expression again. In this case, the predicate $(x =_{par} w) \cdot (y =_{par} z)$ *is obtained.*

That is, two connection primitives are equivalent if and only if the corresponding parameters are also equivalent. The designer may go on with this process to configure the most suitable data model for a given problem.

☐

In this example we can see that the proposed framework is robust enough to handle incomplete, even inconsistent, data definitions. The specifier can analyze intuitively the consequences of the defined semantics.

4 CONCLUSIONS

In this paper we have proposed a framework to help the designer to define, handle and gather properties of Abstract Data Type specifications. As the system behaviour depends on the data definitions, the proposed tool may also be used to assist the specifier in other design tasks targeted to the system as a whole.

This technique is based on a set of transformation rules and may be applied to other contexts where data expression transformation is needed (e.g. theorem proving or functional programming). In our case, this framework has been implemented in LOLA for the analysis and transformation of LOTOS data expressions.

All the information related to the transformation process is present in the same expression being transformed. Consequently, this process can be configured to obtain the most suitable data expressions for a given problem. This can be done choosing an adequate strategy to select targets from the active expression.

The transformation process can be halted at any point returning an expression equivalent to the initial one. The resulting expression (a boolean predicate) may be seen as a set of requisites that must be satisfied by the initial expression to be valid in a given theory. This process makes use of all information available in an ADT specification (even an incomplete one) to offer to the specifier different views of the semantics being defined.

5 ACKNOWLEDGMENTS

This work has been partially funded by the regional government of Galicia, Spain, under contract number XUGA32204B95.

REFERENCES

Eertink, H. and Wolz, D. (1992) Symbolic Execution of LOTOS Specifications. in *Formal Description Techniques V: FORTE'92* (eds. M. Diaz and R. Groz), North-Holland.

Fernández, M. J. (1997). *Contribución al Tratamiento de Datos en LOTOS*, Tesis Doctoral (PhD thesis), Dep. de Tecnologías de las Comunicaciones, Universidade de Vigo, Spain. (In spanish)

Hussman, H. (1985) Unification in conditional-equational theories. *Lecture Notes in Computer Science*, **204**, Springer–Verlag.

ISO (1988) *LOTOS: a Formal Description Technique based on the Temporal Ordering of Observational Behaviour*, ISO International Standard 8807, TC97/SC21.

ISO (1997) *Revised working draft on enhancements to LOTOS*. Technical report, ISO/IEC JTC1/SC21/WG7.

ITU-T (1993) *SDL: Specification and description language*, CCITT Recommendation Z.100.

Logrippo, L., Faci, M. and Haj-Hussein, M. (1992) An introduction to LOTOS: Learning by examples, in *Computer Networks and ISDN Systems*, **23**, 325–342.

Martelli, A. and Montanari, U. (1982) An efficient unification algorithm. in *ACM Transactions on Programming Languages and Systems*, **4(2)**, ACM.

Quemada, J., Fernández, A. and Mañas, J.A. (1987) LOLA: Design and Verification of Protocols using LOTOS, in *Ibercom, Conference on Data Communications*, Lisbon, Portugal.

Mitchell, J.C. (1996) *Foundations for Programming Languages*, Foundations of Computing. The MIT Press.

Turner, K. J. (1993) *Using formal description techniques – An introduction to Estelle LOTOS and SDL*. Wiley, New York.

Vissers, C., Scollo, L., van Sinderen, W. and Brinksma, E. (1990) *On the use of specification styles in the design of distributed systems*, technical report, University of Twente. Faculty of Informatics, Entschede, The Netherlands.

BIOGRAPHY

Manuel J. Fernández graduated from the Universidade de Santiago de Compostela, Spain with a *Ingeniero de Telecomunicación* degree in 1990, and from Universidade de Vigo, Spain with a Doctor in Telecommunications degree in 1.997.

Martín Llamas received the *Ingeniero de Telecomunicación* (1986) and Doctor in Telecommunications (1994) degrees from the Universidad Politécnica de Madrid, Spain.

Both joined the Telecommunication Engineering faculty of the Universidade de Vigo, Spain. In addition to teaching, they are involved in research in the areas of Formal Description Techniques and Computer Based Training.

23

Modelling Digital Logic in SDL

G. Csopaki[a], and K. J. Turner[b]

[a]Department of Telecommunications and Telematics, Technical University of Budapest
H-1521 Budapest, Hungary (Email: csopaki@ttt.bme.hu)

[b]Department of Computing Science and Mathematics, University of Stirling
Stirling FK9 4LA, Scotland (Email: kjt@cs.stir.ac.uk)

Abstract

The specification of digital logic in SDL (Specification and Description Language) is investigated. A specification approach is proposed for multi-level descriptions of hardware behaviour and structure. The modelling method exploits features introduced in SDL-92. The approach also deals with the specification, analysis and simulation of timing aspects at any level in the specification of digital logic.

Keywords: Digital Logic, Hardware Design, SDL (Specification and Description Language), Timing Analysis.

1. Introduction

1.1 Application of SDL

This paper addresses the specification and validation of digital logic components and digital systems using SDL-92 (Specification and Description Language [1,2,3]). SDL was developed by CCITT/ITU-T in the context of telecommunications. ETSI (European Telecommunications Standards Institute) has defined a methodology for preparing standards and reports that describe certain kinds of system (communications protocols or services) whose behaviour is mainly discrete and consists of actions in response to signals or other inputs. The main technique in this methodology for defining behaviour (and associated structure) is SDL. Data is defined using SDL combined with ASN.1 (Abstract Syntax Notation 1 [4]). An SDL description has companion descriptions using ASN.1 and MSCs (Message Sequence Charts [5]). The result has no informal text and is consistent, unambiguous and precise. A combined description can be validated with software tools.

SDL has been widely used to specify and validate communications protocols and communications systems. Nonetheless, SDL is a general-purpose language that is appropriate for many other applications. For example, it has been used in general software design and for real-time systems. Surprisingly, the use of SDL to describe hardware seems to have been rather limited. Hardware description with SDL has evolved in studies of hardware-software co-design [6, 7]. This paper investigates how SDL can be used for the specification and

Formal Description Techniques and Protocol Specification, Testing and Verification
T. Mizuno, N. Shiratori, T. Higashino & A. Togashi (Eds.) © 1997 IFIP. Published by Chapman & Hall

analysis of digital systems. In fact it is permissible to realise system components using hardware or software, so co-design is also possible.

1.2 Digital Logic Design

Digital logic design is partly top-down and partly bottom-up. The general approach is top-down, but designers exploit available components and designs. Components are specified and manufactured to defined interfaces, or sometimes standards. A well-defined specification is the basis of verification and validation. Such a specification should completely cover the required functional behaviour, should quantify time delays inside components, and should exhibit a clear structure of components and their interconnections.

Top-down logic design proceeds in the following way. The user begins by describing the expected behaviour of a functional unit and its timing characteristics. This black-box description is called a functional specification, and refers only to inputs and outputs. The user now has to describe the proposed design structure using functional specifications of its components and their interconnections. At this point the behaviour and properties of the functional unit can be investigated, perhaps by simulation using sample input data sequences. If the proposed design fulfils its requirements, the procedure can be repeated recursively for the subsidiary components. The design process is complete when components are ready-made ones with known characteristics. Design efficiency is improved when verified or validated functional specifications are stored in a library for use by all designers.

1.3 Hardware Description Languages

Digital logic and digital system design are well developed topics. Many textbooks explain the operation of logic gates and other functional elements, and explain how to combine them into larger circuits and systems. Digital logic design uses elements that are available as hardware components from the manufacturer's catalogue [8].

Hardware Description Languages have been extensively studied. Languages such as VHDL (VLSI Hardware Description Language [9]), CIRCAL (Circuit Calculus [10]), CARS [11] and many others have been used to specify and analyse digital hardware. Like SDL, LOTOS (Language Of Temporal Ordering Specification [12]) was developed for describing communications systems. The inspiration for the work reported in this paper was the LOTOS-based approach called DILL (Digital Logic in LOTOS [13]).

DILL supports the hardware engineer when translating a circuit schematic into a LOTOS specification. The DILL library contains a variety of pre-defined components such as might be obtained off-the-shelf. Translation into DILL allows properties of a circuit to be investigated. Once the specification has been verified or validated, it can be realised as actual hardware. The designer must be familiar to some extent with LOTOS so as to combine behaviour expressions, but this is reasonably straightforward and does not require an in-depth knowledge of LOTOS.

The approach of this paper is named ANISEED (Analysis In SDL Enhancing Electronic Design) – somewhat similar to DILL except that the specification language is SDL. Since SDL is widely used in industry and is well supported by commercial tools, it is hoped that the approach will be attractive to electronics engineers. The designer needs only a basic knowledge of SDL in order to describe and analyse circuits. Libraries in the form of SDL packages supply ready-made circuit components and design structures. These present solutions in a form that is familiar to the electronics engineer.

2. Digital Logic Specification with SDL

2.1 Description of Behaviour

The behaviour of a functional unit is given by an SDL description. Process types are used to represent generic components; actual components are instances of these. Component descriptions are stored in a library as SDL-92 named packages. When the generic definition of a component is instantiated, its parameters are set to describe the characteristics of the particular instance. Parameters will include the names of input and output signals, as well as timing characteristics such as propagation delays.

It is highly desirable to have descriptions of logic designs at different levels of abstraction. For example, a low-level description might be appropriate for fabrication or validation of a component. However, once the low-level design has been proven it may be used as a building brick in higher level designs. That is, it may be treated as black box whose internal structure is unimportant at the chosen level of abstraction. A component may thus have several descriptions in a library, to be selected by the designer as appropriate.

The approach deals only with discrete signals, but it handles continuous signals implicitly by modelling their changes (i.e. the signal edges). Hardware signals are modelled as SDL signals with two parameters: the time when the signal is generated, and the binary logic value. The time value of an input signal records when it was generated. The time value is used to determine the time of possible output signals (according to the time delay inherent in a component). The logic value of a signal may be a single bit, but for generality a vector of bits may be used. This caters for common situations such as an address bus or control bus that is several bits wide. In a high-level description, a signal could even carry an SDL data structure, allowing for an abstract description of the interface to a component.

Time delays are often significant in the design of digital logic – especially in asynchronous circuits. It is important that the designer be able to state propagation delays and timing restrictions explicitly. Timing information appears in process parameters and in signals. A process type can be instantiated with different timing values to reflect technology variations or differences due to fan-out of a component. The unit of time in an SDL description is at the discretion of the specifier. Integer time values are used in this work, with a typical interpretation being nanoseconds.

The wires of a circuit are normally considered to carry signals instantaneously between components. Of course this is not strictly true, but the transmission time over a wire is usually negligible compared to the reaction time of a component. In high-speed circuits, a wire can be modelled as a delay if necessary. In digital hardware, the wires between components usually carry signals only in one direction. However bidirectional signals are possible, for example over a bus. The SDL processes representing components are connected by zero-delay channels representing the wires. As usual, channels can be unidirectional or bidirectional.

Process types may have many input and output gates. Figure 1 shows a component described as a functional unit by an SDL process. The input signals are IE_1 to IE_i and the output signals are OE_j to OE_n. The signals carry a time (TIN or TOUT) and a value (V_1 to V_n).

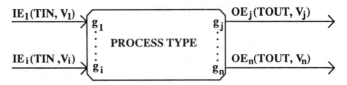

Figure 1: Framework for A Functional Specification

2.2 Description of Structure

A circuit design usually employs a number of components. Processes are therefore combined in an SDL block structure. As a block type, a structure can be stored in an SDL package for use in its own right. Connections between processes go between the gates of process types. This is necessary because in practice an output signal can be sent to the inputs of several other components. The SDL gate construct allows an output gate to send a signal to one or more input gates. The design structure embedded in a block can itself be stored in a library for later use. This is important because components as well as combinations are then available to the designer. An example structure is shown Figure 2, where a two-input NAND gate, a two-input AND gate and a D (Delay) flip-flop are combined in a higher-level design. The components are represented as processes (instantiated process types) and the connections are specified with signal routes (SR1-SR8). Block B1 has four inputs (g_1-g_4) and two outputs (g_5-g_6).

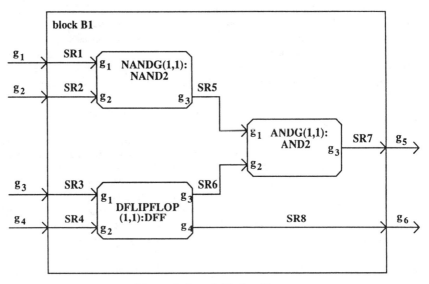

Figure 2: Sample Design Structure

SDL-92 packages are used for design libraries. These can store functional specifications or design structures. Packages are likely to be organised into groups recognisable to a hardware designer. For example there could be packages containing different kinds of gates, latches, flip-flops, counters, registers and so on. Because package elements are parameterised, they represent a variety of instances (implementations) of components and structures

3. Modelling Functional Units

The behaviour of functional units is given by process types. In general, signals carry time and value parameters. As a degenerate case, the time parameter of a signal may be omitted when timing characteristics are not significant. This is appropriate for synchronous logic, where output signals are enabled by clock pulses. In synchronous circuits, component delays can be ignored since it is assumed that the reaction time of a component is faster than the clock rate. But in an asynchronous (unclocked) circuit, exact knowledge of component delays may be necessary to avoid race conditions. Correct operation in the presence of timing constraints may be checked through simulation or through proof of correctness. For adequate modelling it should be possible to:

- give a functional specification of behaviour at any level of abstraction
- define a design structure as a collection of components (process types given in packages) and their interconnection
- have multi-level descriptions, i.e. treat a component as a black box or as a structure containing a finer level of detail
- specify timing characteristics at any level of abstraction.

3.1 Logic Sources and Sinks

Some aspects of logic design require special process types in the SDL. Sometimes it is necessary to specify a source of logic 0 or 1, say to tie an input to a specific level. This is a nullary logic function, specified by process types ZERO and ONE that provide logic 0 and 1 respectively. It may also be necessary to specify a source of other constant values (e.g. some binary input vector). The CONSTANT process type provides a constant output given by its parameter value. Logic sources generate their constants signals at simulation time zero.

If the output of a component is not connected to anything, process output signals have to be consumed but not used. The ABSORB process type is ready to accept and absorb any signal. Note that this differs from standard hardware design: if an output of a component is unused, the engineer simply does not connect anything to it. However, the corresponding SDL process must have a route for output signals to follow (even if nothing is done with them). With a little pre-processing, the use of ABSORB could be made invisible to the specifier. Nonetheless, it could be argued that it is desirable to force an explicit choice of what to do with each output. If an output is accidentally left unconnected, it is useful that a check of the corresponding SDL should point out the error.

3.2 Basic Logic Gates

Logic gates are specified as process types with signal names and gate delays as parameters. Logic gates carry signals that are simple binary (bit) values. A real logic gate exhibits a propagation delay from a change in input to the subsequent output; this reflects the physical processes involved in moving charges into or out of the gate.

Figure 3 illustrates a two-input AND gate. Figure 3 (a) shows the conventional logic symbol for AND2 and its idealised timing diagram. Transitions between logic 0 (low) and logic 1 (high) are imagined to take place instantaneously. In practice, electrical signals never have a sharp edge so Figure 3 (b) is more typical of what happens. Transitions between logic levels are shown as sloping lines here since they are never instantaneous in reality. In Figure

3 (b), TPD1 is the delay for the output C to become valid level 1 after input A changes to 1. TPD0 is the delay taken for the output C to become valid level 0 after input B changes to 0. In real devices, these delays may be different because they require charges to flow in opposite directions. If the engineer does not require to make this distinction, a single delay could be used for output of a 1 or a 0.

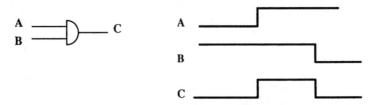

Figure 3 (a): Symbol for AND2 Gate and Idealised Timing Diagram

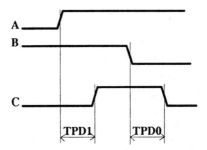

Figure 3 (b): Realistic Timing Diagram for AND2 Gate

An SDL/GR representation of the AND2 gate is given in Figure 4, using timers to specify the propagation delays. The variable OUTN represents the computed value that will be output following the delay time. Only if this differs from the current gate output OUT is it necessary to generate a signal. The delays TPD1 and TPD0 are given actual parameter values during instantiation. Inside the process, delays are handled by timers. When the relevant timer expires, the new output value (OUT) and the time when it is generated (TOUT) are carried by the output signal. An interesting case arises if a further input occurs while the delay timer is running. This corresponds to an input changing before the gate can react to a previous change of input. In such a case, the new output value is calculated. If this differs from the output that is pending, the revised output is scheduled instead after the required delay.

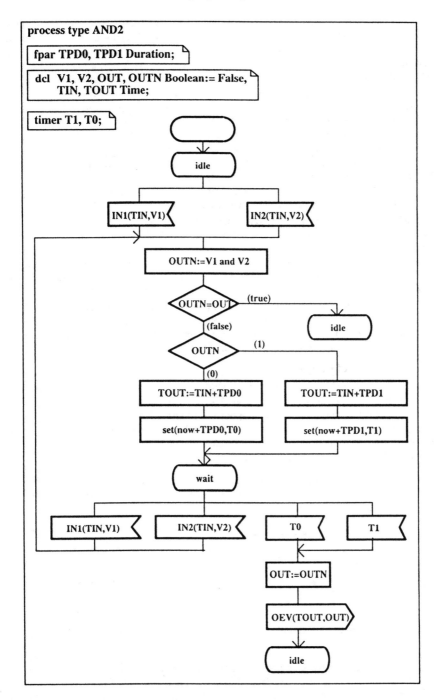

Figure 4: SDL/GR Description of AND2 Gate

Many gates can be modelled in the same way as AND2; a different number of inputs would simply require an alteration to the input signals in the idle and wait states. To compute the new output value, the standard Boolean type and its operators are used. Most of the operators likely to be needed are available in SDL (**or, xor, and, not,** arithmetic operators, relational operators). However, new operators such as **nand** could be defined using the normal data typing features of SDL. In fact it is possible to define a generic logic gate where the number of inputs and the Boolean function are determined by parameters. A generic gate is simply a process type with extra parameters that allow it to become a specific kind of gate – a two-input AND, four-input NOR or whatever.

The initial state of real hardware may be indefinite following power-on. This is not modelled explicitly in the SDL description, since no outputs are generated until inputs occur. However, this is not a limitation since in practice the circuit designer is forced to ensure a well-defined initial state by explicit sequences of signals. For the SDL description, this simply means that the processes will be initialised by specific input signals.

Real logic gates have a fan-out (the maximum number of other gates that can be connected to an output). Real logic gates also have a fan-in (the maximum number of inputs). These are component limitations that can be checked by static analysis of the SDL description. Since fan-out and fan-in have an effect on the delays introduced by gates, the designer can take them into account by choosing appropriate values for the process parameters TPD1 and TDP0.

The wires connecting components introduce a propagation delay that is usually insignificant but can be critical in high-speed circuits. The wires can be considered as components as well, represented by simple delay elements. A process type DELAY is defined for this purpose, with actual transmission delay as a parameter.

3.3 Specification of Delay Flip-Flop

A D (Delay) flip-flop stores its input (D) when a clock pulse occurs. The data will appear on the output (Q) after the next clock pulse. A D flip-flop is thus like a single-bit memory. It is useful for data storage and similar applications. Other kinds of flip-flop include the RS (Reset-Set) and JK types. Figure 5 (a) shows the logic symbol for a D flip-flop and its idealised timing diagram. At point A the D input rises to logic 1. At point B the flip-flop is clocked, so the D input is latched and transferred to the Q output.

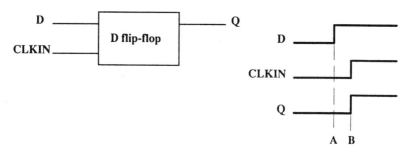

Figure 5 (a): Symbol for D Flip-Flop and Idealised Timing Diagram

Figure 5 (b) illustrates the likely timing behaviour of a real D flip-flop. The time between points A and B (tAB) is called the data setup time. This is measured from the point at which the D input reaches logic 1 and the point at which the clock reaches logic 1. The time is the minimum for which data must be stable at the input of the flip-flop before it is clocked. At point C, the logic 1 input has started to go low. The time between B and C (tBC) is called the data hold time. It is the minimum time for which the data must be held stable after the flip-flop has been clocked. As a result of clocking the flip-flop, its Q output changes state at point D. The time tBD is the maximum time taken for the output to become valid following a clock pulse.

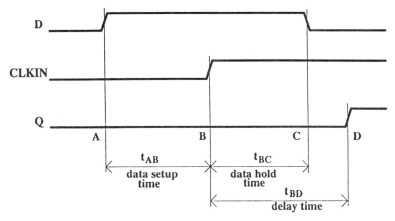

Figure 5 (b): Realistic Timing Diagram for D Flip-Flop

An SDL description of a D flip-flop is given in Figure 6. This follows a similar pattern to the AND2 gate given Figure 4, except that the delays are those particular to a D flip-flop.

3.4 Higher-Level Functional Units

Higher level functional units are frequently used in the top-down design process. These functional units are black-boxes with inputs and outputs and no internal structure. Complex functions can easily be described using the ADT facilities of SDL. Interestingly, it is unimportant whether such functions are realised as hardware or software. This makes it possible to describe mixed hardware-software systems within the same framework. If the designer wishes to specify functional behaviour at an abstract level, it is usually irrelevant whether the realisation is in hardware or software.

Consider, for example, a floating point arithmetic unit. This could be realised as hardware (a co-processor) or as software (by emulation of missing hardware instructions). The designer merely has to specify the interfaces of the functional unit, including input and output data (structures) and timing constraints. At this level of abstraction, a functional unit can be a hardware or software element (and indeed both realisations may be available). Hardware/software functional units can cooperate through signals sent via channels and signal routes. For an abstract description, a signal could carry a data structure such as the content of a data register or address register. Timing characteristics can be given even if the realisation will be in software.

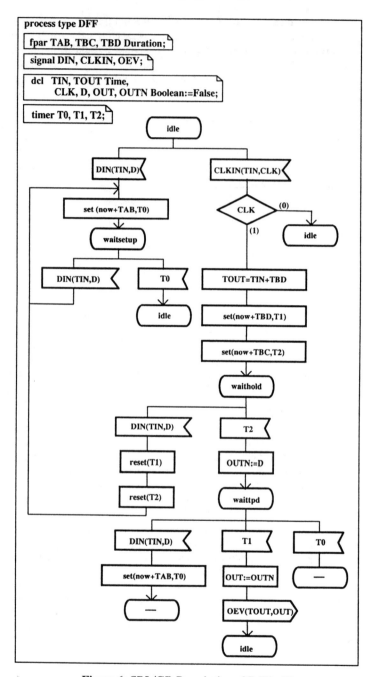

Figure 6: SDL/GR Description of D Flip-Flop

As a concrete example, Figure 7 shows a BCD (Binary Coded Decimal) full adder as a high-level functional unit. AIN and BIN each convey four input bits representing BCD digits from 0 to 9. CIN and COUT are one-bit carry-in and carry-out signals. SOUT conveys BCD sum output in four bits.

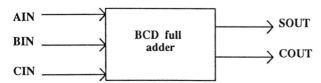

Figure 7: BCD Full Adder as A Black Box

The functional behaviour of this high-level description is given in Figure 8. The value TPDS is the addition delay time – the time from an input change to the change in SOUT. The value TPDCO is the carry delay time for a change in COUT.

The BCD full adder could easily be realised as software, but a hardware implementation will be considered here. The design (internal details of the black box) can be found in any standard textbook on computer logic. Readers unfamiliar with the conventional design should note that it uses the 'trick' of adding 6 (binary 0110) to the first input A. The BCDF block in Figure 9 comprises processes connected via channels between the gates. A, B and SOUT are four-bit hexadecimal values, while CIN and COUT are just one bit. ADD1, ADD2 and ADD3 are instances of the BCDADDER four-bit binary full adder. INV1 is an instance of the FINV four-bit output special inverter. (Of course, BCDADDER and FINV would be found in the component libraries.) The design also requires the special ZERO, ABSORB and CONSTANT process types introduced in Section 3.1. Z1 is used to feed logic 0 to the carry input of ADD1 and ADD2. SIX is used to feed the constant bit vector 0110 to the second input of ADD1. Since the carry outputs of ADD1 and ADD3 are not required, they are absorbed by A1.

4. Validation and Simulation

Validation of a circuit design represented in SDL is conducted through simulation. Although SDL has a formal basis and could thus be used for verification, in practice it is usual to treat verification of SDL descriptions as being exhaustive simulation. Simulation may be an interactive walk through every significant path in order to gain confidence in the design. Step-by-step simulation allows concentration on fewer signals. Another method is to store pre-defined input sequences in a file that is read by the simulator. This is particularly suitable for regression testing, when a design change must be shown to have respected previous behaviour. Simulation results can be presented using MSC sequences, which gives the designer a convenient graphical overview of the behaviour.

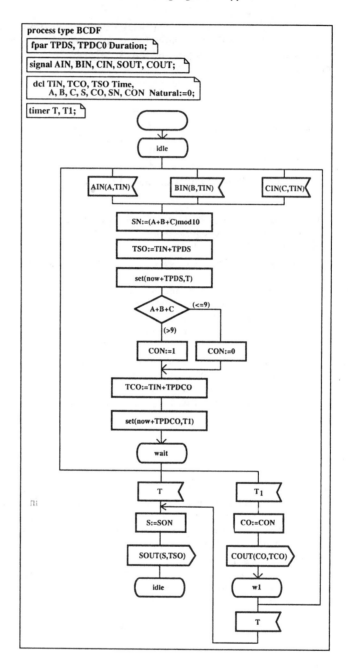

Figure 8: SDL/GR High-Level Description of BCDF Process Type

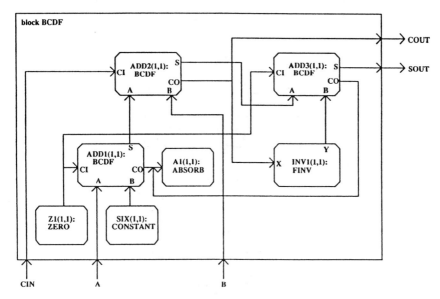

Figure 9: Structure of BCDF Adder

Commercial simulators such as SDT [14] and ObjectGeode [15] can be used for functional analysis of a circuit design. However, they are unsuitable for simulation of timing characteristics because of the special need for event scheduling. A timing simulator has to deal properly with the time parameter in signals. In particular, incoming signals have to be stored in input queues in increasing order of the times at which the signals were generated. This requires the following approach that is rather similar to discrete event simulation:

- Each process needs one input event queue, but events are stored in these queues in increasing order of time parameter.
- Channels have zero delay.
- Initially all input events are evaluated for simulator time zero. This applies especially to process types like ZERO and ONE that generate their constant signals at this point.
- The simulator clock is now set for the next significant event – the earliest time for which any signal is scheduled. All events that are due at this time are evaluated at this point. Note that the order in which events are scheduled is irrelevant because real hardware always exhibits a non-zero delay. This guards against an incorrect order of scheduling events.
- Simulation ceases when no more signals in queues need to be scheduled or when some predefined stop condition is met.

Figure 10 summarises the timed simulation algorithm. As with a standard SDL simulator, inputs could come from a file, outputs could be stored in a file, and outputs could be represented in MSC form. However, the point of simulation is to investigate timing characteristics of a circuit design. This might include the study of race conditions, the determination of minimum and maximum reaction times, and the construction of time sequence diagrams such as those found in Figure 3 (b) and Figure 5 (b).

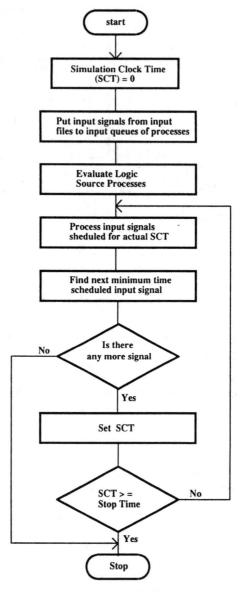

Figure 10: General Overview of Timing Simulator Algorithm

5. Conclusions

An approach for describing digital logic in SDL has been given. From the point of view of the circuit designer, minimal knowledge of SDL is required in order to describe and analyse circuits. Libraries in the form of SDL packages supply ready-made circuit elements and design structures. These present solutions in a form that is familiar to the electronics engineer. Components may be described at several levels of abstraction, and may even be realised as hardware or software.

The approach allows behaviour to be described at several levels from the component level up to the system level. Signals carry values that may be simply bits but could be data structures (vectors of bits). Especially for asynchronous logic, signals may also carry time values that reflect when they were generated. This allows description and analysis of the timing characteristics of a design. A strategy for discrete event simulation of timed circuits has been proposed, allowing timed behaviour to be investigated as well as functional behaviour. Timed simulation can also produce timing diagrams and allow critical time constraints to be checked.

Advantages of the approach include: precise design notation, automated analysis of design, checking equivalences of high/low level designs, use of design component library, and use of a formal design notation.

The work so far shows promise, and has demonstrated that digital logic design in SDL is feasible, following standard engineering practice. Future work will include:

- developing extended libraries of components and design structures
- equivalence checking between different levels of description
- analysing performance issues
- validating timing constraints
- investigating automatic derivation of tests from descriptions
- studying hardware-software co-design in more depth
- constructing tools to support circuit description and analysis, particularly the simulation of timed behaviour.

Acknowledgements

Gyula Csopaki was supported by the British Council during the work described in this paper, and was hosted by the Department of Computing Science and Mathematics, University of Stirling, UK.

References

[1] A. Olsen, *et.al. Systems Engineering Using SDL-92*. North Holland, 1994.

[2] ITU-T. *Specification and Description Language (SDL)*, Recommendation Z.100, International Telecommunications Union, Geneva, 1994.

[3] A. Sarma. Introduction to SDL-92, *Computer Networks and ISDN Systems*, 28 (12), 1996.

[4] L. Vehaard. An introduction to Z.105, *Computer Networks and ISDN Systems*, 28 (12), 1996.

[5] ITU-T. *Message Sequence Chart*, Recommendation Z.120, International Telecommunications Union, Geneva, 1993.

[6] V. Levin, O. Basbugoglu, E. Bounimova and K. Inan. A bilingual specification environment for software/hardware co-design, Proc. *International Symposium on Computer and Information Systems XI*, Middle-East Technical University, Ankara, Turkey, 1996.

[7] J. Peeters, M. Jadoul, E. Holz, M.Wasowski, D. Witaszek and J.-P. Delpiroux. HW/SW co-design and the simulation of a multimedia application, Proc. *7th European Simulation Symposium*, 1995.

[8] A. Clements. *Microprocessor Systems Design*, PWS-Kent Publishing Company, 1992

[9] S. Carlson. *Introduction to HDL Based Design Using VHDL*, Synopsis Inc.

[10] G. A. McCaskill and G. J. Milne. Hardware description and verification using the CIRCAL system. Research Report HDV-24-92, University of Strathclyde, 1992.

[11] G. Csopaki. Hardware description language for specification of digital systems. *Periodica Polytechnica*, 33 (2), 1989.

[12] ISO. *Information Processing – Open Systems Interconnection – LOTOS – A Formal Description Technique based on the Temporal Ordering of Observational Behaviour*, ISO/IEC 8807, International Organization for Standardization, Geneva, 1989.

[13] K. J. Turner and R. O Sinnott. DILL: Specifying digital logic in LOTOS, in *Formal Description Techniques VI*, Elsevier Science B.V., 1994.

[14] Telelogic AB. SDT 3.1: *Tutorial on SDT Tools*, Malmø, Sweden, 1996.

[15] Verilog. *ObjectGeode Simulator Reference Manual*, Toulouse, France, 1996.

24

A Methodology for the Description of System Requirements and the Derivation of Formal Specifications

Atsushi Togashi, Fumiaki Kanezashi, Xiaosong Lu
Department of Computer Science,Shizuoka University
5-1, 3 Johoku, Hamamatsu 432, Japan
phone: +81-53-478-1463   fax: +81-53-475-4595
togashi@cs.inf.shizuoka.ac.jp

Abstract

A methodology for the description of system requirements and the derivation of formal specifications from system requirements are presented. We will specifically deal with the issues (1) mathematical treatment of system requirements and their relationship with formal specifications represented as state transition systems, (2) a sound and complete system with respect to a system requirement, i.e. a standard system of the system requirement specified as a unique model of the system requirement, (3) derivation of standard systems from system requirements, (4) a support system and an application example, and (5) some comparative discussions on the methodology with partial logical Petri Nets, Production systems, and so on.

Keywords

Function Requirement, System Requirement, Formal Specification, State Transition System, Standard System, Logical Petri Net.

1 INTRODUCTION

For a complex and sophisticated system, operational descriptions might be too tedious to handle for rapid prototyping and analysis of a system's behavior. In such cases, it is more convenient to express the system on a higher level, somehow in a functional manner. This approach yields formal specifications that emphasize the system's general behavioral properties rather than its operational details. Moreover, it has a practical significance if the desired

Formal Description Techniques and Protocol Specification, Testing and Verification
T. Mizuno, N. Shiratori, T. Higashino & A. Togashi (Eds.)   © 1997 IFIP. Published by Chapman & Hall

description can be derived or synthesized in a systematic way from the user requirements on system functions.

This paper proposes a new methodology for the description of system requirements and the synthesis of formal specifications from system requirements. The formal specifications can be taken as models of the system requirements. More generally, the main objective is to be able to derive an implementable or operational system description from a given high-level description on system functions. The proposed methodology can be fully automated, hence may/can improve both productivity and quality of system development. We have implemented a support system based on our approach and applied several practical system designs such as a telephone service, a communication protocol, a cable TV system, etc.

In the literature on communicating systems, Formal Description Techniques (FDT), e.g. SDL [5], Estelle [3] and LOTOS [6], have been proposed as high-level specification languages. The conventional state machine oriented approaches such as SDL and Estelle and algebraic approach such as LOTOS are suitable for the purpose of description and investigation of the total behavior of systems. But, these approaches might be not suitable for rapid prototyping and flexible software development. Because we must enumerate and/or determine all system behaviors from an early stage of system design. Our objective is to give theoretical foundations and proposal of a flexible approach on the synthesis of formal specifications from user requirements written in an early stage of system design.

From objectives, our work has some connection with an STR (State Transition Rule) method, which is a specification method based on a production system proposed by Hirakawa and Takenaka in [10]. But, the methodology proposed here differs from their approach mainly in theoretical discussions such as soundness and completeness and formal treatment, rather than practical methodology for description and use. Another related work is a synthesis of communicating processes from temporal logic specification by Manna and Wolper in [11]. Their approach is based on tableau-like method and completely different form ours from technical point of view. Besides those works, no other related works could be found in the literature.

The outline of this paper is as follows: In section 2 after giving some preliminaries, we deal in detail with the issue of system requirements and formal specifications. In section 3, we discuss the key notions, soundness and completeness. Section 4 provides an equivalent transformation on system requirements with the result of determinacy on the resulting transition systems. Section 5 gives an automatic transformation technique from system requirements to formal specifications. Section 6 gives an overview of the support system with an application example followed by the discussions in section 7 and the concluding remarks in section 8.

2 REQUIREMENTS AND FORMAL SPECIFICATIONS

Requirements of a system can be described as expression based on propositional logic. To begin with we will give some preliminaries on propositional logic needed for the description of a system requirement. Let $\mathcal{P}$ be a set of *atomic propositions*. Each atomic proposition describes a specific property of the intended system under the target of design. A *partial interpretation* I is a partial mapping $I : \mathcal{P} \to \{\text{true, false}\}$, where **true** and **false** are the truth values of propositions. If the truth value of a proposition f under I is defined to be **true** then we say that I *satisfies* f, denoted by $I \models f$. $I \not\models f$ denotes that the truth value of f is defined to be **false** and we say I *does not satisfy* f. These can be defined inductively as follows:

(1) $I \models A$ ($I \not\models A$) if I is defined on A and $I(A) = \text{true}$ ($I(A) = \text{false}$), where $A \in \mathcal{P}$.
(2) $I \models \neg f$ ($I \not\models \neg f$) if $I \not\models f$ ($I \models f$).
(3) $I \models f \wedge g$ ($I \not\models f \wedge g$) if $I \models f$ and $I \models g$ ($I \not\models f$ or $I \not\models g$).
(4) $I \models f \vee g$ ($I \not\models f \vee g$) if $I \models f$ or $I \models g$ ($I \not\models f$ and $I \not\models g$).

Note that truth value of a proposition under an interpretation is not always defined since we are concerned with partial interpretations. For propositions f and g, $f \Rightarrow g$ denotes the assertion that for any partial interpretation I, $I \models f$ implies $I \models g$.

Definition 21 Let f and g be propositions.
(1) f is *consistent* if $I \models f$ for some partial interpretation I.
(2) f is *inconsistent* if f is not consistent.
(3) f is *dependent* on g if either $g \Rightarrow f$ (in positive) or $g \Rightarrow \neg f$ (in negative).
(4) f is *independent* of g if f is not dependent on g.   □

A *literal* is an atomic proposition A of the negation of an atomic proposition $\neg A$. Let γ, γ' be consistent conjunctions of literals. It is clear from the definition that $\gamma \Rightarrow \gamma'$ iff $L(\gamma) \supset L(\gamma')$, where $L(\gamma)$ denotes the set of all literals appearing in γ. This implies the following proposition.

Proposition 21 *Let γ be a consistent conjunction of literals. An atomic proposition A is independent of γ iff A does not appear in γ at all neither in positive nor in negative. The negative literal $\neg A$ is independent of γ iff A is independent of γ.*   □

A system can be essentially specified by its fundamental functions and their related constraints for execution. To be more precise, a system function may be invoked by a specific input provided that its pre-condition to be satisfied before execution can hold in the current state. Then, the function is executed, possibly producing some appropriate output. After the execution the current state is changed into the new one. In the new state, other functions (including the same function as well) can be applicable. Taking account into this intuition of system specifications, a function requirement is formally defined in the next definition.

Definition 22 A *function requirement* is a tuple $\rho = \langle id, a, f_{in}, o, f_{out} \rangle$, where

(1) id is a *name* of the function;
(2) a is an *input symbol* of the function;
(3) f_{in} is a *pre-condition* of the function to be satisfied before execution, which is represented as a consistent proposition using atomic propositions in $\mathcal{P}$;
(4) o is an *output symbol* of the function;
(5) f_{out} is a *post-condition* of the function to be satisfied after execution, which is represented as a consistent conjunction of literals by atomic propositions in $\mathcal{P}$. □

For simplicity, in what follows we omit the names and the output symbols from the description of function requirements because they do not play the central roles on the theoretical treatment in this paper. A function requirement $\rho = \langle a, f_{in}, f_{out} \rangle$ is often abbreviated as $\rho : f_{in} \overset{a}{\Rightarrow} f_{out}$.

Definition 23 A *system requirement* is a pair $\mathcal{R} = \langle R, \gamma_0 \rangle$, where R is a set of function requirements and γ_0 is an *initial condition* represented as a consistent conjunction of literals in $\mathcal{P}$. □

In this paper, state transition systems are considered as formal specifications. In the literature, a state transition system is an underling structure of Formal Description Techniques, e.g. SDL [5], Estelle [3] and LOTOS [6], and used to give the operational semantics of concurrent processes in process calculi [12], based on the paradigm of SOS (Structural Operational Semantics) by Plotkin [14].

Definition 24 A *state transition system* is a quadruple $M = \langle Q, \Sigma, \rightarrow, q_0 \rangle$, where Q is a set of *states*, Σ is a set of input symbols, $\rightarrow$ is a *transition relation* defined as $\rightarrow \subset Q \times \Sigma \times Q$, and q_0 is an *initial state*. □

The transition relation defines the dynamical change of states as input symbols may be read. For $(p, a, q) \in \rightarrow$, we normally write $p \overset{a}{\rightarrow} q$. Thus, the transition relation can be written as $\rightarrow = \{ \overset{a}{\rightarrow} \mid a \in \Sigma \}$. $p \overset{a}{\rightarrow} q$ may be interpreted as "in the state p if a is input then the state of the system moves to q". Now, we assume that for an atomic proposition A and for a state $q \in Q$ it is pre-defined whether or not A holds (is satisfied) in q if the truth value of A in q is defined. $q \models A$ indicates that the truth value of A in q is defined and

A holds in q. Let us define the partial interpretation associated with a state q in M, denoted by $I(q)$, in such a way that

$$I(q)(A) = \begin{cases} \textbf{true} & \text{if } q \models A \\ \textbf{false} & \text{if } q \not\models A \ (q \models \neg A) \\ \text{undefined} & \text{otherwise} \end{cases}$$

for all atomic propositions A. Thus, a state transition system can be treated as a Kripke structure [2], where the interpretation of atomic propositions vary over states. Let

$$Sat(q) = \{\, l \mid \text{the truth value of a literal } l \text{ is defined in } q \text{ and } q \models l \,\}.$$

Proposition 22 $q \models f$ *iff* f *is implied from* $Sat(q)$ — *every interpretation satisfying* $Sat(q)$ *also satisfies* f, *for each proposition* f.

Proof: The proof is by structural induction on propositions f.   □

By the completeness of propositional logic, we have that $q \models f$ iff $Sat(q) \vdash f$, f is provable from $Sat(q)$.

Two states p and q in M are *logically equivalent* iff $I(p) = I(q)$. A transition system M is *logically reducible* if there exist distinct logically equivalent states in M. Otherwise, the system is *logically irreducible*. To the rest of this paper, unless stated otherwise, a transition system means a logically irreducible system. Thus, $p = q$ iff $I(p) = I(q)$ $(Sat(p) = Sat(q))$. By this assumption, note that a state q in a (an irreducible) transition system M can be equivalently represented as a consistent set X of literals, where $q \models A$ $(q \models \neg A)$ iff $A \in X$ $(\neg A \in X)$.

3   SOUNDNESS AND COMPLETENESS

Definition 31 A state transition $t = \langle p \xrightarrow{a} q \rangle$ satisfies (is *correct w.r.t.*) a function requirement $\rho : f_{in} \xRightarrow{b} f_{out}$, denoted as $t \models \rho$, if the following conditions hold:

(1) $p \models f_{in}$, $a = b$, and $q \models f_{out}$.
(2) The partial interpretations $I(p)$ and $I(q)$ are identical if atomic propositions independent of f_{out} are only concerned.   □

The condition (1) means the precondition and the postcondition must hold in the current state and the next state, respectively. The condition (2) states that for an atomic proposition A independent of f_{out}, $p \models A$ iff $q \models A$. This means that the truth value of independent atomic propositions *w.r.t.* the postcondition remain unchanged through the state transition.

Example 31 Consider the system requirement

$$\mathcal{R}_1 = \langle \{\rho_1 : A \overset{a}{\Rightarrow} \neg A, \ \rho_2 : B \overset{b}{\Rightarrow} A\}, \ A \wedge B \rangle$$

and the transition system M_1 given in (a) in Figure 1. Now, consider the transition $t_1 = \langle q_0 \overset{a}{\rightarrow} q_1 \rangle$ and the function requirement $\rho_1 : A \overset{a}{\Rightarrow} \neg A$. Since $q_0 \models A$ and $q_1 \models \neg A$ the condition (1) in Definition 31 holds for t_1 *w.r.t.* ρ_1. The atomic proposition independent of $\neg A$ is B. Since the truth values of B in q_0, q_1 are defined and $q_0 \models B$, $q_1 \models B$ the condition (2) in Definition 31 holds. Thus, the transition t_1 satisfies the function requirement ρ_1. In the exactly same way, we can easily check that the transitions $q_0 \overset{b}{\rightarrow} q_0$, $q_1 \overset{b}{\rightarrow} q_0$ satisfy the function requirement $\rho_2 : B \overset{b}{\Rightarrow} A$. □

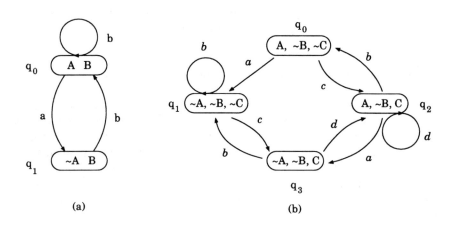

(a)   (b)

Figure 1 Transition Systems M_1 and M_2

Example 32 As a more involved example, let us consider the system requirement

$$\mathcal{R}_2 = \langle \{ \quad \rho_1 : A \overset{a}{\Rightarrow} \neg A \wedge \neg B, \quad \rho_2 : \neg A \wedge \neg B \vee A \wedge C \overset{b}{\Rightarrow} \neg C,$$
$$\rho_3 : \neg C \overset{c}{\Rightarrow} C, \qquad \rho_4 : C \overset{d}{\Rightarrow} A \},$$
$$A \wedge \neg B \wedge \neg C \rangle$$

and the transition system M_2 given (b) in Figure 1. In the same way as in Example 31, it is easily checked that:

- the transitions $q_0 \overset{a}{\rightarrow} q_1$, $q_2 \overset{a}{\rightarrow} q_3$ satisfy ρ_1;
- the transitions $q_1 \overset{b}{\rightarrow} q_1$, $q_2 \overset{b}{\rightarrow} q_0$, $q_3 \overset{b}{\rightarrow} q_1$ satisfy ρ_2;
- the transitions $q_0 \overset{c}{\rightarrow} q_2$, $q_1 \overset{c}{\rightarrow} q_3$ satisfy ρ_3;
- the transitions $q_2 \overset{d}{\rightarrow} q_2$, $q_3 \overset{d}{\rightarrow} q_2$ satisfy ρ_4. □

Let γ be a consistent conjunction of literals. We define a partial interpretation $I(\gamma)$ based on γ by

$$I(\gamma)(A) = \begin{cases} \textbf{true} & \text{if } A \text{ appears positive in } \gamma \\ \textbf{false} & \text{if } A \text{ appears negative in } \gamma, \\ \text{undefined} & \text{otherwise} \end{cases}$$

for all atomic propositions A.

Definition 32 A state transition system $M = \langle Q, \Sigma, \rightarrow, q_0 \rangle$ is *sound* with respect to a system requirement $\mathcal{R} = \langle R, \gamma_0 \rangle$ if the following conditions are satisfied:

(1) $I(q_0) = I(\gamma_0)$;
(2) for any transition t in M there exists a function requirement $\rho \in R$ such that $t \models \rho$.   □

Note that the transition systems M_1 in Example 31 and M_2 in Example 32 are sound with respect to the system requirements $\mathcal{R}_1$ and $\mathcal{R}_2$, respectively.

Definition 33 Let $M = \langle Q, \Sigma, \rightarrow, q_0 \rangle$ and $M' = \langle Q', \Sigma, \rightarrow', q_0' \rangle$ be state transition systems in common input symbols. A *homomorphism* from M into M' is a mapping $\xi : Q \rightarrow Q'$ such that

(1) $\xi(q_0) = q_0'$.
(2) if $p \xrightarrow{a} q$ in M, then $\xi(p) \xrightarrow{a} \xi(q)$ in M'.
(3) $p \models f$ implies $\xi(p) \models f$, for all states p in M and for all propositions f.
    □

The third condition (3) in the above definition can be equivalently relaxed:
(3') $p \models l$ implies $\xi(p) \models l$, for all states p in M and for all literals l.

If a homomorphism $\xi : Q \rightarrow Q'$ is a bijection, a one-to-one and onto mapping, and the inverse function ξ^{-1} is a also homomorphism from M' to M, then ξ is called an *isomorphism*. If there is an isomorphism from M to M', then M and M' are *isomorphic*.

Definition 34 Let M be a sound state transition system with respect to $\mathcal{R}$. M is called *complete* with respect to $\mathcal{R}$ if, there is a homomorphism ξ from M' into M for every sound state transition system M' with respect to $\mathcal{R}$.   □

Definition 35 A sound and complete transition system with respect to $\mathcal{R}$ is called a *standard system* (*model*) of $\mathcal{R}$.   □

Theorem 31 *Let M, M' be standard systems of $\mathcal{R}$, then M and M' are isomorphic.*   □

Let $M(\mathcal{R})$ denote a unique standard system of $\mathcal{R}$ up to isomorphism.

4   TRANSFORMATION AND DETERMINACY

Without loss of generality, a proposition f can be equivalently expressed as a *disjunctive normal form* $\gamma_1 \vee \cdots \vee \gamma_n$, where γ_i are conjunctions of literals. Now, consider the following transformation rules on sets of function requirements:

rule 1 $R \cup \{\gamma_1 \vee \cdots \vee \gamma_n \overset{a}{\Rightarrow} \gamma\} \Rightarrow R \cup \{\gamma_1 \overset{a}{\Rightarrow} \gamma, \ldots, \gamma_n \overset{a}{\Rightarrow} \gamma\}.$

rule 2 $R \cup \{\gamma_1 \wedge A \wedge \gamma_2 \overset{a}{\Rightarrow} \gamma\} \Rightarrow R \cup \{\gamma_1 \wedge A \wedge \gamma_2 \overset{a}{\Rightarrow} \gamma \wedge A\}$
where neither A nor $\neg A$ appears in γ.

rule 3 $R \cup \{\gamma_1 \wedge \neg A \wedge \gamma_2 \overset{a}{\Rightarrow} \gamma\} \Rightarrow R \cup \{\gamma_1 \wedge \neg A \wedge \gamma_2 \overset{a}{\Rightarrow} \gamma \wedge \neg A\}$
where neither A nor $\neg A$ appears in γ.

Lemma 41 *We have the following results on the transformation rules:*

(1) *A transition t is correct w.r.t. a function requirement $\gamma_1 \vee \cdots \vee \gamma_n \overset{a}{\Rightarrow} \gamma$ iff it is correct w.r.t. some function requirement $\gamma_i \overset{a}{\Rightarrow} \gamma$, for some i.*

(2) *A transition t is correct w.r.t. a function requirement $\gamma_1 \wedge A \wedge \gamma_2 \overset{a}{\Rightarrow} \gamma$ iff it is correct w.r.t. the function requirement $\gamma_1 \wedge A \wedge \gamma_2 \overset{a}{\Rightarrow} \gamma \wedge A$, where neither A nor $\neg A$ appears in γ.*

(3) *A transition t is correct w.r.t. a function requirement $\gamma_1 \wedge \neg A \wedge \gamma_2 \overset{a}{\Rightarrow} \gamma$ iff it is correct w.r.t. the function requirement $\gamma_1 \wedge \neg A \wedge \gamma_2 \overset{a}{\Rightarrow} \gamma \wedge \neg A$, where neither A nor $\neg A$ appears in γ.*

Proof: Obvious from the transformation rules.   □

Let $\mathcal{R} = \langle R, \gamma_0 \rangle$ be a system requirement. Let $\hat{\mathcal{R}} = \langle \hat{R}, \gamma_0 \rangle$ denote the resulting system requirement by applying the above transformation rules to $\mathcal{R}$ as much as possible. We call $\hat{\mathcal{R}}$ the *canonical form* of $\mathcal{R}$.

Theorem 41 *Let $\mathcal{R}$ be a system requirement. Suppose that state transition systems M and $\hat{M}$ are standard systems of $\mathcal{R}$ and $\hat{\mathcal{R}}$, respectively, then M and $\hat{M}$ are isomorphic .*   □

Example 41 If we apply the above transformation rules to the requirement $\mathcal{R}_2$ in Example 32, we obtain the following requirement $\hat{\mathcal{R}}_2$.

$$\hat{\mathcal{R}}_2 = \langle \{ \quad \rho_1 : A \overset{a}{\Rightarrow} \neg A \wedge \neg B, \quad \rho_2 : \neg A \wedge \neg B \overset{b}{\Rightarrow} \neg C \wedge \neg A \wedge \neg B,$$
$$\rho_2 : A \wedge C \overset{b}{\Rightarrow} \neg C \wedge A, \quad \rho_3 : \neg C \overset{c}{\Rightarrow} C,$$
$$\rho_4 : C \overset{d}{\Rightarrow} A \wedge C \}, \qquad A \wedge \neg B \wedge \neg C \rangle$$

By Theorem 41, both requirements have the isomorphic standard transition systems.   □

Definition 41 Let M be a transition system. M is called *deterministic* if there are no transitions $p \overset{a}{\rightarrow} q_1$ and $p \overset{a}{\rightarrow} q_2$ for any states p, q_1, q_2 and for any input symbol a such that $q_1 \neq q_2$.   □

Proposition 41 *Let $\mathcal{R}$ be a system requirement. If there are no functions $\rho_1 : f_1 \overset{a}{\Rightarrow} f_1'$, $\rho_2 : f_2 \overset{a}{\Rightarrow} f_2'$ with the input symbol in common such that $f_1 \wedge f_2$ is consistent, then the standard system of $\mathcal{R}$ is deterministic.*

Proof: Suppose the standard system $M(\mathcal{R})$ is nondeterministic, then there exist transitions $t_1 = \langle p \overset{a}{\to} q_1 \rangle$, $t_2 = \langle p \overset{a}{\to} q_2 \rangle$ for some states p, q_1, q_2 and for some input symbol a such that $q_1 \neq q_2$. Let $\rho_1 : f_1 \overset{a}{\Rightarrow} f_1'$, $\rho_2 : f_2 \overset{a}{\Rightarrow} f_2'$ be the functions such that $t_1 \models \rho_1$, $t_2 \models \rho_2$. Then, $p \models f_1$ and $p \models f_2$. Hence, $f_1 \wedge f_2$ is consistent. $\square$

5   SYNTHESIS OF FORMAL SPECIFICATION

Our target is to derive a sound and complete state transition system M from a given system requirement $\mathcal{R} = \langle R, \gamma_0 \rangle$. Now, we state a transformation $\mathcal{T}$ from $\mathcal{R}$ into M. Let us define a transition system $\mathcal{T}(\mathcal{R}) = \langle \Gamma, \Sigma, \to, q_0 \rangle$, where

(1) $\Gamma = \{\gamma \mid \gamma$ is a consistent conjunction of literals in $\mathcal{P}\}$
(2) $\Sigma = \{a \mid \rho : f_{in} \overset{a}{\Rightarrow} f_{out} \in R\}$
(3) $\gamma \overset{a}{\to} \gamma'$ iff there exists a function requirement $\rho : f_{in} \overset{a}{\Rightarrow} f_{out} \in R$ such that
   (a) $I(\gamma) \models f_{in}$.
   (b) $I(\gamma') \models f_{out}$.
   (c) If an atomic proposition A is independent of f_{out}, then $I(\gamma) \models A$ iff $I(\gamma') \models A$.
(4) $q_o = \gamma_0$.

The partial interpretation associated with a state γ in $\mathcal{T}(\mathcal{R})$ is defined as $I(\gamma)$. In other words, the states correspond possible partial interpretations for all atomic propositions in $\mathcal{P}$. It is trivial from the construction that $\mathcal{T}(\mathcal{R})$ is irreducible.

Theorem 51 *The state transition system $\mathcal{T}(\mathcal{R})$ derived from a requirement description $\mathcal{R} = \langle R, \gamma_0 \rangle$ by $\mathcal{T}$ is a standard system of $\mathcal{R}$.*

Proof: Soundness: This direction is clear from the construction of the transition system $\mathcal{T}(\mathcal{R})$.

Completeness: Let $M = \langle Q, \Sigma, \to, q_0 \rangle$ be a sound state transition system with respect to $\mathcal{R}$. Let define a mapping $\xi : Q \to \Gamma$ by $\xi(q) = \gamma$ for $q \in Q$, where γ is a consistent conjunction of literals such that $I(q) = I(\gamma)$. The mapping ξ is well defined.

Now, we will show that ξ is a homomorphism from M into $\mathcal{T}(\mathcal{R})$. It can be easily checked that $\xi(q_0) = \gamma_0$ since M is a sound transition system and the initial state q_0 in M satisfies only literals appearing in γ_0. Let $p \overset{a}{\to} q$ be any transition in M. Suppose $\rho : f_{in} \overset{a}{\Rightarrow} f_{out}$ be the function requirement in R satisfied by this transition. So, we have $p \models f_{in}$ and $q \models f_{out}$. Thus,

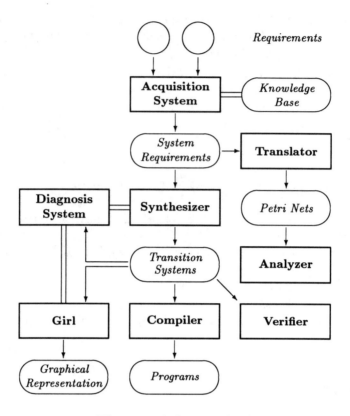

Figure 2 A Support System

$\xi(p) \models f_{in}$ and $\xi(q) \models f_{out}$, by the definition of ξ. The statement "$\xi(p) \models A$ iff $\xi(q) \models A$, for all atomic proposition A independent of f_{out}" can be implied by the statement "$p \models A$ iff $q \models A$, for all atomic proposition A independent of f_{out}". Therefore, we have a transition $\xi(p) \xrightarrow{a} \xi(q)$ in $\mathcal{T}(\mathcal{R})$. By the definition of ξ, $p \models f$ implies $\xi(p) \models f$ for all proposition f. Hence, ξ is a homomorphism from M into $\mathcal{T}(\mathcal{R})$.   □

6   SUPPORT SYSTEM AND APPLICATION EXAMPLE

The outline of a support system for the development of (communication) software is briefly stated. The system consists of *Acquisition System* of system requirements with a help of Knowledge Base, *Synthesizer* of transition systems as formal specifications from system requirements, *Compiler* to C^{++} programs (executable codes) from transition systems (not fully implemented), *Diagnosis System* of system requirements with respect to transition systems (not fully implemented), *Verifier* of specifications via Temporal Logic (not fully implemented), *Translator* of system requirements to partial logical Petri Nets, and *Girl* – Visualizer of transition systems on the X-window system —. Figure 2 shows the system structure of our support system.

As a more real example, we will apply our method to a small portion of a simplified CATV system. The terminal of the CATV system is connected with the host computer, we can take several services on TV programs by controlling the buttons of the remote switch of the terminal. A system requirement of the CATV system is briefly stated: Power button enables power on-off of the system alternatively at any time ([power on/off] function). By pushing the channel-up, channel-down, or ten-key button, we can select the next, previous, or intended channel directly, respectively ([channel-change] functions). As the usual TV systems, the CATV system has muting facility ([mute on/off] function). Force tuning and buzzering functions are the characteristics of the CATV system ([force-tune] and [buzzer on/off] functions). According to the brief description of the system, a system requirement of the CATV system is described by the the following system requirement:

initial_condition : $\neg muteon \wedge \neg force \wedge \neg poweron \wedge \neg buzzer$

power_off : $poweron \wedge \neg force \wedge \neg buzzer \overset{power}{\Longrightarrow} \neg poweron$

power_on : $\neg poweron \overset{power}{\Longrightarrow} \neg muteon \wedge poweron$

channel_up : $poweron \wedge \neg force \wedge \neg buzzer \overset{chup}{\Longrightarrow} -$

channel_down : $poweron \wedge \neg force \wedge \neg buzzer \overset{chdw}{\Longrightarrow} -$

channel_change : $poweron \wedge \neg force \wedge \neg buzzer \overset{tenkey}{\Longrightarrow} -$

mute_on : $\neg muteon \wedge \neg buzzer \wedge \neg force \wedge poweron \overset{mute}{\Longrightarrow} muteon$

mute_off : $muteon \wedge \neg buzzer \wedge \neg force \wedge poweron \overset{mute}{\Longrightarrow} \neg muteon$

force_tune : $\neg poweron \overset{ftune}{\Longrightarrow} force \wedge poweron$

force_tune : $poweron \overset{ftune}{\Longrightarrow} force$

force_cancel : $force \wedge \neg buzzer \wedge poweron \overset{power}{\Longrightarrow} \neg force$

buzzer_on : $\neg buzzer \overset{buzzer}{\Longrightarrow} buzzer$

buzzer_off : $buzzer \overset{anykey}{\Longrightarrow} \neg buzzer$

In the above description the symbol "–" indicates its own precondition of a function. So, e.g. the channel-up function

channel_up : $poweron \wedge \neg force \wedge \neg buzzer \overset{chup}{\Longrightarrow} -$

is the abbreviation of the regular description

channel_up : $poweron \wedge \neg force \wedge \neg buzzer \overset{chup}{\Longrightarrow} poweron \wedge \neg force \wedge \neg buzzer.$

This means that there are no state change by the channel-up function. The formal specification derived from the requirements is depicted in Figure 3, which is the real output (eps file) of the support system sated in the previous section. In the output function names are used instead of input symbols as labels of transitions.

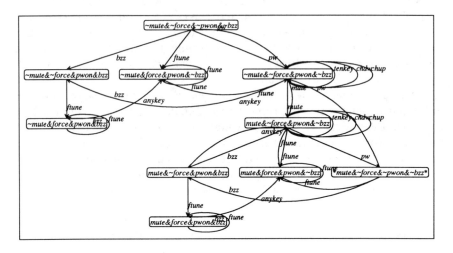

Figure 3 The Derived Formal Specification of the CATV System

7 DISCUSSIONS

The derived state transition system $T(\mathcal{R})$ from a system requirement $\mathcal{R}$ can be proved to coincide with the reachability graph of a Partial Logical Petri Net. A Partial Logical Petri Net, where inhibited arcs (inhibitor arcs) are allowed in both inputs and outputs of transitions, and two kinds of tokens are provided. The Partial Logical Petri Net is an straight extension of a Logical Petri Net proposed by Song and et al [17].

Definition 71 (Partial Logical Petri Net)

A *Partial Logical Petri Net* is a tuple $PN = \langle P, T, I, O, M_0 \rangle$, where

(1) P is a set of *places*;

(2) T is a set of *transitions*;

(3) $I = \langle I_p, I_n \rangle$ is a pair of *input functions* $I_p, I_n : T \to 2^P$ such that $I_p(t) \cap I_n(t) = \emptyset$ for all $t \in T$;

(4) $O = \langle O_p, O_n \rangle$ is a pair of *output functions* $O_p, O_n : T \to 2^P$ such that $O_p(t) \cap O_n(t) = \emptyset$, for all $t \in T$;

(5) $M_0 : P \to \{0, 1, *\}$ is an *initial marking*.   □

A Partial Logical Petri Net can be represented as a bipartite graph in the almost same way as a usual Petri Net [13]. However, in a Partial Logical Petri Net, we have the following extensions and restrictions.

- There are two kinds of arcs, called *positive arcs* and *negative arcs*. If $p \in I_p(t)$ ($p \in O_p(t)$), we make a positive arc, depicted as $\rightarrow$, from p to t (from t to p). If $p \in I_n(t)$ ($p \in O_n(t)$), we make a negative arc, depicted as $\multimap$, from p to t (from t to p).
- There are two kinds of tokens, a *positive token* • and a *negative token* ∘ which represent truth constant **true** and **false**, respectively.
- Marking functions are restricted to the functions with the range $\{0, 1, *\}$, where 0, 1, and $*$ means that the associated condition with the place is "not satisfied", "satisfied", and "undefined", respectively.

The graphical representation of a Partial Logical Petri Net is given in Figure 4 (a).

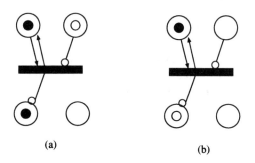

(a)   (b)

Figure 4 Partial Logical Petri Nets

In a marking M, a transitions t is *fireable* (*executable*) if the following conditions are satisfied:

(1) $M(p) = 1$ for all $p \in I_p(t)$.
(2) $M(p) = 0$ for all $p \in I_n(t)$.

If t is fireable, then t suddenly fires and the marking is changed into the marking M' defined by

$$
M'(p) = \begin{cases}
0 & \text{if } p \in O_n(t) \\
1 & \text{if } p \in O_p(t) \\
* & \text{if } p \in (I_n(t) \cup I_p(t)) \cap O_n(t)^c \cap O_p(t)^c \\
M(p) & \text{otherwise}
\end{cases}
$$

The transition in the net (a) in Figure 4 is fireable. After firing, the marking is changed into the one (b) in the figure.

Let $\mathcal{R} = \langle R, \gamma_0 \rangle$ be a requirement in canonical form. We can obtain a Partial Logical Petri Net $\langle P, T, I, O, M_0 \rangle$ from $\mathcal{R}$ as follows:

1. $P = \mathcal{P}$: Places correspond atomic propositions.
2. $T = R$: Transitions correspond function requirements.
3. Let $\rho : A_1 \wedge \cdots \wedge A_n \wedge \neg B_1 \wedge \cdots \wedge \neg B_m \overset{a}{\Rightarrow} C_1 \wedge \cdots \wedge C_j \wedge \neg D_1 \wedge \cdots \wedge \neg D_k$
be a function, where capital letters denote atomic propositions. Then, define input functions $I = \langle I_p, I_n \rangle$ and output functions $O = \langle O_p, O_n \rangle$ by

$$I_p(\rho) = \{A_1, \ldots, A_n\}$$
$$I_n(\rho) = \{B_1, \ldots, B_m\}$$
$$O_p(\rho) = \{C_1, \ldots, C_j\}$$
$$O_n(\rho) = \{D_1, \ldots, D_k.\}$$

4. The initial marking M_0 is defined by

$$M_0(A) = \begin{cases} 0 & \text{if } A \text{ appears negative in } \gamma_0 \\ 1 & \text{if } A \text{ appears positive in } \gamma_0 \\ * & \text{otherwise} \end{cases}$$

Example 71 If we apply the above transformation to the canonical form in Example 41 of the requirement in Example 32, we obtain the Partial Logical Petri Net in Figure 5. The resulting reachability graph of the net coincide with the transition system (b) in Figure 1. This can be guaranteed in general by the next proposition. □

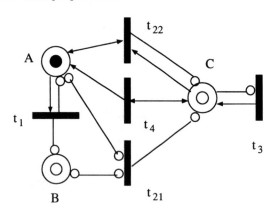

Figure 5 The transformed Partial Logical Petri Net

Proposition 71 *Let M be a standard system of a requirement* $\mathcal{R}$. *Then, the reachability graph of the Partial Logical Petri Net derived from* $\hat{\mathcal{R}}$ *is isomorphic to M.* □

The derived transition system $\mathcal{T}(\mathcal{R})$ can be characterized by Production Systems as well. To be more precise, if $\mathcal{R}$ is a requirement in canonical form, then each function requirement $\rho : f_{in} \overset{a}{\Rightarrow} f_{out}$ can be regarded as a production rule $f_{in} \rightarrow f_{out}$. Then, we have the following result.

Proposition 72 *Let* $\mathcal{R}$ *be a requirement in canonical form. If we take a function requirement* $\rho : f_{in} \overset{a}{\Rightarrow} f_{out}$ *as a production rule* $f_{in} \rightarrow f_{out}$, *then the state transition system of the resulting production system is isomorphic to the standard transition system* $\mathcal{T}(\mathcal{R})$. □

8 CONCLUDING REMARKS

A formal methodology for the description of system requirements and the synthesis of formal specifications from them have been presented. We have specifically dealt with the issues (1) mathematical treatment of system requirements and their relationship with formal specifications represented as state transition systems, (2) sound and complete systems, i.e. standard systems, (3) derivation of standard systems from system requirements, (4) a support system and an application example, and (5) some discussions on partial logical Petri Nets, Production systems, and so on . The proposed framework provides theoretical and practical tools for system design.

To conclude the paper, we state some further comments on our methodology.

Extension to Predicate Logic   The underlying logic of this paper may be easily extended to first order predicate logic. For example, the function of channel_up in the CATV system is expressed more precisely by the function requirement

$$\text{channel\_up} : poweron \land \neg force \land \neg buzzer \land ch(x) \overset{chup}{\Longrightarrow} poweron \land ch(x+1)$$

In the above description, the first order variable x is quantified universally.

Branching Time Temporal Logic   A function requirement $\rho : f_{in} \overset{a}{\Rightarrow} f_{out}$ can be expressed as a proposition $\Box(f_{in} \supset \langle a \rangle f_{out})$ in an extended branching time temporal logic.

REFERENCES

[1] Chellas, B.F., Modal Logic: An Introduction, Cambridge University Press, 1980.

[2] ISO., Estelle: A Formal Description Technique based on the Extended State Transition Model, ISO 9074, 1989.

[3] ISO., *Information Processing Systems – Open System Interconnection – LOTOS – A Formal Description Technique based on the Temporal Ordering of Observational Behavior*, IS 8807, 1989.

[4] CCITT., *SDL: Specification and Description Language*, CCITT Z.100, 1988.

[5] Bolognesi, T., Brinksma, Ed., Introduction to the ISO Specification Language LOTOS, in *the Formal Description Technique LOTOS*, Elsevier Sci. Pub., pp.23–73, 1989.

[6] Emerson, E.A., Temporal and Modal Logic, *Handbook of Theoretical Computer Science*, Elsevier Science Publishers B.V., pp.995–1072, 1990.

[7] Gotzhein, R., Specifying Communication Services with Temporal Logic, *Protocol Specification, Testing and Verification*, XL, pp.295–309, 1990.

[8] van Glabbeek, R.J., *The Linear Time – Branching Time Spectrum*, Lecture Notes in Comput. Sci. **458**, Springer-Verlag, 1990.

[9] Hirakawa, Y., Takenaka, T., Telecommunication Service Description using State Transition Rules, Proc. 6th Int. Work. Software Specification and Design, pp.140–147, 1991.

[10] Manna, Z., P. Wolper, Synthesis of Communicating Processes from Temporal Logic Specifications, *ACM Trans. on Programming Languages and Systems*, **6-**, 1, pp.68–93,1984.

[11] Milner R., *Communication and Concurrency*, Prentice-Hall, 1989.

[12] Murata, T., Petri Nets: Properties, Analysis and Applications, IEEE Proc. Vol.77, No.4, pp.541–580, 1989.

[13] Plotkin, G.D., *A Structural Approach to Operational Semantics*, Computer Science Department, Aarhus University, DAIMI FN-19, 1981.

[14] Shapiro E.Y., *Algorithmic Program Debugging*, Ph.D. Thesis, The MIT Press, 1982.

[15] Shiratori, N., Sugaware, K., Kinoshita, T., Chakraborty, G., Flexible Networks: Basic Concepts and Architecture, IEICE Trans. Commun., Vol.E77-B, No.11. pp.1287–1294, 1994.

[16] Song, K., Togashi, A., Shiratori, N., Verification and refinement for system requirements, *IEICE Trans. on Fundamentals of Elec., Comm. and Comput. Sci.*, Vol. E78-A, No.11, pp.1468–1478, 1995.

[17] Togashi, A., Usui, N., Song, K., Shiratori, N. A derivation of System Specifications based on a Partial Logical Petri Net, Proc. of ISCAS95, 1995.

25
On the Influence of Semantic Constraints on the Code Generation from Estelle Specifications

Ralf Henke, Andreas Mitschele-Thiel
Universität Erlangen-Nürnberg, Lehrstuhl für Informatik VII,
Martensstraße 3, 91058 Erlangen, Germany, Phone/Fax: +49
9131 85 7932 / 7409, email: mitsch@informatik.uni-erlangen.de

Hartmut König
Brandenburgische Technische Universität Cottbus, Institut für
Informatik, Postfach 101344, 03013 Cottbus, Germany, Phone/
Fax: +49 355 692236, email: koenig@informatik.tu-cottbus.de

Abstract
Implementations automatically derived from formal descriptions often do not fulfill the performance requirements of real-life applications. There are several reasons for this. In the paper, we discuss for the FDT Estelle how semantical constraints can influence the efficiency of the generated code. In the first part of the paper we show that certain language features may have a restraining effect on the performance of the implementation. The second part of the paper investigates how the activity thread model, a technique known from manual protocol implementation, can be applied to automatically derive efficient implementations from Estelle specifications. The activity thread model reduces the communication overhead. It is known to be efficient as the server model usually applied. We analyze the prerequisites to apply the model and present measurements comparing the performance achievable with the technique. The measurements are given for different implementations of XTP and the XDT protocol.

Keywords
Estelle, semantics, automated protocol implementation, activity thread model.

1 MOTIVATION

The performance of communication systems is decisively determined by the protocols used. Experience with practical protocol implementations shows that protocol performance depends as much, and usually more, on the implementation than on the design [Wats87][Clar89]. Communication protocols are usually implemented by hand. The implementations are typically based on informal standards provided by ISO, ITU-T or other organizations. The techniques employed for hand-coded implementations have been continuously improved. Examples range from the simple elimination of copy operations between protocol layers by data referencing to more sophisticated optimizations as application level framing and integrated layer processing (ILP) [Clar90][Ahlg96]In addition, the exploitation of various kinds of parallelism to derive efficient protocol implementations has been investigated, but it has not brought the expected efficiency gain.

Product implementations of communication protocols that are automatically derived from formal descriptions are less reported, although approaches for the computer-aided derivation of implementations have been investigated for more than ten years. So far, the use of automatically derived implementations is limited due to their lack of efficiency. Thus, they are mainly used for prototyping or for applications where optimal performance is not crucial.

The main obstacle for the successful application of computer-aided implementation techniques (and probably of formal description techniques in general) is the insufficient efficiency of the generated code. There are several reasons for this [Held95]:

(1)   The implementation model prescribed by the transformation tool is elaborated during tool design. It does not take into account the context, in which the implementation is used.

(2)   Formal description techniques are based on formal semantics. The semantics highly influence the design of the implementation model. Optimizations can only be introduced by ensuring correctness and conformance criteria.

(3)   So far, automated protocol implementations are mainly focused on one-layer implementations (due to their orientation on prototyping).

In recent years the derivation of efficient implementations from formal descriptions has been investigated in several papers. They have brought interesting experimental results, especially for the FDTs based on extended finite state machines (EFSM) as Estelle and SDL. However, the question whether FDT-based implementations can compete with hand-coded ones in performance has not finally been answered yet. The approaches pursued in these papers were different. The influence of specification styles on the runtime efficiency is among others discussed in [Gotz96]. In [Held95] experiments with an experimental Estelle-C compiler are reported that applies optimized algorithms for implementing certain features of the Estelle semantics, e.g. for the synchronisation of module instances for the selection of fireable transitions. It also proposes a variable implementation model for a better adjustment to the given implementation environment. Several approaches were dedicated to the exploitation of the protocol inherent parallelism [Stei93],[Fisc94],[Plat96]. The results obtained show that the efficiency of the derived implementation can be improved by using parallelism. However, the potential of parallelism in protocols is often too small and the additional overhead for synchronisation and communication too high to achieve a considerable efficiency gain. More promising results are expected from an integrating handling of layers and data operations. [Leue96] describes an algorithm for the derivation the *Common Path* from SDL specifications as a base for deriving implementations using the ILP approach. A compiler for deriving ILP implementations from Esterel is reported in [Brau96]. For standardized FDTs such compilers are still under development. However, recent research [Ahlg96] has pointed out that the application of ILP is not recommended in every case.

As long as there are no practicable solutions, practice goes other ways. For the SDL Cmicro-Code-Generator [Tele96], for example, the use of certain SDL constructs (object-oriented features, enabling conditions, continuous signals, *import/export, view/reveal* etc.) is entirely prohibited and the usage of other constructs (*save, create, output to parent/sender* ..) is recommend to avoid. Thus, the implementation overhead caused by the semantics of these language elements is reduced.

Considering these approaches, we see two principle ways to improve the performance of automatically generated implementations:

(1) to support the implementation process (restricted use of certain language features, new language concepts that provide better implementations, appropriate specification styles), and

(2) by improving the implementation techniques (mapping strategy, hardware/software codesign, variable implementation models).

In this paper, we consider aspects of both ways for the standardized FDT Estelle [ISO89]. In particular, we discuss how semantic constraints can influence the efficiency of the generated code. In the first part of the paper we discuss how certain language features may have a restraining effect on the performance of the implementation. The second part of the paper investigates the application of more efficient mapping strategies in the implementation models of the transformation tools. It discusses the use of different process models (server model, activity thread model). We show how the more efficient activity thread model can be used and discuss the constraints for its application to Estelle. The efficiency gain received by applying this technique is shown for implementations of the protocols XDT and XTP.

2 INFLUENCE OF THE SEMANTIC MODEL OF ESTELLE

The derivation of implementations from a formal description requires measures to guarantee the conformance between specification and implementation, i.e. to assure that the verified design is correctly implemented. The straight-forward approach to do this is by exactly implementing the semantics of the respective FDT [Held95], [Gotz96]. This reduces the probability of implementation errors and decreases validation efforts. On the other hand, the semantics of the FDTs require an additional implementation effort that result in an overhead compared to hand-coded implementations. In some cases this overhead could be avoided, because it is not needed for the description of the problem. For example, the dynamic generation of a module in Estelle requires the scheduling of the module according to the parent/children priority principle. To reduce this overhead it is necessary to be familiar with the impact of certain language features on a later implementation. In this section, we give two examples.

2.1 Module attributes

The module attributes play a significant role in an Estelle specification. They define the position and the behaviour of the modules within the module hierarchy, thus determining their relations to other modules as well as the manner how their transitions are selected for firing. The selection depends on the requirements of the application to be specified, but sometimes different choices are possible, i.e. it is the subjective decision of the specifier. The attributes selected may have an influence on the efficiency of the derived implementation. The reason for this is that according to the operational semantics of Estelle the modules are differently handled when selecting the next transitions for firing.To demonstrate this effect we have implemented the XTP protocol [Prot92] based on the Estelle specification of [Budk93] with different attributes. For this experiment, the Estelle specification was embedded into a test environment. This test environment contains at the second level of the module hierarchy the modules: *Application, XTP* and *Virtual Medium (*in the following we call these modules main modules). We have considered the following 4 variants:

- Variant A *(systemprocess)*:specification *non-attributed*, main modules *system-process*, child modules *process*
- Variant B *(process)*: specification *systemprocess*, main modules *process*, child modules *process*
- Variant C *(systemactivity)*: specification *non-attributed*, main modules *systemactivity*, child modules *activity*
- Variant D *(activity)*: specification *systemactivity*, main modules *activity*, child modules *activity*

For the experiments, we used the connection-oriented unconfirmed data transfer service of XTP. We did not observe any changes of the service for these four attributations, i.e. no violation of the semantics of the specification was detected for this service. Due to the high complexity of the specification the influence of these changes of the attributes for the whole specification was not studied in detail for this experiment. The execution times measured for these alternative attribute variants are represented in Figure 1.

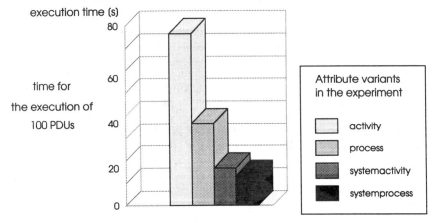

Figure 1   Impact of Estelle attributes on the execution time of an implementation.

The specifications were implemented by means of the PET/DINGO tool [Sije93] on a SUN SPARC 10 with the SunOS 4.1.3. operating system. The modules were mapped as follows: all modules up to the second hierarchy level on operating system processes, all modules below on procedures of the respective parent modules. The results of the measurements point out that the performance loss caused by the semantic constraints (compare the attribute variants *process* with *systemprocess* and *activity* with *systemactivity*). The nondeterministic execution of transitions in modules attributed with *activity* causes a performance loss compared to the *process* variant. Note that the objective of these experiments was to demonstrate the influence the selection of an attribute may have on the implementation. The results depend of course on the implementation tool used and cannot be generalized. However, they point out that specifiers and implementors should prove their decisions on the influence they have on the further development steps.

2.2   Description and execution of transitions

The selection of the transitions for firing is defined by the operational Estelle semantics [ISO89]. It implements the parent/children priority principle and determines the execution of the specification. The used algorithm is very complex. It consists of two passes, one for determining the firable transitions and one for determining the executable ones. As shown in [Held95], up to 90% of the runtime can be spent on this selection. This overhead can be decreased by about 50% by an optimized algorithm that avoids the buffering of the fireable transition and thus needs only one pass.

The approach for selecting transitions in Estelle results in further consequences for the description and execution of transitions, which influence the execution time:

(1) The successor state must be known before executing a transition.

(2) All transitions of a module are considered for determining the
       firable transitions.

In SDL, for comparison, this is not required. The successor state is determined during the execution of the transition and all processes run concurrently. We illustrate the consequences of these differences for the execution time by means of the following example:

An event E1 is supposed to be executed in state A. For this, the data belonging to this event E1 have to be decoded and the checksum has to be calculated. If the checksum is correct, the entity changes in another state, otherwise it remains in the same state and demands the repetition of the transmission. The related specifications in Estelle and SDL are depicted in Figure 2.

The execution of event E1 (t_{trans}) requires the firing of 2 transitions in Estelle, whereas in SDL only one transition is needed. In addition, the transitions in Estelle have to be selected for execution according to the above mentioned algorithm (t_{sel}). In SDL the first firable transition is executed (on average $t_{sel}/2$). Thus, we get under the assumption that the time for the execution of a transition t_{trans} is equal in Estelle and SDL the following relation for the selection and the execution of a transition:

$$t_{estelle} = 2t_{sel} + 2t_{trans} \; > \; t_{sdl} = t_{sel}/2 + t_{trans}$$

The Estelle specification needs in this example $t_{trans} + 1{,}5t_{sel}$ more time than the SDL specification. The difference is even larger when the transition belongs to a child module and the parent module executes arbitrary often (n times) transitions before the child module can fire the second one (parent/children priority principle). This situation is even more extreme for a child module of an *activity*-module, because this module has to be nondeterministically selected between all child modules of the parent module. For this case, we get the following relation:

$$t_{estelle} = 2t_{sel} + 2t_{trans} + n(t_{sel} + t_{trans}) \; > \; t_{sdl} = t_{sel}/2 + t_{trans}$$

```
trans                                        state A
  from state_A to same                         input E1
    when E1 (data)                               call decoding (data);
      begin     decoding (data);                 task cs := checksum (data);
              cs := checksum (data); end         decision cs:
trans                                            (false): (* request event E1 again *)
  from state_A to same                              nextstate _;
    provided cs = false                          (true): (* request next event *)
      begin (* request event E1 again *)            nextstate B;
      end
trans
  from state_A to state B
    provided cs = true
      begin (* request next event *)
      end
```

Figure 2   Comparison of an Estelle and an SDL specification.

2.3   Summary of the discussion

The derivation of an implementation from a formal specification is determined by the formal semantics of the given FDT. The discussion above has shown that features of the semantic model of Estelle (e.g. transition selection rule, parent/children priority principle, static definition of the successor state) may have an adverse effect on the performance of an implementation. The specifier may in part benefit from this knowledge by applying these features with care. On the other hand, a general renunciation of the dynamic creation of modules would strongly restrict the descriptive power of Estelle. The application of asynchronous modules as proposed in [Bred94] could reduce the efficiency loss.

Note that our discussion does not touch the requirement of the independence of the specification from the implementation. It will only show that certain language features due to their semantics can adversely influence an automated implementation, if this is intended. This knowledge may help the specifier in choosing alternative representations. It can also be used for replacing parts of the specification by semantically equal representations when preparing and refining the specification for implementation.

3   INFLUENCE OF THE IMPLEMENTATION MODEL

Formal description techniques possess formal semantics that guarantee the exact interpretation of the specification. Automatically deriving implementations from formal descriptions requires transformation rules, which preserve the semantics of the specification, i.e. the conformance of the implementation with the specification. This adds overhead to the implementation, which is not present with implementations manually derived from informal descriptions. Besides these transformation rules have to be defined during tool development, i.e. they cannot take into account the respective implementation context. The set of these transformation rules forms the implementation model of the tool [Held95]. It defines the structure of the implementation, the interfaces, the process model and the relation to the implementation environment. The decisive factor is the process model, which describes the manner how the specification is mapped on the process structure of the operating system. In manual coding two principle approaches are applied: the server model and the activity thread model [Svob89].

Applying the server model, each system module (*systemactivity* or *systemprocess*) of the Estelle specification is implemented as a server that processes events (e.g. interactions or spontaneous transitions). Due to the semantic similarities between EFSM-based FDTs and the server model, current FDT compilers resort to the server model, because it allows a straight-forward mapping of the FDT semantics. All Estelle compilers known to us apply the server model.

The activity thread model implements the entities as a set of procedures. Each procedure implements a transition of the FSM. An incoming event activates the respective procedure, which immediately handles the event and when producing an output calls the respective procedure of the next entity. The sequence of the inputs and outputs (input→output→input ... input→output ...) results in a sequence of procedure calls - the *activity thread*. (Note that the term *activity thread* does not refer to a thread of the operating system. It denotes the execution path an event takes through the protocol stack.)

The server model exhibits extra overhead, which is not present with the activity thread model, e.g. overhead for asynchronous communication including queuing operations and overhead for process management. The problem with the activity thread model is that it is based on a semantic model that differs considerably from the semantic models of EFSM-based FDTs. We have shown in [Henk97] that the activity thread model can be applied to derive more efficient implementations form SDL specifications. In the following we want to discuss whether this model can be applied for automatically deriving implementations from Estelle specifications. We first give an overview how the transformation can be done in principle. Thereafter we discuss the constraints of this mapping.

3.1   Derivation of Activity Thread Implementations

In order to apply the activity thread model rather than the server model to derive code from protocol specifications, the following differences between the two models have to be taken into account:
- The active elements in the server model are the protocol entities. In general, each entity is implemented by a thread (or process) of the operation system or the runtime environment. The execution sequence of the events is determined by the scheduling strategy of the operating system or the runtime environment. Communication is asynchronous via buffers.
- The activity thread model is event-driven. The execution sequence of the events is determined by the sequence of inputs and outputs as given in the specification. Thus, the activity thread approach supports an optimal execution sequence of the events. Communication is synchronous, i.e. no buffers are employed. Because of that, the model is well suited for the layer-integrating implementation of protocols.

Estelle specifications do not allow a direct mapping on activity threads. In the following we present the principle of a mapping strategy, which comes close to the
activity thread principle:
(1) Each Estelle module is transformed into a reentrant procedure that contains the
    executable code. With each procedure call, a single Estelle transition is executed.
    For each instantiation of an Estelle module, an Instance Control Block (ICB)
    is created. The ICB contains the following information:
    • all internal information (e.g. state information, local variables),
    • the addresses of the interaction points,
    • the address of the reentrant procedure.
    Each event is implemented by a structure containing the following data:
    • the event type,
    • the reference to the parameters of the event,
    • the next procedure to be called (address of the respective ICB).
(2) Each *output*-statement of an Estelle module is replaced by the call of the respec-
    tive procedure, which implements the Estelle module that receives the event.
(3) For the whole specification, the following frame procedure is generated:

```
procedure AT (incoming_event)
        next_event := incoming_event;
        while not_empty(next_event) call_next_procedure (next_event);
```

The procedure *AT* is called by an incoming event. The procedure *AT* calls the procedures implementing the respective Estelle modules according to the structure of their
output-statements. Output events sent to another Estelle module are stored in the variable next_event. The loop of the procedure *AT* is terminated when next_event is
empty, i.e. the transition does not contain an output event. The termination of the
loop also completes the activity thread, and the next incoming event can be handled.
This frame procedure is necessary to map the state-oriented presentation of Estelle
specifications onto the event-driven representation characterizing activity thread implementations. (Otherwise the specification has to be rewritten in an event-oriented
style, which requires a new validation of the specification.)
    The mapping strategy outlined above is depicted in Figure 3. It shows how the
Estelle specification is mapped on the main procedure *AT*, which controls the execution of the activity thread, and the output event on procedure calls.

3.2   Problems and Constraints with Activity Threads

Applying the mapping strategy described above, some problems with the Estelle semantics arise. In the following we provide solutions how to preserve the semantics
of the specification and discuss constraints existing for the application of the technique, respectively. We first discuss two problems, which are of general nature to
this approach. After that we consider the problems, which specifically concern the
Estelle semantics.
    The general problems are:
    - transitions performing an action after an output, and
    - transitions with multiple outputs.

Estelle specification   Activity Thread Implementation

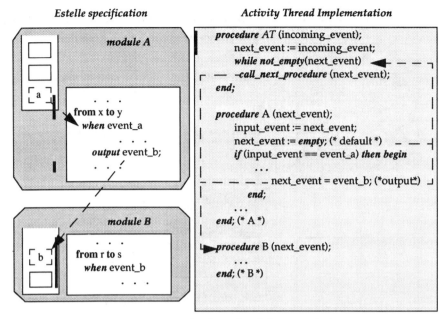

Figure 3 Basic strategy for the derivation of activity threads from Estelle specifications.

- **Transitions performing an action after an output**

The simple replacement of an output statement by the call of the respective procedure (implementing the Estelle module receiving the event) may violate the atomicity of transitions. An example for a possible violation of the atomicity of a transition is given in Figure 4. The figure shows two communicating modules, L2 and L3. For example, let us assume that L3 implements a routing function. An event e1, received by module L2 is routed to module L3. From L3 it is retransmitted to L2 and output as e2 to the medium. Within L2, the variable *dir* is employed to store the direction, in which the last event has been transmitted. The example shows that statements after the *output*-statement may produce a different behaviour as specified.

Our approach is to defer the procedure call until the transition has been completed. To ensure this the output event are stored in the variable next_signal of the frame procedure *AT* in section 3.2. The procedure *call_next_procedure* calls the respective procedure after the transition has been completed. From the viewpoint of the communicating modules this represents a transformation of synchronous communication to asynchronous communication. This is because the procedure implementing the receiving module is no longer called by the sending module itself. Instead, the procedure is called by the procedure *AT*.

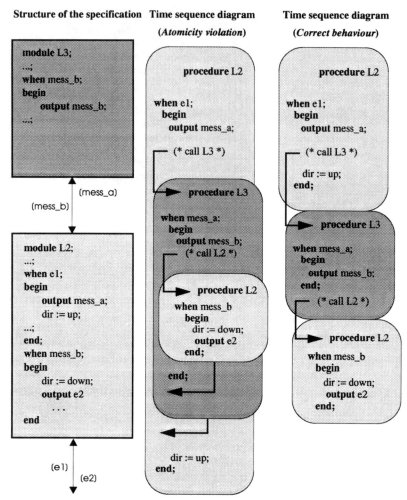

Figure 4   Example for the possible violation of the atomicity of a transition.

- **Transitions with multiple outputs**

A transition may contain several *output*-statements. In this case, a sequential call of the procedures, which implement the Estelle modules the events are sent to, is not always possible. Figure 5 shows an example for this. In the figure, the sequential execution of the two outputs of module P2 results in an overtaking of mess_c by mess_e. In other words, mess_e, which results from mess_d, reaches module P4 before mess_c. This clearly violates the semantics of the Estelle specification.

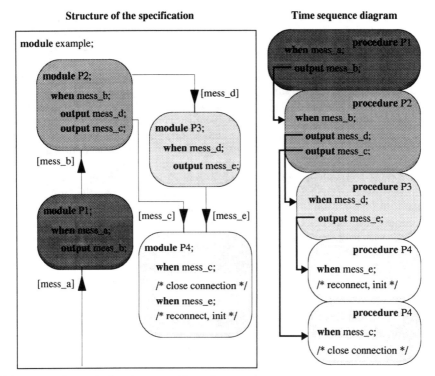

Figure 5   Example for the overtaking of events.

An approach to handle this problem is to split the activity thread in several activity threads. Thus, an additional activity thread is created for each additional output. To handle these additional activity threads, we have devised two process models, namely the *basic activity thread* model (BAT model) and the *extended activity thread* model (EAT model). The BAT model handles the additional activity threads quasi-parallely, whereas the EAT model processes them truly concurrently. As shown in [Henk97] BAT model implementations have in general proved to be more efficient, because they avoid additional communication and organizational overhead. Therefore, we focus on the BAT model in this paper.

For BAT model implementations the frame procedure *AT* is modified as follows:

```
procedure BAT (incoming_event)
      put incoming_event in AT_list
      while (list_not_empty(AT_list)) do
              for_all_elements AT in AT_list do begin
                    next_event := event_of(AT);
                    next_event := call_next_procedure (next_event);
                    if empty(next_event) then remove AT from AT_list;
              end
```

BAT model implementations handle only one incoming event per time, i.e. incoming events are always processed sequentially. No interleaving of the processing of incoming events is possible. Subsequent incoming events are blocked until the previous events have been processed.

Multiple outputs in a transition are handled by the creation of additional activity threads. One additional activity thread is created for each *output*-statement within a transition. These additional activity threads are managed by the AT_list. The management of the activity threads in the AT_list guarantees an execution order that is in accordance with the semantics of the specification. The AT_list is used to schedule the additional activity threads. Note that after the arrival of an incoming event, the AT_list contains exactly one activity thread. Several activity threads only emerge when a transition with multiple outputs is executed. In this case, the activity threads in the AT_list are scheduled according to the Round Robin principle. Only one procedure, i.e. one transition, is executed per iteration of the loop. Thus, the activity threads are executed in a quasi-parallel manner. When an activity thread terminates, it is removed from the AT_list. The main procedure *AT* returns after the last activity thread has terminated.

Now we discuss the semantic problems that are more specifically related to the Estelle semantics.

• **Parent/children priority principle**

With the activity thread model, the sequence of execution of the Estelle modules (schedule) is determined by the output executed by the respective transitions. Conversely, the Estelle semantics employs a complex algorithm to schedule the next transitions to be executed. This contradiction between the activity thread model and the semantics of Estelle can be only solved if the Estelle specification follows one of the following rules:
  - within a subsystem, i.e. a system module with all of its descendants, at most one transition may be executable at a time or
  - the child modules of *systemprocess* or *process* modules may not offer an executable transition at the same time any of its predecessor modules offers an executable transition.

This ensures that an executable transition of a child module can be actually executed instead of its parent, which would not be the case if the parent/children priority principle were applied.

In order to follow these rules, different approaches exist:
  - If a system module has descendants, only the leaves of the hierarchy may be active. In case of descendants of type *systemactivity* or *activity*, only one child may exist (see Figure 6a).[*]
  - The Estelle specification contains only system modules or equivalently, there are only asynchronous modules as proposed in (Bred94) (see Figure 6b).

---

[*] The activity thread model does not allow a nondeterministical selection of the next procedure to be called.

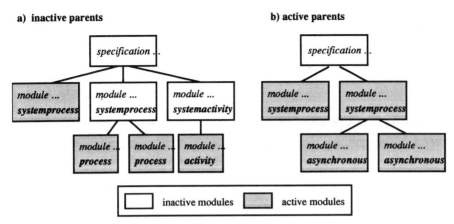

a) inactive parents

b) active parents

Figure 6 Examples of Estelle specifications.

In case the transition has a simple structure, i.e. input→action(s)→output, the transformation strategy described above can be applied. Thus, each output can be directly translated to a procedure call. transitions without output.

- **Spontaneous transitions**

Estelle allows spontaneous transitions, i.e. transitions that are not triggered by an event in a *when*-clause. Transitions with no incoming event cannot be implemented by an activity thread, since there is no event to be mapped on a procedure call. Spontaneous transitions have, therefore, to be removed during the refinement of the specification when preparing for implementation, or an alternative specification strategy has to be applied. If spontaneous transitions are used to model internal events, such an replacement can be usually easily done. The replacement of delayed transitions, however, is more complicated. In this case another specification strategy has probably to be considered. A possible solution could be to introduce timer modules, which are implemented using the *delay*-clause, as they are needed for real-life protocol implementations. Start and reset of the timer as well as a timeout can then be handled as incoming events and mapped on procedure calls.

4 QUANTITATIVE COMPARISON OF THE APPROACHES

In order to evaluate the performance improvements achievable with activity threads, we have manually applied our approach to derive code from XTP (Prot92) and the XDT protocol (eXample Data Transfer) [Koen96] XDT is an example protocol, which is used for teaching purposes in protocol engineering. It is a connection-oriented data transfer protocol supporting implicit connection establishment and data transfer based on the *go-back-N* principle. The XDT specification comprises about 350 lines. The Estelle specification of XDT follows the rules given in section 3.2. Thus, it supports the derivation of code according to the BATmodel.

In order to derive code from XTP, two specification variants have been used. The original XTP specification (rel. 3.6) we used comprises about 7900 lines (Budk93).

In order to apply the BAT model to derive code from the XTP specification, the Estelle specification has been adapted to follow the rules given in section 3.2. The adapted XTP specification is called aXTP in the following. Since the adaptation of the complete XTP specification to suit the BATmodel would have been an enormous effort, aXTP only supports the unconfirmed connection-oriented data transfer service. All other services have been removed from the Estelle specification. The resulting aXTP specification comprises 1000 lines. Deriving the aXTP specification from the original XTP specification, the spontaneous transitions have been replaced by introducing additional events in the input transitions. Note that the semantics of the unconfirmed connection-oriented data transfer service has not been changed by the modifications.

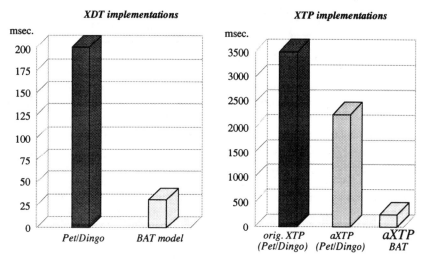

Figure 7   Quantitative comparison of the server model and the activity thread model.

For our experiments, we have derived code according to the BAT and the server model. For the derivation of the implementations according to the server model, we have used the PET/DINGO tools (Sije93). All the measurements were taken for a single connection of the respective protocol. The implementation platform has been a SUN SPARC 10 running the *Solaris* operating system (rel. 2.5). For the measurements, a test environment has been implemented that comprises the service user, which generates the data stream and measures its duration, and a virtual medium (i.e. an Estelle module) for realizing the communication between the protocol entities. The size of the PDUs used in the experiments was 512 bytes for XDT and 64 bytes for XTP. The XTP service we used was the connection-oriented unconfirmed data transfer. The measurements do not take into account the initialization phase. We have measured the times to process 50 PDUs by both peer entities. The results of the measurements are given in Figure 7. For XTP, the measurements for the original version as well as the aXTP version are given.

The measurements show that the activity thread implementations achieve a considerable speedup compared to the respective server model implementations. Compared to the server model implementation, the BATmodel implementation of aXTP exhibits a larger speedup as the respective implementation of the XDT protocol (i.e. a factor of 13.8 for aXTP and a factor of 6.7 for XDT). The difference results from the larger number of Estelle modules employed by aXTP, i.e. 17 versus 5 modules of XDT. This results in a larger number of queuing operations needed for the server implementation of aXTP.

5 CONCLUSIONS

The performance of implementations automatically derived from formal specifications depends on many factors. The inefficiency of current implementations is in part caused by the overhead involved with the implementation of the FDT semantics. In the paper, we have described how these semantical constraints can influence the performance of implementations derived from Estelle specifications. This impact concerns both the application of certain language features and the use of efficient mapping strategies.

In Estelle, additional overhead is caused by the hierarchical approach to select an executable transition. We have presented measurements for XTP, which show that a speedup can be achieve if the attributing principle employed in the specification is chosen with care. In addition, we have analyzed the times to select an executable transition and have compared them with the respective times needed in SDL. The comparison shows that the semantics of Estelle add some extra overhead, which is not present in SDL.

In the main part of the paper, we have investigated the application of the activity thread implementation strategy elaborated in [Henk97] to Estelle. Our measurements for the two protocols XTP and XDT as well as the measurements made for SDL in [Henk97] show that with the activity thread approach a considerable speedup can be achieved. However, compared with SDL, which has only some minor semantic constraints to this approach, the application of the technique to Estelle requires severe limitations, which reduce the descriptive power of the language (as, for instance, a renunciation on the parent/children priority principle). Therefore, we currently focus on the implementation of a prototype compiler for SDL. On the other hand, automated implementation techniques from formal descriptions will be only applied in practice, if the efficiency of the generated code is close to the hand-coded one. Experiment reports like this one may help that FDTs move towards a better support of efficient implementation techniques.

6 REFERENCES

Ahlg96] Ahlgren B., Bjorkman,M.;Gunningberg P.: Integrated Layer Processing can be hazardous to your performance. In Dabbous,W.; Diot,C. (eds.): Protocols for High-Speed Networks V. Chapman&Hall, 1996, 167-181.

[Brau96] Braun, T. et al.: ALFred - An ALF/ILP Protocol Compiler for Distributed Application Automated Design. Rapp. de Recherche No. 2786, INRIA, 1996.

[Bred94] Bredereke J., Gotzhein R.: Increasing the concurrency in Estelle. In Tenney R. L., Amer P. D., Uyar M. Ü. (eds.): *Formal Description Techniques VI*, Elsevier Science Publishers,1994, pp. 26-29.

[Budk93] Budkowski S. (ed.): Formal Specification, Validation and Performance Evaluation of the Xpress Transfer Protocol (XTP). INT Evry (France), Research report No 931004, 1993.

[Clar89] Clark D.D. et al.: An Analysis of TCP-Processing Overhead. *IEEE Communications Magazine*, June 1989.

[Clar90] Clark D.D., Tennenhouse D.L.: Architectural considerations for a new generation of protocols. *ACM SIGCOMM*, 1990, pp. 200-208.

[Fisc94] Fischer S., Hofmann B.: An Estelle Compiler for Multiprocessor Platforms. In Tenney R. L., Amer P. D., Uyar M. Ü. (eds.): *Formal Description Techniques VI*, Elsevier Science Publishers,1994,pp.171-186.

[Gotz96] Gotzhein R. et al: Improving the Efficiency of Automated Protocol Implementation Using Estelle. Computer Communications 19 (1996), pp. 1226-1235.

[Held95] Held T.; Koenig H.: Increasing the Efficiency of Computer-aided Protocol Implementations. In Vuong S., Chanson S. (eds.): *Protocol Specification, Testing and Verification XIV*, Chapman & Hall, 1995, pp. 387-394.

[Henk97] Henke R., Koenig H., Mitschele-Thiel A.: Derivation of Efficient Implementations from SDL Specifications Employing Data Referencing, Integrated Packet Framing and Activity Threads. Accepted for *SDL Forum 97*, Evry, Sept. 1997 (to be published by Elsevier Publishers).

[ISO89] ISO IS 9074: Estelle - A Formal Description Technique Based on an Extended State Transition Model, 1989.

[Koen96] Koenig, H.: eXample Data Transfer. Technical Report 4/96, Department of Computer Science, BTU Cottbus, 1996.

[Leue96] Leue,S.,Oechslin,P.: On Parallelizing and Optimizing the Implementation of Communication Protocols.IEEE/ACM Trans. on Networking, 4(1), Febr. 1996.

[Plat96] Plato R., Held T., König H.: PARES - A Portable Estelle Translator. In Dembinski P., Sredniawa M. (eds.): *Protocol Specification, Testing and Verification XV*, Chapman & Hall, 1996, pp. 383-399.

[Prot92] Protocol Engines Inc.: Xpress Transfer Protocol Definition. Revision 3.6, January 1992.

[Sije93] Sijelmassi R., Strausser B.: The PET and DINGO Tools for Deriving Distributed Implementations from Estelle. Computer Networks and ISDN Systems 25 (1993), pp. 841-851.

[Stei93] Steigner C., Joostema R., Groove C.: PAR-SDL: Software design and implementation for transputer systems. Transputer Applications and Systems '93, R. Grebe et al. (Ed.), vol. 2, IOS Press.

[Svob89] Svobodova L.: Implementing OSI Systems. IEEE Journal on *Selected Areas in Communications*, 7(7),1987, pp. 1115-1130.

[Tele96] Telelogic Malmö AB. SDT Cmicro Package - Technical Description, Telelogic, 1996.

[Wats87] Watson R.W., Mamrak S.A.: Gaining Efficiency in Transport Services by Appropriate Design and Implementation Choices. ACM Transactions on Computer Systems. 5(2),1987, pp. 97-120.

## PART EIGHT

Industrial Usage Reports

26
Using a Formal Description Technique to Model Aspects of a Global Air Traffic Telecommunications Network

J. H. Andrews
N. A. Day
Dept. of Computer Science, University of British Columbia
Vancouver, BC, Canada V6T 1Z4
tel (604)822-3061 fax (604)822-5485
{jandrews, day}@cs.ubc.ca

J. J. Joyce
Hughes Aircraft of Canada Limited
#200 - 13575 Commerce Parkway
Richmond, BC, Canada V6V 2L1
tel (604)279-5721 fax (604)279-5982
jjoyce@ccgate.hac.com

Abstract

Aspects of a draft version of the Aeronautical Telecommunications Network (ATN) Standards and Recommended Practices (SARPs) under development by ISO-compliant committees of the International Civil Aviation Organization (ICAO) have been mathematically modelled using a formal description technique. The ATN SARPs are a specification for a global telecommunications network for air traffic control systems. A version of Harel's statecharts formalism embedded within a machine readable typed predicate logic has been used as a formal description technique to construct this mathematical model. Our model has been 'typechecked' to partially validate the internal consistency of the

Formal Description Techniques and Protocol Specification, Testing and Verification
T. Mizuno, N. Shiratori, T. Higashino & A. Togashi (Eds.)   © 1997 IFIP. Published by Chapman & Hall

specification. The work described in this paper has already uncovered some problems in the draft SARPs, and will provide a basis for follow-on efforts to apply formal analysis methods such as model-checking and symbolic execution to aspects of the ATN SARPs. The success of this approach suggests that typed predicate logic is useful as a syntactic and semantic foundation for specialized Formal Description Techniques (FDTs).

Keywords
Formal description techniques, practical experience, extensions of FDTs, state transition systems, typed predicate logic, OSI application layer, verification and validation.

1   INTRODUCTION

This paper describes the modelling of aspects of the Aeronautical Telecommunications Network (ATN) using the formalism known as 'statecharts' (Harel, 1987) and predicate logic. This effort was performed by workers at Hughes Aircraft of Canada Limited (HACL), the University of British Columbia (UBC) and the University of Victoria (UVic). It is part of the FormalWare project, jointly funded by the BC Advanced Systems Institute, HACL, and MacDonald Dettwiler.

The ATN is a global system under development which will allow aircraft and ground stations to exchange data for the purpose of air traffic control. The various software components of the ATN reside in aircraft or ground station computers, and interact with human users and with each other to perform this data exchange. The communications protocols used by the software components are defined in ICAO documents referred to as Standards and Recommended Practices (SARPs) (SARP, 1996).

This modelling effort consisted of writing textual descriptions of components of the ATN using a formal description technique. There were two goals for this effort: first, to help validate that the SARPs protocols are safe (do not lead to deadlocks or livelocks, for instance); and second, to provide a formal description of the SARPs which can potentially act as a basis for validating implementations of the ATN. The first phase of the effort consisted of writing and typechecking an extensive draft of the model, and doing some informal validation; some problems in the draft SARPs were identified as a result of this work. This paper reports on the first phase, which was done in November and December 1996.

Among the more novel aspects of this work is the use of typed, predicate logic as a foundation for a more specialized formal description technique, namely a version of Harel's statecharts formalism.   Although the 'semantic embedding' of specialized notations within typed, predicate logic is reasonably well-known by formal methods researchers, the effort reported in this paper provides some evidence of the practical benefits of this approach in addition to the more theoretical benefits such as clarifying the semantics of specialized notations. By placing statecharts within a general-purpose environment, we were also able to integrate parts of the specification written in predicate logic itself.

This paper is organized as follows. Section 2 gives background on the tools and methodologies used. Section 3 explains the overall strategy for the modelling effort. Section 4 describes the simplifying assumptions that were made in creating the model to 'abstract out' implementation details. Section 5 discusses the assumptions that had to be made in order to deal with problems identified in the draft SARPs. Section 6 presents the results of the effort. Section 7 discusses the effort planned for future phases of the project. Finally, in Section 8 we review some lessons learned from this effort with respect to the use of formal description techniques.

2   TOOLS AND METHODOLOGIES USED

This section describes the tools and methodologies used in the modelling effort. The statecharts formalism (Harel, 1987) was used to describe the system in terms of parallel state decomposition and state-transition diagrams. 'S' (Joyce, 1994) is a formal description notation which we used to express statechart descriptions as well as other parts of the specification that are more suitably described in predicate logic. 'Fuss' is a typechecking tool for S specifications.

There were a number of factors which contributed to our not using more commonly-known tools such as the Concurrency Workbench (Cleaveland, 1989) or those available for Estelle, LOTOS and SDL (Turner, 1993). First, the specification that we were working from is a combination of text in paragraphs and an informal state transition model given in tables, with explicitly-named states, making the statecharts notation particularly suitable. Second, we wanted to do model checking, rather than simply discrete event simulation, to demonstrate properties of our formal models. Third, many of the available tools for model checking work from the system as a single finite state machine. The nature of the ATN means that the system expressed as a flat finite state machine would have millions of states. We did not want to exclude the possibility that the structure of a hierarchical specification can be exploited to reduce the size of the state space in analysis. Finally, the conditions under which state changes take place in the ATN are relatively complex, and we needed a general logical notation to allow us to express them naturally and accurately. This is where it was advantageous to use predicate logic itself.

Since we did not find this particular constellation of needs to be met by any one tool, we felt that it was most advantageous to us to use and extend a set of tools and methodologies with which members of the team had expertise. Analysis methods such as model checking are under development within this framework.

2.1 Statecharts

Statecharts are described by their inventor, David Harel, as a 'visual formalism' (Harel, 1987). There is precedent in the air traffic control industry for using such formalisms; TCAS II (Traffic Alert and Collision Avoidance System) was formally specified using the Requirements State Machine Language, a notation which is closely related to statecharts (Leveson, 1994).

In the statecharts formalism, a system is described in terms of states and transitions between those states. In this sense, it is like the 'state transition diagram' formalism. However, a statechart state can be more than a state in a state transition diagram.

A statechart state is either a 'basic state', an 'AND-state', or an 'OR-state'. Basic states correspond to the states of a state transition diagram. AND-states represent parallel composition and OR-states represent hierarchical state transition diagrams.

Although there are advantages to graphical representations of statecharts, especially for presentation, we decided to produce the initial model in a textual form, for reasons including portability and ease of integration.

Unfortunately, a number of semantically different versions of statecharts have emerged since the original informal description of the semantics of statecharts given by Harel. For this work, we have used a particular version of this formalism which has a formal semantics defined in a machine-readable format (Day, 1993). Whereas the behaviour of some statechart based tools are not clearly specified, these explicit semantics are used to directly initialize general-purpose analysis tools.

2.2 S - a machine readable notation based on typed, predicate logic

We used the formal description notation called S to represent statecharts textually. This made it possible to integrate the statechart parts of the model easily with predicate logic, as a means of specifying the details of complex state transitions, and with the text-based approach to requirements management followed at HACL. Through the use of parameterization, we were able to reduce the size of the specification and make it easily extensible without   complicating the semantics of statecharts.

S allows us to declare elements such as types, constants, functions and predicates that are left 'uninterpreted'. This contrasts with software (as well as some 'simulation-oriented' formal description techniques) where ultimately everything must be refined, either manually or by means of a compiler, into bits and executable machine code.

Like several other formal description notations, S is based on typed predicate logic. However, in contrast to such notations as Z (which requires the use of an intermediate mark-up notation or other means of handling the specialized symbols and graphical presentation format of Z), it uses a more readable syntax for non-formal methods experts, which emphasizes letters and punctuation characters rather than symbolic characters. It also tends to use common English words (like 'function' and 'select') as keywords, rather than the technical terms (like 'lambda' and 'epsilon') used in other specification languages. Furthermore, the ASCII based syntax of S simplifies the mechanics of integrating formal descriptions into engineering documentation in contrast to notations which involve specialized symbols or graphical presentation formats.

2.3 Fuss

Fuss is a typechecking program for S, roughly corresponding to 'lint' for C. S is a strongly typed specification language, in the sense that the formal and actual parameters of a function call must be of exactly the same type, with no 'typecasting' allowed. In addition, functions can take other functions as parameters, types can be declared in terms of other types, and functions can be declared as taking different patterns of types. This expressiveness makes a rich hierarchy of types available to the user. The Fuss tool checks that the user's specification is well-typed, and also implements a 'type inference' algorithm in the style of the programming language ML, which infers (wherever possible) a precise type for any object for which we have not given an explicit type.

3   MODELLING STRATEGY

The overall strategy used was to model each software entity within the ATN, and each module and 'status' state within an entity, as a statechart state. This section first describes the structure of the ATN as presented in the SARPs, and then discusses how the various aspects of that structure were modelled using statecharts and S.

3.1 Structure of the ATN

The ATN is specified in the SARPs as a set of components interacting via messages. The components are not required to be implemented as separate processes or even as separate objects at the code level, but their behaviour must be consistent with the message-passing model.

Each component we modelled is further specified as a state transition system. Each transition between states is associated with a triggering event and a condition which must be true if the transition is to be followed when the event

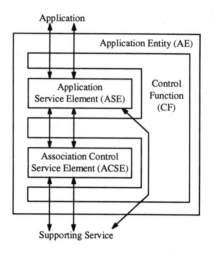

Application

Figure 1: Internal structure of Application Entity

occurs. Each transition also has an associated action which is performed when the transition is followed. Typically, the trigger has to do with the message received and/or current variable settings, and the action is to send a message and/or change the variable settings.

The top-level components of the ATN that human users interact with are referred to simply as the *applications* or *user applications*. These do not communicate directly with each other; rather, they use a number of *Application Entities (AEs)* found in the OSI application layer to provide them with communication services. The four types of AEs modelled in the present effort are the *ADS* (Automated Dependent Surveillance), *CM* (Context Manager), *CPDLC* (Controller Pilot Data Link Communication), and *FIS* (Flight Information Service). There are two versions of each type of AE, a *ground* version which resides in ground stations and an *air* version which resides in aircraft. The AEs communicate with each other via the *supporting service*.

As shown in Figure 1, each AE in turn consists of three entities. The *Application Service Element (ASE)* performs the duty of receiving messages from the application and translating them into OSI-standard messages. The *Association Control Service Element (ACSE)* allows its AE to form associations with other ('peer') AEs. The *Control Function (CF)* mediates all communication amongst the ASE, the ACSE, the application and the supporting service. Each type of AE contains a unique type of ASE, but the CF and the ACSE are the same across all types of AEs.

The SARPs consist of on the order of 1000 pages of text, containing detailed specifications of the four types of ASEs and the CF, along with requirements on the lower OSI layers and various less formal guidelines documents. The ACSE is described in a separate 40-page document, ISO 8650 (ISO, 1994).

3.2 Entities and Status States

The entities modelled were the CF, the ACSE, the CM and ADS ASEs, and the supporting service. The CF and the ASEs are all specified in the SARPs in terms of tables which informally describe a state transition system, as is the ACSE in its specification. The supporting service can also be expressed as a simple state transition system. Each entity was therefore modelled as a statechart OR state, and the resulting models were put together as an AND-state. The decomposition of the task into one state per entity also allowed the work to be distributed to workers and integrated more smoothly.

The CF, the ACSE, the CM ASE, and each module of the ground ADS ASE can be in one of several 'status' states (idle, associated, awaiting response, etc.) as defined in the SARPs. Each of these status states was modelled by a basic statechart state, and these basic states were put together in an OR-state to define the overall module or entity.

Figure 2 illustrates this top-level state decomposition with an example system. Dashed boxes within the large box represent the substates of the overall AND-state; the internal state transition structure of these substates has not been illustrated. The example consists of an air and ground CM AEs, the supporting service over which they communicate, and an 'environment' state Env which will be used to stand in for the applications using the AEs.

Most of the components have parameterized names to eliminate duplicate specifications of the same behaviour. This is similar to the use of procedure calls

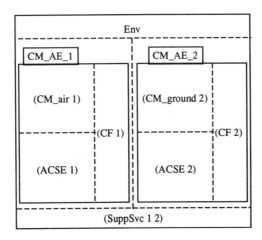

Figure 2: Statechart structure of example system

in a programming language. Thus, (CF 1) is the name of the CF state belonging to the first AE, and (SuppSvc 1 2) is the name of a supporting service connecting the first and second AEs. The text just gives a specification for (CF i) and refers to (CF 1), (CF 2), etc. without requiring new text to be written.

3.3 Transitions

Each module or entity in the SARPs is in one particular status state at any given time. It makes transitions between one status state and another depending on messages it has received, which are modelled as events, and on the results of tests that it makes. Each of these transitions was modelled by a transition in the statechart of the module or entity.

A large number of transitions is given in the SARPs for each entity, so most of the modelling effort went into formally defining these transitions.

3.4 Shared declarations

A file, called 'sc.s' and found in Figure 3, of shared declarations and definitions supports the modelling of statecharts in general. In sc.s, types are defined for statecharts, state names, transitions, and transition names. State names are declared separately from the state specifications. Messages are modelled as events. sc.s also defines term constructors which can be used to build up a statechart definition from basic states and transitions.

To support the task of modelling the SARPs, another S file named 'atncommon.s' was developed to contain declarations for the state names of the top-level ATN entities, and also for the ISO-standardized message types (e.g., 'A-ASSOCIATE request') used by all modules.

These two S files are 'included' by the S files containing the statecharts models in much the same way that a '.h' file may be included by a software module written in C. They provided a common foundation for the half dozen individuals directly involved in the authoring of statechart models, allowing them to work with considerable independence during the initial phase of this project.

4   SIMPLIFYING ASSUMPTIONS MADE

One of the primary benefits of creating a formal model of a software/hardware system is that we can focus on high-level aspects of the system that we are interested in studying, and 'abstract out' implementation details. This process of abstraction consists largely of making simplifying assumptions about the system in order to clearly isolate the aspects we are modelling.

```
%% Type declarations

%% Basic types
: stateName;
: event;
: simpleEvent;   %% Used for "messages"
: action;
: transName;

%% Transition type
: trans == transName # stateName # event # action # stateName;

%% Statechart type
: sc := OrState :stateName :stateName :(sc)list :(trans)list
      | AndState :stateName :(sc)list
      | BasicState :stateName;

%% Constructor declarations

%% Expressions
InState: stateName -> bool;

%% Events
En: stateName -> event;          %% Entering a state
Ex: stateName -> event;          %% Exiting a state
Ev: simpleEvent;                 %% Atomic event name
EvCond: event -> bool -> event;  %% Event and condition
(_ And_e _): event -> event -> event;   %% Both events
(_ Or_e _): event -> event -> event;    %% One or the other event
Tm: event -> num -> event;       %% Event at given time
%% Receipt of message with data from stateName
(:A) Receive: stateName -> simpleEvent -> A -> event;

%% Actions
No_action: action;               %% Null action
Gen: simpleEvent -> action;      %% Generate message
(:A) (_ Asn _): A -> A -> action;        %% Assign var a value
Both : action -> action -> action;       %% Do both actions
%% Broadcast of message with data to all substates of stateName
(:A) Send: stateName -> simpleEvent -> A -> action;
```

Figure 3: 'sc.s' declarations for statecharts in S

In our case, we were primarily interested in studying issues to do with the sequences of messages sent between the various ATN entities. We wanted to examine whether the message protocol as defined in the SARPs is safe (that it does not lead to deadlocks or livelocks, that it is complete and consistent, and so on). We also wanted to provide a formal definition of the high-level structure of the ATN and its protocols, which could be used as a basis for developing and testing the actual software. The simplifying assumptions we made reflect this focus:

• We assumed that the supported service was stable and error-free.

• We assumed that the translation of the data by the various entities did not affect the safety properties of interest to us, and therefore did not need to be modelled.

- Timers are often specified in the SARPs for such purposes as timing out dropped connections. We modelled timeouts of timers as messages sent from the environment, which could be sent at any time, rather than actually modelling time. For instance, a statement in the SARPs which specifies that a timeout will occur 30 seconds after a particular event will be modelled more generally as a timeout that could occur at any time.

Note that the simplifying assumptions were made not because we felt unable to manage the extra details; rather, they were made because in our judgement the extra details were not relevant to the properties we are trying to validate. We may discover that we cannot validate some property because some necessary detail is missing. If this does happen, we will then add the missing detail to the model. However, as long as our simplifying assumptions hold, then any property derivable from the formal description should also be true for the SARPs.

5   DISAMBIGUATING ASSUMPTIONS AND PROBLEMS WITH THE SARPS

In contrast to simplifying assumptions as just discussed, we also found it necessary to make additional assumptions which we have called 'disambiguating assumptions'. From a logical point of view, these assumptions are less 'safe' than simplifying assumptions in that we are not merely shaving away irrelevant detail. With disambiguating assumptions, we are adding necessary detail to the model that may or may not have been intended by the authors of the SARPs.

The effort reported here has revealed some ambiguities and lack of clarity concerning the handling of error conditions (for instance, when messages are received out of sequence) in the draft SARPs. The SARPs give somewhat ambiguous recommendations about what to do in a given error situation. For example, in the specification of the CF, the only substantial passage concerning the behaviour of the CF when a message is received out of sequence is the following: 'The error handling shall result in the association being aborted, if one exists, and a notification being given to the Application user.' This passage makes no mention of the fact that the CF must inform three different entities (the ASE, the ACSE, and the peer AE) of the abort of the connection, it does not describe the sequence or format for these messages, and it does not specify how notification of the abort is to be given to the Application user.

Because the SARPs were ambiguous, the people writing the formal specifications were not able to come up with a model which corresponded unambiguously to the SARPs. Implementations would have the same difficulty. It is observed by experts in software safety that software intensive systems often perform well when operating under normal conditions, but not when operating under unusual or error conditions. There is the potential in the SARPs that protocol errors which go undetected during validation will cause silent aborts of connections, error cascades, or similar problems. For instance, when the ACSE detects a malformed message, it is supposed to send an abort request both to its user and to its peer; but as soon as the CF passes on the first abort message, it goes into a state in which all subsequent abort messages from the ACSE are treated as protocol errors. Because of this, there is the possibility of a further error report from the CF, and the possibility that the peer will not know about the abort of the association. This in turn will at least cause other error conditions, and may have more serious consequences such as error cascades.

Hence we have had to make some disambiguating assumptions about what the SARPs mean. Our group developed an interim strategy for dealing with error conditions to allow the development of the formal model to continue. Since the modelling effort was done, a new version of the SARPs was released around December, 1996. The particular problem noted in the last paragraph remains in the latest version; we have communicated our concerns to the ICAO committee responsible for the development of the SARPs. When these problems are resolved, it should be reasonably straightforward to modify our model based on the resolutions provided.

6   RESULTS OF EFFORT

Component	# of states	# of trans	# of vars	Prior worker knowledge	Worker hours	Lines of S text
CF	5	96	2	Very high	24	680
ACSE	7	123	2	High	20	400
CM	13	201	2	High	44	1790
Ground ADS	22	232	1	Moderate	60	2850
SuppSvc	1	10	0	High	5	50
Misc Support					32	150
Total	48	662	7		185	5920

Figure 4: Results of effort

Figure 4 shows the results of the effort, in terms of the number of hours spent and number of lines of S text produced, according to the number of states, transitions and variables in the given component and the prior worker knowledge of the S formalism. The '# of states' in the column is the number of statechart states. It is not a measure of the complexity of the state space for analysis, but rather a rough measure of the inherent complexity of the specification. The number of hours also includes the time taken to perform static checks for completeness and consistency and integration. All workers were graduate students or faculty; 'very high' prior knowledge means knowledge of S in

```
% A normal transition for the ACSE (no conditions).
% Normally called as "inMessage.(ACSE_TRANS ...)" in
% order to emphasize message.
ACSE_TRANS i sourceState outMessage
           (destState: stateName -> stateName)
           inMessage :=
  ( (PTrans ((ACSE i).sourceState) inMessage),
    ((ACSE i).sourceState),
    (Receive (CF i) inMessage (ACSEData i)),
    (Send (CF i) outMessage (ACSEData i)),
    ((ACSE i).destState)
  );
```

Figure 5: A customizing declaration in S (from ACSE model)

particular, 'high' means knowledge of typed logic but not S in particular, and 'moderate' means only knowledge about first order logic.

Figures 5 and 6 show some sample text from the resultant specification. Figure 5 shows a typical 'customizing' declaration in S; like a declaration of an auxiliary function in a programming language, this declaration allows the rest of the specification to be more compact. The function 'ACSE_TRANS' maps five parameters, 'i', 'sourceState', 'outMessage', 'destState' and 'inMessage' to an instance of a transition denoted by a 5-tuple of the form (transition label, source state, event/condition, action, destination state). Figure 6 shows a typical section of the specification of the ACSE which lists the transitions from a particular state. In the definition of 'Transitions_From_Awaiting_AARE', 'ACSE_TRANS' is used within a let-definition to introduce a local name for a function called 'TRANS_CELL'.   In the let-definition of 'TRANS_CELL', the function

```
Transitions_From_Awaiting_AARE i :=
    /* From Awaiting_AARE state (STA1) */
    let Error_Cell := (ACSE_error i Awaiting_AARE) in
    let TRANS_CELL := (ACSE_TRANS i Awaiting_AARE) in

    [ /* Making connection */
      A_ASSOCIATE_req      . Error_Cell;
      A_ASSOCIATE_rsp_pos . Error_Cell;
      A_ASSOCIATE_rsp_neg . Error_Cell;
      P_CONNECT_ind       . Error_Cell;
      P_CONNECT_cnf_pos    .
          (TRANS_CELL A_ASSOCIATE_cnf_pos Associated);
      P_CONNECT_cnf_neg    .
          (TRANS_CELL A_ASSOCIATE_cnf_neg Idle);
      /* Releasing connection normally */
      A_RELEASE_req        . Error_Cell;
      A_RELEASE_rsp_pos    . Error_Cell;
      A_RELEASE_rsp_neg    . Error_Cell;
      P_RELEASE_ind        . Error_Cell;
      P_RELEASE_cnf_pos    . Error_Cell;
      P_RELEASE_cnf_neg    . Error_Cell;
      /* Releasing connection abnormally */
      A_ABORT_req          . (TRANS_CELL P_U_ABORT_req   Idle);
      P_U_ABORT_ind        . (TRANS_CELL A_ABORT_ind     Idle);
      P_P_ABORT_ind        . (TRANS_CELL A_P_ABORT_ind   Idle)
    ];
```

Figure 6: Typical section of ACSE model

'ACSE_TRANS' is partially evaluated when it is applied to two values, 'i' and 'Awaiting_AARE', as arguments for the first two of the five parameters of 'ACSE_TRANS'. This yields a local function, 'TRANS_CELL' which is used to denote transitions that always originate from the state 'Awaiting_AARE'. 'TRANS_CELL' is parameterized by the remaining three parameters of 'ACSE_TRANS', namely, 'outMessage', 'destState' and 'inMessage'. This use of functions results in a more concise, and potentially easier to understand description. Our use of the S notation provides much the same expressiveness as a general-purpose functional programming language.

Models have been completed for the CF, the ACSE, the air and ground CM ASEs, the supporting service, and part of the ground ADS ASE, incorporating five out of the seven ADS modules defined in the SARPs. The amount of effort required to integrate any new ASEs into the model should be minimal.

We have successfully integrated the CM ASE models with the CF, ACSE, and supporting service, to the level of typechecking. The resulting statechart specification models an air CM AE (consisting of an air CM ASE, a CF, and an ACSE) talking to a ground CM AE (consisting of a ground CM ASE, a CF, and an ACSE) via the supporting service (see Figure 2). When later ASE models are developed, they should be able to be easily added to the specification and re-use much of the existing specification through parameterization.

The integrated system has passed the typechecking of the Fuss tool. This indicates that most interface errors have been eliminated, although it does not allow us to conclude that the models are completely correct.

Some static checks have been performed for the ACSE and CF. These are of two types: completeness checks and consistency checks. The completeness checks are intended to ensure that for each message received by a component of the model, there is at least one transition that will be followed regardless of global variable settings. The consistency checks are intended to ensure that not more than one transition can be followed for each combination of messages received and global variable settings. These checks were carried out by visual inspection. It would be useful to have a tool to do this analysis. Previous efforts at checking the completeness and consistency of state-based models (Heimdahl, 1996; Heitmeyer, 1996) rely on a tabular specification of the transition triggers.

7   FUTURE EFFORT

Two additional phases of the project are planned for the future. Phase 2 consists primarily of effort to adapt/develop a model checking tool as necessary to demonstrate properties of the statechart model. Phase 3 consists of effort by the team as a whole to do the more extensive validation. We also expect our model to be maintained in order to track changes in the SARPs.

At the present time, the only tool available to support the modelling effort is the Fuss typechecker. The second author's PhD thesis research examines how to analyze specifications consisting of integrated components in different notations (such as statecharts and predicate logic), and how to automatically analyze specifications at a high level of abstraction. The SARPs statechart model serves as test data for this effort; the other workers will interact with her to clear up any problems that may arise from the models.

As part of a research collaboration involving two universities and two industrial organizations, the work described in this paper is being used as the basis for a variety of research oriented investigations. Members of the FormalWare project are using, or are expected to use, this example as a case study for the development of methods and software tools for purposes such as automatic test case generation, symbolic execution and possibly code generation. In many cases, the parsing and typechecking functionality of Fuss is used as a front-end for the implementation of software tools which use S as input.   This is easily achieved since Fuss is designed specifically to support user developed extensions which access the internal representation of an S specification created by Fuss.

8   LESSONS LEARNED

The work described in this paper represents the results of using an integrated approach to specifying a model. Using a general-purpose formal description notation (S) as the basis of the entire project, we built models of the components of the ATN based on the statecharts formalism, and laid the groundwork for building analysis tools using S as input. We conclude that this integrated approach has indeed been useful.

8.1 Usefulness of a general-purpose formal description notation

Using a general-purpose formal description notation has been valuable. The alternative would be to use a specialized notation for state-based applications, such as pure statecharts. Our approach allowed us to integrate a state-based formalism with predicate logic to express the complex conditions on transitions. We were also able to use uninterpreted constants to maintain a level of abstraction, and parameterization to reduce duplication.

A future goal of the project is to extend the range of current analysis methods to integrated requirements specifications given in multiple notations and at a high level of abstraction, such as those containing uninterpreted constants. In this paper we have demonstrated that a general-purpose notation can serve as a foundation for expressing specialized notations and integrating notations. Future analysis work will take advantage of the fact the semantics of the specialized notation can also be expressed in the same framework (Day, 1993). This means that we are not locked into a specialized notation for specification and analysis.

8.2 Usefulness of S in particular

S has been particularly appropriate as a general-purpose formal description notation because (a) it is strongly typed and has an associated typechecker, Fuss; (b) it is machine readable; (c) it is more human-readable than many more symbolic notations; and (d) its power and generality allow a good deal of flexibility in how the model components are expressed.

Another important benefit of using typed, predicate logic was the ability to build a layer of infrastructure (i.e., the S file 'atncommon.s' mentioned earlier) on top of the specialized FDT which tailors our use of statecharts specifically to the purpose of modelling aspects of the SARPs.

Finally, our choice of S as the foundation for our approach made it possible to use Fuss 'off the shelf ' for this integration task.

8.3 Classification of assumptions

This work also led to a better appreciation of the distinction between the role of 'simplifying assumptions' and other kinds of assumptions made in the development of a formal representation, such as the 'disambiguating assumptions' made to address aspects of the SARPs which were found to be ambiguous or unclear.   There is a natural tendency to regard any kind of modelling simplification as something that may undermine the validity of results derived from the model. But we have used the term 'simplifying assumptions' to describe aspects of our formal representation which, in effect, increase the generality of these results rather than undermining their validity.

9   ACKNOWLEDGEMENTS

In addition to the authors, a number of individuals contributed to this work, including: from HACL, Ayman Farahat, Alec MacKay, Ofelia Moldovan, Greg Saccone and Robert Taylor; from UBC, Kendra Cooper, Michael Donat, Ken Wong; and from UVic, Dilian Gurov and Bruce Kapron. Michael Donat provided helpful comments and corrections on earlier versions of this paper.

10   REFERENCES

Cleaveland, R., Parrow, J. and Steffen, B. (1989)   A Semantics-Based Verification Tool for Finite State Systems, in *Proc. 9th IFIP Symposium on Protocol Specification, Testing and Verification*, North-Holland.

Day, Nancy (1993) A model checker for statecharts, M.Sc. thesis, Department of Computer Science, University of British Columbia, Technical Report 93-35.

Harel, David (1987)   Statecharts: A visual formalism for complex systems. *Science of Computer Programming*, **8**, 231-274.

Heimdahl, Mats P.E. and Leveson, Nancy G. (1996)   Completeness and consistency in hierarchical state-based requirements. *IEEE Transactions on Software Engineering,* **22(6)**, 363-377.

Heitmeyer, Constance L., Jeffords, Ralph D. and Labaw, Bruce G. (1996) Automated consistency checking of requirements specifications.   *ACM Transactions on Software Engineering and Methodology*, **5(3)**, 231-261.

ISO (International Organization for Standardization) (1994)   ACSE Protocol, ITU-T Rec. X.227 -- ISO/IEC 8650-1: Edition 2. Available in electronic format via anonymous FTP at URL 'ftp://ftp.stel.com/ pub/atnp2/iv/P2'.

Joyce, J., Day, N. and Donat M. (1994) S: A machine readable specification notation based on higher order logic, in *7th International Workshop on Higher Order Logic Theorem Proving and Its Applications*, 285-299.

Leveson, Nancy G., Heimdahl, Mats P. E., Hildreth, Holly and Reese, Jon D. (1994)   Requirements Specification for Process-Control Systems.   *IEEE Transactions on Software Engineering*, **20(9),** 684-107.

SARP (1996) Aeronautical Telecommunication Network Panel. Draft. Available via anonymous FTP at URL 'ftp://ftp.stel.com/ pub/atnp2'.

Turner, K. J. (ed) (1993) Using Formal Description Techniques: An Introduction to Estelle, LOTOS and SDL. Wiley.

11   BIOGRAPHIES

Jamie Andrews is an Assistant Professor in the Department of Computer Science, University of Western Ontario. This work was carried out while he was a post-doctoral fellow at the University of British Columbia (UBC) working with Paul Gilmore and Jeff Joyce.

Nancy Day is a PhD student in the Department of Computer Science at UBC. She expects to complete her dissertation near the end of 1997.

Jeff Joyce is a Research Scientist at Hughes Aircraft of Canada and an Adjunct Professor in the Department of Computer Science at UBC.

27

An experiment in using RT-LOTOS for the formal specification and verification of a distributed scheduling algorithm in a nuclear power plant monitoring system

L. Andriantsiferana, J.-P. Courtiat, R.C. De Oliveira
LAAS / CNRS
7, avenue du Colonel Roche - 31077 Toulouse Cedex - France
E-mail: {andrian,courtiat,cruz}@laas.fr

L. Picci
Electricité de France - Direction des Etudes et Recherches
6, Quai Watier - 78400 Chatou - France
E-mail: {laurence.picci}@der.edfgdf.fr

Abstract: The paper relates an industrial experiment performed jointly by LAAS-CNRS and Electricité de France (EdF in short) for assessing the application of a formal method to the reverse engineering of (a part of) a fault-tolerant monitoring system designed for the control room of French N4 nuclear power plants. More specifically, the experiment is devoted to the formal specification and verification of the distributed scheduling algorithm managing the hot redundancy between the two computers composing the system, a single fault hypothesis being assumed for this function. The formal method used for the experiment is RT-LOTOS, a temporal extension of the LOTOS standard Formal Description Technique (FDT in short). The main motivation behind the experiment was to get a better understanding of the fault-tolerant features of the scheduling algorithm by means of both simulation and formal verification.

Keywords: Formal Specification and Verification, Distributed and Real-Time Systems, Reverse Engineering, LOTOS & RT-LOTOS

Formal Description Techniques and Protocol Specification, Testing and Verification
T. Mizuno, N. Shiratori, T. Higashino & A. Togashi (Eds.)   © 1997 IFIP. Published by Chapman & Hall

1   Introduction

The paper describes an experiment dealing with the formal specification and validation of a fault-tolerant monitoring system designed for the control room of French N4 nuclear power plants. The monitoring system has been designed to be transparent to a single failure, and it is composed of two computers in hot redundancy. Both machines, master and slave, process the same application inputs and monitor their internal errors. The master is in charge of application processes scheduling and emission of application messages. In case of a single error the faulty computer is isolated and the other one becomes master. A distributed scheduling algorithm has been devised for implementing this hot redundancy scheme.

Although non-critical for the safety of the nuclear power plant, the monitoring system has been recognized by EdF as representative enough for starting and supporting an experiment aiming at assessing the use of formal methods in the reverse engineering of a part of this system, namely the scheduling algorithm. Main expectations on the project achievements were twofold: (i) to assess the feasibility of the reverse engineering process [CDH+96] starting from an analysis of the monitoring system source code, written in Ada, which has been implemented by a third party following the (informal) requirements of EdF (ii) to better understand and assess the fault-tolerant capabilities of the scheduling algorithm under several faulty conditions.

Three main requirements have been expressed by EdF for selecting a particular formal description technique for this study:

- To have executable specifications to facilitate the reverse engineering process.
- To represent the physical distribution of the monitoring system components, these components running asynchronously the one with respect to the others.
- To specify a large and complex system made of several components; the method should therefore provide facilities for composing large specifications from simpler and possibly reusable components.

The previous requirements led to the choice of a process algebra. Assuming also that explicit time constraints had to be expressed, the availability of a complete environment, the RTL software tool, for validating (simulating and verifying) formal specifications was the main reason which led to the choice of RT-LOTOS (Real-Time LOTOS, a temporal extension of the LOTOS FDT).

The paper is organized as follows: Section 2 presents the monitoring system informal description. Section 3 presents how the reverse engineering process has been carried out. Section 4 describes the main capabilities of the RTL software tool. Section 5 presents some of the achieved validation results. Finally, some conclusions are drawn in a last section.

2   System informal specification

This section presents a brief and informal description of a part of the fault-tolerant monitoring system designed for the control room of French N4 nuclear power plants. This informal description relies on different documents (informal text, and diagrams) provided by EdF for describing the system functionality. The textual documentation includes an overview of the monitoring system and an informal description of its functional modules. The diagrams illustrate the system functional decomposition, and include many state graphs, as well as Ada tasks flow charts.

2.1   System Overview

For plant availability reasons, the monitoring system has been designed to tolerate a single failure; it is made up of two computers, called CPC_1[1] and CPC_2 in hot redundancy. One of these computers is considered to be in the *master* mode, since it is in charge of scheduling the application processes running on both computers and of supervising the messages exchanged by the application processes with their environment; the other computer is in the *slave* mode. Single failure occurrence in any of the two computers leads to the isolation of the faulty computer (it enters the *isolated* mode), the other computer entering the *single_master* mode.

Synchronization messages are exchanged between both computers through bidirectional High-Speed Data channels (HSD in short). These messages make it possible for a computer to notify its mode alteration to the other. Other messages are exchanged through a dedicated network (the N2 network, not explicited in Figure 1) between the computers and the environment. These messages are: (i) the stimuli received from the environment by both computers, and (ii) the event notifications sent by the computers to the environment.

The monitoring system functional decomposition is presented in Figure 1 where five main modules (*Processes, Stimuli Manager, Scheduler, Service Manager* and *Watchdog*) have been defined.

2.2   The Environment

The environment sends stimuli to the stimuli managers in both computers, each stimulus conveying information to be used for scheduling the application processes. This information includes the stimulus priority, the stimulus creation date, the identification of the application process to which the stimulus relates and the stimulus type (*Activation_Stimulus, Resume_Stimulus* or *Allocation_Stimulus*). The master computer sends consistently, through a HSD channel, the stimulus it elected (i.e. the one corresponding to an execution

---

[1]Central Processing Computer

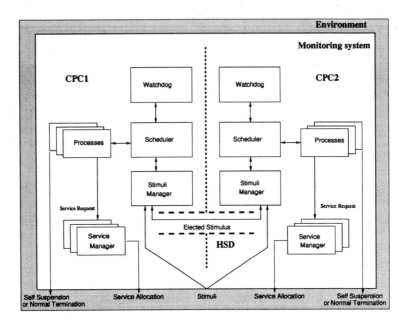

Figure 1: Functional Decomposition

thread activated on the master computer) to the slave computer in order to implement the hot redundancy scheme.

The event notifications sent to the environment are the following: (i) *Normal_Termination* sent by an application process when it terminates, (ii) *Self_Suspension* sent by an application process when it suspends itself, and (iii) *Service_Allocation* sent by the service manager when it allocates a service requested by an application process.

The environment basic behavior may be described as follows; for any application process:

- An *Activation_Stimulus* is sent after some random delay, following the reception of a *Normal_Termination* event notification of this process
- A *Resume_Stimulus* is sent after a fixed delay, following the reception of a *Self_Suspension* event notification of this process
- An *Allocation_Stimulus* is sent after a fixed delay, following the reception of a Service_Allocation event notification of the service manager.

2.3   System Functional Decomposition

2.3.1   The Processes

The application processes run on both computers in order to implement the hot redundancy scheme. Any application process is composed of two synchronizing Ada tasks: (i) a task called *Process* implementing application functions (it may suspend itself either at some predefined suspension point or when it requests some service from the service manager), and (ii) a task called *Interface* interfacing the process with the scheduler.

2.3.2 The Stimuli Manager

Each stimuli manager manages the stimuli by processing two queues. The *Candidate_Queue* contains all the stimuli received from the environment awaiting to be elected; they are sorted by priority and age. The *Elected_Queue* contains elected stimuli that have not yet been scheduled; on the master computer, it contains only the last elected stimulus; on the slave computer, it contains the stimuli elected by the master not yet scheduled by the slave.

The master elects the oldest stimulus among those with the highest priority, i.e. the stimulus located at the head of its *Candidate_Queue*. Whenever the slave receives a stimulus elected by the master, it puts it at the end of its *Elected_Queue* and removes it from its *Candidate_Queue*.

2.3.3 The Scheduler

The scheduling mechanism elects one stimulus at a time and the scheduler starts one execution thread in a process; then, the scheduler idles and waits for the process termination or suspension, which finally leads to removing the stimulus from the *Elected_Queue*.

2.3.4 The Service Manager

A service is a procedure that has only one executable instance, for instance a disk access. Several services are available. Requests to these services are managed by the *Service Manager* which is not further detailed here.

2.3.5 The Watchdog

The *Watchdog* task waits for the occurrence of events (either internal or external) that may lead to modifying the scheduling operating mode. Only internal events corresponding to failures detected inside the monitoring system have been considered. These failures are: (i) the *dis-symmetry* failure when both computers do not receive the same stimuli from the N2 network, recovery being made by the *Isolate* action, (ii) the CPU *overflow* failure, recovery being made by the *Isolate* action, and (iii) the *HSD* channel failure, recovery being made by the *HSD_Failure* action.

By the *Isolate* action, the computer, where the failure has been detected, orders its peer, through a HSD channel, to switch to the *single_master* operating mode; it then disconnects itself from the N2 network and the HSD channels.

By the *HSD_Failure* action, the master switches its operating mode to *single_master* and orders its peer, through the N2 network, to switch to the *isolated* operating mode; the slave then waits some time for a message coming from the master; if there is no message, then it switches to the *single_master* operating mode.

2.4   Required Properties

Two types of properties may be identified depending whether one considers the nominal behavior (without any internal failure) or a single failure situation.

2.4.1   Case of the nominal behavior

The property characterizing the correct expected behavior of the scheduling algorithm is expressed as follows:

Property 1 *The stimuli sequence elected on the slave computer is identical to the one of the master, with the possible exception of the last stimulus elected on the master (which may not yet have been scheduled by the slave).*

2.4.2   Case of a single failure

Let us consider a single failure situation together with the assumptions that the failures are sudden and total, and that the failure detection mechanism is fully reliable.

The property to be verified corresponds to the correctness of the operating mode switching. Assuming an initial configuration, where CPC_1 is the *master* and CPC_2 the *slave*, the two following properties may be stated:

Property 2 CPC_1 *switches from the* master *to the* isolated *operating mode, while* CPC_2 *switches from the* slave *to the* single_master *operating mode, in case of an internal failure detected in* CPC_1.

Property 3 CPC_1 *switches from the* master *to the* single_master *operating mode, while* CPC_2 *switches from the* slave *to the* isolated *operating mode, in case of an internal failure detected in* CPC_2.

2.5   Conclusion

The specification of the monitoring system depends on several parameters, among them : the number of application processes to be scheduled on each computer, the number of services available on each computer, the number of sequential threads per application process, and several timing parameters like the transit delays across the N2 network and the HSD channels, the duration of each application process thread and the duration of each available service.

3 System formal specification

The purpose of this section is to illustrate the main features of the design method that has been used for translating the informal specification introduced in section 2 into a RT-LOTOS formal specification.

The design method is essentially based on the LOTOS design methodology developed within the European LotoSphere project [BvdLV95]. The key concept of the approach is the *design trajectory*. A design trajectory is made up of several design steps. Starting from an initial high-level specification expressed in LOTOS, the execution of each design step leads to refining the specification by using so-called *transformations*. Two of these transformations, known as the *functionality decomposition* and the *functionality rearrangement* [BvdLV95], are particularly useful for building step by step complex specifications. The same design method may be applied to RT-LOTOS since the difference between RT-LOTOS and LOTOS stands essentially at the level of the elementary action offering, but not at the level of the composition operators.

Applying this design approach has been relatively easy since the starting point of the re-engineering process was a functional architecture of the monitoring system (see the Ada tasks functional decomposition and the associated state graphs introduced in Section 2).

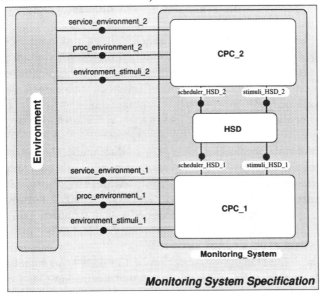

Figure 2: High-level specification architecture

3.1   The High-Level Specification

From the system functional architecture presented in Figure 1, two main entities may be identified, namely the monitoring system and its external environment. The monitoring system is itself composed of three modules corresponding respectively to the two CPC computers and to the bidirectional HSD channels. This high-level functional architecture may be immediately translated into the high-level RT-LOTOS specification architecture presented in Figure 2, where RT-LOTOS process instances[2] are represented by boxes and where synchronization gates are represented by small circles labeled by their name.

Both computers may potentially have the same behavior; they correspond therefore to two instances of the same process definition, namely *process CPC_Computer[···](mode:mode_type)*; formal parameter *mode* permits the definition of the operating mode of each instance (*master, slave, ···*) [3].

3.2   Refining the High-level Specification

The next step consists in performing the refinement of process *CPC_Computer* following the functional decomposition of Figure 1. A RT-LOTOS process has been associated with each functional module, and the processes have been composed in parallel with a mandatory synchronization on their common gates, as illustrated in the architecture depicted in Figure 3.

Stimuli election is performed by synchronizing processes *Scheduler* and *Stimuli_Manager* on gate *stimuli_scheduler*. Process scheduling results from the synchronization of processes *Scheduler* and *Processes* on gate *scheduler_proc*. Service requests result from the synchronization of processes *Service_Manager* and *Processes* on gate *proc_service*. Finally, mode switching results from the synchronization of processes *Scheduler* and *Watchdog* on gate *watchdog_scheduler*.

In the informal specification, it is easy to distinguish the behavior and the data parts of the system. The behavior part results in a composition of RT-LOTOS processes using the parallel composition, the choice, ··· operators. The data part describes the values (messages) exchanged between processes through the synchronization gates. Every message structure (stimulus, event notification, mode) defined for the monitoring system has been translated into a particular data type.

In standard LOTOS, the description of the data type signatures is completed by the definition of equations, expressed in the Act-One formalism, for providing the type semantics. For many reasons, related to the non-obvious industrial applicability of Act-One, only the data type signature is expressed

---

[2]Depending on the context the term process will define either a process definition or a process instance

[3]We assume an initial configuration where *CPC_1* is the *master* and *CPC_2* the *slave*

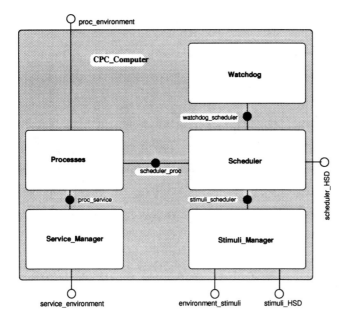

Figure 3: CPC_Computer process architecture

in RT-LOTOS, the meaning of the data type operations being provided by C++ or Java classes defined within a user library accessible from the RTL software tool (itself written in C++).

3.3 Failure specification

Modeling (internal) failures of the monitoring system consists basically in introducing new behaviors in the formal specification of the system that lead, after some random delay, to the occurrence of an event characterizing a failure detection (remember that the failure detection mechanism is assumed to be fully reliable). Such an event will activate the associated recovery mechanism (i.e. the operating mode switching) described in section 2.

3.4 Conclusion

The resulting specification comprises around fifty processes for a total of approximatively one thousand RT-LOTOS lines (without the data type implementation). Each leaf process (defined at the bottom of the process hierarchy) is rather simple and corresponds merely to a state machine of few symbolic states. This appears to be one of the most visible interest of LOTOS-based approaches: the description of highly complex concurrent behaviors by the stepwise composition of processes always becomes simpler when going the hierarchy down.

4   The RTL tool environment

The validation techniques implemented within the RTL tool may roughly be classified into two main categories:

- *Verification techniques* for formally proving some property; the purpose here is to analyze a complete (finite) model of the (RT-)LOTOS specification; these techniques become not feasible when such a model cannot be computed (either because it is infinite or just because it is too big with respect to the size of the available RAM memory).
- *Simulation techniques* for observing some possible traces of the specification global behavior; the purpose here is to understand the global behavior of the specification better and to gain a certain level of confidence on the validity of some property (i.e. no trace has violated the property during the - as many and as long as possible - simulation runs)

4.1   RTL Verification Capabilities

The verification method implemented in the RTL software tool consists in translating a RT-LOTOS specification into a timed automaton model, on which reachability analysis is performed. The general way to proceed is not original, but the specific method implemented in RTL presents several advantages: (i) it permits to minimize the number of clocks in each control state of the timed automaton, thanks to the definition of the DTA (Dynamic Timed Automata) model, and (ii) reachability analysis is performed *on the fly* when generating the DTA model from the RT-LOTOS specification (see [CdO95] for details).

Both advantages are important from a practical point of view, since the complexity of verification algorithms developed for timed automata depends directly on the number of clocks [YL93]. The DTA model, initially developed for taking into account non regular RT-LOTOS processes, has proven to be very efficient since it has permitted to drastically reduce the number of clocks to be defined in each control state of the model (from more than 20 clocks to around 0 to 5 clocks per control state for the present case study). Reachability analysis does not furthermore require the underlying (untimed) LOTOS behavior being finite, since it is performed on the fly.

4.2   RTL Simulation Capabilities

Besides its verification capabilities, RTL also provides several simulation capabilities. Although simulation cannot in any case be used for formally proving a property, it appears as particularly useful for debugging a complex specification and/or gaining a good level of confidence on the satisfaction of some property. The observer approach (discussed in the next section) may still be used within a simulation framework, and becomes an interesting testing technique.

The trade-off between verification and simulation is very easy to understand. Depending on the available RAM memory (100 M-bytes in our case) and on the average size of a state representation in memory, one can easily estimate the maximal number of states that can potentially be produced before entering the swap zone. Many enhancements have been performed in RTL for drastically reducing the memory size of the state representations. For the present case study, this has led to having a 28 k-bytes representation of a state be decreased to 2.2 k-bytes. As a consequence, around 45,000 different states can potentially be developed for this specification.

The trade-off is therefore between (i) the simulation of the complete specification that has been produced by the re-engineering process, and (ii) the verification of a simplified specification derived from the original specification.

5 Validation of the scheduling algorithm

This section presents some results related to the simulation and verification of the monitoring system scheduling algorithm. It is organized as follows:

- Simulation results are detailed first with the purpose of illustrating the use of RTL (i) for performing the initial debug of the complex specification produced by the re-engineering process, and (ii) for gaining a certain level of confidence on the validity of the required properties
- Verification results are presented next on a simplified formal specification in order to be able to master the size of the induced state space; under these simplification assumptions, the desired properties of the scheduling algorithm have been formally proven.

5.1 Simulation of the Scheduling Algorithm

5.1.1 Using Simulation for Debugging the Specification

Simulation has extensively been used for debugging the specification. It has been particularly useful for identifying undesirable deadlock situations due to the incorrect specification of RT-LOTOS processes synchronizations.

Debugging has essentially been achieved by the display of simulation event traces. From these event traces, scheduling diagrams (see Figure 4) have been produced. These diagrams display the interleaving of the processes executions threads in both computers, and permit therefore to analyze the behavior of the scheduling algorithm.

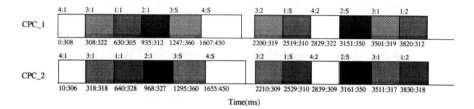

Figure 4: Scheduling Diagram

5.1.2   Validating the System Behavior by Simulation

The monitoring system specification is composed of many processes and gates
where data values are exchanged. Pattern scanning and processing languages
(like Awk and Perl) have been used for displaying relevant parameters of the
monitoring system as a function of time, starting from raw data extracted
from the simulation traces.

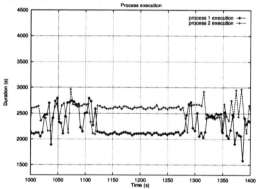

Figure 5: Process execution duration

By this way, with a minimal effort, it becomes possible to observe several
parameters like: the time required for executing a process, the load of the
stimuli queues, the number of elected stimuli per period of time $\cdots$.

As an illustration, let us consider Figure 5 which shows the execution time
of processes 1 and 2 on the master computer (CPC_1). Assuming that the
priority of process 1 is higher than the one of process 2, it clearly appears that
the execution time of process 1 is less than the execution time of process 2.

5.1.3   Expressing a Property by an Observer

A classical verification technique is related to the so-called *observer* (or tester)
approach [BvdLV95]. Basically, observers are modules synchronizing themselves with the specification on some internal gates, and checking on-line
whether some particular condition characterizing the violation of some prop-

erty arises; in case of such a violation, the observer offers a specific *error* action. Proving that the property checked by the observer is valid consists therefore in showing that the *error* action is not reachable. The advantage of the technique is that it can easily be implemented; its main drawback is that it is less powerful than model checking.

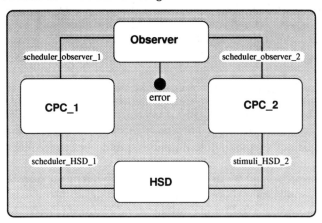

Figure 6: Architecture of the system with an observer

Regarding *Property 1* intensive simulations of the monitoring system specification, including the observer, have been performed with different sets of parameter values (principally time parameters). Assuming the time parameter configuration provided by EdF, no action *error* has been detected in these simulations [4].

Then, temporal values, characterizing some delay of the slave computer when running the application processes, were selected. Since the slave computer elected stimuli queue has by definition a limited capacity, a too important delay with respect to the master computer causes stimuli to be lost, leading to the occurrence of action *error*. In this way, it has been possible to identify a set of parameter values leading to an incorrect behavior of the scheduling algorithm. These parameter values have been analyzed by EdF and the third party software company in charge of the implementation of the scheduling algorithm. Several changes have been made in the monitoring system in order to overcome the (potential) error situation identified by these simulations.

In a similar way, observer processes have been developed for validating Property 2 and Property 3 of section 2. No *error* occurrence has been reported, validating consequently the mode switching mechanism.

---

[4]Note that, within a simulation framework, the error action is specified as a goal for the simulation kernel; if, in some state, error is enabled, then it will necessarily be fired

5.2   Formal Verification of the Monitoring System

Validation by simulation can obviously not be considered as a formal proof, since it does not cover the complete specification state space. However, the simulation results have already provided some level of confidence on the specification quality and on the validity of the desired properties.

Using observers, the verification principle is simple, and corresponds to a standard reachability analysis. The same observers have been used for simulation and verification.

5.2.1   Simplification of the system specification

Verification by reachability analysis faces the classical state explosion problem [Hol93]. Several simplifications have been made on the specification:

- The number of parallel components has been reduced by decreasing the number of application processes in each computer (from 4 for the simulation to only 2) and the number of services that may be requested by the application processes.
- The specification has been simplified by withdrawing any behavior which did not directly affect the property to be verified.
- The internal architecture of the specification has been simplified, by replacing a composition of processes by an equivalent unique process.
- The value domain of some parameters has been reduced, and parameters that do not directly interfere with the scheduling algorithm have been removed.

5.2.2   Formal verification of Property 1

Various verification-oriented specifications have been derived from the initial formal specification (the one which has been intensively simulated). Each verification-oriented specification includes the observer process used for verifying the relevant property. Results related to the formal proof of Property 1 are summarized below.

- Case I: RT-LOTOS processes involved in the mode switching mechanism have been removed, and only two application processes have been considered, without any service. Equal values have been considered for the durations of the execution threads in both computers (i.e. $t\_exec\_min = t\_exec\_max$), and a latency of $250ms$ has been specified for the communications between the environment and the computers ($N2\_max - N2\_min = 250$); the transmission delay of the HSD channels has finally be neglected (i.e. $dHSD = 0$). Under these assumptions, the complete reachability graph has been constructed (see details in Table 1) with action *error* being *not reachable* in this graph, proving therefore formally Property 1 for this configuration.

Cases	I	II	III
N2 min	500	500	500
N2 max	750	750	750
dHSD	0	0	10
t_exec_min (CPC_1/CPC_2)	500/500	500/750	500/500
t_exec_max(CPC_1/CPC_2)	500/500	500/750	510/510
DTA states	1695	594	730
0-clock DTA states	315	115	110
1-clock DTA states	381	144	169
2-clock DTA states	547	191	231
3-clock DTA states	302	111	142
4-clock DTA states	140	33	72
5-clock DTA states	10	-	6
Classes/Arcs	1949/5338	689/1600	1508/3345
Memory used (KB)	34300	28764	23544

Table 1: Property 1 verification results

- Case II: The same specification has been considered plus the additional assumption that the slave computer (*CPC_2*) has an important (processing) deterministic delay (i.e. $250ms$) with respect to its master computer. This situation leads to the violation of Property 1.
- Case III: The same specification has been considered but with new timing parameters. The processing delay of the slave computer has been removed, and a latency of $10ms$ has been introduced for characterizing the variability of the thread duration. The transmission delay of the HSD channels has been established to its nominal value, i.e. $10ms$. With these assumptions, the proof of Property 1 has been successful.

Many other verifications have been performed with different parameter sets. Due to the lack of space, they are not reported here.

6   Conclusion

The specification phase has been much more simple than initially expected; the re-engineering process has greatly been facilitated by the existence of Ada flow charts, state diagrams, ... The use of a LOTOS-based approach has also greatly simplified the specification development. This is largely due to the LOTOS general parallel composition operator with multi-way synchronization, which permits the specification of complex behaviors by the composition of much more simple ones.

The simulation phase has brought much more results than initially expected; many simulations have been conducted for debugging the initial specification, and then for validating the scheduling algorithm behavior with numerous parameter configurations. Error situations have been reported for some parameter values, and have been analyzed in depth by our industrial partners. The use of the observer approach within such simulation framework has proven to be very simple and efficient.

The verification phase has been as difficult as initially expected; several improvements, not detailed here, on the RTL tool have been made during the project, most of them for reducing the number of bytes required for coding in memory a RT-LOTOS state; although verification results have been obtained only for simplified configurations of the monitoring system (when reducing the number of application processes, the number of threads or services, ...) the validity of the proposed approach has been validated on a complex industrial application.

One important return of experience is the successful trade-off achieved between simulation and verification. Both have been carried out consistently and in cooperation, and not in isolation: (i) the verification-oriented specifications have been derived following a strict methodology from the complete formal specification (ii) the observer approach has been used for both the simulation and the reachability analysis. As a consequence, errors in simulation have been better understood by analyzing reachability graphs, and vice versa reasons preventing the convergence of the reachability graph minimization have been better understood thanks to the simulation.

References

[BvdLV95] T. Bolognesi, J. van de Lagemaat, and C. Vissers, editors. *LOTO-Sphere: Software Development with LOTOS.* Kluwer Academic Publisher, 1995.

[CDH+96] J.-P. Courtiat, P. Dembinski, G. Holzmann, L. Logrippo, H. Rudin, and P. Zave. Formal methods after 15 years: status and trends. To appear in Computer Networks and ISDN Systems, 1996.

[CdO95] J.-P. Courtiat and R.C. de Oliveira. A Reachability Analysis of RT-LOTOS Specifications. In *Proc. 8th Intern. Confer. on Formal Description Techniques (FORTE'95)*, Montreal, Canada, October 1995. Chapman & Hall.

[Hol93] G.J. Holzmann. Design and validation of protocols: a tutorial. *Computer Networks and ISDN Systems*, 25:981–1017, 1993.

[YL93] M. Yannakakis and D. Lee. An efficient algorithm for minimizing real-time transition systems. In *CAV '93*, volume 697 of *LNCS*, pages 210–224. Springer-Verlag, 1993.

28
Intelligent Protocol Analyzer with TCP Behavior Emulation for Interoperability Testing of TCP/IP Protocols

Toshihiko Kato, Tomohiko Ogishi, Akira Idoue and
Kenji Suzuki
KDD R&D Laboratories
2-1-15, Ohara, Kamifukuoka-shi, Saitama 356, Japan
E-mail : {kato, ogishi, idoue, suzuki}@hsc.lab.kdd.co.jp

Abstract
Recently, the TCP/IP protocols are widely used and it is mentioned that, in some cases, throughput is limited due to problems such as network congestion. To solve such problems, the details of communication need to be examined. In order to support such examination, we are developing an 'intelligent' protocol analyzer which can estimate what communication has taken place by emulating the behaviors of the TCP protocol entity in a pair of communicating computers. Since modern TCP includes some internal procedures for the flow control, such as the slow start algorithm, this analyzer can emulate these procedures as well as the state transition based behaviors of the basic TCP. This paper describes the overview of the analyzer, the detailed design of the TCP behavior emulation function and the implementation results.

Key words
Protocol testing, Interoperability testing, TCP/IP, Protocol Behavior Emulation, Industrial usage report

Formal Description Techniques and Protocol Specification, Testing and Verification
T. Mizuno, N. Shiratori, T. Higashino & A. Togashi (Eds.) © 1997 IFIP. Published by Chapman & Hall

1   INTRODUCTION

Recently, the TCP/IP protocols[1,2] are widely used in various computer communications.   Here, most users of computers use the communication functions installed in operating systems or commercial software products as they are, and do not pay attentions to their details.   So, the protocol testing is not focused on for the TCP/IP protocols.   For example, the conformance testing is not performed for TCP/IP software products, and the only tools for the interoperability testing are commercially available protocol analyzers which just monitor networks.

Actually, most of the commercial TCP/IP software products can communicate with each other without the conformance testing.   However, it is mentioned that the TCP/IP protocols, especially TCP (Transmission Control Protocol), have some problems such that the performance is degraded due to the congestion avoidance algorithms[2,3], and due to the incompatibility between send and receive socket buffers[4].   When such problems occur, the details of communications need to be examined to detect the problem sources.   Those problems occur in the actual communications, and therefore the interoperability testing needs to be performed for the TCP/IP protocols.   As mentioned above, commercial protocol analyzers such as [5] are used for this purpose, but they do not have enough functionality. They only have the functions to capture PDUs (Protocol Data Units) transmitted over networks, to analyze only their formats, and to present those results.   The analysis of protocol behaviors and the investigation of problem sources need to be performed manually by TCP experts.

In order to support of the interoperability testing of TCP/IP communications, we have proposed an 'intelligent' protocol analyzer which can estimate what communication has taken place[6].   This analyzer maintains the specification of the state transition based behaviors of TCP and emulates the behaviors of the TCP protocol entity in communicating computers.   Since modern TCP includes some internal procedures for the flow control, such as the slow start algorithm, this analyzer can emulate these procedures as well.   Based on the design described in [6], we have finished the implementation of the analyzer as software running in workstations.   This paper describes the overview of our intelligent protocol analyzer, the mechanism of the TCP behavior emulation and the implementation results.

2   OVERVIEW OF INTELLIGENT PROTOCOL ANALIZER

(1)   As depicted in Fig. 1, our intelligent protocol analyzer is attached to a LAN and observes it.   It captures PDUs transmitted over the LAN, and analyzes PDU formats and emulates the TCP behavior of a specific pair of computers, e.g. computers A and B in the figure.

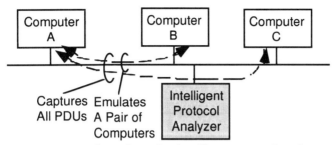

Figure 1   Network configuration using intelligent protocol analyzer.

(2)   The analyzer function is implemented as a software running in UNIX workstations.   Figure 2 depicts the software structure of the analyzer. It consists of the capture module and the TCP emulation module.   The capture module captures PDUs transmitted over the LAN, analyzes their format and parameter values according to the TCP/IP protocols, and saves those results in the PDU log.   The TCP emulation module selects PDUs which a pair of computers sent and received into the event sequence log for individual computers, and emulates the TCP behaviors of these computers according to the event sequence logs.

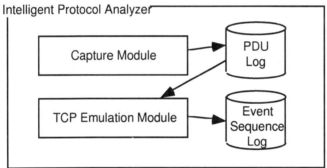

Figure 2   Software structure of intelligent protocol analyzer.

(3)   The event sequence log contains TCP protocol events, each of which is a "sent TCP segment" or a "received TCP segment".   The TCP emulation module maintains the state transition specification for received events and sent events, and processes each event according to the following procedure.

●   When the event is a received TCP segment, it looks up the specification for received events, and performs a corresponding state transition.   If it sends out a segment, the module checks a sent TCP segment in the event sequence and emulates the received and sent segments.

●   When the event is a sent TCP segment, the TCP emulation module looks up the specification for sent events and checks whether the TCP protocol entity can send out the segment.   If there is no input to generate the segment, it decides that the TCP protocol entity has some protocol errors.

3   DETAILS OF TCP EMULATION

3.1   Overview of TCP Behaviors

Behaviors of Basic TCP

TCP provides a reliable data transfer over an unreliable network layer protocol, IP. In order to transfer data, it establishes a connection which is identified by source and destination ports together with IP addresses in the IP header.   The connection establishment is performed by the three way handshake using SYN, SYN+ACK and ACK segments.   Data transferred over a TCP connection is identified by use of 32 bit sequence number, and acknowledged by 32 bit acknowledgment number. The flow control is performed by use of 16 bit window size advertised by a receiver. In the end of the data transfer, a TCP connection is released by the exchange of FIN and ACK segments by both side.

Internal Procedures of Modern TCP

Most of Modern TCP software products contain four internal procedures to perform a flow control with congestion control and retransmission; slow start, congestion avoidance, fast retransmit, and fast recovery[3].   The basic ideas are ;
- In order to solve the problem that that some segments are lost by the network congestion in the case that there is a slower link between sender and receiver, the sender uses a congestion window, called cwnd, which control the sending rate internally in the sender, and a slow start threshold size, called ssthresh.
- In order to improve the throughput under moderate congestion, especially for large windows, a sender retransmits a segment which is consider to be lost detected by the reception of duplicate ACKs and, after that, it invokes congestion avoidance.

   The procedures are summarized as follows.
(1)   The sender never send more than the minimum of cwnd and the advertised window from the receiver.
(2)   Congestion is indicated by a timeout or the reception of duplicate ACKs.   If the congestion is indicated by a timeout, the sender sets ssthresh to one-half of current window size (the minimum of cwnd and the advertised window), and sets cwnd to one segment.
(3)   For duplicate ACKs, that is, when the sender receives three duplicate ACKs, it considers that these ACKs indicate a lost segment, sets ssthresh to one-half of the current cwnd, and retransmits the missing segment.   After that, when the first ACK arrives that acknowledges new data, the sender sets cwnd to

ssthresh.

(4) Every time when new data is acknowledged by the receiver, cwnd is increased in the following way. If cwnd is less than or equal to ssthresh, TCP is in slow start and cwnd is incremented by one segment every time an ACK is received. The sender starts by transmitting one segment and waiting for its ACK. When that ACK is received, cwnd is incremented from one to two, and two segments can be sent. When each of those two segments is acknowledged, cwnd is increased to four. This opens the window exponentially. If cwnd is greater than ssthresh, congestion avoidance is being performed and cwnd is incremented by segsize * segsize / cwnd each time an ACK is received, where segsize is the segment size. This is a linear growth of cwnd, and the increase in cwnd is at most one segment for each round-trip time.

3.2 Principles of TCP Emulation

(1) The TCP emulation module maintains a specification of TCP behaviors in response to sent and received segments for individual states of a TCP connection. For a sent segment, it specifies what occurs in a TCP protocol entity to send the segment. For a received segment, it specifies what occurs when the segment is processed. The specification includes both the basic TCP behaviors and the internal procedures of modern TCP.

(2) In order to specify the basic TCP behaviors, the TCP emulation module maintains the state and internal variables corresponding to those defined in RFC 793[1].

(3) The invocation of the internal procedures is estimated from the input and output of TCP segments. For example, the invocation of the slow start algorithm is estimated when a connection is established or when it detects a DATA segment retransmission caused by timeout (not duplicated ACKs).

(4) In order to specify the internal procedures, the TCP emulation module maintains a status which specifies what procedure is being performed, and the internal variables associated with them.

(5) The TCP emulation module estimates the behavior of a TCP protocol entity by tracing a state transition, based on the specification, for events saved in the event sequence log on an event by event basis.

3.3. Specification of TCP Behavior

State and Internal Variables for Basic TCP

The state takes the following values representing the behavior of basic TCP for the connection establishment and the data transfer phases:

CLOSED, SYN_SENT, SYN_RCVD, and ESTABLISHED.

The internal variables include the followings:

send sequence variables such as

SND.NXT : the send sequence number to be sent next,

SND.UNA : the least send sequence number which is unacknowledged,

SND.WND : the advertised window offered by the remote side,

SND.MAX : the highest sequence number which has been already sent,

SND.WL1 : the sequence number of the latest received segment which updated SND.WND,

SND.WL2 : the acknowledgment number of the latest received segment which updated SND.WND,

ISS : the initial send sequence number, and

MSS : the maximum segment size, and

receive sequence variables such as

RCV.NXT : the receive sequence number to be received next,

RCV.WND : the advertised window offering to the remote side,

RCV.UNA : the least receive sequence number which is unacknowledged,

RCV.ADV : the maximum receive sequence number which can be received by the advertised window, and

IRS : the initial receive sequence number.

Status and Internal Variables for Internal Procedures

The TCP emulation module maintains the status on the internal procedures, STATUS, which indicates what algorithm is being emulated. It takes the values: NORMAL, SS (slow start), CA (congestion avoidance), and FR (fast retransmit). It also maintains the following internal variables associated with the slow start and congestion avoidance algorithms and the fast retransmit and fast recovery algorithms:

variables associated with the slow start and congestion avoidance

CWND : the estimated congestion window,

SSTHRESH : the estimated slow start threshold, and

variables associated with the fast retransmit and fast recovery

D_ACK : number of the received duplicated ACK.

State Transitions of TCP

By use of the state and internal variables described above, the specification of TCP behavior for the TCP behavior emulation is defined as follows. Figure 3 shows the state transitions of TCP for the connection establishment and data transfer phases.

A state transition for one event and for one state is associated with one or more possibilities. A possibility includes condition, output and requirements for its parameters (only for a received event), next state, and variable update. The

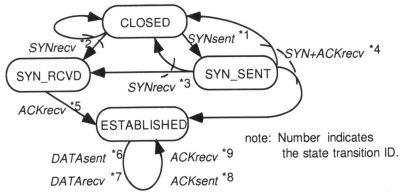

Figure 3 TCP state transition for connection establishment and data transfer.

following specifies details for some state transitions in the figure.

State Transition 1 : *SYNsent* in CLOSED
    next state: SYN_SENT;
    variable update: ISS = sequence number (SEQ) in *SYN*; SND.UNA = ISS;
      SND.NXT = ISS+1; SND.MAX = SND.NXT; RCV.WND = window size
      (WND) in *SYN*; MSS = maximum window size (MSS) in *SYN*;
      CWND = MSS; SSTHRESH = 65535; STATUS = SS;
State Transition 4 : *SYN+ACKrecv* in SYN_SENT
    1) output: *ACK* with SEQ = SND.NXT
                 and acknowledgment number (ACK) = SEQ in *SYN+ACK* +1;
    next state: ESTABLISHED;
    variable update: IRS = SEQ in *SYN+ACK*;
      SND.UNA = ACK in *SYN+ACK*;
      SND.WND = WND in *SYN+ACK*; RCV.NXT = IRS+1;
      RCV.WND = WND in *ACK*;   RCV.UNA = ACK in *ACK*;
      RCV.ADV = RCV.UNA + RCV.WND;
      MSS = min (current MSS, MSS in *SYN+ACK*);
    2) output: *RST* with SEQ = ACK in *SYN+ACK*;   next state: CLOSED;
State Transition 6 : *DATAsent* in ESTABLISHED
    1) condition: SND.UNA <= SEQ in *DATA* < *SND.NXT*
    next state: ESTABLISHED;
    variable update: SSTHRESH =
          max (2*MSS, 1/2 * min (CWND, SND.WND));
    CWND = MSS; STATUS = SS;
    /* The invocation of the slow start is detected and the interval variable
    SSTHRESH and CWND are estimated. */
    2) condition: SND.NXT <= SEQ in *DATA* < SND.UNA + SND.WND
    next state: ESTABLISHED;

variable update: SND.NXT = SEQ in *DATA* + length (LEN) of *DATA*;
    RCV.NXT = ACK in *DATA*; RCV.WND = WND in *DATA*;
    if (STATUS != NORMAL && SEQ in *DATA* + LEN of *DATA*
        > SND.UNA + CWND)   STATUS = NORMAL;
    /* Check whether the internal procedures are used at every DT sending. */
State Transition 9: *ACKrecv* in ESTABLISHED
    1) condition: D_ACK == 2 and ACK in *ACK* == SND.UNA
    output: *DATA* with SEQ in *DATA* == SND.UNA;
    next state: ESTABLISHED;
    variable update: SSTHRESH =
            max (2*MSS, 1/2 * min (CWND, SND.WND));
        CWND = SSTHRESH+3*MSS; D_ACK = 0; STATUS = FR;
        /* This corresponds to the fast retransmit. */
    2) condition: D_ACK < 2 and ACK in *ACK* == SND.UNA
    next state: ESTABLISHED;
    variable update: if (STATUS != FR) D_ACK = D_ACK +1;
        else CWND = CWND+MSS;
    3) condition: SND.UNA < ACK in *ACK* <= SND.NXT
    next state: ESTABLISHED;
    variable update:
        SND.UNA = ACK in *ACK*; SND.WND = WND in *ACK*; D_ACK = 0;
        if (STATUS == FR) CWND = SSTHRESH; STATUS = CA;
        /* This is the start of congestion avoidance following the fast retransmit. */
        if (STATUS == SS)
            if (CWND <= SSTHRESH) CWND = CWND+MSS;
            else CWND = CWND+MSS*MSS/CWND; STATUS = CA;
        if (STATUS == CA)
            CWND = CWND+MSS*MSS/CWND;
            if (CWND > 65535) STATUS = NORMAL; CWND = 65535;
        /* These two are the slow start and congestion avoidance. */

3.4   Emulation Based on Specification

By use of the specification defined in the previous subsection, the TCP emulation module traces the behaviors of TCP protocol entity.   The algorithm is summarized as follows.

(1)   The TCP emulation module reads out an event from the event sequence log one by one.

(2)   If the event is a sent TCP segment, the module looks up the corresponding transition, and selects a possibility considering values of segment parameters and internal variables.   If a correct one is found, the module emulates it, including changing the state to the next state and updating the internal variables.

If such a possibility does not exist, the TCP emulation module considers that there may be some protocol errors, and resets the state and internal variables to restart the emulation.

(3)   If the event is a received TCP segment, the TCP emulation module looks up the state transition for the current state and the received segment.   If the transition does not send out any outputs, then the transition is emulated.

If the transition sends out some outputs, the TCP emulation module looks for the next sent event in the event sequence log.   If the next sent event is the correct output for the transition being traced, then the TCP emulation module reads out the sent event, and emulates the current received event and the sent event.

If the correct output is not found, the TCP emulation module supposes that there are no outputs for the event.   If there is a possibility with no output for the transition, then the possibility is emulated.   If there are no possibilities which do not send out any output for the transition, then the TCP emulation module considers that the corresponding frame is lost in the network and that the current received event did not take place in this protocol entity.

According to this algorithm, the behaviors of computer A in Fig. 4 are emulated in the following way.

(i)   First, the TCP emulation module reads out an event *SYNsent* and selects **State Transition 1** for this event.   The new state is SYN_SENT.

(ii)   Next, the module reads out an event *SYN+ACKrecv* and looks up **State Transition 4**.   Since this transition includes some outputs, then the module looks for the next sent event in the event sequence log, and finds an event *ACKsent.* This output conforms to possibility 1) in **State Transition 4**, *SYN+ACKrecv* and *ACKsent* are emulated.   The new state is ESTBLISHED.

(iii)   Then, the module reads out an event *DATAsent* and looks up **State Transition 6**.   According to the value of sequence number, possibility 2) is selected and the check of the slow start in this possibility is also passed.

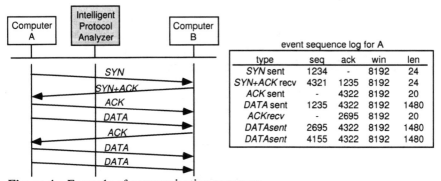

Figure 4   Example of communication sequence.

(iv) Then, the module reads out an event *ACKrecv* and looks up **State Transition 9**. According to the value of acknowledgment number, possibility 3 is selected. Since current STATUS is SS, the value of CWND is updated from 1460 to 2920.

(v) Similarly with step (iii), the following two *DATAsent* is emulated according to possibility 2) of **State Transition 4**.

4   IMPLEMENTATION RESULTS

4.1   Program Structure

According to the detailed design described in section 3, we have implemented the intelligent protocol analyzer. The analyzer is implemented as a software running over SUN workstations with Solaris 2.5. As depicted in Fig. 5, the software works together with the functions in the kernel space. The Capture module uses the DLPI (Data Link Provider Interface) module in the kernel in order to read frames through the Ethernet device driver. It relies on the BUF module in the kernel to assign timestamps to captured frames. Asking the kernel to assign timestamps makes the assigned timestamp values accurate.

The software of the intelligent protocol analyzer consists of capture module, TCP emulation module and GUI module. This structure corresponds to that described in Fig. 2. The capture module uses a dedicated thread for reading frames from the network, reading thread in the figure. The rest of this module stores the information of captured frames in the PDU log. The TCP emulation module performs the protocol emulation for a specified pair of computers according to the method described in the previous section. The GUI module provides a graphic window interface for operators of the analyzer. This module uses Tcl/Tk for implementing GUI.

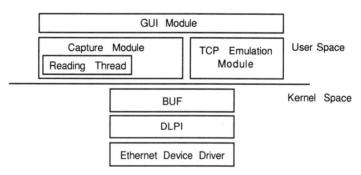

Figure 5   Program modules of intelligent protocol analyzer.

4.2   User Interface

Figure 6 shows an example of a window for TCP emulation.   In the top part of the window, there are some buttons for display control commands.   The middle part of the window gives a list of communication sequence between a pair of computers, which are indicated by 'SYSTEM A' and 'SYSTEM B'.   For one computer, there are two rectangles assigned, the left of which indicates sent or received segments and timpstamp, and the right of which indicates the parameter values of the segment and the estimated value for the state and the internal variables.   The bottom part of the window shows the full information of the selected line in the rectangles of the middle part.

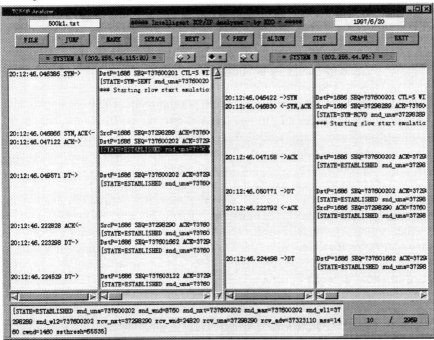

Figure 6   An example of graphical user interface.

4.3   Example of TCP Estimation

This subsection gives examples of the results of TCP emulation generated actually by the intelligent protocol analyzer.   When SYSTEM A sends a SYN segment, the analyzer gives the following estimation.

```
    A 20:12:46.046386 SYN->
        DstP=1686 SEQ=737600201 CTL=S WIN=24820 MSS=1460
    +   [STATE=SYN-SENT snd_una=737600201 snd_wnd=???
```

```
+    snd_nxt=737600202 snd_max=737600202 snd_wl1=0
+    snd_wl2=0 rcv_nxt=??? rcv_wnd=24820 rcv_una=???
+    rcv_adv=??? iss=737600201 irs=0 mss=1460 cwnd=1460 ssthresh=65535]
+    *** Starting slow start emulation. ***
```

Note that the internal variables are written in lower case and that the state names are different from those in section 3.   When a SYN segment is sent, some internal variables such as snd_wnd and rcv_nxt are not determined in the initiator side and denoted as '???'.   The emulation of the slow start in this side is started from this timing, and cwnd is set to 1460 (one MSS) and ssthresh is set to 65535.

Next, the SYN segment sent by SYSTEM A is received by SYSTEM B and a SYN+ACK segment is replied.   The analyzer emulates this sequence and produce the following results.

```
B 20:12:46.046422 ->SYN
    DstP=1686 SEQ=737600201 CTL=S WIN=24820 MSS=1460
B 20:12:46.046830 <-SYN,ACK
    SrcP=1686 SEQ=37298289 ACK=737600202 CTL=SA
+    WIN=8760 MSS=1460
+    [STATE=SYN-RCVD snd_una=37298289 snd_wnd=24820
+    snd_nxt=37298290 snd_max=37298290 snd_wl1=0
+    snd_wl2=0 rcv_nxt=737600202 rcv_wnd=8760
+    rcv_una=737600202 rcv_adv=737608962 iss=37298289
+    irs=737600201 mss=1460 cwnd=1460 ssthresh=65535]
+    *** Starting slow start emulation. ***
```

By this emulation, all the internal variables have the estimated values in the responder side and the emulation of the slow start is also started.

When SYSTEM A receives the SYN+ACK segment, it replies an ACK segment which is then received by SYSTEM B.   The results of emulation for those are as follows.

```
A 20:12:46.046866 SYN,ACK<-
    SrcP=1686 SEQ=37298289 ACK=737600202 CTL=SA
+    WIN=8760 MSS=1460
A 20:12:46.047122 ACK->
    DstP=1686 SEQ=737600202 ACK=37298290 CTL=A WIN=24820
+    [STATE=ESTABLISHED snd_una=737600202 snd_wnd=8760
+    snd_nxt=737600202 snd_max=737600202
+    snd_wl1=37298289 snd_wl2=737600202 rcv_nxt=37298290
+    rcv_wnd=24820 rcv_una=37298290 rcv_adv=37323110
+    mss=1460 cwnd=1460 ssthresh=65535]
B 20:12:46.047158 ->ACK
    DstP=1686 SEQ=737600202 ACK=37298290 CTL=A WIN=24820
+    [STATE=ESTABLISHED snd_una=37298290 snd_wnd=24820
+    snd_nxt=37298290 snd_max=37298290 snd_wl1=737600202
+    snd_wl2=37298290 rcv_nxt=737600202 rcv_wnd=8760
```

+   rcv_una=737600202 rcv_adv=737608962 mss=1460

+   cwnd=1460 ssthresh=65535]

Since the communication monitored is the data connection for ftp, SYSTEM A then begin transmitting data. Now the current cwnd is 1460 in the initiator size, and only one data segment is sent. The sending and receiving of the data segment are emulated in the following way.

A 20:12:46.049571 DT->

    DstP=1686 SEQ=737600202 ACK=37298290 CTL=PA

+   WIN=24820 LEN=1460

+   [STATE=ESTABLISHED snd_una=737600202 snd_wnd=8760

+   snd_nxt=737601662 snd_max=737601662

+   snd_wl1=37298289 snd_wl2=737600202 rcv_nxt=37298290

+   rcv_wnd=24820 rcv_una=37298290 rcv_adv=37323110

+   mss=1460 cwnd=1460 ssthresh=65535]

B 20:12:46.050771 ->DT

    DstP=1686 SEQ=737600202 ACK=37298290 CTL=PA

+   WIN=24820 LEN=1460

+   [STATE=ESTABLISHED snd_una=37298290 snd_wnd=24820

+   snd_nxt=37298290 snd_max=37298290 snd_wl1=737600202

+   snd_wl2=37298290 rcv_nxt=737601662 rcv_wnd=8760

+   rcv_una=737600202 rcv_adv=737608962 mss=1460

+   cwnd=1460 ssthresh=65535]

Then SYSTEM B sends an ACK segment for the received data segment as follows. We can estimate that this ACK is sent by 200 msec timeout from the timestamp for this segment.

B 20:12:46.222792 <-ACK

    SrcP=1686 SEQ=37298290 ACK=737601662 CTL=A WIN=8760

+   [STATE=ESTABLISHED snd_una=37298290 snd_wnd=24820

+   snd_nxt=37298290 snd_max=37298290 snd_wl1=737600202

+   snd_wl2=37298290 rcv_nxt=737601662 rcv_wnd=8760

+   rcv_una=737601662 rcv_adv=737610422 mss=1460

+   cwnd=1460 ssthresh=65535]

When SYSTEM A receives this ACK segment, it increments its cwnd from 1460 to 2920 and sends two data segments. These behaviors in SYSTEM A are emulated as follows.

A 20:12:46.222828 ACK<-

    SrcP=1686 SEQ=37298290 ACK=737601662 CTL=A WIN=8760

+   [STATE=ESTABLISHED snd_una=737601662 snd_wnd=8760

+   snd_nxt=737601662 snd_max=737601662

+   snd_wl1=37298290 snd_wl2=737601662 rcv_nxt=37298290

+   rcv_wnd=24820 rcv_una=37298290 rcv_adv=37323110

+   mss=1460 cwnd=2920 ssthresh=65535]

A 20:12:46.223298 DT->

```
        DstP=1686 SEQ=737601662 ACK=37298290 CTL=A
  +     WIN=24820 LEN=1460
  +     [STATE=ESTABLISHED snd_una=737601662 snd_wnd=8760
  +     snd_nxt=737603122 snd_max=737603122
  +     snd_wl1=37298290 snd_wl2=737601662 rcv_nxt=37298290
  +     rcv_wnd=24820 rcv_una=37298290 rcv_adv=37323110
  +     mss=1460 cwnd=2920 ssthresh=65535]
  A 20:12:46.224529 DT->
        DstP=1686 SEQ=737603122 ACK=37298290 CTL=PA
  +     WIN=24820 LEN=1460
  +     [STATE=ESTABLISHED snd_una=737601662 snd_wnd=8760
  +     snd_nxt=737604582 snd_max=737604582
  +     snd_wl1=37298290 snd_wl2=737601662 rcv_nxt=37298290
  +     rcv_wnd=24820 rcv_una=37298290 rcv_adv=37323110
  +     mss=1460 cwnd=2920 ssthresh=65535]
```

When SYSTEM B receives these two data segments, it sends one ACK segment immediately according to the rule that every other data segment is acknowledged[2]. When SYSTEM A receives this ACK segment, then it increments its cwnd by one segment size from 2920 to 4380 in the following way.

```
  A 20:12:46.227211 ACK<-
        SrcP=1686 SEQ=37298290 ACK=737604582 CTL=A WIN=8760
  +     [STATE=ESTABLISHED snd_una=737604582 snd_wnd=8760
  +     snd_nxt=737604582 snd_max=737604582
  +     snd_wl1=37298290 snd_wl2=737604582 rcv_nxt=37298290
  +     rcv_wnd=24820 rcv_una=37298290 rcv_adv=37323110
  +     mss=1460 cwnd=4380 ssthresh=65535]
```

5 DISCUSSIONS

(1) It is considered that the intelligent protocol analyzer is used effectively for the detailed analysis of TCP/IP communications. Especially, it is helpful to analyze the behavior of TCP internal procedures for flow control. For example, the analyzer can estimate the numbers of invocation of the slow start and congestion avoidance algorithms. It can also estimate the number of DATA segment retransmission. Since TCP/IP is rather a mature protocol, it is considered that available protocol software does not include so many protocol errors. The intelligent protocol analyzer is used to analyze the details of communication in the case of problems such as the throughput degradation.

(2) It is possible that the intelligent protocol analyzer performs a wrong estimation of event sequence and a wrong TCP emulation. For example, the estimation of event does not take account of the buffering delay in routers. When such buffering delay is larger than propagation delay and transmission time, estimated

event sequence by the analyzer may be wrong in the actual processing order. In order to cope with such cases, it is required to reorder event sequence when any protocol error is detected during the emulation. It is possible to apply the rule based programming to implement such reordering based on heuristic algorithm[7].
(3) By analyzing the behaviors of commercial TCP/IP software by the analyzer, we have obtained some information such as:
- that the TCP software in the Window 95 operating system does not support the fast retransmit and fast recovery algorithms, and
- that the TCP software in Solaris 2.5 might not conform to the congestion avoidance after the fast retransmit algorithm.

6 CONCLUSION

In this paper, we have described the intelligent protocol analyzer which supports the interoperability testing of TCP/IP protocols. It provides the function which can estimate what communication has taken place by emulating the behaviors of the TCP protocol entity in a pair of communicating computers. Since modern TCP includes some internal procedures for flow control, such as the slow start algorithm, our analyzer can emulate these procedures as well as the state transition based behaviors of the basic TCP. We have also presented the implementation results of the analyzer.

The Intelligent protocol analyzer is effective in analyzing TCP/IP communication, including counting how many times the slow start algorithm is invoked and how many DATA segments are retransmitted by the timeout retransmission and the fast retransmit algorithm.

7 ACKNOWLEDGMENT

The authors wish to thank Dr. H. Murakami, Managing Director of KDD, for his continuous encouragement to this study.

8 REFERENCES

[1] DARPA Internet Program Protocol Specification (1981) Transmission Control Protocol. *RFC 793*.
[2] Stevens, W.R. (1994) TCP/IP Illustrated, Vol. 1 : The Protocols. Addison Wesley.
[3] Stevens, W.R. (1997) TCP Slow Start, Congestion Avoidance, Fast Retransmit, and Fast Recovery Algorithms. *Request for Comments: 2001*.
[4] Moldeklev, L. and Gunningberg, P. (1995) How a Large ATM MTU Causes

Deadlock in TCP Data Transfer. *IEEE Trans. On Networking*, **3**.

[5] Tekelec (1992) Chameleon User's Manual.

[6] Kato, T., Ogishi, T., Idoue, A. and Suzuki, K. (1997) Design of Protocol Monitor Emulating Behaviors of TCP/IP Protocols. To appear in *Proc. of 10th International Workshop on Testing of Communicating Systems*.

[7] Kato, T., Ogishi, T., Idoue, A. and Suzuki, K. (1996) Design of Protocol Interoperability Testing System based on Rule Base Programming. in *IPSJ SIG Notes*, **DPS 77-5**.   (in Japanese)

9   BIOGRAPHY

Toshihiko Kato is the senior manager of High Speed Communications Lab. in KDD R&D Labs.   Since joining KDD in 1983, he has been working in the field of OSI, formal specification and conformance testing, distributed processing, ATM and high speed protocols.   He received the B.S., M.E. and Dr. Eng. Degrees of electrical engineering from the University of Tokyo, in 1978, 1980 and 1983 respectively.   From 1987 to 1988, he was a visiting scientist at Carnegie Melon University.   Since 1993, he has been a Guest Associate Professor of Graduate School of Information Systems, in the University of Electro-Communications.

Tomohiko Ogishi is a member of High Speed Communications Lab. in KDD R&D Labs.   Since joining KDD in 1992, he worked in the field of computer communication.   His current research interests include the protocol testing on TCP/IP communication.   He received the B.S. Degree of electrical engineering from the University of Tokyo in 1992.

Akira Idoue is a research engineer of High Speed Communications Lab. In KDD R&D Labs.   Since joining KDD in 1986, he worked in the field of computer communication.   His current research interests include implementation of high performance communication protocols and communication systems.   He received the B.S. and M.E. Degrees of electrical engineering from Kobe University, Kobe, Japan, in 1984 and 1986 respectively.

Kenji Suzuki is the senior manager of R&D Planning Group in KDD R&D Labs.   Since joining KDD in 1976, he worked in the field of computer communication.   He received the B.S., M.E. and Dr. Eng. Degrees of electrical engineering from Waseda University, Tokyo, Japan, in 1969, 1972 and 1976 respectively.   From 1969 to 1970, he was with Philips International Inst. of Technological Studies, Eindhoven, The Netherlands as an invited student.   He received Maejima Award from Communications Association of Japan in 1988, Achievement Award from the Institute of Electronics, Information and Communication Engineers in 1993, and Commendation by the Minister of State for Science and Technology (Persons of scientific and technological research merit) in 1995.   Since 1993, he has been a Guest Professor of Graduate School of Information Systems, in the University of Electro-Communications.

29

Eight years of experience in test generation from FDTs using TVEDA

Roland Groz, Nathalie Risser
FRANCE TELECOM CNET DTL/MSV,
2 avenue Pierre Marzin, F-22307 Lannion Cedex, FRANCE
Tel: +33 2 96 05 11 11, Fax: +33 2 96 05 39 45,
E-mail: (groz,risser)@cnet.francetelecom.fr

Abstract

This paper relates France Télécom experience with automatic test generation from FDT. We describe our approach and the evolution of tools and practice over 8 years. Both technical and non-technical causes for evolution are analysed. We provide data on two dozens of industrial applications of FDT-based test generation in our organization.

Keywords

test generation, conformance testing, FDT-based protocol engineering, FDT tools, Estelle, SDL, TTCN

1   INTRODUCTION

Conformance and acceptance testing of network equipment is a major concern for a

Formal Description Techniques and Protocol Specification, Testing and Verification
T. Mizuno, N. Shiratori, T. Higashino & A. Togashi (Eds.)   © 1997 IFIP. Published by Chapman & Hall

telecommunication operator. Integrating network elements that come from different sources can be reliable only if it is possible to ensure that they will interoperate in conformance with their interface definitions. Actually, a major part of our R&D centre (CNET for France Télécom) is dedicated either to specification of network interfaces (protocols and services) or to validating network products from telecommunication manufacturers through testing. A key point that should be kept in mind is that the staff involved in testing activities is at least as numerous as those involved in specification.

FDTs had been initially designed to enhance the quality of specifications. Until the end of the previous decade (1980s), CNET invested in FDT tools (for Estelle and SDL) mainly on the specification side: editing, simulation, verification, and a tiny interest in code generation. The goal was to provide support for better (more reliable) ways of designing protocols and services.

Somehow in the end of the 1980s, we started to realize (!) that FDTs could provide a good basis for supporting some aspects of testing. We think that we just followed a general trend : when we look at papers presented in the PSTV (and later in FORTE) conferences, we feel that although testing was present from the beginning, the FDT sub-community broadened its activities to testing in the beginning of the 1990s (of course, there were pioneers before that time; also, IWPTS was started in 1988 by B. Davis, S. Chanson and S.T. Vuong). The item on Formal Methods in Conformance Testing [ISO 95] was started in ISO and CCITT at the same period.

Anyway, the use of FDTs to support test activities (mainly test generation in our case) has proved a key factor for the extension of FDT practice in our organization. In fact, our tool TVEDA has been the driving force for the use of SDL within CNET. Thanks to the results gained from using automatic test generation with TVEDA, interest in specifying in FDT has risen drastically, a feat which a decade of promoting verification techniques had never achieved.

This paper provides an overview of the development and use of our test generation technique over the past 8 years (1989-1996). Section 2 introduces the background that motivated the development of such a method, the approach we followed and the evolution of our tool. Section 3 presents a collection of data on our case studies. Section 4 explains how the expectations of customers of test generation techniques have evolved. Section 5 concludes on the main lessons we get from those 8 years of experience.

2   MOTIVATIONS AND TECHNICAL APPROACH

2.1   Background

Our work on test generation was triggered by a request from the ISDN test division in CNET. At that time (1989), our own division had already been working for 8 years on protocol verification techniques (first with Petri nets and FSM-based approaches, later on with Estelle, simulation and model-checking techniques).

People in the ISDN test division felt relatively comfortable with their test execution environment : they had convenient ISDN testers (viz. machines), they were applying IS 9646 (standardized in 1989)[ISO 94] and were relatively happy starting to use TTCN. Therefore, they did not expect anything from us on test architecture (a hot topic of the early PSTV workshops) or test specification (it is a well known fact that TTCN emerged as an ad hoc language - sorry, notation! - designed by test experts for test experts who had grown wary and weary of FDT inadequacies).

They asked for help on two topics:

1. Optimizing the order of test runs (i.e. the order of test cases). That first problem was addressed by a different group in our division (using expert-systems and constraint logic programming).

2. Automating the design of test suites, i.e. automatic test generation. The main concern at that time was not only the fact that writing test suites was a lengthy and costly job (approximately 1 man-year for an ISDN protocol), but also that the maintenance of test suites was costly, because a new test suite had to be redesigned each time a new version of the protocol was standardized. It was felt that even if automatic test generation was costly for the first test suite generation, it could prove cost-effective for the maintenance of it. This second problem could be solved based on our FDT background.

At the same time, we knew that test experts:

- were rather sceptical towards formal descriptions
- found that the existing theoretical work on test generation from FSM was wholly inapplicable to their purpose.

2.2   State of the art in 1988 and inadequacies

When we started in 1988-1989, papers had been published on test generation for protocols. Most of those papers dealt with FSM, and were either based on previous work on the FSM model, such as the DS[Gonenc 70] or W[Chow 78] methods, or adapted for protocols, such as Wp[Fujiwara 91], UIO[Sabnani 88], UIOv[Vuong 89] and UE[Cavalli 92] methods.

Such methods produce tests in the form of (usually just one) long test sequences. If that sequence is correctly accepted by an implementation, then the test is passed, otherwise the implementation is considered non-conformant. This type of test suite is of no use for a test lab, because the information it provides is too simplistic :

1. Most of the elements that make a test suite useful (in fact just usable) are missing like test suite structuring and test identification.

2. Conformance testing for software based protocol implementations is very different from acceptance checks on chips (which was the initial framework for DS & W methods).

Also, a major drawback for the theoretical FSM methods is the length of the sequences they generate (in the hundreds or thousands of events even for toy FSMs of less than 10 states). Small bounds on the implementation are unrealistic; therefore test labs prefer using convenient test suites of a limited size rather than would-be exhaustive test sequences whose exhaustiveness is based on unrealistic assumptions. In [Groz 95], we discuss more thoroughly the research issues in test generation.

2.3   Our approach : empirical and pragmatic

Our analysis of the expectations of test experts and their assessment of the state of the art in automatic test generation method resulted in our choice of approach which we qualified as both empirical and pragmatic [Phalippou 90a]. Our starting point was the request from our test experts to provide them with a test suite that was as close as possible to an existing test suite they used for the LAPD protocol (viz. one from the european CTS-WAN project). LAPD is the link layer protocol for ISDN D-channel.

Our approach was empirical because we observed the human mental process (as could be inferred from its inputs and outputs) to elicit the underlying algorithmic principles for test generation. By design, the tool should directly meet the current needs of the LAPD european test centres, but above all it should be generic enough to provide valuable tests for other protocols than just LAPD.

It was also pragmatic because we concentrated on those parts where the added value of automatic test generation could be rapidly felt and did not need too much research or development. Our initial goal was to generate 90% of the existing (human generated) test suite, considering that the 10% left could be left to human ingenuity. Given the high cost of test design, computerizing 90% of the task should prove very cost-effective. In the case of LAPD, we achieved this goal by leaving out, in the first experiment, the generation of test preambles, postambles and test constraints (which altogether amounted to only about 50 pages of TTCN out of the 800 total). Details are reported in [Phalippou 90b]

The further development of TVEDA confirmed our initial intuitions based on a case study and the generalization of its observed principles.

First, we proved that the test parameters that are fed into the algorithms of TVEDA were sufficiently generic to reproduce another completely different test suite for the same protocol, namely the ICOT north-american test suite [Phalippou 91]. Then we applied our tool to more and more specifications, as will be seen in section 3.

2.4   Tool development

The evolution of the tool is summarized on figure 1.

Following a preliminary study on the algorithms, the first version (called V0) was developed in a few weeks in the autumn of 1989, and later on slightly enhanced in 1990 and 1991. It was developed on the Prolog-based environment that we had developed for our Estelle simulation & verification tool called Veda. Rapid prototyping of our ideas was made easy because we found in this environment all the building blocks needed (parsing, grammar transformers etc.) and the language used was high-level (Prolog + a specific language for describing grammar transformations called Spécode, see [Monin 89]).

It was later decided to move to a CASE software environment called Concerto. TVEDA V1 was therefore functionally very close to V0 except for the software base, which implied a complete reprogramming of our prototype into the object-oriented Lisp language of Concerto called Ulysse. From then on, TVEDA V0 could be considered as a throw-away prototype. The advantage of using Concerto was the ability to integrate easily with the various environments that existed inside Concerto, in particular for TTCN and SDL. The disadvantage was that software development of TVEDA was made more difficult, much slower and time consuming. On top of that, we had to follow Concerto versions (evolution). With today's technology, we would probably choose a different approach, keeping our tool separate and interfering with other tools through a CORBA based Open ToolBus architecture. In V2, the front-processing was redesigned so that both Estelle and SDL specifications could be used as inputs; to that end, an internal common EFSM representation in

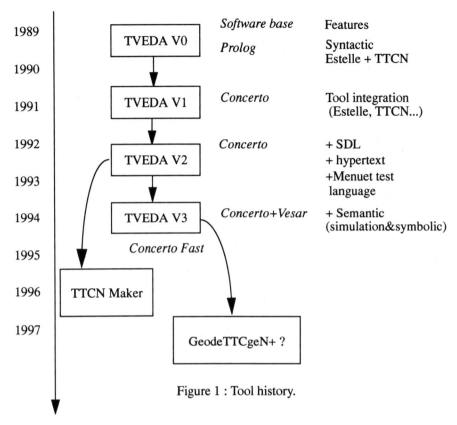

1989	TVEDA V0	*Software base*	Features
		Prolog	Syntactic
1990			Estelle + TTCN
1991	TVEDA V1	*Concerto*	Tool integration
			(Estelle, TTCN...)
1992	TVEDA V2	*Concerto*	+ SDL
			+ hypertext
1993			+Menuet test
			language
1994	TVEDA V3	*Concerto+Vesar*	+ Semantic
			(simulation&symbolic)
1995	*Concerto Fast*		
1996	TTCN Maker		
1997	GeodeTTCgeN+ ?		

Figure 1 : Tool history.

Lisp objects was redefined. Later on, the same approach was taken to produce tests either in TTCN or for an in-house test language and environment called Menuet.

Following our pragmatic approach of developing an unsophisticated tool that could produce 90% of the TTCN code (to be completed by a test expert), the first versions of TVEDA (V0 until V2) only implemented the *syntactic method*, which we sometimes also call (test) *skeleton* strategy. This means that only syntactical translations are made on the Estelle or SDL code. Therefore, the tests produced cannot chain several transitions, and correspond to the *single transition test method* paradigm (each test = preamble to start state + 1 transition + postamble from final state). However, this syntactic approach is intrinsically limited : it cannot compute values in constraints, and it cannot generate preambles and postambles.

In fact, our pragmatic approach was only of the «first things first» kind. From the beginning (in 1988-1989), we had planned that, as a second step, we would enrich our tool with semantic computations. Two approaches have been tried and implemented [Clatin 95].

a. «Symbolic computation» [Phalippou 94], in fact a mixture of symbolic execution and of a technique called abstract interpretation [Cousot 77]. This approach turned out to be too limited to process real specifications.

b. «Simulation», in fact coupling TVEDA to a model-checker for Estelle, called Vesar. Thanks to our front-processing architecture, reachability can be computed for both SDL and Estelle specifications (the internal common EFSM representation is decompiled into Estelle for Vesar). This approach proved workable, thanks to well-designed heuristics to adapt the specification to test generation purposes and to prune the reachability graph.

With the new version, the test suites were complete: unlike previous versions, they included test declarations: PCO, ASP, PDU and constraint values and the test cases were complete with the description of preambles and postamble. Simultaneously, the test purposes evolved from one test for one transition to a set of tests going through a set of chosen branches of transitions. These test purposes could be expressed (and phrased) according to format expressions depending on parameters of the tested path.

Implementing this major addition in test generation strategy resulted in TVEDA V3, started in 1993, but which was only finalized in early 1996.

In the meantime, CNET and a consortium of companies led by Sema Group had responded to a European Commission call for tender based on the then existing technology, i.e. TVEDA V2. The corresponding project, called INTOOL/CATG took place from the spring of 1995 to the spring of 1997. The outcome is a commercial tool called TTCN Maker [Desécures 97], based on a new technology called Fast which Sema Group developed for its Concerto environment.

Talks are now in progress with Vérilog and INRIA to integrate TVEDA V3 in the Geode environment along with test case computation algorithms from the TGV tool [Fernandez 96]. In that case, the reachability analysis would be done directly with Geode tool, instead of the Estelle-based Vesar tool. We hope to be able to lift the «single SDL process» current restriction of TVEDA. The main drawback will be the loss of the ability to input Estelle specifications.

3   EXPERIENCE IN TEST GENERATION

3.1   Protocols and services experimented

In table 1, we mention most relevant automatic test suite generations performed

Table 1 : Collected applications.

Application name	Size	Effort	Nb of tests	Cover age	Tests produced
INRES	428		11 [#]	100%	OK
LAPD [*]	9500/800	3 m-m	765	90%	OK + 37d
ISDN D $_u$ [*]	4500/700	2 m-m	681	90%	OK + 39 d
ISDN D $_n$ [*]	10500/1000	5 m-m	2321	90%	OK + 38 d
ISDN D $_{R6 n}$	1800		131	90%	OK

Table 1 : Collected applications.

Application name	Size	Effort	Nb of tests	Cover age	Tests produced
MAP	1800/100	3 m-m	198 [#]		not relevant
RTSE [*]	5400/200	1.5 m-m			inadequate
SDLC	500		34 48 16[#]	100%	~ OK
TP	4200	0.5 m-m	10 [#]		not relevant
HOLD	5000	0.5 m-m	23 [#]	80%	not relevant
SS1	1673	/	76 [#]	90%	OK
SS2	1960	/	10 [#]	90%	OK
SS3	1792	1.5 m-m	106 [#]	90%	OK
SS4	1757	/	28 #	90%	OK
SS5	1801	/	172	90%	OK
P1 X400[*]	7600/115	3 m-m			failure
SDH mngt	348	0.5 m-m	6 [#]	100%	OK
SSCOP	2100/95	3 m-m	213 [#]	90%	in use
PME[*]	6000				inadequate
Q2931-n	3300		158 [#]	90%	OK
IN S1	689	0.5 m-d	57 [#]	81%	in use
IN S2	2800	2 m-d	96 [#]	87%	in use
IN S3	1100	1 m-d	96 [#]	100%	in use
IN S4	1100	1 m-d	115 [#]	100%	in use

using TVEDA until 1996, regardless of the versions and strategies of the tool. Actually, other specifications were «tested» on TVEDA, but those we mention here correspond to applications for which a client team expressed interest in test generation (except INRES, a well known toy example which is presented here just for reference).

The specifications are presented in approximate *chronological order* (again, except INRES). The chronology refers to the time when the test suite considered in the subsequent columns was generated with TVEDA (not the time when the formal specification was written).

3.2   Organization of the table

The first column lists the names of the applications (protocols or services) supplied to TVEDA. The specifications are often written in LDS, the specifications annotated with an asterisk have been described in Estelle.

The specifications called «SSx» correspond to five military ISDN Supplementary Services represented by five different specifications (undetailed for confidentiality reasons).

Similarly, names of the form «IN Sx» correspond to (undetailed) Intelligent Network Services.

The expression «SDH mngt» is a shorthand for an experiment with a part of a protocol involved in SDH management.

The small letters «n» and «u» in indicate which side a specification describes: n for network and u for user.

The second column «Size» gives the number of lines of the specification including the comments; the optional second number (following a «/») is the number of extended transitions of the specification when it is written in Estelle.

The third column «Effort» indicates the formalization effort (in manpower : m-m stands for man-month) to get the specification in a formal language. The figures represent the effort spent for each specifier regardless of his or her experience in formalization, in the domain or in the formal language. Moreover, although most specifications were entirely written for test generation, others which already existed were simply completed (or made more formal, in the case of existing «pseudo-SDL» specifications). Basically, when the effort is over 1 man-month, it corresponds to a formal specification written entirely from a natural language. Efforts counted in man-days correspond to adaptations of existing formal descriptions.

The fourth column «nb of tests» gives the number of automatically generated tests if it was possible to generate them and in this case if they were significant. For the SDLC Protocol, three test suites were generated from one single specification. When the «reachability tree analysis» strategy was used to get the test suite, the symbol: # indicates it, otherwise when the syntactic method was used, no symbol appears.

The fifth column «Coverage» alternatively presents two coverage rate depending on the two TVEDA strategies. Both are instances of what we presented as «specification coverage» in [Groz 96]. The syntactic strategy is linked to the following rate: number of tests divided by the number of (extended) transitions. The «reachability tree analysis» strategy is linked to the following rate: number of tests divided by the number of branches of transitions (the difference may be significant for SDL).

The sixth column «Tests produced» gives a snapshot appraisal of the results of the experiments. The following verdicts are used, listed from the worst to the best :

1. failure (TVEDA was unable to generate any test)
2. inadequate (the tests produced were not considered by test experts to be satisfactory for this type of protocol)

3. not relevant (sensible tests could be produced, but not really considered satisfactory by experts for various reasons)
4. OK (test experts considered the tests were OK to be used in testing); the phrase «+ X d», used for earlier experiments with TVEDA, refers to a test suite which could be considered OK once completed by a test expert for those parts which TVEDA was unable to produce automatically at that time (mainly declarations); in that case, X stands for the number of days needed for completion;
5. in use (the test team actually uses the automatically generated test suite instead of one written by a test expert).

3.3 Fields addressed

The following applications correspond to ISDN protocols or services : the three specifications for the ISDN D-channel protocol (i.e. layer three of this protocol), LAPD (layer 2), and all the services referring to ISDN SS (mind : in ISDN words, TP stands for «Terminal Portability», not «Transport Protocol» or «Transaction Processing» as in the OSI world!).

Broadband communication is covered by two ATM-related protocols : SSCOP [Dinsenmeyer 97] and Q2931.

Intelligent Network services are somehow similar (in nature) to ISDN supplementary services.

RTSE (CCITT X.410) and P1 (CCITT X.409) are OSI layer 7 (application) protocols related (at least originally) to Message Handling Systems (X400 series of standards).

Communication with mobiles is represented by the MAP protocol of the GSM standard for wireless (cellular) telephones.

Finally, we find a few unrelated protocols : SDLC (data networks), SDH management, PME (smart card application).

As can be seen, TVEDA has proved to be applicable to a large variety of fields. Also, the size of the specifications is quite typical of the protocols in those fields, and show that there is no particular limit on the size of the applications handled.

3.4 Writing formal specifications

One of the most important lessons from this experience is the low cost in writing formal specifications. Until about 5 years ago, there was a widespread belief in CNET (and we think elsewhere as well) that writing formal specifications was a heavy task, which implied specialized manpower (preferably experts in FDTs) and required an effort at least as important as writing (from scratch) the corresponding specification in natural language. Therefore, the task of writing formal specifications was doubted to be cost effective. Our results show that test generation alone may be sufficient to justify writing a formal description.

In our opinion, this low cost in writing FDT specifications can be attributed to the following factors :

1. Estelle and SDL are FSM based techniques easily understood by rank-and-file engineers in an R&D centre such as CNET (in a majority of cases, especially the older ones, specifications were written by novice users of FDTs).

2. They are supported by efficient tools (the key factor in this respect is not so much in editing as in validation and simulation).

3. Writing a formal specification from an informal but precise one is more a matter of translation than really designing a specification.

4. Protocols and basic services have a rather low complexity : most of the inherent complexity of a network lies in its architecture and layering; once you get down to a single element (one protocol module), it is a relatively straightforward automaton. Also, the size results more from the enumeration of transitions for different states and types of messages than from embedded structures or functions, so that the process of writing specifications is a highly repetitive one.

3.5   Incomplete test suites

Until 1994, TVEDA was used to generate test suites with the syntactic method (see Tool development page 4). Therefore the final customer had to finalize the test suite. In the early period, exhaustive assessments had been done on the Link Layer for ISDN D Protocol, so we give in this paper the figures for those assessments, even though the current release of TVEDA is more powerful.

After 1995, no completion effort is required any longer.

3.6   Non significant results

Two main reasons explain why some results are inadequate or not relevant.

Firstly, for TVEDA, one SDL process or one Estelle body represent the black box and consequently the Implementation Under Test. All its interfaces can be considered as PCOs, or at least can have indirect effects on test generation. Consequently, all specifications which describe a protocol or a service distributed on several processes or bodies or with interfaces which mix messages from PCOs and uncontrolled messages will not have a significant test suite. Some specifications which already existed before being handled by TVEDA do not always take into account these TVEDA assumptions, so the tests generated from them are inadequate. Such tests are only satisfactory according to the structure of the specification. This is the case for the following specifications : MAP, PME, TP.

The case of RTSE is slightly different, because the primary goal for writing this formal specification (in Estelle) was not to generate a test suite, but to validate an existing one. Test validation was conducted by feeding test scenarios into an Estelle simulation of the specification with the tool EWS (Estelle WorkStation [Ayache 89]). This was quite successful, as 16 errors where quickly found in the test suite which was a draft CCITT recommendation, for a total investment (formalization included) of less than 2 man-months.

Secondly, TVEDA interprets the specification and some parameters supplied by the user on the specification. If the specification describes the protocol or the service with too many details or at an inappropriate level of abstraction, sometimes TVEDA cannot generate any test suite, or in most cases, TVEDA will generate test suites with the same inappropriate level of abstraction. The specifiers generally corrected their specification to obtain a significant test suite. The cost of such corrections has been included in the «formalization effort» column of our table.

3.7   Evolution of the use of tests generated : from case studies to operational

tests

Results of test generations vary from failure to «in use». In fact, different kinds of results were expected from our experiments. Some of them were purely case studies for test generation; in other instances, an operational test suite was expected.

Obviously, the LAPD experiment was a pure case study, since the tests already existed and were in use. In accordance with the objectives of our experiment, TVEDA might have been used for later versions of the protocol, but this did not occur (LAPD has remained stable).

Later, specifications were used to judge the capabilities of TVEDA to be applied to different fields. Less and less, issued test suites were compared with manually written already existing test suites. More and more, automatically generated test cases were compared to parts of specification they test. The relevance of test purposes was analysed according to the usual testing practice, depending on the kind of protocol or the network field (ISDN, MHS etc.). The specifications of ISDN D protocol for network side or for user side, RTSE, P1, MAP... were concerned.

Things have evolved, and most generated test suites are now put into the field. This is the case both for conformance testing, e.g. for protocols such as SSCOP [Dinsenmeyer 97], or for other forms of acceptance testing (e.g. test suites given to industrial implementers) for instance in the case of Intelligent Network services.

4   SHIFTING FROM PRODUCTIVITY TO QUALITY CONCERNS

4.1   Exhaustive assessment of preliminary ISDN results

This first experiment was conducted on the LAPD protocol. Rapidly set up in 1989, this experiment had practically shown the benefits that could be gained from an automatic test generation tool based on our empirical approach [Phalippou 90b].

The manually written test suite consisted of 646 tests. A test expert for this field is supposed to write an average of three test cases during one day. So eleven months are required to write such a test suite.

Three months were necessary to get the specification in the Estelle formal language (including first level validation of this Estelle code). Approximately, one week allowed to adjust TVEDA parameters to obtain a significant result.

In order to have a proper assessment of the quality of the automatically generated test suite, we decided to have it audited by an independent expert (from a company outside CNET). In 1993, this expert made a thorough assessment, comparing each test case from the TVEDA test suite with the CTS-WAN tests. The conclusion was that two months would have to be spent to get an operational test suite from the automatically generated test suite. This proved that the cost of test generation with TVEDA was only 50% of the cost without such a tool.

This 50% gain had been estimated for V2 of TVEDA. Although we did not pay an external auditor for the following versions, it is clear that most of the «completion time» (2 months in the case of LAPD) was no longer needed for the newer versions. This means that the cost would now go down to 30% which means a threefold improvement.

Nevertheless, the test suites still had to be completed. Some tests were missing:

principally test cases which contain errors on timers, or on identifiers. Identifiers did not follow the usual rules agreed by ISDN test experts. All errors could be corrected.

Anyway, this first experiment showed that cost effectiveness could already be achieved. Of course, since the tool had been built for LAPD, this was not convincing enough, and we needed more experiments.

4.2   Enlargement of test generation field

So our next task, between 1991 and 1993, consisted in experimenting TVEDA with significant specifications from various fields. Several experiments began with Estelle specifications. These specifications had significant sizes, between 4000 and 10000 lines, to assess the capabilities of TVEDA. In parallel, there were requests for accepting specifications written in SDL, the formal standard language of ITU. So TVEDA was enriched with SDL input capacity.

As can be seen on table 1, TVEDA proved efficient for lower layer protocols. However, the syntactic approach was wholly unfitted to upper layer protocols such as RTSE or P1. For this type of service, typical test purposes refer not only to the messages sent in one transition but above all to global procedures as a set of exchanged messages or functionality of the service. The state concept is completely missing in these test purposes. More over, TVEDA V1 and V2 indicated only a syntactical form of constraint values, the form roughly derived from the specification. In the experiment for RTSE, the description of a constraint was as important as the description of the test case, so real constraint values had to be computed to get useful test suites.

4.3   From complete test cases to an increasing test quality

Between 1993 and 1995 V3 was designed to provide an alternative strategy (namely the semantic one) to circumvent the problems encountered on upper layer protocols. The tool generally satisfied its customers, apart from its major (still in force) restriction: only one SDL process or one Estelle body can be targeted as the black box to test. If this restriction is not observed, the test suite may not be completely significant. Apart from this restriction, the tests produced by TVEDA are now considered to be reliable, and in particular more reliable than tests written by an expert; actually, writing tests is error-prone, and automatic test generation avoids this drawback. Therefore, although the first advantage of the TVEDA approach was cost effectiveness, the quality of tests has now become a prominent issue (in favour of TVEDA). For its customers, TVEDA brought benefits mainly in two directions:

1. firstly for big size specifications, an enforcement to reduce global time to produce test suites
2. secondly for smaller specifications, an improvement of the quality of automatically generated test suites (thanks to the use of simulations and reachability analysis).

4.4   TVEDA customers

Between 1993 and 1995, our typical customers would rely on the TVEDA development team to run the tool on their specifications and provide them with a ready-to-use test suite. The demand of customers increased because TVEDA V2 and the first elements of V3 had proved the benefits of automatic test generation.

A training course allowed the motivated customers to practise TVEDA and become active users. Since 1996, the prototype TVEDA is now run in CNET on a pool of workstations with the proper software licences, with a direct access for each user.

Who is the typical customer of TVEDA ? There are different cases, corresponding to the successive stages of the life cycle of a specification. One person may define the specification to be considered, another may write the formal specification, another may be responsible for defining the test architecture and parameters, and the test suite may be expected by yet another group (not to mention the ensuing life cycle of the tests).

4.5   How to integrate TVEDA in the new test practices?

As TVEDA is available for customers, their concern lies now in how TVEDA can be integrated in the software life cycle. A practical experiment is today running to assess new approaches of testing with different tools of test generations -now some test generation tools are commercially available-. Another experiment tries to interleave specification steps and test generation steps to compute the global time benefit.

Nevertheless, an astonishing (sic) consequence obviously appears. Automatic test generation requires a formalized specification! Once the formalization is done, the customer gets for free a form which allows him not only to prepare tests but also several fruitful operations: validation, properties verification, animation...

In fact, our experience in CNET has been that before the existence of TVEDA, virtually no group would invest on formal specification. Writing a formal specification was seen as a waste of time, because crude levels of verification (cross-reading) could be done on the informal specification. Also, informal specifications evolved all the time and the formal specifications would be hard to maintain.

With TVEDA, this investment can be easily justified because tests must be written, and writing them is tedious and error prone. Time reduction and the increase in the quality of tests and consequently of the products has proved a powerful incentive for the use of FDTs.

5   CONCLUSION

When TVEDA was started in 1989, no commercial tool was available. Things have evolved now. In the meantime, tools were first developed for computer aided (but not fully automatic) test generation from SDL with STED [Ek 93] (the current commercial name is now TTCN Link), and TTCgeN. Based on TVEDA, a new generation of tools that perform automatic generation of test suites appear. They are fully automatic because they generate test purposes, instead of relying on a human expert to design significant test scenarios. The first commercial tool is TTCN Maker [Desécures 97].

CNET now considers that the development of such tools can be done by tool vendors. We feel that our work on TVEDA has been pioneering in that direction. Here are the main lessons that we get from those 8 years of research and development in test generation.

1. The momentum for the widespread use of FDTs within CNET was provided by test generation applications. We feel that the main reason for that are psychological. People working on the specification side are not eager to use methods that may question their design practice; on the other hand, writing tests is not a rewarding task, it is boring and despised, therefore there is a strong interest in making it smarter with the use of tools.

2. Only 1 year was needed to develop a prototype to prove the feasibility and cost effectiveness of our approach. But 5 further years of case studies and promoting were necessary before test suites produced by TVEDA were actually used by pilot customers. Technology transfer takes longer than technology development, because it takes time to for groups of people to change their practice and move from polite interest to practising.

3. Automatic test generation is feasible, cost effective, and can be integrated in product development.

6   ACKNOWLEDGEMENTS & DRAMATIS PERSONÆ

Although this paper has been written by two actors in the history of TVEDA, the list of contributors is so long that it would by far exceed the capacities of an authors' line on the first page. In this last section, we try to mention the most prominent actors.

Philippe Cousin, then at CNET (and who moved to the European Commission not long after) provided the initial trigger in the form of the problems mentioned in section 2.1. Roland Groz and Marc Phalippou put TVEDA on its track by defining the initial approach (empirical and pragmatic), and planning the two step response (syntactic, then semantic). Marc Phalippou did most of the Prolog implementation of TVEDA V0, with some help from Roland Groz and Jean-François Monin. Claude Hervé from Cap Gemini provided the initial analysis of the CTS-WAN test suite thus contributing to the definition of the algorithms, and Philippe Riou from the same company wrote the LAPD specification in Estelle. Didier Bouësnard, managing director at Cap Gemini Rennes deserves a special mention for his inspired bet and faith in the future of FDT and test generation and his support of TVEDA related activities: our first contacts date back to as early as 1988.

The implementation of TVEDA V1, V2 and V3 was done by the permanent staff who joined the TVEDA project (Martine Brossard-Guerlus, Nathalie Risser, Laurent Boullier, Isabelle Dinsenmeyer and Marylène Clatin) and by students (Gaëtan Offredo, François Simonneau , Rabe Harou, François-Alexandre Fauchier, Roxane Arsaut). Richard Thummel from CENA (french Air control R&D centre) did the feasibility study for the «simulation» part of TVEDA V3 (interface with Vesar) and proposed many interesting solutions. This «simulation» part was implemented by Marylène Clatin with the help of Roland Groz, while Marc Phalippou developed the «symbolic computation» module.

TVEDA also benefited from collaborations we established with other laboratories in test generation. On the international side, CNET collaborated with NTT [Arakawa 92] and BTRL [Boullier 94]. On the MAP protocol, we compared our approach with that of INT [Anido 96]. We also participated in a french consortium [Doldi 96] on a military contract with Claude Jard, Thierry Jéron from IRISA Jean-Claude Fernandez from IMAG [Fernandez 96]), Didier Bouësnard, Sylvie Le Bricquir, Nathalie Texier from Cap Gemini Rennes and Laurent Doldi fromVérilog.

Finally, we also participated in the european consortium for INTOOL/CATG [Desécures 97]. Throughout those years, we also collaborated with colleagues working on test environments in CNET in Paris, especially with Anne Rouger on SDL test selection strategies [Boullier 93] and Michel Martin on Menuet.

On the customer side, Patrice Desclaud, who headed a project on specification and test methodology, was active in promoting TVEDA; in particular, he commissioned the audit on ISDN test suites mentioned in section 4.1; this audit was done by Gérard Guillerm. We are indebted to our customers, who have become enthusiastic users of TVEDA; among those, we would like to mention Serge Gauthier, Jacques Kerbérénès, Marie-Paule Simon and Laurent Dinsenmeyer who have accompanied TVEDA since the beginning of its external uses in 1995.

Finally, we would like to remind that the heart, main contributor and conductor of TVEDA has been Marc Phalippou, who headed the project from 1991 to its end in 1996. Marc was so successful on this project that he has now moved to other strategic projects for France Télécom.

7   REFERENCES

R. Anido, A. Cavalli, T. Macavei, L.P. Lima, M. Clatin, M. Phalippou, *Engendrer des tests pour un vrai protocole grâce à des techniques éprouvées de vérification*, CFIP'96, Rabat, october 1996.

N. Arakawa, M. Phalippou, N. Risser, T. Soneoka, *Combination of conformance and interoperability testing*, in Proceedings of FORTE'92, Lannion, France, 1992.

J.M. Ayache, J. Berrocal, S. Budkowski, M. Diaz, J. Dufau, A.M. Druilhe, N. Echevarria, R. Groz, M. Huybrechts: *Presentation of the Sedos Estelle Demonstrator project*; The Formal Description Technique Estelle, North-Holland (Elsevier), 1989, pp 423-437.

L. Boullier, M. Phalippou, A. Rouger, *Experimenting test selection strategies*, proceedings of SDL forum 93, Darmstadt, October 1993.

L. Boullier, B. Kelly, M. Phalippou, A. Rouger, N. Webster. *Evaluation of some test generation tools on a real protocol example*, in Proceedings of the IWPTS VII - Int Workshop on Protocol Test Systems, Tokyo, Japan, 1994.

A. Cavalli, Sung Un Kim, *Automated protocol conformance test generation based on formal methods for LOTOS specifications*, proceedings of the 5th International Workshop on Protocol Test Systems, Montréal, September 1992.

T. Chow, *Testing Software Design Modelled by Finite-State Machines*, IEEE Transactions on Software Engineering, vol. SE-4, n. 3, may 1978.

M. Clatin, R. Groz, M. Phalippou, R. Thummel. *Two approaches linking a test generation tool with verification techniques*, in Proceedings of the IWPTS VIII - Int Workshop on Protocol Test Systems, Evry, France, 1995.

P. Cousot, R. Cousot, *Abstract interpretation: a unified lattice model for static analysis of programs by construction or approximation of fixed points*, in Proceedings of the 4th ACM Symposium on Principles of Programming Languages, pp 238-252, 1997.

L. Doldi, V. Encontre, J-C. Fernandez, T. Jéron, S. Le Bricquir, N. Texier, M. Phalippou, *Assessment of automatic generation methods of conformance test suites in an industrial context*, proceedings of IWTCS'96, Darmstadt, September 1996.

[A. Ek, J. Ellsberger, A. Wiles, *Experiences with computer aided test suite generation*, in Proceedings of IWPTS VI - Int Workshop on Protocol Test Systems, Pau, France, 1993.

E. Desécures, L.Boullier, B.Péquignot, *The INTOOL/CATG Europen project; Development of an industrial tool in the field of Computer Aided Test Generation*, submitted to IWTCS IX- Int Workshop on Test System, Seoul, Korea, 1997.

I. Dinsenmeyer, S. Gauthier, L. Boullier, *L'outil TVEDA dans une chaîne de production de tests d'un protocole de télécommunication*, in Proceedings of CFIP'97, Liège, Belgium, 1997.

J.C. Fernandez, C. Jard, T. Jéron, C. Viho, *Using on the fly techniques for the generation of test suites*, in Proceedings of CAV'96 (Conference onf Computer Aided Verification), August 1996, Rutgers University, USA.

S. Fujiwara, G. Bochmann, F. Khendek, M. Amalou, A. Ghedamsi, *Test Selection Based on Finite State Models*, IEEE Transactions on Software Engineering, vol. 17, n. 6, juin 1991.

G. Gonenc, *A method for the design of fault-detection experiments*, IEEE Trans. Comput. vol. C-19, june 1970.

R. Groz, M. Phalippou, *La génération automatique de tests est-elle possible?*, in Proceedings of CFIP'95, Rennes, France, 1995.

R. Groz, O. Charles, J. Renévot, *Relating Conformance Test Coverage to Formal Specifications*. Proceedings of FORTE 96, Kaiserslautern, Germany.

ISO, *Information Technology, Open Systems Interconnection, Conformance Testing Methodology and Framework*, International Standard IS-9646, 1994. Also available as recommendations X.290 (for Part 1 of ISO document), X.291 (Part 2)... from ITU-T.

ISO/IEC/JTC1/SC21/P.54 - ITU-T SG10 Q.8, *Formal Methods in Conformance Testing*, Working Draft, March 1995. To become Recommendation Z.500 from ITU-T.

J.F. Monin, *Programmation en logique et compilation de protocoles: le simulateur Véda*, PhD thesis, Université de Rennes No 292, 01/1989.

M. Phalippou, R. Groz, *From Estelle Specifications to Industrial Test Suites, Using an Empirical Approach*, in Proceedings of the IWPTS III - Int Workshop on Protocol Test Systems, Washington, USA, 1990.

M. Phalippou, R. Groz, *Evaluation of an empirical approach for computer-aided test cases generation*, in Proceedings of the FORTE'90, Madrid, Spain, 1990.

M. Phalippou, M. Brossard: *ICOT vs CTS LAPD Test Suites*; Position statement presented at IWPTS'91, Liedschendam, Octobee 1991.

M. Phalippou, *Test sequence generation using Estelle or SDL structure information*, in Proceedings of FORTE'94, Bern, Switzerland, October 1994.

K. Sabnani, A. Dahbura, *A Protocol Testing Procedure*, Computer Networks and ISDN Systems, vol. 15, n. 4, 1988.

S. Vuong, W. Chan, M. Ito, *The UIOv-Method for Protocol Test Sequence Generation*, proceedings of the second International Workshop on Protocol Test Systems, Berlin, october 1989.

Distributed Object Consistency in Mobile Environments

Distributed Object Consistency in Mobile Environments

Henry Chang
Mobility Software, IBM T.J. Watson Research Center
Route 134, Yorktown Heights, NY 10598, USA
phone: +1-914-945-1831   fax: +1-914-945-4297
hychang@watson.ibm.com

Abstract

Distributed object has emerged as a popular paradigm for constructing large scale distributed applications. In both internet and intranet environments, major infrastructures such as Java RMI, Corba ORB, and Active-X are being used as the new basis of client/server applications. These infrastructures are available in multiple platforms and support a rich set of functionality which includes remote object invocation, component-based application composition, and mobile code migration. While the object paradigm has reduced the complexity of writing distributed applications, it did not reduce the difficulty of creating end-to-end solutions that guarantee robustness and reliability. Distributed consistency is a problem that exemplify the complexity.

In mobile environments, the performance of a distributed object application is strongly affected by the latency and speed of the network, where remote dial-in via a slow link and occasional connection are the normal mode of operations. In order to overcome the mobile communication constraints, object caching and replication have been used pervasively to reduce network traffic and improve the efficiency of interaction of objects.

Object consistency is the critical issue of an object caching/replication system, where versions of objects are distributed and may get diverged. Conflicts arise when synchronization happens. To avoid loss work, distributed applications must be designed to prevent conflicts or to recover gracefully from the conflicts. This is an area where formal methodology may help. This talk will first give an overview of the distributed object technologies. and survey object consistency and conflict resolution protocols in systems such as Bayou at Xerox Park, Lotus Notes, Orable, and GoldRush at IBM.

Formal Description Techniques and Protocol Specification, Testing and Verification
T. Mizuno, N. Shiratori, T. Higashino & A. Togashi (Eds.)   © 1997 IFIP. Published by Chapman & Hall

## PART NINE

Concurrent Systems

30

Self-independent Petri Nets for Distributed Systems*

Yong SUN[†], Shaoying LIU[*], Mitsuru OHBA[*]
†Department of Computer Science
The Queen's University of Belfast
Belfast BT7 1NN, Northern Ireland
Email: Y.Sun@qub.ac.uk
*Faculty of Information Sciences
Hiroshima City University
Asaminami-ku, Hiroshima, 731-31, Japan
Email: {shaoying, ohba }@cs.hiroshima-cu.ac.jp

Abstract

We propose that asynchronous mechanism for communication should be considered as a primitive for semantic models of distributed systems in contrast to the synchronous mechanism (hand-shaking) of CSP (Hoare, 1985) and of CCS (Milner, 1989). By doing so, we present an approach to producing a deadlock-free system, and give a denotational model for CCS in a non-interleaving fashion.

Keywords

asynchronous communication, self-independence, Petri nets, CCS, pre-order, partial order, complete partial order (cpo), denotational semantics

1 Introduction and Overview

Hand-shaking is an abstract and ideal synchronous mechanism for communication. It is adopted by Hoare in CSP (1985) and by Milner in CCS (1989) as a primitive to describe communications between different components in distributed systems. However, there is a synchronization problem associated with the implementation of this mechanism (Lamport et al, 1982; Sun & Yang, 1996). Asynchronous communication is inevitable in the implementation of distributed and concurrent systems. The conventional way to cope with this problem is to simulate asynchronous communication using the hand-shaking mechanism. However, such a simulation is often awkward and inconvenient. Furthermore, the hand-shaking mechanism may result in deadlock; in our proposed asynchronous communication this will not occur. For instance,

$$\alpha?x.\beta?y.B(x,y)\|\beta!e_1.\alpha!e_2.C$$

*Work is supported in part by the Ministry of Education of Japan under Joint Research Grant-in-Aid for International Scientific Research (08044167), by Hiroshima City University under Hiroshima City University Grant for Special Academic Research (International Studies) SCS-FM (A440), and by NIDevR of the Queen's University of Belfast.

Formal Description Techniques and Protocol Specification, Testing and Verification
T. Mizuno, N. Shiratori, T. Higashino & A. Togashi (Eds.)   © 1997 IFIP. Published by Chapman & Hall

in CSP and

$$(\alpha x.\beta y.B(x,y)|\bar{\beta}e_1.\bar{\alpha}e_2.C)\backslash \alpha\beta$$

in CCS are two examples of deadlocked processes (or deadlocked agents). They both describe that the left agent of the operator $\|$ (or $|$) synchronizes with the right agent. The left agent performs an α action (receiving) followed by a β action (receiving) and then behaves like $B(x,y)$. The right agent performs a β action (sending) followed by an α action (sending) and then behaves like C. Since the order of the actions in one agent does not match that in the other agent, deadlock is caused in synchronous communication. Such occurrences should not arise in the design of distributed systems. If a complete inference rule system existed which could be used to decide whether a deadlock exists in a communicating system, then we would use this system to exclude all deadlocks. Unfortunately, such a system cannot be constructed since the decision problem for deadlock is undecidable.

In order to solve the problems mentioned above, we propose that an asynchronous mechanism for communication be used as a primitive for semantic models of distributed systems, say CCS. Thus, as we will show, hand-shaking communication becomes a special case of asynchronous communication. Because of the asynchronous communication, every process (or agent) is self-independent which prevents deadlocks in distributed systems. That is, deadlock-freedom is a property inherited from the asychronous communication rather than resulting from an intricate matching of the orders of communicating actions among distributed processes (or agents).

In this paper, we first replace the synchronous mechanism by the asynchronous mechanism using specialized Petri nets (Petri, 1976; Peterson, 1981). This new kind of Petri net is called a *self-independent* Petri net. Such Petri nets provide a denotational model for CCS in a non-interleaving fashion (or so-called true concurrency). However, we omit the interpretation of CCS terms in the model since it is the same as that in the CCS semantics model. The reader may consult Goltz and Mycroft (1984) or Winskel (1983) for details.

The model to be presented is described through labelled Petri nets in which all constructable nets are self-independent (or deadlock-free). The class of all constructable nets provides an appropriate and sufficiently abstract domain in which the fixpoints of semantic equations can be evaluated when the isomorphic nets are regarded as being identical. Technically, the result can be regarded as an improvement of the following:

1. Winskel's (1983) event structures in which non-determinism is non-symmetric (i.e. the semantics of $B_1 + B_2$ is not the same as the semantics of $B_2 + B_1$);

2. Goltz and Mycroft's (1984) labelled Petri nets which can not systematically deal with recursive processes (i.e. they have a difficulty in giving a semantics to B_2 in $\{B_1 \Leftarrow \alpha.Nil|\gamma.B_1, B_2 \Leftarrow B_1|\beta.B_2\}$);

3. Boudol and Castellani (1987) follow Winskel's approach to solve the problem of non-symmetric of non-determinism, but their results are limited to finite structures.

In summary, we demonstrate that asynchronoous systems have better deadlock free properties. We will give a class of Petri nets which preserve deadlock freeness. Such nets can be used as a CCS semantics model (Milner, 1989).

The remainder of this paper is organized as follows. Section 2 introduces Self-independent Petri Nets (SPNs). Section 3 describes constructable SPNs. In section 4, we discuss isomorphic classes of SPNs. Finally, we compare the newly introduced notation with existing notations, give some conclusions, and sketch some future work.

2 Introduction to Self-independent Petri Nets (SPNs)

A labelled Petri net (LPN or simply PN) N can be denoted as

$$< S, E, T, L >$$

where S is a set of states (or places), E is a set of events, T is a set of transitions which is a subset of $(S \times E) \cup (E \times S)$, L is a labelling from events E to labels Lab. For the sake of simplicity, we let N stand for $S \cup E$.

For $x \in N$, the pre- and post-set of x are defined, respectively, as

1. $^{\bullet}x =_{def} \{y \in N| < y, x > \in T\}$ and as

2. $x^{\bullet} =_{def} \{y \in N| < x, y > \in T\}$.

A root-set of net N, $^{\circ}N$, is defined as $^{\circ}N =_{def} \{x \in N|^{\bullet}x = \emptyset\}$. $|X|$ stands for the cardinality of the set X. The roots of N (i.e. the elements of $^{\circ}N$) can be considered as being always initially marked, although markings are not discussed at all in this paper. This treatment can be viewed as presupposing that all roots (states) are always marked initially and that multi-tokens are linearized into multi-nets, i.e. every initial state represents one token, see the example in Figure 1.

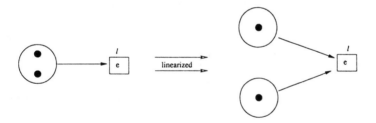

Figure 1: linearizing of tokens

For Petri nets, there are three phenomena which need to be classified.

1. The events e_1 and e_2 in Figure 2 are to be exclusively triggered, i.e. either but not both can be triggered.

   In other words, these two events are not independent of each other. This phenomenon is commonly referred to as non-determinism.

2. The states s_1 and s_2 in Figure 3 are independent of each other, i.e. both states s_1 and s_2 are triggered by the event e at the same time.

   This phenomenon is commonly referred to as concurrency or parallelism.

Figure 2: non-determinism

Figure 3: parallelism

3. The event e in Figure 4 can only be triggered by both the states s_1 and s_2 (but not by only one of them).

   This phenomenon is commonly referred to as synchronization.

To exploit these phenomena further, we introduce a parallel operator | over nets. Consider the net N_1 in Figure 5 and the net N_2 in Figure 6.

Then, their composition $N_1|N_2$ is as shown in Figure 7.

In these diagrams (see Figures 5, 6 and 7), each horizontal arrow can roughly be viewed as an execution (or evolution) of a process, and non-horizontal arrows are interactions between processes. The events labelled by τs (say e_4 and e_5) are successful communicating actions, and the events labelled differently (say e_1 and e_2) indicate unsuccessful communicating actions. More specifically, N_1 in $N_1|N_2$ at s_2 does not know whether the action performed on port ℓ is a result of the event e_1 or a result of the event e_4 (or e_5), unless it checks the receiving message. That is, the result of an action has to be examined before N_1 understands whether it is a successful action. In other words, an empty message implies that the action was unsuccessful. Among successful actions, N_1 needs to be informed by N_2 which event produced the received message, say event e_4 or event e_5. Therefore, every communicating action takes place only subject to the performer's

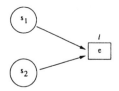

Figure 4: synchronization

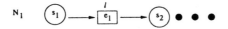

Figure 5: net N_1

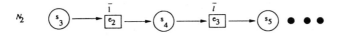

Figure 6: net N_2

wish regardless of:

1. who a message sender or a message receiver is, and

2. who the communicating partner is.

Synchronization is the result of a *"pure coincidence"* (i.e. the actions labelled by l and its partner $\bar{l}$ can be executed independently). Therefore, synchrony is a special case of asynchrony.

We obtain two extra advantages from such asynchrony. These are:

1. every process is *self-independent*;

2. the message loss in communication is a natural consequence of the asynchronous communication.

Because of the latter, systems are able to tolerate message losses, which is an important feature in distributed computing systems. Actually, we say that a relation $\|$ on nets is the *independent* relation on nets, i.e. $\| : (S \times S) \cup (E \times E) \cup (S \times E) \cup (E \times S)$ is the least relation on net N which satisfies the following six conditions:

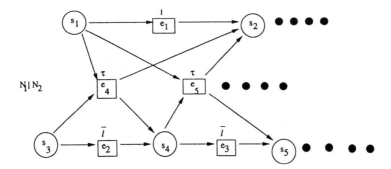

Figure 7: composition $N_1 | N_2$

1. $\forall s_1, s_2 \in {}^\circ N. s_1 \neq s_2 \Rightarrow s_1 || s_2;$

2. $\forall e \in E. \forall s_1, s_2 \in e^\bullet. s_1 \neq s_2 \Rightarrow s_1 || s_2;$

3. $\forall e_1, e_2 \in E. (({}^\bullet e_1 \cap {}^\bullet e_2 = \emptyset) \wedge (\forall s_1 \in {}^\bullet e_1. \forall s_2 \in {}^\bullet e_2. s_1 || s_2)) \Rightarrow e_1 || e_2;$

4. $\forall e_1, e_2 \in E. e_1 || e_2 \Rightarrow (\forall s_1 \in {}^\bullet e_1. s_1 || e_2) \wedge (\forall s_2 \in {}^\bullet e_2. e_1 || s_2);$

5. $\forall s \in S. \forall e \in E. ((\forall s' \in {}^\bullet e. s || s') \vee (\forall s' \in {}^\bullet e. s' || s)) \Rightarrow (s || e \wedge e || s);$

6. $\forall s \in S. \forall e \in E. (s || e \Rightarrow \forall s' \in e^\bullet. s || s') \wedge (e || s \Rightarrow \forall s' \in e^\bullet. s' || s).$

The first condition states that the distinct initial states are independent of each other; the second condition states that the distinct states triggered by a same event are independent of each other; the third condition expresses the fact that the distinct events triggered by totally independent states are independent of each other; the fourth condition indicates that an event and a state are independent of each other if the state is triggered by another event independent of this event; the fifth condition states that an event and a state are independent of each other if all the states triggering the event are independent of the state; the sixth condition indicates that a state (say s) is independent of another state (say s') if the other state (s') is triggered by an event which is independent of the state (s). With the understanding of the independence relation on nets, we are interested in those nets which have the following six properties:

1. $\forall x, y \in N. ({}^\bullet x = {}^\bullet y) \wedge (x^\bullet = y^\bullet) \Rightarrow x = y;$

2. $\forall e \in E. \exists s_1, s_2 \in S. s_1 T e \wedge e T s_2;$

3. $||$ is irreflexive;

4. $\forall e \in E. \forall s_1, s_2 \in {}^\bullet e. s_1 \neq s_2 \Rightarrow s_1 || s_2;$

5. $\forall s \in S. \forall e_1, e_2 \in {}^\bullet s. \neg(e_1 || e_2);$

6. $\forall s \in S - {}^\circ N. \bigcap \{{}^\bullet e | e \in {}^\bullet s\} \neq \emptyset.$

The first property states that the nets we are interested in have no redundancy; the second property expresses the fact that every event has a cause and a result in the nets; the third property states that either an event or a state cannot be independent of itself; the fourth property states that the distinct states triggering the same event must be independent of each other; the fifth property indicates that the distinct events triggering the same state cannot be independent of each other; the last property asserts that if a non-initial state is triggered by a collection of events, these events must share a common triggering state.

If, for any net N, the independence relation $||$ on the nets has the above six properties, we say that N is a *self-independent* (labelled) Petri net, or a SPN. Actually, the fourth property of SPNs excludes the nets such as that in Figure 8 and the fifth property excludes the nets such as that in Figure 9. These exclusions are reasonable, i.e. *non-determinism* should not lead to *synchronization*; and *concurrency* (or *parallelism*) should not be reduced to *non-determinism*.

It is worth mentioning that $||$ is a reversed version of # (the conflict relation) in Neilsen et al (1980), but SPNs are richer than the nets described there since they (i.e. SPNs) are

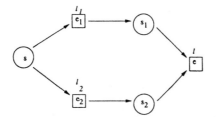

Figure 8: non-determinism → synchronization

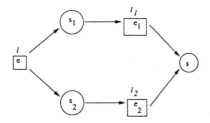

Figure 9: parallelism → non-determinism

not necessarily cycle-free. However, the existence of a cycle in a net is closely related to a deadlock in the net. An exmaple with a cycle is shown in Figure 10. Fortunately, this does not happen in the constructable nets with which this paper is concerned.

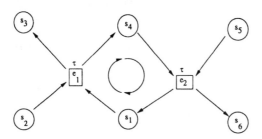

Figure 10: a circle in a net (a deadlock)

We are not interested in the trivial SPN, i.e. $< \emptyset, \emptyset, \emptyset, \emptyset >$ is not included inside our SPNs. Also, since we intend to regard isomorphic nets as identical, we treat all the singleton nets as the same net and write it as $< \{\bullet\}, \emptyset, \emptyset, \emptyset >$. For simplicity, we assume that the roots of each net are states, i.e. $^\circ N \subseteq S$. This assumption is implied by the second property of SPNs. From now on, we concentrate on Self-independent Petri Nets (SPNs).

We now introduce a partial order $\sqsubseteq$ on SPNs such that $N_1 \sqsubseteq N_2$ iff the following three conditions hold:

1. $N_1 \subseteq N_2$ (i.e. $S_1 \subseteq S_2$, $E_1 \subseteq E_2$, $T_1 \subseteq T_2$, $L_1 \subseteq L_2$);

2. $^\circ N_1 \subseteq {}^\circ N_2$;

3. $\forall x, y \in N_1 . x T_2 y \Rightarrow x T_1 y$.

The third condition implies that the projection of N_2 to N_1 coincides with N_1 (i.e. $N_2\lceil_{N_1} = N_1$). This means that the prefix closure implies the partial order. Hence, the least upper bound $\sqcup N_k$ of an ω-chain $\{N_k\}$ is $\sqcup N_k = \cup N_k$, and the least element $\perp$ is $< \{\bullet\}, \emptyset, \emptyset, \emptyset >$. Therefore, our SPNs form a *cpo* (complete partial order) (Plotkin, 1985).

3   Constructable SPNs

Let ℓ range over *Lab*, ℓ and $\bar{\ell}$ represent an input label and an output label respectively. Note that $^-$ is the complementary function over *Lab*. We give constructions for SPNs below.

1. (bottom) $\perp = < \{\bullet\}, \emptyset, \emptyset, \emptyset >$;

2. (sequence or prefix) $\ell.N =_{def} < \{s\} \cup S, \{e\} \cup E, \{< s, e >\} \cup \{< e, s' > | s' \in {}^\circ N\} \cup T, L \cup \{< e, \ell >\} >$, where $N \neq \emptyset$, and s and e are not in N;

3. (non-deterministic composition) $N_1 + N_2 =_{def} < S, E_1 \cup E_2, T, L_1 \cup L_2 >$, where $N_i \neq \emptyset$ $(i = 1, 2)$ are disjoint, $S =_{def} (S_1 - {}^\circ N_1) \cup (S_2 - {}^\circ N_2) \cup ({}^\circ N_1 \times {}^\circ N_2)$ and $T =_{def} \{< s, e > | s =< s_1, s_2 > \wedge (< s_1, e > \in T_1 \vee < s_2, e > \in T_2) \wedge s_i \in {}^\circ N_i$ $(i = 1, 2)\} \cup T_1\lceil_{(S_1 - {}^\circ N_1) \times E_1 \cup E_1 \times (S_1 - {}^\circ N_1)} \cup T_2\lceil_{(S_2 - {}^\circ N_2) \times E_2 \cup E_2 \times (S_2 - {}^\circ N_2)};$

4. (parallel composition) $N_1 | N_2 =_{def} < S_1 \cup S_2, E_1 \cup E_2 \cup prod, T_1 \cup T_2 \cup conc, L_1 \cup L_2 \cup comm >$, where N_1 and N_2 are disjoint, and $prod =_{def} \{e =< e_1, e_2 > \in E_1 \times E_2 | L_1(e_1) = \overline{L_2(e_2)} \neq \tau\}$, $comm =_{def} \{< e, \tau > | e \in prod\}$, and $conc =_{def} \{< s_1, e >, < s_2, e >, < e, s_1' >, < e, s_2' > | e =< e_1, e_2 > \in prod \wedge s_1 \in {}^\bullet e_1 \wedge s_2 \in {}^\bullet e_2 \wedge s_1' \in e_1^\bullet \wedge s_2' \in e_2^\bullet\};$

5. (synchronizer) $N\backslash\ell$ will be defined later;

6. (recursive operator) μ or fix (fixpoint operator).

To motivate the definition of the synchronizer (say $N\backslash\ell$), we display an example. Consider Figure 7 in the previous section. The intuitive meaning of $(N_1|N_2)\backslash\ell$ is to enforce that all the communications on port ℓ must be synchronized. The effect of such enforcement is demonstrated in Figure 11. By referring to Figure 7, we know that the events e_1 and e_2 are cut off after synchronization in Figure 11.

Also, there is no event which follows the state s_4, i.e. the events e_3 and e_5 and the state s_5 are cut off as well. In contrast, there may be events which follow the state s_2.

In order formally to capture synchronization, we have to introduce another concept, called *well-rooted* SPNs.

Let $N^{[0]} =_{def} {}^\circ N$ $(\subseteq S)$, $N^{[2i+1]} =_{def} \{e \in E | {}^\bullet e \subseteq \bigcup_{k=0}^{i} N^{[2k]}\}$, $N^{[2i+2]} =_{def} \{s \in S | {}^\bullet s \cap N^{[2i+1]} \neq \emptyset\}$, and $N^* =_{def} < S^*, E^*, T^*, L^* >$ where $S^* = \bigcup N^{[2i]}$, $E^* = \bigcup N^{[2i+1]}$,

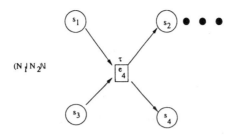

Figure 11: the net of $(N_1|N_2)\backslash\ell$

$T^* = T\lceil_{(S^* \times E^*) \cup (E^* \times S^*)}$ and $L^* = L\lceil_{E^*}$. We say that a SPN, N, is *well-rooted* iff $N = N^*$.

Informally, the idea of well-rooted nets is similar to the idea of a well-known property in set theory, viz. well-foundness. N is well-rooted means that if $x \in N$, then x can be reached by a(n) (arbitrary) finite number of steps (or executions or evolutions) from the roots $^\circ N$. In other words, a well-rooted N says that every state in N is reachable. We consider here only the well-rooted SPNs.

Lemma 2.1: *The property of well-rootedness is an invariant of the constructions of prefixes, non-deterministic compositions and parallel compositions.*

For a well-rooted N, we define $N\backslash\ell =_{def} < S', E', T', L' >$, where $N^{(0)} =_{def} {}^\circ N$, $N^{(2i+1)} =_{def} \{e \in E|^\bullet e \subseteq \bigcup_{k=0}^{i} N^{(2k)} \wedge L(e) \neq \ell, \bar{\ell}\}$, $N^{(2i+2)} =_{def} \{s \in S|^\bullet s \cap N^{(2i+1)} \neq \emptyset\}$, $S' = \bigcup N^{(2i)}$, $E' = \bigcup N^{(2i+1)}$, $T' = T\lceil_{(S' \times E') \cup (E' \times S')}$, and $L' = L\lceil_{E'}$. Thus, the result of Lemma 2.1 can be extended to include synchronizers.

Lemma 2.2: *The well-rooted property is an invariant of the constructions mentioned in Lemma 2.1 with the extra constructions of synchronizers.*

Also, we have that the well-rooted property is closed under least upper bounds. Formally,

Lemma 2.3: *Let $\{N_i\}$ be an ω-chain of SPNs, if N_i is well-rooted for every i, then $\sqcup N_i$ is well-rooted, also.*

Therefore, the result of Lemma 2.2 can be extended to include all constructions of the nets.

Theorem 2.4: *The well-rooted property is an invariant of all the constructions of the nets.*

The key point of the proof for Theorem 2.4 is that all the constructions are monotonic and continuous. Therefore, the fixpoints always exist when we apply μ (or fix) to all the possible combinations of constructions, and the other constructions naturally follow from the previous lemmas.

It is not hard to prove that all constructable nets are *self-independent*. Furthermore, if we regard every root of a net as a process and syntactically exclude the instance of constructions like $N_1 + (N_2|N_3)$, then every process is self-independent in constructable nets. That is, every token can move along constructable nets without getting stuck if we consider each root to be initially marked. This is the origin of the name of *self-*

independent Petri nets comes from.

4   Isomorphic Classes of SPNs ([SPN])

In order that the non-deterministic compositions of $N_1 + N_2$ and of $N_2 + N_1$ can be regarded as identical, we must consider isomorphic classes of SPNs. First, let us give a definition for net isomorphism. N_1 is *isomorphic* to N_2, written as $N_1 \cong N_2$, iff there is a bijection $\phi : N_2 \to N_1$ such that

1. $s \in S_2 \Leftrightarrow \phi(s) \in S_1$,

2. $e \in E_2 \Leftrightarrow \phi(e) \in E_1$,

3. $< x, y > \in T_2 \Leftrightarrow < \phi(x), \phi(y) > \in T_1$, and

4. $L_2 = L_1 \circ \phi$.

It is not difficult to see that the isomorphic relation between SPNs is an equivalence relation. Then, we define the correpsonding equivalence class over the SPNs as: $[N] =_{def} \{N' \in SPN | N' \cong N\}$. We use [SPN] to denote the set of all the equivalence classes.

Lemma 3.1: *The following three definitions are equivalent: for all N_i $(i = 1, 2)$*

1. $[N_1] \trianglelefteq_1 [N_2]$ iff $\exists N_2' \in [N_2].N_1 \sqsubseteq N_2'$;

2. $[N_1] \trianglelefteq_2 [N_2]$ iff $\exists N_1' \in [N_1].N_1' \sqsubseteq N_2$;

3. $[N_1] \trianglelefteq_3 [N_3]$ iff $\exists N_1' \in [N_1].\exists N_2' \in [N_2].N_1' \sqsubseteq N_2'$.

We naturally want to extend the partial order $\sqsubseteq$ from SPNs to $[SPN]$s, denoted by $\trianglelefteq$. Because of Lemma 3.1, we can relate the order $\trianglelefteq$ to any one of $\trianglelefteq_i$ $(i = 1, 2, 3)$ as we require. However, $\trianglelefteq$ is not, in general, a partial order but a pre-order. This is due to symmetricity, i.e. $[N_1] \trianglelefteq [N_2] \trianglelefteq [N_1]$ does not imply $[N_1] = [N_2]$.

In order to present the non-partial-order problem more clearly, let $\preceq$ be a partial order in an arbitrary X and $\sim$ be an isomorphism on X. Then, we say that a partial order $\preceq$ in X has a *well-extended* property (on X over $\sim$) if

$$\forall x_1, x_2 \in X.x_1 \preceq x_2 \wedge x_1 \sim x_2 \Rightarrow x_1 = x_2.$$

It is obvious that $\trianglelefteq$ is a partial order if $\sqsubseteq$ has the well-extended property. Unfortunately, $\sqsubseteq$ does not, in general, have the well-extended property. Therefore, it is not hard to conclude that $\trianglelefteq$ can not be a partial order in SPNs in general. An example to show that $\sqsubseteq$ does not have the well-extended property is shown below, see Figure 12.

Let $N_i = < S_i, E_i, T_i, L_i > (i = 1, 2)$ and

1. $S_i = \{s^i\} \cup \{s_{n,k}^i | ((i = 1) \wedge (n \geq 1) \wedge (1 \leq k \leq 2n)) \vee ((i = 2) \wedge (n \geq 1) \wedge (1 \leq k \leq 2n + 1))\}$;

2. $E_i = \{e^i\} \cup \{e_{n,k}^i | ((i = 1) \wedge (n \geq 1) \wedge (1 \leq k \leq 2n - 1)) \vee ((i = 2) \wedge (n \geq 1) \wedge (1 \leq k \leq 2n))\}$;

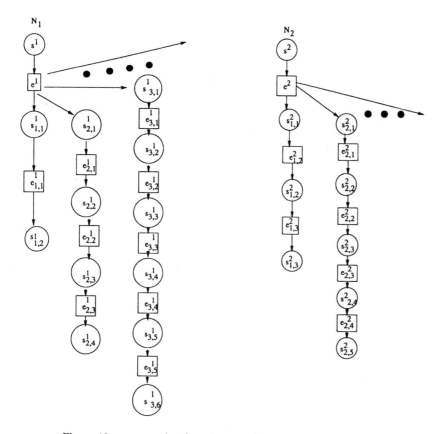

Figure 12: an example of not having the well-extended property

3. $T_i = \{< s^i, e^i >\} \cup \{< e^i, s^i_{n,1} > | n \geq 1\} \cup \{< s^i_{n,k}, e^i_{n,k} >, < e^i_{n,k}, s^i_{n,k+1} > |((i = 1) \wedge (n \geq 1) \wedge (1 \leq k \leq 2n - 1)) \vee ((i = 2) \wedge (n \geq 1) \wedge (1 \leq k \leq 2n))\}$;

4. $L_i = \{< e^i, \alpha >\} \cup \{< e^i_{n,k}, \beta > |((i = 1) \wedge (n \geq 1) \wedge (1 \leq k \leq 2n - 1)) \vee ((i = 2) \wedge (n \geq 1) \wedge (1 \leq k \leq 2n))\}$.

It is obvious that N_1 is not isomorphic to N_2. However, we still have to check whether or not there exist N'_1 and N'_2 as required in $\trianglelefteq_3$ in Lemma 3.1. This is done by considering an embedding of N_1 into N_2 and conversely N_2 into N_1.

Thus, in order that $\trianglelefteq$ be a partial order, we need to restrict the SPNs to a certain sub-collection. This sub-collection must be rich enough to accommodate CCS. The restriction we derive is the *finitely-branched* condition for SPNs, which is now introduced.

Let N be a SPN. N is said to be *finitely-branched* (or to be a fSPN) if N has the following three properties:

1. for roots $|{}^{\circ}N| < \infty$;

2. $\forall e \in E.|e^{\bullet}| < \infty$; and

3. for events $\forall i \geq 0.|N^{[2i+1]} - N^{[2i-1]}| < \infty$ where $N^{[-1]} = \emptyset$.

These three conditions restrict SPNs to fSPNs such that the increase of the numbers of branches for each execution (or evolution) is limited to arbitrary finite numbers.

Since a finite set is identical to another finite one if one is a subset of the other and if there exists a 1-1 mapping between the elements of the two sets, fSPNs naturally hold the well-extended property. We formally express this result as a lemma.

Lemma 3.2: $\sqsubseteq$ *is well-extended in fSPNs.*

Therefore,

Theorem 3.3: $\trianglelefteq$ *is a partial order in [fSPN].*

We now know that the finitely-branched conditions are sufficient to guarantee that $\trianglelefteq$ be a partial order. However, the necessity of these conditions needs to be demonstrated. On the other hand, we can show that they are not too restrictive in the sense that there are examples where the symmetricity does not hold. Such an example has been given earlier in demonstrating that the well-extended property does not hold in general. We will give another example in the proof of Lemma 3.4. In this sense, we claim that the finitely-branched conditions are necessary for $\trianglelefteq$ to be a partial order.

Lemma 3.4: *There are* $[N_1]$ *and* $[N_2]$ *such that* $\exists N_1' \in [N_1]$ *and* $\exists N_2' \in [N_2]$ *and they satisfy the following three conditions:*

1. $N_1 \sqsubseteq N_2'$, i.e. $[N_1] \trianglelefteq_1 [N_2]$;

2. $N_2 \sqsubseteq N_1'$, i.e. $[N_2] \trianglelefteq_1 [N_1]$; and

3. N_1 and N_2 are not isomorphic to each other, i.e. $[N_1] \neq [N_2]$.

Proof

The example about the well-extended property can be used here. However, this only shows that $\forall e \in E.|e^\bullet| < \infty$ is necessary. Removing the prefix from the example, it will be shown that $|{}^\circ N| < \infty$ is necessary. For the other case, we provide another example. Assume that $\exists j.|N^{[2j+1]} - N^{[2j-1]}| \not< \infty$ and $|N^{[2j+2]} - N^{[2j]}| \not< \infty$ (the conjunction is because of the nets' characteristics of no redundancy), say $j = 0$. Readers are encouraged to draw a graphical representation of the following example, since a diagram can aid understanding.

Let $N_i = < S_i, E_i, T_i, L_i > (i = 1, 2)$, where

1. $S_i = \{s^i\} \cup \{s_k^i | k \in Nat - \{0\}\}$;

2. $E_i = \{e_{k,n}^i | ((i = 1) \wedge (1 \leq n \leq 2k - 1)) \vee ((i = 2) \wedge (1 \leq n \leq 2k))\}$;

3. $T_i = \{< s^i, e_{k,n}^i >, < e_{k,n}^i, s_k^i > | ((i = 1) \wedge (1 \leq n \leq 2k - 1)) \vee ((i = 2) \wedge (1 \leq n \leq 2k))\}$; and

4. $L_i = \{< e_{n,k}^i, \alpha^i > | ((i = 1) \wedge (1 \leq n \leq 2k - 1)) \vee ((i = 2) \wedge (1 \leq n \leq 2k))\}$.

It is obvious that N_1 is not isomorphic to N_2 (see Figure 13).

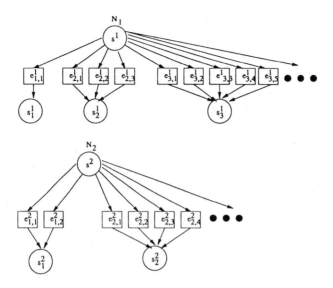

Figure 13: non-isomorphic nets

However, we still need to check whether there exist N_1' and N_2' as required. This is achieved by considering an embedding N_1 into N_2 and conversely N_2 into N_1. $\square$

For an ω-chain $\{[N_k]\}$ in [fSPN], its least upper bound is:

$$\sqcup[N_k] = [\bigcup N_k']$$

where $\forall k . N_k' \in [N_k] \wedge N_k' \sqsubseteq N_{k+1}'$. So, [fSPN] forms a cpo.

Naturally, we can extend the constructions of the last section (Section 3) to [fSPN] as follows:

1. $\alpha.[N] = [\alpha.N]$

2. $[N_1] + [N_2] = [N_1 + N_2]$

3. $[N_1]|[N_2] = [N_1|N_2]$

4. $[N]\backslash\alpha = [N\backslash\alpha]$.

Since these constructions on [fSPN] are defined by their counterparts in fSPNs, their monotonicities and continuities follows easily. However, there is no trivial way to extend recursive definitions to isomorphic classes. For example, $x \Rightarrow x|\alpha.\bot$ where x is an identifier for nets. That is, μ is not closed in fSPNs. So we have further to restrict constructions to certain instances.

By the application of the *well-constructed* restriction, a unary construction F (or functor) of fSPNs is said to be *well-constructed* if

$$\exists K . \forall k \geq K . |^\circ F^k([< \{\bullet\}, \emptyset, \emptyset, \emptyset >])| = |^\circ F^{k+1}([< \{\bullet\}, \emptyset, \emptyset, \emptyset >])| < \infty.$$

The well-constructed condition for nets excludes the possibility of increasing the number of roots of a constructable net to an infinite number, i.e. a well-constructed F cannot increase the number of roots of a net to an arbitrary large value by recursively applying F. So we have that

Theorem 3.5: *For any well-constructed F, $\mu x.(\lambda x.F)$ is closed in [fSPN].*

The proof is omitted. However, the key point of the proof is to show that a series of numbers of root states (to the least element) converges by recursively applying F, and so the least fixed point of F is an fSPN. A similar theorem (to Theorem 3.5) for mutual recursive functors $\vec{F}$ can also be established, but we do not give its details.

Since it can be easily checked that other constructions are closed in [fSPN], the net constructions in section 3 under the well-constructed restriction are also closed in [fSPN]. Therefore, it is safe to interpret CCS terms in the well-constructed [fSPN], i.e. the well-constructed [fSPN] is an adequate denotational semantics model for CCS provided that the CCS terms are well-guarded.

5　Discussion and Conclusions

From the categorical point of view, the constructions + and | are consistent with their sum and product in the Petri nets category (Winskel, 1987) except that | is a restricted product which only affects the events with complementary labels. This shows that the intuition of the asynchrony from implementation has a categorical background. Also, it is worth mentioning that the asynchrony is by its nature deadlock-free.

We should point out that the synchrony and the asynchrony in this paper are different from those of Milner (1984). His synchrony implies the existence of a universal clock and his asynchrony implies the non-existence of such a clock.

In this paper, we propose that the asynchronous mechanism for communication should be considered as a primitive for semantic models of distributed systems in contrast to the synchronous mechanism (hand-shaking) of CSP (Hoare, 1985) or of CCS (Milner, 1989). By doing so, we present an approach to producing a deadlock-free system, and give a denotational model for CCS in a non-interleaving fashion by means of self-independent Petri Nets.

Another advantage from the asynchrony not having been exploited in this paper is to incorporate broadcasting communication into the model. This will be the subject of future investigation.

Comparing constructable nets (or cSPNs for short) with fSPNs, we understand that the collection of cSPNs is more restrictive than the collection of fSPNs. In the light of this observation, the condition for net constructions to be both finitely-branched and well-constructed may be over restrictive. This issue requires more attention.

Acknowledgement

The first author would like to thank G. Plotkin for his insight of SPNs. We would like to thank Prof. M. Clint of the Queen's University of Belfast and four anonymous referees for their constructive comments on early version of this paper and Y.Sato of Hiroshima City

University for his support in using computer facilities. This work is supported in part by the Ministry of Education of Japan under Joint Research Grant-in-Aid for International Scientific Research FM-ISEE (08044167), by Hiroshima City University under Hiroshima City University Grant for Special Academic Research (International Studies) SCS-FM (A440), and by NIDevR of the Queen's University of Belfast. We would like to express our special gratitude to these organizations for their support.

References

[1] G. Boudol and I. Castellani (1987), "On the Semantics of Concurrency: partial order and transition systems", TAPSOFT'87, Lecture Notes in Computer Science, Vol. 249, Springer-Verlag, 1987.

[2] U. Goltz and A. Mycroft (1984), "On the Relationship of CCS and Petri Nets", Lecture Notes in Computer Science, Vol. 172, Springer-Verlag, 1984.

[3] C. A. R. Hoare (1985), "Communicating Sequential Processes", Prentice-Hall International, 1985.

[4] L. Lamport, R. Shostak, and M. Pease (1982), "The Byzantine General Problem", ACM Trans. on Prog. Lang. and Syst., Vol.4, No.3, July 1982.

[5] Robin Milner (1984), "Calculi of Synchrony and Asynchrony", Journal of Theoretical Computer Science, 1984.

[6] Robin Milner (1989), "Communication and Concurrency", Prentice-Hall International, 1989.

[7] M. Neilsen, G. Plotkin and G. Winskel (1980), "Petri Nets, Event Structures and Domains, Part 1", Journal of Theoretical Computer Science, 1980; also appears in internal report of Department of Computer Science, University of Edinburgh, CSR-47-79, November 1979.

[8] J. L. Peterson (1981), "Petri Nets Theory and the Modelling of Systems", Prentice-Hall, Englewood Cliffs, N. J., 1981.

[9] C. A. Petri (1976), "General Net Theory", *Communication Disciplines*, ed. B. Show, Proc. Joint IBM and University of Newcastle Seminar, 1976.

[10] G. D. Plotkin (1985), "Domain Theory", Lecture notes for the postgraduates of the Department of Computer Science, University of Edinburgh, 1985/6.

[11] Yong Sun and Hongji Yang (1996), "Communication mechanism independent protocol specification based on CSP: a case study", in the proceedings of the 22nd Euromicro conference: Beyond 2000: Hardware and Software Design Strategies, IEEE Computer Society, 2-5 September, Pregue, Czech Republic, 1996.

[12] G. Winskel (1983), "Event Structure Semantics for CCS and Related Languages", Lecture Notes in Computer Science, Vol.140, Springer-Verlag, 1983.

[13] G. Winskel (1987), "Petri Nets, Algebras, Morphisms, and Compositionality", Journal of Information and Computation, Vol.72, 1987.

6   Biography

Yong Sun was born at Nanning, Guangxi Province, P R China on 1st September 1961. He received his BSc, MSc and PhD (all in Computer Science) from Peking University in 1982, the Institute of Computing Technology (Chinese Academy of Sciences, Beijing) in 1985, and the University of Edinburgh in 1992, respectively. And he has over 20 publications in refereed journals and conferences. He is a member of IEEE Computer Society, the Association of Computing Machinery, the European Association of Theoretical Computer Science. At the present, he is a Lecturer in Computer Science of the Queen's University of Belfast, Northern Ireland.

Shaoying Liu was born in Shaanxi Province, P R China on 25th April 1960. He received his BSc and MSc in Computer Science in 1982 and 1987 respectively from Xi'an Jiao-Tong University, P R China, his PhD in Formal Methods of Software Engineering in 1992 from the University of Manchester, United Kingdom. He has served as general chair, co-program chair, and program committee members for many international conferences and has over 30 publications in refereed journals and conferences. He is a member of IEEE Computer Society, IEICE Japan and the Order of International Fellowship (MOIF). At present, he is an Associate Professor in Computer Science of Hiroshima City University in Japan.

Mitsuru Ohba is a Prefessor of Computer Science Department, Hiroshima City University since April 1994. Before that, he was a senior researcher of IBM at Systems Engineering Laboratory, Tokyo, Japan. He received both BSc and MS from Aoyama Gakuin University, Tokyo, Japan, in 1971 and 1973 respectively. His present interests are in collaboration in a distributed environment, software testing and application of software metrics for software process improvements.

31
Combining CSP and Object-Z: Finite or Infinite Trace Semantics?

Clemens Fischer
Universität Oldenburg, FB Informatik, Abt. Semantik,PO Box 2503,
26129 Oldenburg, Germany, fischer@informatik.uni-oldenburg.de

Graeme Smith
Technische Universität Berlin, FB Informatik, Sekr. FR 5-6,
Franklinstr. 28/29, 10587 Berlin, Germany, graeme@cs.tu-berlin.de

Abstract

In this paper we compare and contrast two alternative semantics as a means of combining CSP with Object-Z. The purpose of this combination is to more effectively specify complex, concurrent systems: while CSP is ideal for modelling systems of concurrent processes, Object-Z is more suitable for modelling the data structures often needed to model the processes themselves. The first semantics, the *finite trace model*, is compatible with the standard CSP semantics but does not allow all forms of unbounded nondeterminism to be modelled (i. e. where a choice is made from an infinite set of options). The second semantics, the *infinite trace model*, overcomes this limitation but is no longer compatible with the standard CSP semantics. Issues involving specification, refinement and modelling fairness are discussed.

Keywords
CSP, Object-Z, concurrent systems, combining FDTs, semantics, refinement

1 INTRODUCTION

CSP [15] is a process algebra developed for the formal specification of concurrent systems. The standard semantics of CSP is the *failures-divergences semantics* [3] where finite traces are used together with refusal sets of events to model a process' external behaviour. This semantics, however, does not support *unbounded nondeterminism* (where a process can choose from an infinite set of options) thus restricting what can be expressed in the language. For example, it is not possible to specify a process which nondeterministically chooses, and then outputs, any natural number.

To overcome this limitation, two "alternative" semantics of CSP have been proposed. The first [18], which we will refer to as the *finite trace model*, doesn't alter the

Formal Description Techniques and Protocol Specification, Testing and Verification
T. Mizuno, N. Shiratori, T. Higashino & A. Togashi (Eds.)   © 1997 IFIP. Published by Chapman & Hall

form of the semantics, i.e. finite traces and refusal sets are used, nor the meaning of the CSP operators, but provides an alternative way of defining these operators which allows some forms of unbounded nondeterminism to be modelled. For example, a process which nondeterministically selects any natural number, outputs this natural number and then stops can be modelled in this semantics. It does not, however, allow the specification of a process which nondeterministically selects any natural number n and then performs a particular event n times. More precisely, because it only uses finite traces to model a process, it cannot distinguish between a process which can undergo any finite sequence of an event a and a process which can also undergo an infinite sequence of a's. The second semantics [20, 19], which we will refer to as the *infinite trace model*, overcomes this problem by introducing the set of infinite traces of the process. It can model all forms of unbounded nondeterminism but its form, and the meaning of some operators, are different from those of the standard failures-divergences semantics.

These alternative semantics of CSP allow it to be combined with state-based specification languages where unbounded nondeterminism arises naturally. For example, the finite trace model has been used as a common semantics for Z [24] and CSP in [10], and Object-Z [9] and CSP in [23] and [11]. The purpose of such combinations is to more effectively specify complex, concurrent systems: the state-based language is used for the specification of data structures for which CSP is not particularly suited. This need for more than one specification language to model complex, concurrent systems has also arisen in the ODP (Open Distributed Processing) standardisation initiative [16] where a combination of Z and LOTOS [2] has been proposed [7].

In this paper, we compare and contrast the two alternative semantics as a means of integrating CSP with Object-Z. The advantage of using Object-Z over Z is that it is object-oriented and therefore has a class structure which encapsulates state and operations. Such a structure is readily identifiable with a process as illustrated in [22]. The lack of a similar structure in Z makes such an integration more difficult. Indeed, [7] requires an intermediate language ZEST [5] which also has a class structure and is (automatically) translatable to Z, in order to define a common semantics with the process algebra part of LOTOS. Also, [10] proposes additional syntax for grouping the Z schemas corresponding to a process.

In Section 2, we provide a brief introduction to Object-Z and, in Section 3, we define the alternative CSP semantics. Section 4 then looks at integrating Object-Z and CSP using each of the semantics. A case study illustrating the difference between the two approaches is presented in Section 5 and rules of refinement in Section 6. In Section 7, we look at how the infinite trace semantics can be used to specify fairness properties.

2 OBJECT-Z

Object-Z [9] is an extension of Z [24] designed to support an object-oriented specification style. It includes a special class construct to encapsulate a state schema with all the operations which may affect its variables. A class is represented syntactically by a named box possibly with generic parameters. In this box there may be local type and constant definitions, at most one state schema and associated initial state schema, and zero or more operation schemas. As an example, consider the following specification of a bounded buffer. This specification is generic since the type T of the items in the buffer is not specified.

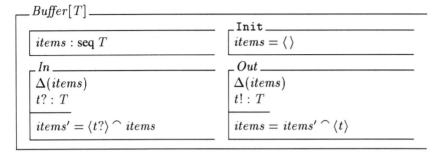

The class has a single state variable *items* denoting the items in the buffer. Initially, the buffer is empty and the operations *In* and *Out* enable items to be input to and output from the buffer, respectively, on a first-in/first-out basis. Each operation schema has a Δ-list of state variables which it may change, a declaration part consisting of input (denoted by names ending in ?) and output (denoted by names ending in !) parameters and a predicate part relating the pre- and post-values (denoted by names ending in ') of the state variables.

In addition, a class may have a *history predicate* enabling the specification of fairness properties. This will be discussed further in Section 7.

3   MODELLING UNBOUNDED NONDETERMINISM IN CSP

In this section, we present the details of the alternative semantics of CSP which enable the modelling of unbounded nondeterminism. We let a denote events, X and Y denote sets of events, s and t denote finite traces and u denote infinite traces.

The *finite trace model* of CSP is presented in [18]. As in the standard failures-divergences semantics [3], a process is modelled by its *failures* $\mathcal{F}$ and *divergences* $\mathcal{D}$. The failures of a process are pairs (s, X) where s is a finite trace that the process may undergo and X is a set of events that the process may refuse to perform after undergoing s. That is, if the process after undergoing s is in an environment which only allows it to undergo events in X, it may deadlock. The divergences of a process are the traces after which the process may undergo an infinite sequence of internal events, i.e. livelock. Failures and divergence are defined in terms of a non-empty alphabet of events Σ as follows.

Failures: $\mathcal{F} \subseteq \Sigma^* \times \mathbb{P}\Sigma$ such that $\mathcal{F}$ is not empty and the following axioms hold.

(1) $(s \frown t, \varnothing) \in \mathcal{F} \Rightarrow (s, \varnothing) \in \mathcal{F}$
(2) $(t, X) \in \mathcal{F} \wedge Y \subseteq X \Rightarrow (t, Y) \in \mathcal{F}$
(3) $(t, X) \in \mathcal{F} \wedge \forall a \in Y \bullet (t \frown \langle a \rangle, \varnothing) \notin \mathcal{F} \Rightarrow (t, X \cup Y) \in \mathcal{F}$

Axiom 1 captures the requirement that the traces of a process are prefix-closed. Axiom 2 states that if a process can refuse all events in a set X then it can refuse all events in any subset of X. Axiom 3 states that a process can refuse any event which cannot occur as the next event.

In addition, the standard failures-divergences semantics [3] has an axiom which states that all events in a set can be refused if all events in its finite subsets can be refused. Hence, a process which selects, and then outputs, any natural number cannot be modelled. Such a process would be able to refuse to output all numbers in any finite set of natural numbers, but not all numbers in the set of all natural numbers.

This axiom was included in the failures-divergence semantics to make the nondeterminism order on processes complete which was necessary to define the fixed point theory. However, the finite trace model, by defining the same fixed point theory in a different way (see [19] for details), removes the need for this axiom and, hence, removes the associated restriction on modelling unbounded nondeterminism.

Divergences: $\mathcal{D} \subseteq \Sigma^*$ such that the following axioms hold.

(4) $s \in \mathcal{D} \Rightarrow s \frown t \in \mathcal{D}$

(5) $s \in \mathcal{D} \Rightarrow (s \frown t, X) \in \mathcal{F}$

These axioms capture the idea that it is impossible to determine anything about a divergent process in a finite time. Therefore, the possibility that it might undergo further events cannot be ruled out. In other words, a divergent process behaves chaotically.

The *infinite trace model*, presented in [20, 19], is an extension of the finite trace model which allows all forms of unbounded nondeterminism to be modelled. In particular, it can distinguish between a process which can undergo any finite sequence of an event a and one which can also undergo an infinite sequence of a's.

In addition to failures and divergences, it also includes a component $\mathcal{I}$ corresponding to the infinite traces of a class. $\mathcal{I}$ is defined in terms of the alphabet of events Σ as follows.

Infinite traces: $\mathcal{I} \subseteq \Sigma^\omega$ such that the following properties hold[*].

(6) $s \frown u \in \mathcal{I} \Rightarrow (s, \varnothing) \in \mathcal{F}$

(7) $s \in \mathcal{D} \Rightarrow s \frown u \in \mathcal{I}$

(8) $(s, \varnothing) \in \mathcal{F} \Rightarrow$
$\exists T \bullet \forall t \in T \bullet (s \frown t, \{a \mid t \frown \langle a \rangle \notin T\}) \in \mathcal{F} \wedge \{s \frown u \mid u \in \bar{T}\} \subseteq \mathcal{I}$

where Σ^ω is the set of infinite traces over Σ and T is a non-empty, prefix-closed set of finite traces and $\bar{T} = \{u \in \Sigma^\omega \mid \forall t < u \bullet t \in T\}$.

Axioms 6 and 7 are straightforward extensions of axioms 1 and 4 respectively. The purpose of axiom 8 is to ensure there are enough infinite traces analogously to the way that axiom 3 ensures there are enough failures. It states that there exists at least one deterministic refinement of a process after it has undergone a trace s (the finite traces of this refinement are given by the set T), and that any infinite trace that this refinement can undergo is in $\mathcal{I}$. The explanation and derivation for this axiom are quite subtle and the interested reader is referred to [19] for details.

---

[*]We adopt the form of axiom 8 from [20] which is equivalent to that in [19] as argued in the appendix of that paper.

4   MODELLING OBJECT-Z CLASSES AS PROCESSES

In this section we present both finite and infinite trace models for Object-Z classes. The former is based on the approaches adopted in [23] and [11]. The latter is a simple extension of these approaches and proofs of the necessary additional axioms are given. As a preliminary, we present the semantics of Object-Z classes from which the finite and infinite trace models are derived.

4.1   Semantics of Object-Z classes

A class in Object-Z can be modelled as a set of values each corresponding to a potential object of the class at some stage of its evolution. Such a semantics is presented in [22] where the value chosen to represent an object is the sequence of states the object has passed through together with the corresponding sequence of operations the object has undergone. This value is referred to as the *history* of the object.

The histories of a class can be derived from its sets of states S, initial states I, operations O and state transitions ST. Object-Z classes can also have fairness properties which restrict the set of histories. These, however, are not considered in this section.

The states of a class are an assignment of values to the class' local constants and state variables as well as any global constants to which the class may refer. Given the set of all possible identifiers Id and the set of all possible values $Value$, the states of a class can be represented by a set $S \subseteq (Id \nrightarrow Value)$ such that the states of a class refer to a common set of identifiers: $s_1 \in S \wedge s_2 \in S \Rightarrow \operatorname{dom} s_1 = \operatorname{dom} s_2$

The initial states of a class are simply a subset of its states: $I \subseteq S$. The operations of a class are instances of the class' operation schemas. They can be represented by the name of an operation schema together with an assignment of values to its parameters. The operations of a class can be represented by a set $O \subseteq Id \times (Id \nrightarrow Value)$ such that an operation name is always associated with the same set of parameters: $(n, p_1) \in O \wedge (n, p_2) \in O \Rightarrow \operatorname{dom} p_1 = \operatorname{dom} p_2$

Finally, the state transitions are represented by an operation together with a pair of states representing a possible pre-state and corresponding post-state of the operation: $ST \subseteq O \times S \times S$.

No conditions are placed on the state transitions allowing an operation to be associated with no transitions (corresponding to having an unsatisfiable precondition) or multiple transitions with the same pre-state (corresponding to nondeterminism).

A history of a class is a non-empty sequence of states together with a sequence of operations. Either both sequences are infinite or the state sequence is one longer than the operation sequence. The first state is an initial state of the class and each pair of consecutive states corresponds to a state transition of the corresponding operation.

Given a class C with states S, initial states I, operations O and state transitions ST, the histories of the class are derived as follows.

$$\mathcal{H}(C) = \{(s, o) \mid (s \in S^\omega \wedge o \in O^\omega \vee s \in S^* \wedge o \in O^* \wedge \#s = \#o + 1) \wedge \\ s(1) \in I \wedge \forall i \in \operatorname{dom} o \bullet (o(i), s(i), s(i + 1)) \in ST\}$$

Note that the interpretation of an operation in Object-Z differs from that in Z in that an operation cannot occur when its precondition is not enabled. In Z, the operation would be able to occur but the outcome would be unspecified. In [11], a slightly different interpretation of an operations is taken allowing both the views of Object-Z and Z to be accommodated. In this paper, however, we adopt the standard view of Object-Z.

The following property of the set of histories of a class (without fairness properties) is given as a lemma without proof.

Lemma 1: An infinite history (s, o) is a history of C if and only if all its prefixes are histories of C:

$$\forall (s, o) \in S^{\omega} \times O^{\omega} \bullet (s, o) \in \mathcal{H}(C) \Leftrightarrow$$
$$\forall n \in \mathbb{N} \bullet (1 \mathinner{.\,.} n + 1 \lhd s, 1 \mathinner{.\,.} n \lhd o) \in \mathcal{H}(C) \qquad \square$$

4.2 A finite trace model of Object-Z classes

In order to model a class as a process, we need to relate operations and events. A number of ways of doing this exist (see [23] and [11] for examples). For the purposes of this paper, we will assume the existence of a function *event* which returns a unique event for each operation of a class. This function also returns identical events for operations from different classes which have the same name and whose parameters have the same basenames (i. e. apart from the ? or !) and values. This allows such common-named operations to synchronise and their matching input and output parameters to be equated. The function *events* is the straight-forward extension of *event* to traces. A class which has unsatisfiable initial conditions will have no histories and hence cannot be given a failures-divergences semantics in an obvious way. Such classes need to be considered separately from other classes.

The failures of a class C with satisfiable initial conditions, and hence at least one history, can be derived from its set of histories as follows: (t, X) is a failure of C if there exists a history of C such that

- the sequence of operations of the history corresponds to the sequence of events in t and
- for each event in X, there does not exist a history which extends the original history by an operation corresponding to that event.

$$\mathcal{F}(C) = \{(t, X) \mid \exists (s, o) \in \mathcal{H}(C) \bullet t = events(o) \wedge$$
$$\forall e \in X \bullet \nexists st \in S, op \in O \bullet$$
$$e = event(op) \wedge (s ^\frown \langle st \rangle, o ^\frown \langle op \rangle) \in \mathcal{H}(C)\}$$

As shown in [23], the failures of C defined in this way satisfy axioms 1 to 3 of the finite trace model.

Since Object-Z has no notion of internal operations nor recursive definitions of operations*, the set of divergences of a class C is empty: $\mathcal{D}(C) = \varnothing$. This trivially satisfies the properties 4 and 5 of the finite trace model.

---

*Although recursive definitions of operations have been suggested for Object-Z (e.g. [8]), we have adopted a more conservative view of Object-Z in this paper.

4.3 An infinite trace model of Object-Z classes

The infinite trace model can be derived from the finite trace model by the addition of a set of infinite traces. Given a class C with operations O these are derived from the histories of C as follows.

$$\mathcal{I}(C) = \{t \mid \exists(s, o) \in \mathcal{H}(C) \bullet o \in O^\omega \wedge t = events(o)\}$$

The proof of axiom 6 of the infinite trace model follows from Lemma 1. The proof of axiom 7 is trivial since the set of divergences of a class is empty. It remains then to prove axiom 8.

Let C be a class with states S, initial states I, operations O and state transitions ST and let $(t, \varnothing)$ be a failure of C. From the definition of the function $\mathcal{F}$, there exists a history (s, o) of C such that $t = events(o)$.

The proof of axiom 8 relies on finding another class D whose initial state is the final state of s and is otherwise a deterministic refinement of class C. Such a class can be constructed as follows.

Let D be a class with states S, initial states $I' = \{last(s)\}$, operations O and state transitions $ST' \subseteq ST$ such that

$$\forall\, op \in O, st_1 \in S \bullet \exists\, st_2 \in S \bullet$$
$$(op, st_1, st_2) \in ST \Rightarrow \exists_1\, st \in S \bullet (op, st_1, st) \in ST'$$

That is, whenever C can perform an operation from a particular state, D can also perform the operation from that state and there is a unique post-state.

The set T whose existence must be found to satisfy axiom 8 is then simply the set of traces of D, i.e. $T = \{t \mid (t, X) \in \mathcal{F}(D)\}$. To prove this, we need to prove the following for T.

(1) T is prefix-closed and non-empty.
(2) $\forall t \in T \bullet (s \frown t, \{a \mid t \frown \langle a \rangle \in T\}) \in \mathcal{F}(C)$.
(3) $\{s \frown u \mid u \in \bar{T}\} \subseteq \mathcal{I}(C)$ where $\bar{T} = \{u \in events(O^\omega) \mid \forall t < u \bullet t \in T\}$.

Proof

(1) The failures of a class are prefix-closed and non-empty.
(2) By axiom 2 of the finite trace model, $\forall t \in T \bullet (t, \varnothing) \in \mathcal{F}(D)$. Hence, by axiom 3 of the finite trace model, $\forall t \in T \bullet (t, \{a \mid (t \frown \langle a \rangle, \varnothing) \notin \mathcal{F}(D)\}) \in \mathcal{F}(D)$ and, therefore, $\forall t \in T \bullet (t, \{a \mid t \frown \langle a \rangle \notin T\}) \in \mathcal{F}(D)$. Also, from the definition of D, it can be shown that $\{(s \frown t, X) \mid (t, X) \in \mathcal{F}(D)\} \subseteq \mathcal{F}(C)$. Therefore, $\forall t \in T \bullet (s \frown t, \{a \mid t \frown \langle a \rangle \notin T\}) \in \mathcal{F}(C)$.
(3) Given $u \in \bar{T}$, let $t_n = 1..n \lhd u$ for all $n \in \mathbb{N}$. Since for all $n \in \mathbb{N}, t_n \in T$, from the definition of $\mathcal{F}$, there exists a history (s_n, o_n) of D such that $t_n = events(o_n)$. Since D is deterministic, each sequence of operations will only occur with one sequence of states in the set of histories of D. Also, since $event$ returns a unique event for each operation of a class, any sequence of events of D will correspond to a unique sequence of operations and hence a unique sequence of states. Therefore, $\forall\, m, n \in \mathbb{N} \bullet m < n \Rightarrow s_m < s_n$.

Let (s, o) be the infinite history such that $\forall n \in \mathbb{N} \bullet s(n) = last(s_n)$ and $o = events(u)$. Then $\forall n \in \mathbb{N} \bullet (1 \mathinner{.\,.} n + 1 \lhd s, 1 \mathinner{.\,.} n \lhd o) \in \mathcal{H}(D)$. Therefore, by Lemma 1, $(s, o) \in \mathcal{H}(D)$ and hence, by the definition of $\mathcal{I}$, $u \in \mathcal{I}(D)$. Also, from the definition of D, it can be shown that $\{s \frown u \mid u \in \mathcal{I}(D)\} \subseteq \mathcal{I}(C)$. Therefore, $\{s \frown u \mid u \in \bar{T}\} \subseteq \mathcal{I}(C)$.   □

5   CASE STUDY

To demonstrate the differences between the finite and the infinite trace models, we develop a management system for process identifiers (PIDs). This system is implemented using a bus where all numbers are transmitted serially. As the number of PIDs is not bounded (a PID can be any natural number) and any free PID can be chosen by the system, the number of communications on the bus – a local operation – is unbounded but finite. This causes the system to diverge in the finite trace model but not in the infinite trace model.

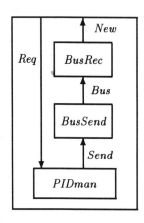

A connection diagram of the system can be found in Fig. 1. Every box in the diagram stands for an Object-Z class. The semantics of the combination is defined by CSP parallel composition and hiding (see below).

The type *PID* is just an abbreviation for a natural number (*PID* $==$ $\mathbb{N}$). The class *PIDman* has two operations: A request from the environment for new PIDs by

Figure 1 Connection Diagram for *PIDbus*

Req is answered with the operation *Send* that transfers a new PID to the bus. Releasing PIDs that are no longer used is not considered in this simple example.

PIDman	
used : $\mathbb{P}\, PID$ alloc : seq PID [alloc are the PIDs which have been allocated to send.]	__Init__ used $= \varnothing$ alloc $= \langle\,\rangle$
__Req__ $\Delta(used, alloc)$ front $alloc' = alloc$ last $alloc' \notin used$ $used' = used \cup \{last\, alloc'\}$	__Send__ $\Delta(alloc)$ $n! : PID$ $alloc \neq \langle\,\rangle$ $n! = head\ alloc \wedge alloc' = tail\ alloc$

The output of the PID manager is transmitted via a serial bus to the environment. The type $STATUS$ is used to code the control flow of the bus and the type $SBIT$ extends a bit with the stop signal S denoting end of transmission.

$STATUS ::= trans \mid wait$

$$SBIT ::= 0 \mid 1 \mid S$$

The status *trans* holds during transmission of bits over the bus. When the system is ready to receive a new *PID* the status *wait* holds. The class *BusSend* receives a natural number and sends it bitwise over the bus.

```
┌─ BusSend ──────────────────────────────────────────────────────┐
│                                    ┌─ Send ──────────────────┐  │
│  ┌──────────────────────────┐      │ Δ(st, v)                │  │
│  │ v : ℕ                    │      │ n? : PID                │  │
│  │ st : STATUS              │      ├─────────────────────────┤  │
│  ├──────────────────────────┤      │ st = wait ∧ v' = n?     │  │
│  │ st = wait ⇒ v = 0        │      │ st' = trans             │  │
│  └──────────────────────────┘      └─────────────────────────┘  │
│                                                                  │
│  ┌─ Init ───────────────────┐                                   │
│  │ st = wait                │                                   │
│  └──────────────────────────┘                                   │
│                                                                  │
│  ┌─ Bus ─────────────────────────────────────────────────────┐  │
│  │ Δ(st, v)                                                   │  │
│  │ b! : SBIT                                                  │  │
│  ├────────────────────────────────────────────────────────────┤ │
│  │ st = trans                                                 │  │
│  │ v = 0 ⇒ (b! = S ∧ st' = wait)                              │  │
│  │ v > 0 ⇒ (b! = v mod 2 ∧ v' = v div 2 ∧ st' = trans)        │  │
│  └────────────────────────────────────────────────────────────┘ │
└──────────────────────────────────────────────────────────────────┘
```

The class *BusRec* receives bits over the bus, decodes the bits to a number and sends the result to the environment. The variable *bc* counts the number of bits already received.

```
┌─ BusRec ───────────────────────────────────────────────────────┐
│                                    ┌─ New ───────────────────┐  │
│  ┌──────────────────────────┐      │ Δ(st, v)                │  │
│  │ st : STATUS              │      │ n! : PID                │  │
│  │ v, bc : ℕ                │      ├─────────────────────────┤  │
│  └──────────────────────────┘      │ st = wait ∧ n! = v      │  │
│                                     │ st' = trans ∧ v' = 0    │  │
│  ┌─ Init ───────────────────┐      └─────────────────────────┘  │
│  │ st = trans ∧ bc = 0 ∧ v = 0 │                               │
│  └──────────────────────────┘                                   │
│                                                                  │
│  ┌─ Bus ─────────────────────────────────────────────────────┐  │
│  │ Δ(v, st, bc)                                               │  │
│  │ b? : SBIT                                                  │  │
│  ├────────────────────────────────────────────────────────────┤ │
│  │ st = trans                                                 │  │
│  │ b? ≠ S ⇒ ( v' = b? * 2^bc + v ∧ st' = trans ∧ bc' = bc + 1)│  │
│  │ b? = S ⇒ ( st' = wait ∧ bc' = 0 ∧ v' = v)                  │  │
│  └────────────────────────────────────────────────────────────┘ │
└──────────────────────────────────────────────────────────────────┘
```

The system *PIDbus* is defined by the following CSP expression (compare Fig. 1):

$$PIDbus = \left(PIDman \underset{\{\!|Send|\!\}}{\|} BusSend \underset{\{\!|Bus|\!\}}{\|} BusRec \right) \backslash \{\!| Send, Bus |\!\}$$

where $C_1 \|_X C_2$ denotes the CSP parallel composition of C_1 and C_2 synchronising on the set of events X. The notation $\{| \, Send \, |\} = \{Send.n \mid n \in PID\}$ abbreviates the set of events of the operation $Send$.

By the CSP hiding operator $\backslash$, the operations $Send$ and Bus are made invisible to the environment. These operations are called *local operations*. The different semantics of hiding in the CSP models let $PIDbus$ diverge in the finite trace model, but not in the infinite trace model:

The divergences of the hiding operator in the finite trace model are:*

$$\mathcal{D}(C \backslash X) \;=\; \{(s \backslash X) ^\frown t \mid s \in \mathcal{D}(C)\} \cup$$
$$\{s ^\frown t \mid \forall n \bullet \exists s_n : \mathcal{T}(C) \bullet s_n \backslash X = s\}$$

where $\mathcal{T}(C) = \{s \mid (s, \varnothing) \in \mathcal{F}(C)\}$ is the set of traces of C. Divergent traces are a result of finite traces on the local operation with no upper bound. There are at least n div $2 + 2$ operations Bus if the PID n is transmitted. As there is no upper bound for PIDs, there is no upper bound for the number of local operations of $PIDbus$. Thus $PIDbus$ diverges after receiving a request for new PIDs.

A divergent trace in the infinite trace model is a result of an infinite trace involving only local operations.

$$\mathcal{D}(C \backslash X) \;=\; \{(u \backslash X) ^\frown t \mid u \in \mathcal{I}(C) \wedge u \backslash X \text{ is finite}\} \cup \{s \backslash X \mid s \in \mathcal{D}(C)\}$$

As we argued above, there is no infinite trace of Bus in $PIDbus$. As well, the other local operation, $Send$, can only occur once during a transmission of a PID. Thus $PIDbus$ is not divergent in the infinite trace model.

6   REFINEMENT

In this section we investigate the differences in the refinement rules for both semantic models.

CSP refinement is based on set inclusion: A process P refines a process Q in the finite trace model if P is more deterministic $\mathcal{F}(P) \subseteq \mathcal{F}(Q)$ and less divergent $\mathcal{D}(P) \subseteq \mathcal{D}(Q)$. We write $Q \sqsubseteq_F P$ if Q is refined by P in the finite trace model.

For the infinite trace model, the additional condition $\mathcal{I}(P) \subseteq \mathcal{I}(Q)$ must hold. We write $Q \sqsubseteq_I P$ to denote this refinement. The symbol $\sqsubseteq$ denotes refinement in both models.

Refinement of Object-Z classes can be defined similar to Z data refinement. A class C refines a class A (both having the same set of operations), if every step of C can somehow be simulated by the class A. The exact definition relies on a retrieve relation $Retr$ that relates the states of C and A. Thus $Retr$ can be written in the form

---

*Note that in [18] a refined version of the hiding operator is given that can cope with some forms of unbounded nondeterminism. But this definition is not widespread in the CSP literature and as the main problem remains, we use the simpler standard definition for hiding in a finite trace model.

$[A.\texttt{State};\ C.\texttt{State}\mid P]$ for some predicate P. ($A.\texttt{State}$ refers to the state schema of class A. Similar notation is used to refer to the other schemas of a class.) We assume in this section that C and A do not have variables with the same name.

The class C is a *forward simulation* of the class A if the following holds for all operations *op*.

1. The initial states are related: $C.\texttt{Init} \Rightarrow \exists A.\texttt{Init} \bullet Retr$
2. The preconditions of related states are equivalent: $Retr \Rightarrow (\text{pre } A.op \Leftrightarrow \text{pre } C.op)$
   where pre $op = \exists \texttt{State}' \bullet op$ is the precondition of an operation.*
3. The state change of $C.op$ can be simulated by the operation $A.op$ for corresponding states: $Retr \wedge C.op \Rightarrow \exists A.\texttt{State}' \bullet Retr' \wedge A.op$

The definition of *backward simulation* is similar. It is omitted here.

The relation between these simulation techniques and CSP refinement was worked out by He [14], Josephs [17], Woodcock and Morgan [26] and Butler [4]. They prove that forward simulation is sound, i.e. if C is a simulation of A then $A \sqsubseteq C$ holds. This is proven by Hallerstede [13] in the finite trace model for a combination of CSP and Z similar to our combination of Object-Z and CSP. This proof is easily extended to the infinite trace model.

Differences occur, however, if the simulation technique is extended to local operations which are not visible for the environment. Thus we investigate extensions of the simulation rules to prove $A \sqsubseteq C \setminus X$ where X is the set of events of the local operations L. Simulation rules for this situation are called *weak simulation* [6].

6.1 Weak simulation in the infinite trace model

The idea of a weak simulation rule is that the abstract state is not changed by a local operation in the concrete class.

Let C and A be classes and $Retr$ a retrieve relation and L the set of local operations of C. Then C is a *weak forward simulation* of A if the following holds for all operations *op* of A:
1. The initial states are related: $C.\texttt{Init} \Rightarrow \exists A.\texttt{Init} \bullet Retr$
2. The preconditions of related states are equivalent if no local operation is enabled.

$$(Retr \wedge \forall l : L \bullet \neg \text{pre } C.l) \Rightarrow (\text{pre } A.op \Leftrightarrow \text{pre } C.op)$$

Note that refusals cannot be observed if a local operation is enabled. Therefore the equivalence of the preconditions must only hold if no local operation is enabled.
3. The state change of $C.op$ can be simulated by the operation $A.op$ for corresponding states: $Retr \wedge C.op \Rightarrow \exists A.\texttt{State}' \bullet Retr' \wedge A.op$

and there exists a schema $term \mathrel{\widehat{=}} [\, C.\texttt{State};\ t : \mathbb{N} \mid P_{term} \,]$ such that $term$ is defined for all states of C and for all local operations $l \in L$ the following holds:
4. l does not change the abstract state:

   $C.l \wedge Retr \Rightarrow \exists A.\texttt{State}' \bullet \Xi A.\texttt{State} \wedge Retr'$
5. l does not diverge: $C.l \wedge term \Rightarrow t' < t \wedge term'$

---

*Note that the definition of pre here is different to that in Z where the outputs are also hidden.

6.2   Weak simulation in the finite trace model

For the finite trace model we have to add an extra condition concerning bounded nondeterminism. Increase of the value t of the schema *term* must be finitely bounded.

Hence, C is a *weak forward simulation* of A if 1.-5. from above and the following extra condition holds

6. There exists a constant $b : \mathbb{N}$ that bounds the increase of t for all operations c:

$$C.c \wedge \Delta term \Rightarrow t' < t + b$$

Furthermore the initial value of t must be bound: $C.\texttt{Init} \wedge term \Rightarrow t < b$

A refined version of this rule that also takes the communicated parameters into account can be found in [10].

6.3   Case study

We use the example from Section 5 to indicate the application of the simulation rules. We have designed class *PIDman* such that it is a specification of the complete system *PIDbus* when *Send* is renamed to *New*, i.e. we want to prove
$PIDman[New/Send] \sqsubseteq_I PIDbus$.

To apply the weak simulation rule, we have to remove the parallel composition in the definition of *PIDbus*. This can be done by a rule for parallel composition from [10]: Parallel composition of classes with disjoint state spaces is basically equivalent to conjunction. Thus CSP parallel composition can be expressed by Object-Z inheritance. According to this rule, the parallel composition of *PIDman*, *BusSend* and *BusRec* is equivalent to the following class that inherits all three classes.

```
┌─ PIDcon ──────────────────────────────────────────────
│ PIDman[n/n!]
│ BusSend[st_S/st, v_S/v, n/n?, b/b!]
│ BusRec[st_R/st, v_R/v, b/b?]
│         [new/old renames a variable, parameter or schema name old to new.]
└
```

Note that the common named schemas of the inherited classes are implicitly conjoined.

With proper renaming, we can apply the rule for weak forward simulation for the infinite trace model to prove

$$PIDman[New/Send, used_A/used, alloc_A/alloc] \sqsubseteq_I PIDcon \setminus \{| \ Send, Bus \ |\}$$

using the schemas *retr* and *term* from Fig. 2. A proof of the side conditions will be published in a technical report. As we expect, the proof of the extra side condition 6 in the finite case fails; there is no bound for the variable t after the operation *Send*.

Note that the case study, although very simple, shows a general principle of using Object-Z and CSP for the development of communicating systems: The class *PIDman* corresponds to the specification of some layer N that is implemented us-

---

Retr _____

$PIDman.\texttt{State} \; [used_A/used, alloc_A/alloc]$
$PIDcon.\texttt{State}$

_____

$(st_S = wait \land st_R = trans) \Rightarrow alloc_A = alloc$
$st_S = st_R \Rightarrow (tail \; alloc_A = alloc \land head \; alloc_A = v_S * 2^{bc} + v_R)$
$(st_S = trans \land st_R = wait) \Rightarrow$
$\qquad (tail \; tail \; alloc_A = alloc \land head \; alloc_A = v_R \land head \; tail \; alloc_A = v_S)$
$used_A = used$

---

term _____

$PIDcon.\texttt{State}$
$t : \mathbb{N}$

_____

$(st_S = wait \land st_R = trans \land alloc \neq \langle \rangle) \Rightarrow t = head \; alloc + 3$
$st_S = st_R = trans \Rightarrow t = v_S + 2$
$st_S = st_R = wait \Rightarrow t = 1$
$(st_S = trans \land st_R = wait) \Rightarrow t = 0$

Figure 2 Retrieve Relation and Termination Schema for the Case Study

ing the functionality of some lower layer $N + 1$, the bus. The two steps in this section, parallel composition and weak simulation, are generally applicable in this situation.

7 SPECIFYING FAIRNESS

In this section, we investigate the specification of fairness properties as a possible application of the infinite traces model. Fairness properties state that an operation which is either repeatedly or continuously enabled must eventually occur. They therefore affect only the infinite, and not the finite, traces of a process.

Consider specifying a buffer which never does an infinite sequence of inputs without outputting. Such a specification may be modelling a finite buffer in which case, an implementation would need to restrict the behaviour so that an input cannot be performed when the capacity of the buffer is reached. Alternatively, it may be modelling a finite buffer which is shared by a number of other clients and whose capacity, therefore, from the point of view of a particular client, changes nondeterministically.

In standard CSP, such fairness properties can be captured by the introduction of an auxiliary "supervising" process but not at a higher level of abstraction. In Object-Z, they can be captured by the use of temporal logic predicates to restrict the set of histories of a class. For example, the buffer described above could be specified as follows.

Buffer₁[T] _____

$Buffer[T]$

_____

$\Box \Diamond (\mathbf{Out} \; \mathbf{occurs})$

The class $Buffer_1$ inherits $Buffer$ from Section 2. Inheritance can be resolved here by replacing $Buffer$ with its definition. The history predicate $\Box\Diamond(Out\ \mathbf{occurs})$ (always eventually Out occurs) restricts the infinite histories of $Buffer_1$ to those histories which do not end in an infinite sequence of In operations. (See [21] for details.)

However, we cannot specify such fairness properties when the class is going to be used in the CSP setting because axiom 8 does not hold. $Buffer_1$ must do any finite sequence of In, but it cannot do an infinite trace of In.

Fairness properties can only be placed on events which are totally under the control of a process and not its environment [1]. Therefore, to specify a fair buffer, we need to model the decision to perform an Out operation as internal to the buffer. For example, the class $Buffer_2$ has an auxiliary variable $next\_out$ modelling the buffer's decision to perform an Out operation next. It can only do this when there is an item in the buffer to output.

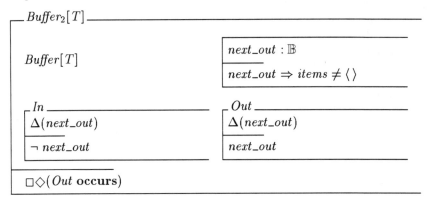

In this case, the fairness condition on Out makes sense. If we place $Buffer_2$ in an environment which continuously offers events corresponding to the operation In but never events corresponding to the operation Out, it will eventually deadlock (when $next\_out$ becomes true). Note that the resulting finite trace with only events corresponding to In is a possible trace of the class. The fairness property only restricts the class' infinite traces. Axiom 8 holds for $Buffer_2$ (although the proof from section 4.3 does not translate directly for this situation).

An other form of operations that are totally under control of a process are local operations. However, fairness constraints on local operations can implicitly put constraints on visible operations and thus making axiom 8 invalid.

The infinite trace model can, therefore, be used as a means to specify fairness properties abstractly in a combined Object-Z/CSP notation. But appropriate restrictions are required to prevent the occurrence of classes which do not fulfill all axioms of the CSP model. The exact form of such restrictions and the practicalities of enforcing them requires further investigation.

8   CONCLUSION

In this paper we have investigated the influence of the finite and the infinite trace models on the combination of Object-Z and CSP. First of all, the underlying idea of deriving a CSP semantics from the histories semantics of an Object-Z class and the proof of the necessary axioms can be carried out without difficulty in both models. In fact both CSP semantics for Object-Z are equivalent if only bounded nondeterminism is considered.

From a theoretical point of view, the infinite trace model seems to be preferable: It can handle all forms of unbounded nondeterminism properly, the rules for weak simulation have fewer side conditions and, thus, are more flexible. It is even possible to handle fairness constraints under certain restrictions; although the exact formulation of the restrictions is a topic for further research.

However, the finite trace model is more widespread in the CSP community. Tools like the CSP model checker FDR [12] or an encoding of CSP in Isabelle [25] are only available for the finite trace model. Although the refinement rule is more complicated, the simulation technique is only complete* in the finite trace model [4, 13]. Furthermore, it is not clear whether the restricted use of fairness constraints in the infinite trace model is a significant advantage since the necessity of modelling internal choice in Object-Z classes can often complicate specifications. This is evident from the example in Section 7.

To decide which model is preferable, the role of unbounded nondeterminism in the application domain of the combination of Object-Z and CSP has to be considered. If unbounded finite sequences of events occur, the infinite trace model is the better choice. However, in real systems such unbounded sequences of events do not occur. For example, there would be a bound for PIDs in any real system. If we used the type $PID == 0 .. maxPID$ (for some natural number $maxPID$) instead of $\mathbb{N}$ in the case study of Section 5, we could safely use the finite trace model. The specification would still be fairly abstract as we wouldn't have to specify the exact value of $maxPID$. Further case studies will show if such restrictions can be found in general.

Acknowledgements
The work of C. Fischer is part of the project UniForM supported by the German Ministry for Education and Research (BMBF) under grant No. FKZ 01 IS 521 B2. G. Smith is supported by a research fellowship granted by the Alexander von Humboldt-Stiftung, Germany.

REFERENCES

[1] M. Abadi and L. Lamport. Composing specifications. In J.W. de Bakker, W.-P. de Roever, and G. Rozenberg, editors, *REX Workshop*, volume 430 of *Lecture Notes in Computer Science*, pages 1–41. Springer-Verlag, 1990.

[2] T. Bolognesi and E. Brinksma. Introduction to the ISO specification language LOTOS. *Computer Networks and ISDN Systems*, 14(1):25–59, 1988.

[3] S.D. Brookes and A.W. Roscoe. An improved failures model for communicating processes. In *Pittsburgh Symposium on Concurrency*, volume 197 of *Lecture Notes in Computer Science*, pages 281–305. Springer-Verlag, 1985.

[4] M. J. Butler. *A CSP Approach To Action Systems*. PhD thesis, University of Oxford, 1992.

[5] E. Cusack and G.H.B. Rafsanjani. ZEST. In *Object-Orientation in Z*, Workshops in Computing, pages 113–126. Springer-Verlag, 1992.

---

*This means that every relation $A \sqsubseteq C$ for classes A and C can be proven by successive application of forward and backwards simulation.

[6] J. Derrick, E. Boiten, H. Bowman, and M. Steen. Weak refinement in Z. In J. Bowen and M. Hinchey, editors, *ZUM'97: The Z Formal Specification Notation*, volume 1212 of *LNCS*, pages 369–388, 1997.

[7] J. Derrick, E.A.Boiten, H. Bowman, and M. Steen. Supporting ODP - translating LOTOS to Z. In *First IFIP International workshop on Formal Methods for Open Object-based Distributed Systems (FMOODS)*. Chapman & Hall, 1996.

[8] J. Dong, R. Duke, and G. Rose. An object-oriented approach to the semantics of programming languages. In G. Gupta, editor, *17th Annual Computer Science Conference (ACSC'17)*, pages 767–775, 1994.

[9] R. Duke, G. Rose, and G. Smith. Object-Z: A specification language advocated for the description of standards. *Computer Standards and Interfaces*, 17:511–533, 1995.

[10] C. Fischer. Combining CSP and Z. Technical report, University of Oldenburg, 1997.

[11] C. Fischer. CSP-OZ: A combination of Object-Z and CSP. To appear in Formal Methods for Open Object-Based Distributed Systems (FMOODS '97), 1997.

[12] Formal Systems (Europe) Ltd. *Failures-Divergence Refinement: FDR 2*, Dec 1995. Preliminary Manual.

[13] S. Hallerstede. Die semantische Fundierung von CSP-Z. Master's thesis, University of Oldenburg, 1997. In German.

[14] J. He. Process simulation and refinement. *Formal Aspects of Computing*, 1(3):229–241, 1989.

[15] C.A.R. Hoare. *Communicating Sequential Processes*. International Series in Computer Science. Prentice-Hall, 1985.

[16] ITU Recommendation X.901-904. *Open Distributed Processing - Reference Model - Parts 1-4*, July 1995.

[17] M.B. Josephs. A state-based approach to communicating processes. *Distributed Computing*, 3:9–18, 1988.

[18] A.W. Roscoe. An alternative order for the failures model. *Journal of Logic and Computation*, 3(2), 1993.

[19] A.W. Roscoe. Unbounded nondeterminism in CSP. *Journal of Logic and Computation*, 3(2), 1993.

[20] A.W. Roscoe and G. Barrett. Unbounded nondeterminism in CSP. In *Mathematical Foundations of Programming Semantics*, volume 442 of *Lecture Notes in Computer Science*, pages 160–193. Springer-Verlag, 1989.

[21] G. Smith. *An Object-Oriented Approach to Formal Specification*. PhD thesis, Department of Computer Science, University of Queensland, 1992.

[22] G. Smith. A fully abstract semantics of classes for Object-Z. *Formal Aspects of Computing*, 7(3):289–313, 1995.

[23] G. Smith. A semantic integration of Object-Z and CSP for the specification of concurrent systems. To appear in Formal Methods Europe (FME '97), 1997.

[24] J.M. Spivey. *The Z Notation: A Reference Manual (2nd Ed.)*. International Series in Computer Science. Prentice-Hall, 1992.

[25] H. Tej and B. Wolff. A corrected failure-divergence-model for CSP in Isabelle/HOL. To appear in Formal Methods Europe (FME '97).

[26] J. C. P. Woodcock and C. C. Morgan. Refinement of state-based concurrent systems. In *Proceedings of VDM Symposium 1990*, volume 428 of *LNCS*, pages 340–351. Springer-Verlag, 1990.

32
Selective mu-calculus: New Modal Operators for Proving Properties on Reduced Transition Systems

Roberto Barbuti
Dipartimento di Informatica. Università di Pisa, 56125 Pisa, Italy.
e-mail: `barbuti@di.unipi.it`

Nicoletta De Francesco, Antonella Santone, Gigliola Vaglini
Dipartimento di Ingegneria dell'Informazione. Università di Pisa, 56126 Pisa, Italy.
e-mail: {`nico,santone,gigliola`}`@iet.unipi.it`

Abstract
In model checking for temporal logic, the correctness of a (concurrent) system with respect to a desired behavior is verified by checking whether a structure that models the system satisfies a formula describing the behaviour. Most existing verification techniques, and in particular those defined for concurrent calculi like as CCS, are based on a representation of the concurrent system by means of a labelled transition system. In this approach to verification, state explosion is one of the most serious problems. In this paper we present a new temporal logic, the *selective mu-calculus*, with the property that only the actions occurring in a formula are relevant to check the formula itself. We prove that the selective mu-calculus is as powerful as the mu-calculus. We define the notion of ρ-bisimulation between transition systems: given a set of actions ρ, a transition system ρ-bisimulates another one if they have the same behaviour with respect to the actions in ρ. We prove that, if two transition systems are ρ-equivalent, they preserve all the selective mu-calculus formulae with occurring actions in ρ. Consequently, a formula with occurring actions ρ can be more efficiently checked on a transition system ρ-equivalent to the standard one, but smaller than it.

Keywords
Mu-Calculus, State Explosion, Abstraction, CCS

Formal Description Techniques and Protocol Specification, Testing and Verification
T. Mizuno, N. Shiratori, T. Higashino & A. Togashi (Eds.)   © 1997 IFIP. Published by Chapman & Hall

1   INTRODUCTION

In model checking for temporal logic, the correctness of a (concurrent) system with respect to a desired behavior is verified by checking whether a structure that models the system satisfies a formula describing the behaviour. Most existing verification techniques, and in particular those defined for concurrent calculi like as CCS [23], are based on a representation of the concurrent system by means of a labelled transition system [8, 12]. In this approach to verification, state explosion is one of the most serious problems: systems are often described by transition systems with a prohibitive number of states. On the other hand, in several cases, it is sufficient to verify a property on a reduced transition system containing only the "parts" which "influence the property". Thus a solution to state explosion is the definition of suitable abstraction criteria by means of which a reduced transition system can be obtained, which abstracts from the parts not concerned with the property to be verified. The works [3, 22, 24, 25, 26, 28, 29] deal with abstractions of transition systems preserving only properties expressible by sub-languages of a general temporal logic language, for example avoiding the use of some operators. The works [1] and [10] present methods for constructing reduced transition systems, where the reduction is based on a temporal logic formula: the reduced system preserves the truth value of the formula. However, [10] refers only to formulae written in a subset of CTL logic, while the method in [1] can be applied only to systems obtained as the composition (product) of smaller ones. In both cases, the reduced transition system is obtained by means of a non-trivial algorithm. Other methodologies exist in which abstraction criteria are issued by the user of the verification environment [6, 7, 12, 13]; although useful in practice, this approach cannot be automated.

Since our aim is to obtain reductions in an automatic way from a formula expressing a temporal property, we consider two main aspects: the first one is the definition of a formalism suitable to express such properties; the second one is the method for extracting, from the definition of a property, the information sufficient to characterize the reduced transition systems. A suitable formalism to express temporal properties could be the modal mu-calculus extended with fixpoint formulae [27]. However, this formalism, although very powerful, cannot be used for easily deducing, from a formula, the reduction which can be performed on the standard transition system to obtain a smaller one on which the formula can be equivalently checked (this point will be discussed extensively in the following).

In order to cope with this problem, we define a different calculus, called *selective mu-calculus*, obtained by replacing the modal operators of the mu-calculus by new "selective modal operators". This new calculus has the same power of the original one: the mu-calculus can be expressed by means of the selective mu-calculus, and viceversa. In addition, each formula written using the selective operators allows us to immediately point out the parts of the transition system that can be disregarded in checking the formula. To formalize this fact, we define the notion of ρ-equivalence between transition systems: given a set of actions ρ, two transition systems T_1 and T_2 are ρ-equivalent iff they present the same behaviour with respect to the actions in ρ. We prove that, if two transition systems are ρ-equivalent, they preserve all the formulae such that the set of actions occurring inside the modal operators of the formulae is a subset of ρ. Thus, to prove a formula, with occurring actions ρ, we check it on a transition system which is ρ-equivalent to the standard one, but which contains only the actions in ρ.

We would like to remark the elegance and the simplicity of our approach: the selective mu-calculus is very easy to understand and to use, being a slight modification of standard mu-calculus. Nevertheless, differently from mu-calculus, its formulae can be proved on reduced transition systems, the structure of which is suggested by the formulae themselves.

After the preliminaries in Section 2 and an informal overview of the approach in Section 3, we define the selective mu-calculus in Section 4. Experimental results are given in Section 5 and Section 6 concludes the work. The proofs of the theorems are only sketched. The complete proofs can be found in [2].

2 PRELIMINARIES

2.1 The Calculus of Communicating Systems

Let us now quickly recall the main concepts about the Calculus of Communicating Systems (CCS) [23]. The syntax of *process expressions* (*processes* for short) is the following:

$$P ::= nil |X| \alpha.P |P + P| \ P|P \ |P \backslash L| P[f]$$

where α ranges over a finite set of actions $\mathcal{A} = \{\tau, a, \overline{a}, b, \overline{b}, ...\}$. The action $\tau \in \mathcal{A}$ is called the *internal action*. The set of *visible actions*, $\mathcal{L}$, ranged over by $l, l' ...$, is defined as $\mathcal{A} - \{\tau\}$. Each action $l \in \mathcal{L}$ (resp. $\overline{l} \in \mathcal{L}$) has a *complementary action* $\overline{l}$ (resp. l). X ranges over a set of *constant* names: each constant X is defined by a constant definition $X \overset{def}{=} P$. We denote the set of process expressions by $\mathcal{E}$.

An *operational semantics* is a transition relation $\longrightarrow_O \subseteq \mathcal{E} \times \mathcal{A} \times \mathcal{E}$, where $\mathcal{E}$ is the set of all the processes. If $(P, \alpha, Q) \in \longrightarrow_O$, we write $P \overset{\alpha}{\longrightarrow}_O Q$. The standard semantics of CCS as defined in [23], will be denoted by $\longrightarrow_S$.

Given an operational semantics $\longrightarrow_O$, if $\delta \in \mathcal{A}^*$ and $\delta = \alpha_1 ... \alpha_n, n \geq 1$, we write $P \overset{\delta}{\longrightarrow}_O Q$ to mean $P \overset{\alpha_1}{\longrightarrow}_O \cdots \overset{\alpha_n}{\longrightarrow}_O Q$. For the empty sequence of actions $\lambda \in \mathcal{A}^*$ we have $P \overset{\lambda}{\longrightarrow}_O P$. With $\mathcal{D}_O(P) = \{Q|P \overset{\delta}{\longrightarrow}_O Q\}$ we denote the set of the *derivatives* of P by $\longrightarrow_O$.

A term P is *image finite* by an operational semantics $\longrightarrow_O$ if each derivative of P by $\longrightarrow_O$ has a finite number of *immediate* derivatives, i.e., for each $Q \in \mathcal{D}_O(P)$, it holds that the set $\{Q' | Q \stackrel{\alpha}{\longrightarrow}_O Q'\}$ is finite.

A *(labelled) transition system* is a quadruple (S, T, R, s_0), where S is a set of states, T is a set of transition labels, $s_0 \in S$ is the initial state, and $R \subseteq S \times T \times S$ is a set of transitions. Given a process P and an operational semantics $\longrightarrow_O$, $O(P) = (\mathcal{D}_O(P), \mathcal{A}, \longrightarrow_O, P)$ is the transition system of P built by means of the relation $\longrightarrow_O$; for example, $S(P)$ is the standard transition system of P. Note that, with abuse of notation, we use $\longrightarrow_O$ for denoting both the operational semantics and the transition relation among the states of the transition system.

2.2   The mu-calculus

We use the modal mu-calculus [21] in a slightly extended form [27] as a branching temporal logic to express behavioural properties. The syntax of the extended mu-calculus is the following, where K ranges over sets of actions and Z ranges over variables:

$$\phi ::= \texttt{tt} \mid \texttt{ff} \mid Z \mid \phi_1 \vee \phi_2 \mid \phi_1 \wedge \phi_2 \mid [K]\phi \mid \langle K \rangle \phi \mid \nu Z.\phi \mid \mu Z.\phi$$

A fixed point formula has the form $\mu Z.\phi$ $(\nu Z.\phi)$ where μZ (νZ) *binds* free occurrences of Z in ϕ and an occurrence of Z is free if it is not within the scope of a binder μZ (νZ). A formula is *closed* if it contains no free variables. The formula $\mu Z.\phi$ is the least fixpoint of the recursive equation $Z = \phi$, while $\nu Z.\phi$ is the greatest one.

The verification of a formula ϕ by a (finite) term P is defined recursively in the following. In the definition, subformulae containing free variables are dealt with using *valuations*, i.e. functions ranged over by $\mathcal{V}$, which assign a subset $\mathcal{V}(Z)$ of processes in $\mathcal{E}$ to each variable Z. Moreover the notion of verification is given also with respect to a semantics $\longrightarrow_O$: the verification of ϕ by P and $\mathcal{V}$ is denoted by $P \models_{\mathcal{V}}^O \phi$. We assume that P is image finite with $\longrightarrow_O$. The transition system $O(P)$ verifies a formula ϕ, written $O(P) \models_{\mathcal{V}} \phi$, if and only if $P \models_{\mathcal{V}}^O \phi$, i.e. the initial state verifies ϕ by $\longrightarrow_O$ and $\mathcal{V}$.

$P \not\models_{\mathcal{V}}^O$ ff

$P \models_{\mathcal{V}}^O$ tt

$P \models_{\mathcal{V}}^O Z$ iff $P \in \mathcal{V}(Z)$

$P \models_{\mathcal{V}}^O \phi \wedge \psi$ iff $P \models_{\mathcal{V}}^O \phi \wedge P \models_{\mathcal{V}}^O \psi$

$P \models_{\mathcal{V}}^O \phi \vee \psi$ iff $P \models_{\mathcal{V}}^O \phi \vee P \models_{\mathcal{V}}^O \psi$

$P \models_{\mathcal{V}}^O [K]\phi$ iff $\forall P'.\forall \alpha \in K.\text{if } P \stackrel{\alpha}{\longrightarrow}_O P' \text{ then } P' \models_{\mathcal{V}}^O \phi$

$P \models_{\mathcal{V}}^O \langle K \rangle \phi$ iff $\exists P'.P' \models_{\mathcal{V}}^O \phi.\exists \alpha \in K.P \stackrel{\alpha}{\longrightarrow}_O P'$

$P \models_{\mathcal{V}}^O \nu Z.\phi$ iff $P \models_{\mathcal{V}}^O \nu Z^n.\phi$ for all natural numbers n

$P \models_{\mathcal{V}}^O \mu Z.\phi$ iff $P \models_{\mathcal{V}}^O \mu Z^n.\phi$ for some natural number n

where $\nu Z^n.\phi$ and $\mu Z^n.\phi$ are defined as:

$$\nu Z^0.\phi = \mathtt{tt} \qquad\qquad \mu Z^0.\phi = \mathtt{ff}$$
$$\nu Z^{n+1}.\phi = \phi\{\nu Z^n.\phi/Z\} \quad \mu Z^{n+1}.\phi = \phi\{\mu Z^n.\phi/Z\}$$

where the notation $\phi\{\psi/Z\}$ indicates the substitution of ψ for every free occurrence of the variable Z in ϕ.

Note that closed formulae do not depend on valuations. Thus, in case of a closed formula ϕ we can simply write $P \models^O \langle K \rangle \phi$ in place of $P \models_\nu^O \langle K \rangle \phi$. Moreover, the verification of a recursive formula by a term P is given considering natural numbers, instead of ordinals, since we consider only image finite terms [27].

In the sequel we will use the following abbreviations (where K range over sets of actions and $\mathcal{A}$ is the set of CCS actions):

$$[\alpha_1,\ldots,\alpha_n]\phi \stackrel{def}{=} [\{\alpha_1,\ldots,\alpha_n\}]\phi; \quad [-]\phi \stackrel{def}{=} [\mathcal{A}]\phi; \quad [-K]\phi \stackrel{def}{=} [\mathcal{A}-K]\phi$$

3 AN INFORMAL OVERVIEW OF THE APPROACH

In this section we present a brief overview of our approach, together with the problems it can solve. For this purpose, we use as an example the following CCS description of an automatic cash dispenser. The dispenser is able to perform two kinds of operations: to provide two different amounts of cash and to give information about a bank account. Each user of the cash dispenser owns a credit card with a personal code that must be supplied before requiring an operation. If the code is correctly inserted, the operation is accepted and executed after the return of the credit card; otherwise, the card is held and no operation is performed. In any case, the dispenser is able to go back to the state in which other requests can be accepted. After having correctly inserted the personal code, the user can ask either for one of two different amounts of money or for an account information. Then the dispenser returns the card and gives either the money or the requested information. In every case, the user must collect the item before the dispenser goes back to the initial state.

$$x \stackrel{def}{=} card.code.(\overline{right}.(cash.(cash_1\_req.\overline{ret\_card}.\overline{cash_1}.collect.x+$$
$$cash_2\_req.\overline{ret\_card}.\overline{cash_2}.collect.x)+$$
$$account.\overline{ret\_card}.\overline{account\_info}.collect.x)+$$
$$\overline{wrong}.\overline{hold}.x)$$

Figure 1 shows $S(x)$, which has 13 states.

Now, let us suppose we want to verify the (mu-calculus) formula ψ_1 below:

$$\psi_1 = \nu Z.([\overline{cash_1}]\mathtt{ff} \wedge [-\overline{right}]Z)$$

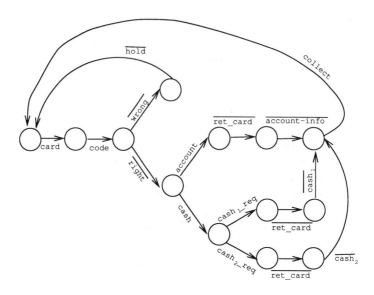

Figure 1

ψ_1 expresses the safety property: "after each action different from $\overline{right}$, an action $cash_1$ cannot be performed".

It is easy to see that ψ_1 it satisfied by the transition system of Figure 1, but it has the same truth value if evaluated on the transition system of Figure 2(a), which is obtained from the transition system of Figure 1 by keeping only the transitions labelled by the actions $\overline{cash_1}$ and $\overline{right}$, and collapsing the states consequently. In fact we can note that, in order to check ψ_1, it is sufficient to observe only the part of the transition system containing these two actions. The problem we want to solve is to devise, given a formula, an automatic way for defining a suitable reduced system on which the formula has the same truth value of the complete system. In other words, given a formula ϕ, we look for a method to individuate those actions labelling transitions which do not alter the value of ϕ. Given such a set of actions, we can eliminate from the transition system the transitions labelled by them, and reduce the system consequently, still preserving the truth value of ϕ.

Consider again ψ_1. We note that the set of actions to be ignored does not coincide with the set of actions not occurring in the formula. In fact, this set contains only the action $\overline{right}$ (recall that $-a$ is a shorthand for $\mathcal{A} - a$), and generates the reduced transition system of Figure 2(b), if interpreted as the set of actions to be ignored. The formula ψ_1 is not satisfied by this transition systems, while it holds on the complete one.

It is important to note that it does not exist a mu-calculus formula expressing the above property and containing only the actions $cash_1$ and $\overline{right}$, which are the only ones relevant for proving the property. Intuitively, the "cycle" $\nu Z.(\dots[-\overline{right}]Z)$ in ψ_1 means "go ahead over non-interesting actions"; thus,

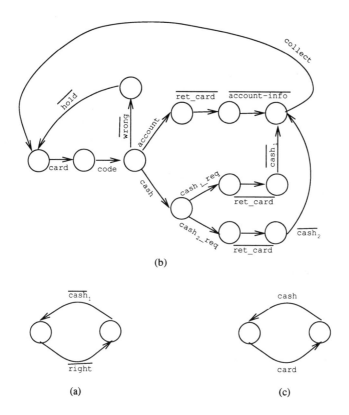

(b)

(a)   (c)

Figure 2

to express the fact that $\overline{right}$ is an interesting action, we need to mention all the other ones.

Consider now the following formula ψ_2, whose informal description is "it holds repeatedly that: there is a finite path leading to a $\overline{right}$ action and, after executing it, there is a finite path leading to a $\overline{cash_1}$ action".

$$\psi_2 = \nu Z.(\mu X.\langle -\overline{right}\rangle X \vee \langle \overline{right}\rangle \text{tt}) \wedge$$
$$(\nu W.[\overline{right}](\mu Y.\langle -cash_1\rangle Y \vee \langle cash_1\rangle Z) \wedge [-\overline{right}]W)$$

All the actions occur in this formula; nevertheless, it can be equivalently checked on the transition system of Figure 2(a). Thus, also in this case, all actions, apart from $\overline{right}$ and $\overline{cash_1}$, can be ignored. The above formulae seem to suggest that the interesting actions are only the ones occurring in the formula both in the form K and $-K$ inside the modal operators. It is sufficient the trivial formula $\psi_3 = [card]\langle cash\rangle \text{tt}$ to realize that this is false. This formula is not satisfied by the transition system of Figure 1 but it is verified by the reduced transition system of Figure 2(c).

The above examples show that it does not exists an intuitive algorithm for extracting the set of actions to be ignored from a mu-calculus formula. On the other hand, they suggest the introduction of new modalities for expressing properties, such that the actions which are relevant for proving a formula are the only ones explicitly mentioned by the modal operators occurring in the formula itself. For instance, we would like to express the property ψ_1 by a formula in which the only occurring actions are $\overline{cash_1}$ and $\overline{right}$. To this purpose, we define the (selective) modal operator $[K]_R$, where K and R are set of actions, such that $[K]_R\,\phi$ is verified by a process which, for every performance of a sequence of actions not belonging to $R \cup K$, followed by an action in K, evolves in a process obeying ϕ. With this new modal operator the property ψ_1 can be expressed by the formula: $\psi_{s1} = [\overline{cash_1}]_{\{\overline{right}\}}\mathtt{ff}$, in which the set of occurring actions is exactly $\{\overline{cash_1}, \overline{right}\}$. The new modality $\langle K \rangle_R\,\phi$ can be is defined analogously.

The idea of the selective mu-calculus is very simple although powerful. Formulae written using the new modalities can be checked equivalently, either on the complete transition system or on the one obtained by disregarding all the actions not occurring in the formula itself. A formula in selective mu-calculus corresponding to ψ_2 is $\psi_{s2} = \nu Z. \langle \overline{right} \rangle_\emptyset\, \mathtt{tt} \wedge [\overline{right}]_\emptyset\, \langle \overline{cash_1} \rangle_\emptyset\, Z$. The actions occurring in this formula "say" that it can be checked on the system of Figure 2(a).

For what regards ψ_3, we obtain the following formula in selective mu-calculus: $\psi_{s3} = [card]_{\{A-card\}}\langle cash \rangle_{\{A-cash\}}\mathtt{tt}$. The occurring actions in this formula are the whole set $\mathcal{A}$; according to the fact that the formula is not checkable on the reduced system of Figure 2(c).

4   THE SELECTIVE MU-CALCULUS

The selective mu-calculus substitutes the modal operators $[K]$ and $\langle K \rangle$ with the selective operators $\langle K \rangle_R$ and $[K]_R$, with $R, K \subseteq \mathcal{A}$, the definition of which is the following:

$$P \models^O_\mathcal{V} [K]_R\,\phi \quad \text{iff} \quad \forall P'.\forall \delta \in (\mathcal{A} - (R \cup K))^*.$$
$$\forall \alpha \in K.\text{if } P \xrightarrow{\delta}_O Q \xrightarrow{\alpha}_O P' \text{ then } P' \models^O_\mathcal{V} \phi$$

$$P \models^O_\mathcal{V} \langle K \rangle_R\,\phi \quad \text{iff} \quad \exists P'.\exists \delta \in (\mathcal{A} - (R \cup K))^*.$$
$$\exists \alpha \in K.P \xrightarrow{\delta}_O Q \xrightarrow{\alpha}_O P' \text{ and } P' \models^O_\mathcal{V} \phi$$

Informally, these new operators require that the formula ϕ is verified after the execution of an action of K, provided that it is not preceded by any action in $R \cup K$. More precisely:

$[K]_R\,\phi$ is verified by a process which, for every performance of a sequence of actions not belonging to $R \cup K$, followed by an action in K, evolves to a process obeying ϕ.

$\langle K \rangle_R\,\phi$ is verified by a process which can evolve to a process obeying ϕ after performing a sequence of actions not belonging to $R \cup K$, followed by an action in K.

The selective mu-calculus is equivalent to the mu-calculus. In fact it is easy to see that the standard mu-calculus operators can be defined by means of the selective operators subscribed by the whole set of actions $\mathcal{A}$:

$$[K]\phi = [K]_\mathcal{A}\,\phi \text{ and } \langle K\rangle\phi = \langle K\rangle_\mathcal{A}\,\phi$$

On the other hand, the selective operators can be expressed in standard mu-calculus as follows:

$$\langle K\rangle_R\,\phi = \mu Z.\langle K\rangle\phi \vee \langle-(R\cup K)\rangle Z \text{ and } [K]_R\,\phi = \nu Z.[K]\phi \wedge [-(R\cup K)]Z$$

Note that the mu-calculus formulae obtained by translating the selective mu-calculus operators have a structure recalling the one of formulae expressing, respectively, weak liveness and safety properties, as classified in [27].

Note also that, the translation from mu-calculus to selective mu-calculus produces formulae in which all the actions ($\mathcal{A}$) occur. This is not necessary in principle: we use this translation only to show how to pass, in a simple way, from one calculus to the other. Of course, it is possible to define more clever algorithms, which base the translation on the structure of mu-calculus formulae, such that the resulting formulae do not contain all the actions $\mathcal{A}$.

Given a set of actions $\rho \subseteq \mathcal{A}$ and a semantics $\longrightarrow_O$, we define a transition relation ignoring all actions in $\mathcal{A} - \rho$.

Definition 1 Given a set of actions $\rho \subseteq \mathcal{A}$ and an operational semantics $\longrightarrow_O$, we define the relation $\longrightarrow_{O\rho}$ in the following way:

for each $\alpha \in \rho$ and $\delta \in (\mathcal{A} - \rho)^*$ $P \xrightarrow{\alpha}_{O\rho} P' \equiv \exists Q.P \xrightarrow{\delta}_O Q \xrightarrow{\alpha}_O P'$.

By $P \xrightarrow{\alpha}_{O\rho} P'$ we express the fact that it is possible to pass from P to P' (according to the operational semantics $\longrightarrow_O$) by performing a (possibly empty) sequence of actions not belonging to ρ and then the action α in ρ. Note that $\longrightarrow_{S\mathcal{A}} = \longrightarrow_S$. Using the $\longrightarrow_{O\rho}$ relation we now give the notions of ρ-bisimulation and ρ-equivalence between transition systems. Informally, two transition systems are ρ-equivalent iff they behave in the same way with respect to the actions in ρ.

Definition 2 (ρ-bisimulation, ρ-equivalence) Let $\rho \subseteq \mathcal{A}$ be a set of actions and $\longrightarrow_O$ and $\longrightarrow_\Omega$ two operational semantics. Let $O(P) = (\mathcal{S}_1, \mathcal{A}, \longrightarrow_O, P)$ and $\Omega(P') = (\mathcal{S}_2, \mathcal{A}, \longrightarrow_\Omega, P')$ the transition systems built for the terms P and P' using the two semantics.

- A *ρ-bisimulation*, $\mathcal{B}$, is a binary relation on $\mathcal{S}_1 \times \mathcal{S}_2$ such that $R\mathcal{B}Q$ implies:
  (i) $R \xrightarrow{\alpha}_{O\rho} R'$ implies $Q \xrightarrow{\alpha}_{\Omega\rho} Q'$ with $R'\mathcal{B}Q'$; and
  (ii) $Q \xrightarrow{\alpha}_{\Omega\rho} Q'$ implies $R \xrightarrow{\alpha}_{O\rho} R'$ with $R'\mathcal{B}Q'$
- $O(P)$ and $\Omega(P')$ are *ρ-equivalent* ($O(P) \approx_\rho \Omega(P')$) iff there exists a ρ-bisimulation $\mathcal{B}$ containing the pair (P, P').

To indicate that two CCS terms P and Q are ρ-equivalent with respect to an operational semantics $\longrightarrow_O$ (i.e. it occurs $O(P) \approx_\rho O(Q)$), we write $P \approx_\rho^O Q$.

Note that $\approx_\mathcal{A}^S$ coincides with Milner's strong equivalence; while $\approx_\mathcal{L}^S$, defined by considering only the visible actions, does not coincide with observational equivalence. In fact, τ actions are completely ignored by $\approx_\mathcal{L}^S$, but this does not occur in the case of observational equivalence. For example, the processes *a.nil* + *τ.nil* and *a.nil* are $\mathcal{L}$-equivalent, while they are not observationally equivalent. On the other hand, *a.nil* + *a.(c.nil* + *τ.nil)* and *a.(c.nil* + *τ.nil)* are observationally equivalent, but they are not $\mathcal{L}$-equivalent. Actually, $\approx_\mathcal{L}^S$ is the same as the $\tau * \alpha$ *equivalence* defined in [14, 17], and implies the *safety equivalence* defined in [5].

Now we can formulate the main theorem of the paper, stating that two transition systems verify a formula ϕ of the selective mu-calculus iff there exists a ρ-bisimulation between them, where ρ contains the set of actions occurring in ϕ. This means that the set of formulae with occurring actions contained in ρ completely characterizes ρ-equivalence, as well as the set of all mu-calculus formulae characterizes strong equivalence [27].

Definition 3 (occurring actions) Given a formula ϕ of the selective mu-calculus, the set $\mathcal{C}(\phi)$ of the actions occurring in ϕ is inductively defined as follows:

- $\mathcal{C}(\mathtt{tt}) = \mathcal{C}(\mathtt{ff}) = \mathcal{C}(Z) = \emptyset$
- $\mathcal{C}(\langle K \rangle_R \phi) = \mathcal{C}([K]_R \phi) = K \cup R \cup \mathcal{C}(\phi)$
- $\mathcal{C}(\phi_1 \vee \phi_2) = \mathcal{C}(\phi_1 \wedge \phi_2) = \mathcal{C}(\phi_1) \cup \mathcal{C}(\phi_2)$
- $\mathcal{C}(\nu Z.\phi) = \mathcal{C}(\mu Z.\phi) = \mathcal{C}(\phi)$

Theorem 4 Let P and Q be two CCS terms and let $\longrightarrow_O$ and $\longrightarrow_\Omega$ be two operational semantics. Suppose that P is image finite by $\longrightarrow_O$ and Q is image finite by $\longrightarrow_\Omega$. For each $\rho \subseteq \mathcal{A}$:

$$O(P) \approx_\rho \Omega(Q) \qquad \text{if and only if}$$
$$P \models^O \phi \Leftrightarrow Q \models^\Omega \phi, \text{ for every } \phi \text{ such that } \mathcal{C}(\phi) \subseteq \rho.$$

Proof Sketch.
(only if) By natural induction on the depth of a formula ϕ of the selective mu-calculus, where the depth of ϕ is the number of nested selective operators ($\langle K \rangle_R$ and $[K]_R$) in ϕ.
(if) By contradiction, i.e. by supposing that $O(P) \not\approx_\rho \Omega(Q)$ and by finding a formula ϕ such that $P \models^O \phi$ and $Q \not\models^\Omega \phi$.

Note that, as well as for mu-calculus and strong equivalence, the *only if* direction in the theorem above holds also for non-image finite terms, while the *if* direction holds only if the terms are image finite [2].

5 USING SELECTIVE MU-CALCULUS TO REDUCE STATE EXPLOSION

The selective mu-calculus has the property that, in each formula, the occurring actions are the only ones relevant to check the formula itself. In this section we discuss how state explosion can be reduced using selective mu-calculus. First of all we state the following proposition, relating transition systems obtained by using different operational semantics defined by O with different sets ρ of actions. We recall that, given a term P and a semantics $\longrightarrow_O$, $O^\rho(P)$ is the transition system generated by the operational semantics $\longrightarrow_{O^\rho}$.

Proposition 5 Given a term P and $\rho, \rho' \subseteq \mathcal{A}$, if $\rho \subseteq \rho'$, $O^\rho(P) \approx_\rho O^{\rho'}(P)$.
Proof Sketch. *By showing that* $\longrightarrow_{(O^{\rho'})_\rho} = \longrightarrow_{O^\rho}$.

If $O = S$ and $\rho' = \mathcal{A}$, the transition system generated by $\longrightarrow_{S^\rho}$ is ρ-equivalent to the one obtained by $\longrightarrow_{S^\mathcal{A}} \equiv \longrightarrow_S$, that is the standard transition system. As a consequence of the above proposition, a strategy to check a property ϕ on $S(P)$ may be that of checking it on $S^\rho(P)$, where $\rho = \mathcal{C}(\phi)$. In fact, in general, $S^\rho(P)$ is smaller than $S(P)$, even if it may be not the minimum one ρ-equivalent to $S(P)$. In order to furtherly reduce the state space, $S^\rho(P)$ can be minimized by known techniques finding the minimum transition system with respect to strong equivalence (see for example [12, 17]).

Example 6 Reconsider the CCS specification of the cash dispenser in Section 3 and let us express some other properties using the selective mu-calculus.

$\psi_1 = [\overline{hold}]_{\{\overline{wrong}\}} \mathbf{ff}$: "the card is not held if the wrong code is not inserted".

$\psi_2 = [right]_\emptyset(\langle cash \rangle_\emptyset \, \mathbf{tt} \vee \langle account \rangle_\emptyset \, \mathbf{tt})$: "if the right code is inserted, it is possible to perform either a cash request or an account information".

$\psi_3 = \nu Z.[card]_\emptyset(Z \wedge [card]_{\{collect, \overline{hold}\}} \mathbf{ff})$: "a card can be inserted only if either the previously inserted card, if any, has been held or the previous operation, if any, has been successfully executed".

Each formula ψ_i, $i \in [1..3]$, can be checked on the transition systems $S^{\rho_i}(x)$ (reduced with respect to strong equivalence), where $\rho_i = \mathcal{C}(\psi_i)$: $\rho_1 = \{hold, \overline{wrong}\}$, $\rho_2 = \{right, cash, account\}$, $\rho_3 = \{card, collect, \overline{hold}\}$. Figure 3 shows $S^{\rho_i}(x)$ (reduced w.r.t. strong equivalence) for each i.

In order to effectively apply the above methodology to processes with any number of states, we need a tool able to build the reduced transition system $S^\rho(P)$, for a CCS term P and a set ρ of actions. We can simulate such a tool by using existing verification environments and standard notions of bisimulations. In fact, we can use the facilities for hiding actions (i.e. renaming some actions as τ), offered by most existing verification environments, and build

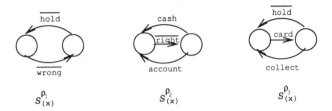

Figure 3

the minimum transition system with respect to a bisimulation ignoring τ actions. To experimentally evaluate the degree of reduction induced by selective mu-calculus, we used a known environment with its notions of bisimulation, i.e. the CADP environment [14, 17]. We applied the following methodology to build a reduced transition system for checking a formula with occurring actions ρ.

1. hide the actions in $\mathcal{A} - \rho$ in the specification, using the hiding facilities of CADP;
2. build the transition system with the -imin option of aldebaran, issuing $\tau^*\alpha$ equivalence reduction.

In order to show that the above strategy is correct, we state the following proposition:

Proposition 7 Let us denote by $H_\rho(P)$ the transition system obtained by $S(P)$ by substituting τ to all actions in $\mathcal{A} - \rho$. We have $H_\rho(P) \approx_\rho S(P)$. Moreover $H_\rho^{min}(P) \approx_\rho S(P)$, where $H_\rho^{min}(P)$ is the minimum transition system $\tau^*\alpha$ equivalent to $H_\rho(P)$.
Proof Sketch. *By Proposition 5 and by transitivity of* $\approx_\rho$.

Example 8 Let us consider the task scheduling system, taken from [23]: n processes wish to perform a task repeatedly, and a scheduler is required to ensure that they begin the task in cyclic order starting with the first process. The different task-performances need not exclude each other in time (for example the second process can begin before the first one finishes), but the scheduler is required to ensure that each agent finishes one performance before it begins the following. The action a_i signals to the i-th process that it can perform the task, whereas b_i signals its completion. The execution of each task is scheduled by a single process:

$$A \overset{def}{=} a.C \quad C \overset{def}{=} c.E \quad E \overset{def}{=} b.D + d.B \quad B \overset{def}{=} b.A \quad D \overset{def}{=} d.A$$

If we define $A_i \overset{def}{=} A[f_i]$, $D_i \overset{def}{=} D[f_i]$, etc., where $f_1 = [a_i/a,\ b_i/b,\ c_i/c,\ \overline{c_n}/d]$, and $f_i = [a_i/a,\ b_i/b,\ c_i/c,\ \overline{c_{i-1}}/d]$, for $1 < i \le n$, an n-task scheduler is:

$$Sched_n \overset{def}{=} (A_1 \mid D_2 \mid \cdots \mid D_n)\backslash \tilde{c}$$

where $\tilde{c}$ denotes the set $\{c_1, \ldots, c_n\}$. The sort of the scheduler is $\{a_i, b_i \mid 1 \le i \le n\}$.

The properties we wish to prove about the scheduler are the following.

1. the start-task actions $a_1, \ldots, a_n$ are performed cyclically starting with a_1;
2. for each i, the start-task action a_i and the end-task action b_i are performed alternately.

A selective mu-calculus formula expressing (1) is:

$$
\begin{aligned}
\phi = \quad &\nu Z. [\tilde{a} - a_1]_{\{a_1\}} \, \text{ff} \wedge [a_1]_\emptyset \\
&\left([\tilde{a} - a_2]_{\{a_2\}} \, \text{ff} \wedge [a_2]_\emptyset\right. \\
&\left([\tilde{a} - a_3]_{\{a_3\}} \, \text{ff} \wedge \ldots \wedge [a_{n-1}]_\emptyset\right. \\
&\left(\left.\left.[\tilde{a} - a_n]_{\{a_n\}} \, \text{ff} \wedge [a_n]_\emptyset \, Z\right) \ldots\right)\right)
\end{aligned}
$$

while the formulae expressing (2) are, for each $1 \le i \le n$, of the form

$$\psi_i = \nu Z.([b_i]_{\{a_i\}} \, \text{ff} \wedge [a_i]_\emptyset([a_i]_{\{b_i\}} \, \text{ff} \wedge [b_i]_\emptyset Z))$$

Note that the property expressed by ϕ is rather weak, since it implies that the a_i's are performed in cyclic order, but it does not imply that each a_i is ever executed.

Table 1 summarizes the experimental results obtained using CADP, showing the number of states of the standard transition systems and of the reduced ones, for some values of the number n of processes. In the table we use the following symbols:

- S_1: number of states of the standard transition system;
- S_2: number of states of the standard transition system minimized using the $\tau^*\alpha$ bisimulation;
- S_3: number of states of $H_\rho^{min}(Sched_n)$, where $\rho = \mathcal{C}(\phi) = \{a_1, \ldots, a_n\}$;
- S_4: number of states of $H_{\rho_i}^{min}(Sched_n)$, where $\rho_i = \mathcal{C}(\psi_i) = \{a_i, b_i\}$, for each $1 \le i \le n$.

Note that we obtain for the scheduler's example a reduction comparable to the one in [7]. The difference is that, while we derive the interesting actions from the formula, in [7] the hiding of the b_i actions is based on informal reasonings and consequently must be proved correct. Actually, our work can be seen as proving a general framework to extensively use practical techniques for process abstraction, driven by temporal logic formulae.

Finally, note that the above methodology cannot be used when $\tau \in \rho$; however, this is not a great limitation because in general it is not important to observe τ.

n	S_1	S_2	S_3	S_4
2	13	8	2	2
3	37	24	3	2
8	3073	2048	8	2
10	15361	10240	10	2

Table 1

6   CONCLUSION

In this paper we present a new temporal logic, the selective mu-calculus, with the property that the actions relevant to check a formula are only the ones occurring in the formula itself.

The degree of reduction we obtain depends on the actions occurring in a formula. This means that there are cases, i.e. when the actions occurring in the formula are almost the whole set $\mathcal{A}$ of actions, for which we do not obtain significant reductions. This occur when checking properties which must hold for every state of the transition system as, for example, deadlock-freeness. In fact our calculus deals with a specific kind of abstraction, namely deleting all paths in which some actions do not occur. Other kinds of abstractions were proposed in the literature, which are general abstractions or cope with a specific property, as, for example, deadlock freeness [7, 11, 28].

The selective mu-calculus is useful in practice because it allows the use a reduced transition system in property verification. Thus all the verification systems which base their behaviour on the analysis of transition systems can profit from the method. In particular, our approach can be integrated with an on-the-fly methodology [9, 15, 16, 19, 20], where on-the-fly means that the system is verified during its generation. Other approaches to model checking fall inside the automata-theoretic framework [4, 18, 30, 31], in which each temporal logic formula is associated with a (either word or tree) automaton accepting exactly all computations that satisfy the (negation of the) formula. To check whether a transition system satisfies a formula, a product is done between the transition system and the automaton describing the formula. Our approach can be also used in conjunction with this methodology, thus obtaining a more efficient verification.

REFERENCES

[1] A. Aziz, T.R. Shiple, V. Singhal, A.L. Sangiovanni-Vincentelli. *Formula-Dependent Equivalence for Compositional CTL Model Checking.* In Proceedings of Workshop on Computer Aided Verification (CAV'94),

LNCS 818, 1994. 324–337.

[2] R. Barbuti, N. De Francesco, A. Santone, G. Vaglini. *Formula-Based Reduction of Transition Systems.* Internal Report IR-3/97, Dipartimento di Ingegneria dell'Informazione, Univ. of Pisa.

[3] S. Bensalem, A. Bouajjani, C. Loiseaux, J. Sifakis. *Property Preserving Simulations.* In Proceedings of Workshop on Computer Aided Verification (CAV'92), LNCS 663, 1992. 260–273.

[4] O. Bernholtz, M.Y. Vardi, P. Wolper. *An Automata-Theoretic Approach to Branching-Time Model Checking.* In Proceedings of Workshop on Computer Aided Verification (CAV'94), LNCS 818, 1994. 142–155.

[5] A. Bouajjani, J.C. Fernandez, S. Graf, C. Rodriguez, J. Sifakis. *"Safety for Branching Time Semantics".* In Proceedings of the 18th International Colloquium on Automata, Languages and Programming. LNCS 510, 1991. 76–92.

[6] G. Bruns. *A Case Study in Safety-Critical Design.* In Proceedings of Workshop on Computer Aided Verification (CAV'92), LNCS 663, 1992. 220–233.

[7] G. Bruns. *A Practical Technique for Process Abstraction.* In Proceedings of International Conference on Concurrency Theory (CONCUR'93), LNCS 714, 1993. 37–49.

[8] R. Cleaveland, J. Parrow, B. Steffen. *The concurrency workbench: operating instructions.* Tech. Notes Sussex University, 1988.

[9] C. Courcoubetis, M. Vardi, P. Wolper, M. Yannakakis. *Memory Efficient Algorithms for the Verification of Temporal Properties.* In Workshop on Computer Aided Verification, DIMACS 90, June 1990.

[10] D. Dams, O. Grumberg, R. Gerth. *Generation of reduced models for checking fragments of CTL.* In Proceedings of Workshop on Computer Aided Verification (CAV'93), LNCS 697, 1993. 479–490.

[11] N. De Francesco, A. Santone, G. Vaglini. *A Non-Standard Semantics for Generating Reduced Transition Systems.* In Proceedings of LOMAPS'96, LNCS 1192, 1996. 370-387.

[12] R. De Simone, D. Vergamini. *Aboard AUTO.* INRIA Technical Report 111, 1989.

[13] R. De Simone, A. Ressouche. *Compositional semantics of ESTEREL and verification by compositional reductions.* In Proceedings of Workshop on Computer Aided Verification (CAV'94), LNCS 818, 1994. 441–454.

[14] J.C. Fernandez, A. Kerbrat, L. Mounier. *Symbolic Equivalence Checking.* In Proceedings of the 5th International Conference on Computer-Aided Verification, LNCS 697, 1993. 85-96.

[15] J.C. Fernandez, L. Mounier. *Verifying bisimulation on the fly.* In Third International Conference on Formal Description Techniques, FORTE'90, Madrid, November 1990.

[16] J.C. Fernandez, L. Mounier. *"On th Fly" Verification of Behavioural Equivalences and Preorders.* In Proceedings of the Third International

Conference on Computer-Aided Verification, LNCS 575, 1991. 181-191.

[17] J.C. Fernandez et al. *"CADP A Protocol Validation and Verification Toolbox"*. In Proceedings of the Third International Conference on Computer-Aided Verification, LNCS 1102, 1996. 437-440.

[18] T.A. Henzinger, O. Kupferman, M.Y. Vardi. *A Space-Efficient On-the-fly Algorithm for Real-Time Model Checking*. In Proceedings of International Conference on Concurrency Theory (CONCUR'96), LNCS 1119, 1996. 514–529.

[19] C. Jard, T. Jéron. *On-Line Model-Checking for Finite Linear Temporal Logic Specifications*. In International Workshop on Automatic Verification Methods for Finite State Systems, LNCS 407, 1989. 189–196.

[20] C. Jard, T. Jéron. *Bounded-memory Algorithms for Verification On-the-fly*. In Proceedings of the Third International Conference on Computer-Aided Verification, LNCS 575, 1991. 192-201.

[21] D. Kozen. *Results on the propositional mu-calculus*. Theoretical Computer Science, 27, (1983). 333–354.

[22] Y.S. Kwong. *On reduction of asynchronous systems*. Theoretical Computer Science 5, 1977. 25–50.

[23] R. Milner. *Communication and Concurrency*. Prentice-Hall, 1989.

[24] D. Peled. *All from one, one for all, on model-checking using representatives*. In Proceedings of the Fifth International Conference on Computer-Aided Verification(CAV'93), LNCS 679, 1993. 409–423.

[25] D. Peled. *Combining Partial Order Reductions with On-the Fly Model-Checking*. In Proceedings of the 6th International Conference on Computer-Aided Verification(CAV'94), LNCS 818, 1994. 377–390.

[26] J. Sifakis. *Property Preserving Homomorphisms of Transition Systems*. In Logics of Programs, LNCS 164, 1983.

[27] C. Stirling. *An Introduction to modal and temporal logics for CCS* In Concurrency: Theory, Language, and Architecture, LNCS 391, 1989.

[28] A. Valmari. *A stubborn attack on state explosion*. In Proceedings of International Conference on Computer-Aided Verification (CAV'90), LNCS 531, 1990. 156–165.

[29] A. Valmari, M. Clegg. *Reduced Labelled Transition Systems Save Verification Effort*. In Proceedings of the International Conference on Concurrency Theory (CONCUR'91), LNCS, 1991. 526–540.

[30] M.Y. Vardi, P. Wolper. *An automata-theoretic approach to automatic program verification*. In Proceedings of the First Symposium on Logic Computer Science, Cambridge, 1986. 322–331,

[31] M.Y. Vardi, P. Wolper. *An automata-theoretic techniques for modal logics of programs*. Journal of Computer and System Science, 32(2):182-21, April 1986.

33

On a Concurrency Calculus for Design of Mobile Telecommunication Systems

Sendai National College of Technology,
1, Kitahara, Kamiayashi, Aoba-ku, Sendai, 989-31
JAPAN.
Telephone: +81-22-392-4761, Fax:+81-22-392-3359.
e-mail:{tando,kato}@info.sendai-ct.ac.jp,
kaoru@cc.sendai-ct.ac.jp

Abstract

Process algebras with name passing can be suitable to describe dynamical changes of connections. To describe mobile communication, however, it is necessary to consider locations at which processes run. We propose a description method to design such systems using a concurrency calculus in this paper. The concept of a field is introduced to model locality. An extension of π-calculus with a field is proposed. A field is given when behaviors of a target system is verified for a particular environment. The aim of the extension is to verify and to test connectivity between processes under various geographical constraints. This method could be design-oriented in this context. Equivalence relations with/without location in this calculus are also discussed.

Keywords

mobile telecommunication, π-calculus, location, field, location bisimulation, location erased bisimulation

Formal Description Techniques and Protocol Specification, Testing and Verification
T. Mizuno, N. Shiratori, T. Higashino & A. Togashi (Eds.) © 1997 IFIP. Published by Chapman & Hall

1 INTRODUCTION

Process algebras are one of the most successful formalisms for specification of concurrent systems. Especially, name passing calculi are useful to describe dynamical changes of connections among processes in mobile systems. π-calculus (Milner et al., 1991 and 1992) is a typical one. Channel names are treated as data that are used to create new channels dynamically and to communicate between processes in π-calculus.

Some issues are raised when π-calculus is applied to mobile telecommunication. Locations and mobility of mobile terminals may be hard to describe using only a name passing technique while behaviors of mobile processes can be influenced by them in many cases. Various approaches have been tried to treat locality in the context of concurrency calculi. Sangiorgi (1994) proposed a method using located processes. A located process is roughly written $l :: P$, where P is a π-calculus process and l is a location at which the process P runs. Location bisimulation based on location transitions, with the form $\overset{a}{\underset{l}{\rightarrow}}$, is an equivalence relation considering not only causality but locations. Amadio et al.(1994 and 1997) focused failures and mobility, and treated detection of failures and mobility of processes using optional functions such as detection of locations and replication of processes. In these approaches, locality is treated within the framework of these languages. Locality may make specification more difficult although locality is needed to be specified.

For simple specification, we take another approach in this paper. Environment is modeled independently of a language, and a calculus consists of π-calculus and the model of an environment in this approach. To model a relation between π-calculus and this model, the concept of a *field* is introduced. A field is considered as a set of constraints on communication, and is set up for each application. In the context of fields, an interesting issue is how the same party of processes behaves for various fields. To discuss an equivalence relation on such behaviors, we introduce two bisimulation relations, i.e. location bisimulation and location erased bisimulation. These bisimulations are almost the same as that of CCS (Milner, 1989) except for focusing on connectivity of processes. The aim of our method is to support design of mobile communication systems. To decide whether processes can similarly behave even if an environment is changed, the above bisimulations are used. Furthermore, these bisimilarities should be invariants when specifications are refined.

The rest of this paper consists as follows. In Section 2, the concept of a field will be introduced. Using a field, we will also define a variant πF of π-calculus in this section. Next, equivalence relations will be discussed in Section 3. Section 4 shows an application of πF to a mobile telecommunication system. Finally, we conclude this paper in Section 5.

2　AN EXTENSION OF π-CALCULUS WITH A FIELD

Processes may be affected by their environment. We model such a situation with a concept of a *field*. π-calculus is extended with a field. To define the extension, we premise the following: (1) Behaviors of processes can be affected by their environment; (2) The environment can not be affected by processes. The first premise is the start point of this paper. We consider environments such as *networks* or *geographical features*, e.g. the configurations of buildings and roads, in this paper. Such environments can be regarded as not changed in the short term that communication processes are running. From the two premises, we take the style that locations are appended to π-calculus processes.

Field and movement
A field presents a set of constraints on communications among processes.

Definition 1 *Field*
Field $\mathcal{F}$ is defined as a pair as follows:
$\mathcal{F} = \langle Loc, RL \rangle$,
where Loc is a set of locations (places at which processes can be located) and $RL \subseteq Loc \times Loc$ is a relation such that a message can be sent from l to m if $(l, m) \in RL$. If $(l, m) \in RL$, then (l, m) is called a road *from l to m, or simply a road.*

Movement of processes can be treated using a higher-order language (see e.g. Milner, 1991). However, we place movement at the outside of the syntax of a language for the simple description and the separation of processes from a field.

Definition 2 *Movement*

Constraints on movement of processes are represented by $MV \subset Proc \times Loc \times 2^{Loc}$, where Proc is the set of all process identifiers. Let $P \in Proc$, $l \in Loc$ and $A \subset Loc-\{l\}$. $(P, l, A) \in MV$ expresses that a process with the process identifier (abbreviated to PID) P may move from the location l to some location $m \in A$. PIDs mean initial process names from which current processes are reduced by the reduction rules in Definition 5.

Example 1 *The following field $\mathcal{F}_1$ is shown as a directed graph in Figure 1. A process at '1' can not communicate with a process at '3', but a process at '2' and '4'. A message can not be sent from '4' to '1' while it can be sent from '1' to '4'. Furthermore, MV represents the situation that a process with the PID R can move '4' to '3'.*

$$
\begin{aligned}
\mathcal{F}_1 &= \langle Loc_1, RL_1 \rangle \\
Loc_1 &= \{1, 2, 3, 4\} \\
RL_1 &= \{(1,4), (1,2), (2,1), (2,3), (3,2)\} \\
MV &= \{(R, 4, \{3\})\}.
\end{aligned}
$$

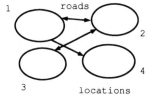

Figure 1 An example of a field: $\mathcal{F}_1$.

Note that no-movements of processes such as $(Q, 4, \{4\})$ must not be written in MV even if other *fixed* processes are located on a field! Readers have to take care that there are no relations between movement of processes and directions of arrows in Fig. 1. The arrows in the figure show transimission directions of data, but not process movement.

πF

An extension of π-calculus with a field is defined here to model behaviors of mobile processes on a particular environment. Syntax of this extension, called πF, is the same as that of the standard poliadic π-calculus(see e.g. Sangiorgi, 1994) except for without matching. In this calculus, locations are appended to π processes as parameters. Such a process with a location is called a *labor*.

In this paper, L, M, $\cdots$ range over the set $\mathcal{LAB}$ of all labors, P, Q, $\cdots$ the set $\mathcal{P}roc$ of all process identifiers, a, b, $\cdots$ the set Ch of all channel names, l, m, $\cdots$ the set Loc of all locations, and $\tilde{b}$, $\cdots$ the set of all channel name vectors.

Definition 3 *Syntax of πF*

$$\begin{array}{llll}
actions & : \alpha & ::= & a(\tilde{b}) \mid \overline{a}[\tilde{b}] \\
processes: & P & ::= & 0 \mid \alpha.P \mid P+Q \mid P|Q \mid \nu b P \mid D\langle \tilde{b} \rangle \\
labors & : L & ::= & \{P\}l \mid L|M \mid \nu b L
\end{array}$$

In the above definition, actions occur through channels, and processes receive ($a(\tilde{b})$) / send ($\overline{a}[\tilde{b}]$) messages (channel names). The definitions of processes represent inaction, action prefix, sum, parallel composition, restriction of a name and constant application respectively. In the definitions of labors, $\{P\}l$ represents that the process P is at the location 'l'. The rest of definitions of labors are corresponding to parallel composition and restriction of a name. If $\tilde{b}$ is empty, brackets [ ] and ( ) will be omitted. D is defined as $D(\tilde{c})$ $\overset{\text{def}}{=} P$, where $\tilde{c}$ is formal parameters. $D\langle \tilde{b} \rangle$ is that formal parameters $\tilde{c}$ is replaced with actual parameters $\tilde{b}$ in D.

Congruence relations on processes are the same as the relations of the standard π-calculus. Additional congruence relations on labors are as follows.

Definition 4 *Congruence of labors*

1. $L|\{0\}l \equiv L$ 2. $L|M \equiv M|L$
3. $\{P|Q\}l \equiv \{P\}l|\{Q\}l$ 4. $\nu a \nu b L \equiv \nu b \nu a L$
5. $\nu a\{P\}l \equiv \{\nu a P\}l$
6. $\nu x(L|M) \equiv L|\nu x M$ *if $x \notin fn(L)$, where $fn(L)$ is the set of all free names in L.*

Labors are reduced according to the following reduction rules. Constraints on communications by locations and movement of processes are reflected in these rules.

Definition 5 *Reduction rules*

$$MOVE: \quad \frac{(P,l,A) \in MV, m \in A}{\{P\}l \xrightarrow{\tau@\{l,m\}} \{P\}m} \qquad PAR: \quad \frac{L \xrightarrow{a@Pls} L'}{L|M \xrightarrow{a@Pls} L'|M}$$

$$RES1: \quad \frac{L \xrightarrow{a@Pls} L', \; x \neq a}{\nu x L \xrightarrow{a@Pls} \nu x L'} \qquad RES2: \quad \frac{L \xrightarrow{a@Pls} L', \; x = a}{\nu x L \xrightarrow{\tau@Pls} \nu x L'}$$

$COMM:$ $\dfrac{(l, l_\lambda) \in RL \ for \ \lambda \in \Lambda, \ \Delta(\neq \emptyset) \subseteq \Lambda}{}$

$$\{\cdots + \overline{a}[\tilde{b}].P\}l \mid \prod_{\lambda \in \Lambda} \{\cdots + a(\tilde{c}).Q_\lambda\}l_\lambda$$

$$\xrightarrow{a@(\{l\} \cup \{l_\lambda | \lambda \in \Delta\})}$$

$$\{P\}l \mid \prod_{\lambda \in \Delta} \{Q_\lambda\langle \tilde{b}\rangle\}l_\lambda \mid \prod_{\lambda' \in \Lambda - \Delta} \{\cdots + a(\tilde{c}).Q_{\lambda'}\}l_{\lambda'}$$

$STRUCT:$ $\dfrac{Q \equiv P, P \xrightarrow{a@Pls} P', P' \equiv Q'}{Q \xrightarrow{a@Pls} Q'}$

$L \xrightarrow{a@Pls} L'$ *represents that* L *is reduced to* L' *by the occurrence of the action through the channel* a, *and processes at locations listed in* Pls *are related to this reduction. In* $COMM$, $\prod_{\lambda \in \Lambda} L_\lambda = L_{\lambda_1} | L_{\lambda_2} | \cdots \ (\lambda_1, \lambda_2, \ldots \in \Lambda)$. Λ *is a communicatable process index set. In this case, the arity (number of names) of both* $\tilde{b}$ *and* $\tilde{c}$ *must be the same.* τ *represents an unobservable action.*

Communications are restricted by a field such that processes can communicate with each other when these processes are located at communicatable locations. $COMM$ allows broadcast or multicast transmission. Cases $\Delta = \Lambda$ and $\Delta \subset \Lambda$ correspond to broadcast and multicast respectively. Movement of processes is represented with $MOVE$, and the communication by a bounded channel name is represented with $RES2$. Observation of communications is emphasized in this reduction rules, so that restricted communications and movement of processes are regarded as internal actions.

Example 2 *Consider processes* P, Q *and* R, *given as follows, running on the field* $\mathcal{F}_1$ *under* MV *in Example 1.*
$P \stackrel{\text{def}}{=} \overline{a}[b]. \ P'$, $Q \stackrel{\text{def}}{=} a(x). \ \overline{x}. \ Q'$, $R \stackrel{\text{def}}{=} b. \ R'$.
Then, let P, Q *and* R *be located at '1', '2' and '4' respectively. This labor may behave as follows:*

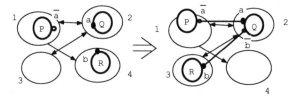

Figure 2 *A behavior of a labor of Example 3.*

$$\{P\}1 \mid \{Q\}2 \mid \{R\}4 \xrightarrow{a@\{1,2\}} \{P'\}1 \mid \{\overline{b}.\ Q'\}2 \mid \{R\}4$$
$$\xrightarrow{\tau@\{3,4\}} \{P'\}1 \mid \{\overline{b}.\ Q'\}2 \mid \{R\}3$$
$$\xrightarrow{b@\{2,3\}} \{P'\}1 \mid \{Q'\}2 \mid \{R'\}3$$

3 EQUIVALENCE RELATIONS

Behaviors of located processes described in πF are a subset of behaviors of the standard π-calculus processes. Even if the same processes are given, behaviors of a party of these processes may be different according to fields. In telecommunication systems, it is required that the same communication services are ensured even in the case that the topology of a network changes.

To discuss the equivalence in such a situation, we define bisimulation relations in this section. At first, a *location bisimulation* is defined in terms of located reduction. In this bisimulation, causality of locations is not considered, different from the location bisimulation of Sangiorgi (1994). Next, a *location erased bisimulation* is defined. This bisimulation is that location information is omitted from the above location bisimulation.

Location bisimulation

A *weak reduction* is introduced by omitting reductions with internal actions. If there exists $m, n \geq 0$, location sets $Pls_{1,i} \subseteq Loc$ ($1 \leq i \leq m$), location sets $Pls_{2,j} \subseteq Loc$ ($1 \leq j \leq n$), such that $L \xrightarrow{\tau@Pls_{1,1}}$ $\ldots \xrightarrow{\tau@Pls_{1,m}} \xrightarrow{a@Pls} \xrightarrow{\tau@Pls_{2,1}} \ldots \xrightarrow{\tau@Pls_{2,n}} L'$ for an observable channel name a ($\neq \tau$) and a location set Pls, we write $L \xRightarrow{a@Pls} L'$.

Now, to compare two labors, we define location bisimulation.

Definition 6 *Location bisimulation*

A binary relation $\mathcal{S} \subseteq \mathcal{LAB} \times \mathcal{LAB}$ over labors is a location bisimulation if $(L, M) \in \mathcal{S}$ implies, for all $a \in Ch$ and all $Pls \in Loc$,

1. *Whenever $L \xRightarrow{a@Pls} L'$, there exists M' such that $M \xRightarrow{a@Pls} M'$ and $(L', M') \in \mathcal{S}$;*

2. *Whenever $M \xRightarrow{a@Pls} M'$, there exists L' such that $L \xRightarrow{a@Pls} L'$ and $(L', M') \in \mathcal{S}$.*

Then, $\approx_l$ is defined as follows, and $L \approx_l M$ if $(L, M) \in \approx_l$.
$$\approx_l = \bigcup \{\mathcal{S} \mid \mathcal{S} \text{ is a location bisimulation}\}$$

Proposition 1

1. $\{\overline{a}[\tilde{b}].P\}l \mid \{x(\tilde{c}).Q\}m \approx_l \{0\}l'$ for any l' if $x \neq a$,

2. $\{\nu a P\}l \mid \{Q\}m \approx_l \nu a(\{P\}l|\{Q\}m)$, if $a \notin fn(Q)$.

Location erased bisimulation

Location information may not always be important in telecommunication. When mobile terminals can move freely, connection or data transfer becomes a more important problem. To consider such a situation, location erased bisimulation is defined.

A notation of *location erased reduction* is used for the definition of location erased bisimulation, i.e. $L \xrightarrow{a} L'$ if $L \xrightarrow{a@P|s} L'$. Weak location erased reduction $L \overset{a}{\Longrightarrow} L'$ is also defined similarly.

Definition 7 *Location erased bisimulation*

A binary relation $S \subseteq \mathcal{LAB} \times \mathcal{LAB}$ over labors is a location erased bisimulation if $(L, M) \in S$ implies, for all $a \in CH$,

1. Whenever $L \overset{a}{\Longrightarrow} L'$, there exists M' such that $M \overset{a}{\Longrightarrow} M'$ and $(L', M') \in S$;

2. Whenever $M \overset{a}{\Longrightarrow} M'$, there exists L' such that $L \overset{a}{\Longrightarrow} L'$ and $(L', M') \in S$.

Then, $\approx$ is defined as follows, and $L \approx M$ if $(L, M) \in \approx$.
$\approx = \bigcup\{S|S$ is a location erased bisimulation$\}$,

We show simple properties on location erased bisimulation.

Proposition 2 $L \approx_l M$ *implies* $L \approx M$.

Proposition 3

1. $\{\overline{a}[\tilde{b}].P\}l \mid \{x(\tilde{c}).Q\}m \approx \{0\}l'$ if $x \neq a$.

2. $\{\nu a P\}l \mid \{Q\}m \approx \nu a(\{P\}l|\{Q\}m)$, if $a \notin fn(Q)$.

A design method using πF

We simply show an overview of a design method of mobile concurrent systems using πF.

1. Processes and a field are described at first.

2. These processes are located on the field.

3. Necessary properties such as connectability and the detection of failure are verified.

4. If necessary, the design will be more refined: (1) Location bisimulation and/or location erased bisimulation are used as invariants in the refinement; (2) More detailed informations are appended into restricted actions of the previous design.

Location erased bisimulation may be an effective invariant in the case of the design of mobile telecommunication systems. Bisimilarities are difficult to be verified in actual systems as reduction trees may be very complex. Thus, a simulator can be an useful tool in the actual design. A simulator checks a subset of all possible action sequences, so that a part of connectivity and failures could be checked.

4 APPLICATION OF πF TO MOBILE TELECOM- MUNICATIONS

An application of πF to mobile telecommunication systems is shown in this section. In mobile telecommunication systems, cellular systems have been employed. Multiplicity of trunks by reuse of frequencies within limited frequency band and the flexibility for increase of users are the reason. The transmission power of a mobile terminal is low, so that a mobile terminal can be connected with only the nearest base station. In cellular systems, each terminal is registered to the area to which the terminal can access every moment (*location registration*). The location of a terminal is traced whenever the terminal moves to other areas (*location tracing control*). A *simultaneous call* technique is used to search a particular terminal in a location registration area (Padgett et al., 1995). We will describe the situation that one calls someone driving a car with a movable terminal from a fixed terminal at home.

A fixed terminal (*Home*) is connected to base stations (*Base$_i$*, $i = 1, 2, 3$) via the relay station (*Station*) with cables, and base stations communicate with a movable terminal (*Car*) with a radio equipment if the terminal is in its area (cell) (see Figure 3).

At first, a field and movement of processes are set up for this system.

$$
\begin{aligned}
\mathcal{F} &= \langle Loc, RL \rangle \\
Loc &= \{h, s, c_1, c_2, c_3\} \\
RL &= \{(h, s), (s, h), (s, c_1), (c_1, s), (s, c_2), (c_2, s), (s, c_3), \\
&\quad (c_3, s)\} \cup \{(p, p) | p \in Loc\}
\end{aligned}
$$

$$MV \quad =\{(Car, c_1, \{c_2, c_3\}), (Car, c_2, \{c_1, c_3\}), (Car, c_3, \{c_1, c_2\})\}$$

'h', 's' and 'c_i' ($i = 1, 2, 3$) are locations at which the fixed terminal, the relay station and base stations are located respectively. MV represents the situation that $Home$, $Station$ and $Base_i$ ($i = 1, 2, 3$) can not move while Car can move among cells 'c_i' ($i = 1, 2, 3$).

Next, these processes are given as follows:

$$Home(n) \quad \overset{\text{def}}{=} \quad \overline{call}[n].CallingHome\langle n \rangle,$$
$$Station \quad \overset{\text{def}}{=} \quad call(n).\overline{search}[n].SearchingStation\langle n \rangle,$$
$$Base_i \quad \overset{\text{def}}{=} \quad search(n).\overline{n}.RelayingBase_i,$$
$$Car \quad \overset{\text{def}}{=} \quad num.ReceivingCar.$$

Car has its own telephone number num. The fixed terminal $Home$ requests for a call to the movable terminal Car with some telephone number. When the relay station $Station$ is requested for a call, the simultaneous call will be done. After each base station receives the number, the corresponding station will connect with Car if Car is in its cell. Then, a connection between $Home$ and Car is established.

We assume that $Home$, $Station$, $Base_i$ and Car are located at 'h', 's', 'c_i' and 'c_1' at the start. This situation is represented as the labor $Call\langle num \rangle$:

$$Call\langle num \rangle \quad \overset{\text{def}}{=} \quad \{Home(num)\}h \mid \{Station\}s \mid \{Base_1\}c_1$$
$$\mid \{Base_2\}c_2 \mid \{Base_3\}c_3 \mid \{Car\}c_1.$$

The labor $Call\langle num \rangle$ may be reduced as follows. Then, a connection between $Home$ and Car will be established.

$$Call\langle num \rangle$$

$\overset{call@\{h,s\}}{\longrightarrow}$

$\{CallingHome\langle num \rangle\}h$
$\mid\{\overline{search}[num].SearchingStation\langle num \rangle\}s$
$\mid\prod_{i=1}^{3}\{search(n).\overline{n}.RelayingBase_i\}c_i$
$\mid\{Car\}c_1$

$\overset{search@\{s,c_1,c_2,c_3\}}{\longrightarrow}$

$\{CallingHome\langle num \rangle\}h$
$\mid\{SearchingStation\langle num \rangle\}s$
$\mid\prod_{i=1}^{3}\{\overline{num}.RelayingBase_i\}c_i$
$\mid\{num.ReceivingCar\}c_1$

$\overset{\tau@\{c_1,c_2\}}{\longrightarrow} \overset{num@\{c_2\}}{\longrightarrow}$

$\{CallingHome\langle num \rangle\}h$
$\mid\{SearchingStation\langle num \rangle\}s$
$\mid\{\overline{num}.RelayingBase_1\}c_1$
$\mid\{RelayingBase_2\langle num \rangle\}c_2$

$$|\{\overline{num}.RelayingBase_3\}c_3$$
$$|\{ReceivingCar\}c_2$$

In this reduction sequences, the second reduction shows the simultaneous call. Each base station waits to connect with Car. Whenever Car connects with one of the base stations, Car can communicate by the same channel name num. The result of this reduction sequences is shown in Figure 3 (b). Communications with the same names are enabled in various situations, so that process descriptions may become simple. This is an advantage of πF.

Next, consider the accident that the cable between the relay station and one of the base stations, e.g. base station 2, is broken down. How will $Call\langle num\rangle$ behave at that time? Roads (s, c_2) and (c_2, s) are removed from RL in this case. Even if Car can not communicate with $Base_2$, Car will communicate with $Base_1$ or $Base_3$ by movement to 'c_1' or 'c_3'. So, those behaviors are location erased bisimilar before and after this accident.

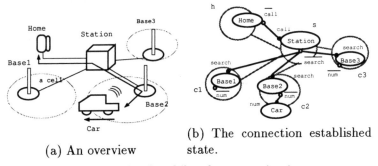

(a) An overview

(b) The connection established state.

Figure 3 An example of mobile telecommunication systems.

5 CONCLUSION

We proposed a concurrency calculus for design of mobile telecommunication systems in this paper. By separation of process description from location information, process description can be simple. Verification of features on processes can be flexibly done for various environments. πF may be suitable for mobile telecommunication systems managed in fluidal environments.

Our future works are as follows. A simulator is necessary to support design of mobile communication systems. Using such a simulator, one will be able to check properties such as connectivity of communication processes on particular environments. In addition, effects of fields to behaviors of processes will be investigated. Furthermore, a method of specification from requirement acquisition will be developed based on πF. We have proposed a method based on topology of description techniques (Ando et al., 1996). We plan to apply this topological method to the specification using πF.

REFERENCES

Amadio,R.M.(1997) An asynchronous model of locality, failure, and process mobility, Rappoit Interne LIM (to appear), and INRIA Research Report 3109.

Amadio, R.M. and Prasad, S.(1994) Localities and failures. *Proceedings of* 14^t*h FST and TCS Conference, FST-TCS'94, LNCS 880*, pp. 205-16. Springer-Verlag.

Ando, T., Takahashi, K. and Kato, Y.(1996) A Topological Framework of Stepwise Specification for Concurrent Systems, *The Institute of Electronics, Information and Communication Engineers Transactions on Fundamentals*, **E79-A**, the Institute of Electronics, Information and Communication Engineers, pp. 1760-7.

Milner, R.(1989) *Communication and Concurrency.* Prentice Hall.

Milner, R.(1991) The poliadic π-calculus: a tutorial, *Technical Report* ECS – LFCS – 91 – 180, Labo. for Foundations of Comp. Sci., Dept. Comp.Sci., Univ. Edinburgh, UK.

Milner, R., Parrow, J. and Walker, D.(1992) A calculus of mobile processes, Pert I and II. *Journal pf Information and Computation*, **100**, pp. 1–77, September.

Padgett, J., Gunther, C. and Hattori, T.(1995) Overview of Wireless Personal Communications, *IEEE Communication Magazine*, **33**, pp.28–41.

Sangiorgi, D.(1994) Locality and non-interleaving semantics in calculi for mobile processes. Technical Report ECS-LFCS-94-282, Laboratory for Foundations of Computer Science, Department of Computer Science, University of Edinburgh, UK.

INDEX OF CONTRIBUTORS

KEYWORD INDEX